PROGRESS IN PLANT PROTOPLAST RESEARCH

CURRENT PLANT SCIENCE AND BIOTECHNOLOGY IN AGRICULTURE

Aims and Scope of the Bookseries

The bookseries is intended for readers ranging from advanced students to senior research scientists and corporate directors interested in acquiring in-depth, state-of-the-art knowledge about research findings and techniques related to plant science and biotechnology. While the subject matter will relate more particularly to agricultural applications, timely topics in basic plant science and biotechnology will be explored as well. Some volumes will report progress in rapidly advancing disciplines through proceedings of symposia and workshops while others will detail fundamental information of an enduring nature that will be referenced repeatedly.

Evans, H.J., Bottomley, P.J. and Newton, W.E. (eds): Nitrogen fixation research progress. 1985. ISBN 90-247-3255-7

Zimmerman, R.H., Griesbach, R.J., Hammerschlag, F.A. and Lawson, R.H. (eds): Tissue culture as a plant production system for horticultural crops. 1986. ISBN 90-247-3378-2

Verma, D.P.S. and Brisson, N. (eds): Molecular Genetics of Plant-Microbe Interactions. 1987. ISBN 90-247-3426-6

Civerolo, E.L., Collmer, A., Davis, R.E. and Gillaspie, A.G. (eds): Plant Pathogenic Bacteria. 1987. ISBN 90-247-3476-2

Summerfield, R.J. (ed): World Crops: Cool Season Food Legumes. 1988. ISBN 90-247-3641-2

Gepts, P. (ed): Genetic Resources of *Phaseolus* Beans. 1988. ISBN 90-247-3685-4

Puite, K.J., Dons, J.J.M., Huizing, H.J., Kool, A.J., Koornneef, M., Krens, F.A. (eds): Progress in Plant Protoplast Research. 1988. ISBN 90-247-3688-9

Progress in Plant Protoplast Research

Proceedings of the 7th International Protoplast Symposium, Wageningen, the Netherlands, December 6–11, 1987

edited by

K.J. PUITE, J.J.M. DONS, H.J. HUIZING,
A.J. KOOL, M. KOORNNEEF, F.A. KRENS

KLUWER ACADEMIC PUBLISHERS
DORDRECHT / BOSTON / LONDON

Library of Congress Cataloging in Publication Data

International Protoplast Symposium (7th : 1987 : Wageningen, Netherlands)
Progress in plant protoplast research : proceedings of the 7th International Protoplast Symposium, Wageningen, the Netherlands, December 6-11, 1987 / edited by K.J. Puite ... [et al.].
p. cm. -- (Current plant science and biotechnology in agriculture ; 7)
Includes index.

1. Plant protoplasts--Congresses. 2. Cell hybridization--Congresses. 3. Plant genetic engineering--Congresses. I. Puite, K. J. II. Title. III. Series.
QK725.I495 1987
581.87'3--dc19 88-701
CIP

ISBN-13: 978-94-010-7754-5 e-ISBN-13: 978-94-009-2788-9
DOI: 10.1007/978-94-009-2788-9

Published by Kluwer Academic Publishers,
P.O. Box 17, 3300 AA Dordrecht, The Netherlands.

Kluwer Academic Publishers incorporates
the publishing programmes of
D. Reidel, Martinus Nijhoff, Dr W. Junk and MTP Press.

Sold and distributed in the U.S.A. and Canada
by Kluwer Academic Publishers,
101 Philip Drive, Norwell, MA 02061, U.S.A.

In all other countries, sold and distributed
by Kluwer Academic Publishers Group,
P.O. Box 322, 3300 AH Dordrecht, The Netherlands.

CONTENTS

PREFACE

In the series of International Protoplast Symposia the Symposium of 1987 was held in Wageningen (The Netherlands). Earlier Symposia took place in Jena (DDR) 1963, Brno (CSSR) 1967, Salamanca (Spain) 1971, Nottingham (UK) 1975, Szeged (Hungary) 1979 and Basel (Switzerland) 1983. This 7th International Protoplast Symposium was organized by K.J. Puite (Secretary), J.J.M. Dons (Treasurer), H.J.Huizing and E.J.L. Hotke-Staal (Local Organizers), the first three persons being scientists, respectively, from the Research Institute Ital, the Institute for horticultural plant breeding IVT and the Foundation for agricultural plant breeding SVP at Wageningen. Scientific Advisers of the Symposium were A.J. Kool, M. Koornneef and F.A. Krens. The International Agricultural Centre IAC served as the Symposium location.

The Organizing Committee decided that the scientific programme of the Symposium should be mainly focussed on protoplast technology of relevance to plant breeding. Therefore research on microbial protoplasts and on secondary metabolites was not included.

About 250 scientists from 27 different countries were welcomed at the meeting. Speakers at Symposium Sessions and authors of Poster contributions were asked to hand over their manuscripts for the Symposium Proceedings already at the meeting, permitting early publication of the Proceedings. These manuscripts give the state of the art of the protoplast research and illustrate the progress since the last Protoplast Symposium.

The Symposium Programme included Lectures, Poster discussions and Workshops. Although the daily programmes went on until 21.30 h each night great interest was shown for an extra workshop on the use of symbols in protoplast research. It was agreed that the symbol for any type of somatic hybrid should be (+). This new symbol would clearly distinguish somatic hybrids from plant chimaeras which are presently represented by the + symbol. The social events e.g. the Symposium dinner at the Doornenburg castle further stimulated the contacts between the participants of the Symposium.

At the closure of the Symposium it was announced that the next Protoplast Symposium will probably be held in Uppsala, Sweden in 1991.

Klaas Puite

LEGENDS TO THE PHOTO IMPRESSION

1. Opening address by Klaas Puite, Secretary of the Organizing Committee.

2,9. Plenary lectures were held in the large conference room of the International Agricultural Centre.

3. Some participants attending the book exhibition.

4. Departure for visits to the National Parc "De Hoge Veluwe" and Amsterdam.

5,7, 12, 15. Over 150 posters were presented during this symposium.

6. The Opening lecture of the congress by Ingo Potrykus.

8, 10, 13. Lunches and dinners in the IAC restaurant.

11. Workshops were held in the afternoon and evening.

14. The workshop on "Problems in protoplast isolation and regeneration" was organized by Frans Krens.

PHOTO IMPRESSION OF THE PROCEEDINGS

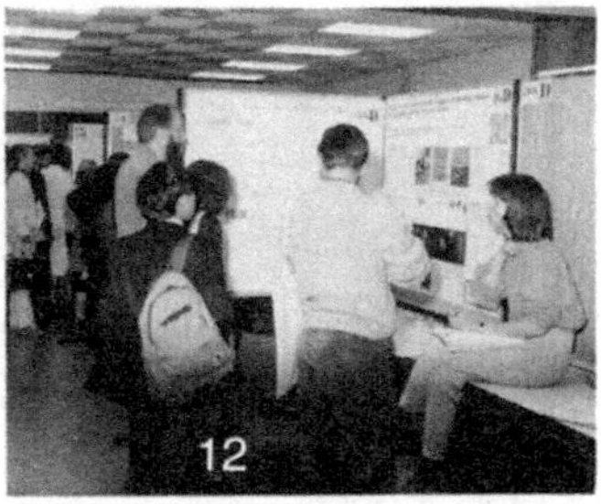

PROGRESS IN PLANT PROTOPLAST RESEARCH

I. POTRYKUS

Institut für Pflanzenwissenschaften, ETH-Zentrum, Ch-8092 Zürich, Switzerland.

1. INTRODUCTION

If we look back at the state of the art of plant protoplast research in 1983, as published in the LECTURE Proceedings and POSTER Proceedings of the 6th International Protoplast Symposium, August 12-16, 1983, Basel and compare that to the present state of the art, the progress made during the past four years is indeed impressive and demonstrates that the interest in and the potential of plant protoplast research is far from slowing down. It is virtually impossible for this introductory lecture to the 7th International Protoplast Symposium to even only mention all the many new achievements in all the many different fields of protoplast research. All the author can try to do is to pick out a few examples from different areas in a quite subjective way which to him represent important achievements. He also hopes that it will be acceptable if he does not exclude work from his own laboratory.

2. PROTOPLAST CULTURE

This topic will be discussed in great detail in subsequent presentations by e.g. M.R. Davey and K.J. Webb. A very important achievement in this area is the development of the MICROCULTURE technology by H.U. Koop (3) [94] ([ ] refers to page no in the Book of Abstracts of this Symposium) and colleagues who set up an efficient protocol for the culture of individual protoplasts in nanoliter drops, which opens the way to many novel experimental questions and which will have a big impact on future developments. An important breakthrough with regard to so far recalcitrant plant species was the simultaneous success in several laboratories in plant regeneration from CEREAL protoplasts e.g. rice (4, 5, 6) maize, and probably wheat.

3. PROTOPLAST FUSION: TECHNIQUES

ELECTROFUSION has been developed to an efficient and routine technique which is in use in numerous laboratories and will be presented in several presentations at this Symposium. There has been considerable progress in the development of SELECTION schemes for the recovery of fusion hybrids as e.g. the development of "universal hybridisers" combining each a dominant and a recessive marker in one fusion partner, thus allowing for save selection in any given combination with wildtype protoplasts (7), or the use of pollen tetrade protoplasts as generally applicable half-selective system [32], or the efficient enrichment for heterokaryocytes via fluorescence-activated-cell-sorting [54, 55]. One of the most interesting developments is, however, the protocol for MICROFUSION & CULTURE as developed by H.U. Koop and G. Spangenberg (8) (see also G. Spangenberg, this Proceedings), which allows for predictable and precise combination of genomes and recovery of defined somatic hybrid clones between and desired wildtype protoplasts.

4. PROTOPLAST FUSION: CONSEQUENCES

In 1983 we saw the first publication on striking recombination between chondriomes in fusion hybrids, as judged from altered mtDNA restriction pattern (9). Since than numerous publications appeared describing this phenomenon in different fusion combinations and the impression is that chondriomes in somatic hybrid cells are very recombinogenic. In contrast it took a real effort to find the first case of recombination between plastidomes (10) and the general impression is that plastidomes recombine only in very rare cases. Somatic hybridisation led to the establishment of several novel plastidome/chondriome combinations and partial loss of chromosomes often led to the recovery of novel asymmetric nuclear hybrids.

5. PROTOPLAST FUSION: APPLICATION

This paragraph is representative for a general change in attitude becoming visible during the last four years: an increasing number of laboratories is aiming at the application of the novel techniques to practical problems in plant breeding. There is the TRANSFER OF MALE STERILITY via transfer and sorting-out of "male sterile" mitochondria as a consequence of protoplast fusion (for more details see G. Pelletier and K. Glimelius, this Proceedings). There is also the TRANSFER OF PATHOGEN RESISTANCE from wild relatives to culture varieties (e.g. J.P. Helgeson, this Proceedings) and the TRANSFER OF HERBICIDE RESISTANCE (e.g. S.M. Jain, this Proceedings). Following the first report on the recovery of a SYNTHETIC Brassica napus (11) at the last Symposium, synthetic hybrids in Brassica are produced in various combinations (e.g. K. Glimelius, this Proceedings). The somatic hybridisation between rice and orchard grass (12) was not only of interest because it was the first case of plantlet regeneration from cereal fusion hybrids, but also because it provided the very important proof that cereal protoplasts indeed can regenerate to plants.

6. TRANSFER OF PARTIAL GENOMES

Deliberate asymmetric genome combination (cybridisation) is usually achieved by the "donor-recipient" method (13), by which the irradiated nucleus of the donor protoplasts is assumed not to contribute to the genome of the recovered fusion hybrids. Some caution may be advisable in the interpretation of such "cybrids" as it has been shown (14) that massive quantities of nuclear DNA of the donor fusion partner can be found in the cybrids. The development of microfusion in microdrops of defined subprotoplasts and protoplasts (15) (see Spangenberg, this Proceedings) offers now a safe method for predictable combinations of nuclear and extranuclear genomes.

7. MUTANTS AND VARIANTS

One of the motivations to search for biochemical mutants in protoplast-derived clones was their value in providing tight selection schemes for somatic hybrids, and there are numerous examples where auxotrophs have been used in complete or half selective fusion combinations. How effective mutants can be used for the characterisation and isolation of plant genes has been beautifully shown by the work of the group in Versailles (see also Caboche, this Proceedings) careful and consequent accumulation and molecular analysis of a great number of nitrate reductase-deficient mutants in N. plumbaginifolia finally led to the isolation of complete genomic DNA clones of the enzyme. A group of mutants, besides many others, where one would hope for an important breakthrough are the temperature sensitive auxine auxotrophs isolated at the FMI Basel (see also Fracheboud, this Proceedings) which might, in the long run, help us better to understand the real function of auxins. Somaclonal variation recieving considerable attention during the past four years, appears not to have produced the expected collection of novel genotypes.

8. STABLE GENE TRANSFER

Gene transfer to protoplasts was one of the most successful areas in protoplast research. Although working against the overwhelming competition of *Agrobacterium*-mediated gene transfer, it could contribute to the general progress in knowledge - in transient and stable gene transfer experiments. It could be shown that encapsidation of marker genes into liposomes and subsequent fusion with protoplasts produced stable integrative transformation (16). Direct gene transfer by incubation of protoplasts with DNA, PEG and additional treatment by electroporation led to recovery of transgenic plants (17); subsequent detailed genetic analysis demonstrated that the foreign hybrid gene had a pattern of inheritance identical to single endogenous genes (18); the efficiency of stable transformation could be increased to values in the range of up to 10 % of protoplast-derived clones by careful optimisation of numerous parameters, either including (19) or excluding electroporation (20, 21). Such high values of integrative transformation were byond the imagination even of the greatest optimists a few years ago, and they have consequences as they make work with non-selectable genes feasible. Also micro-injection of DNA into protoplasts has been developded (22) and optimised to reach values of 40 % (23). It also has been shown that there is no "host-range" problem and that protoplasts from graminaceous species can be transformed as those from herbaceous dicots (24, 25). Mixed application of two independent genes, e.g. a selectable and a non-selectable one disclosed that on top of the high transformation there is the phenomenon of co-transformation at frequencies higher than 50 % (the maximum case so far is 88 %), which makes it very easy to recover transformants with any gene by just mixing it with a selectable marker gene and screening through a number of selected clones for the presence of the non-selected gene (26). *In situ* hybridization of radio-actively labelled DNA probes of marker genes have been used to visualize the presence and the position of genes in metaphase chromosomes, and this could be used to map foreign genes on specific chromosome arms of specific chromosomes (27). A further important result is the demonstration that single copy genes can be transferred from total plant genomes without their prior isolation, by treating protoplasts with sheared total genomic DNA, as has been exemplified in an experiment with a transgenic tobacco plant containing one stable locus for kanamycin resistance (T. Müller, pers. communication). Direct gene transfer leads to the integration at one (76 %), two independent or linked, three or more independent or linked chromosomal loci; there are, however, also a number of cases of inheritance for which the straight forward interpretation would be integration into the chondriome or plastidome; there are, however, no molecular data to support this interpretation. Detailed genetic analysis through to the eighth sexual generation has shown that foreign genes in transgenic plants can be, at least, as stable as original host genes. A few interesting cases of instability have, however, also been found, which show that one can not rely a priori at the absolute stability of transformed genes (Potrykus, pers. communication). The integration of foreign genes into host genomes is obviously, and independent of the method used, random. It would, of course, be of some advantage if it was possible, to direct the integration to a specific predetermined site in the genome. Recent work on homologeous recombination and on gene targeting has probably opened the route to this goal: In work with complementing 5' and 3' deletion mutants of a selectable marker gene on separate plasmids it has been shown that homologeous recombination is active in somatic cells and can be optimised by provision of appropriate substrate (M. Baur, pers. communication). On the basis of these studies it was possible to design experiments, again with the use of 5' and 3' deletion mutants of a selectable marker gene, this time, however, with part of the system stably integrated in a host genome, to test whether and to proof that foreign DNA can be placed precisely (by one base pair) into a predetermined position in the host genome (J. Paszkowski, pers. communication). Also these experiments have important consequences as they most probably enable now to work towards gene replacement, towards directed mutagenesis via gene inactivation, and to study the (hypothetical) position effect on gene expression.

9. TRANSIENT GENE TRANSFER

Incubation of protoplasts with genes and treatment with electroporation (28) or magnesium/PEG (I. Negrutiu, pers. communication) not only leads to stable integrative transformation, but also to "transient gene expression" (see also Bates, this Proceedings). This phenomenon has been used in numerous experiments for rapid analysis of genes of hybrid gene constructs as well as for detailed analysis of promoter and enhancer functions, as for the analysis of the protein coding region, including studies on gene splicing. This possibility for rapid analysis of gene functions in homologeous and heterologeous systems has profoundly contributed to the overwhelming progress in the undersanding of gene regulation during the very recent years. The same also applies to the similar studies on transient functions with caulimo viruses and gemini viruses.

10. CELL SORTING

The enrichment for specific protoplast types from mixed populations has shifted away from enrichment based on differential buoyant density and sedimentation characteristics in step density gradients to the use of "fluorescence activated cell sorting" which has been developed from the first attempts at the time of the previous Symposium to a routine and efficient technique for laboratories having access to one of the expensive automated fluorescence activated cell sorters (see also K. Glimelius, this Proceedings, and e.g. [54], [55], [61]. Automated cell sorting has great merits in the area of analysis of large populations and has been shown to be efficient in the enrichment for fusion heterokaryocytes, enabling recovery of somatic hybrids of wildtype protoplasts, thus providing experimental independency from selection protocols based on the availability of mutants. In this second area of application the manual combination of two wildtype protoplasts in microdrops and their susequent electrofusion and microculture as described in the contribution by G. Spangenberg (this Proceedings) appears to be a valuable alternative for safe recovery of somatic fusion hybrids independent of any selection scheme.

11. PHYSIOLOGY & FUNDAMENTAL RESEARCH

Because of the authors personal involvement the review was discussing so far only progress in the area of protoplast culture and genetic manipulation. There are, however, other fields which are as important, which cannot be discussed in more depth in this review and the reader is referred to the contributions at the Symposium.

REFERENCES

1. Potrykus, I. et al.: Protoplasts 1983, Poster Proceedings, Experientia Suppl. Vol. 45, Birkhäuser, Basel 1983.

2. Potrykus, I. et al.: Protoplasts 1983, Lecture Proceedings, Experientia Suppl. Vol. 46, Birkhäuser, Basel 1983.
3. Koop, H.U. and Schweiger, H.G., J. Plant Physiol. 121: 245-257 (1985)
4. Fujimura, T. et al., Plant Tissue Culture Lett. 2: 74-75 (1985)
5. Toriyama, R. et al., Theor. Appl. Genet. 73: 16-19 (1986)
6. Kyozuka, J. , Mol. Gen. Genet. 206: 408-413 (1987)
7. Brunold, C. et al., Mol. Gen. Genet. 208: 469-473 (1987)
8. Koop, H.U. and Schweiger, H.G., Europ. J. Cell Biol. 39: 46-49 (1985)
9. Pelletier, G. et al., Mol. Gen. Genet. 191: 244-250 (1983)
10. Maliga, P. et al., in Green CE et al. (eds) Plant Tissue and Cell Culture. Alan Less, New York, pp 265-274 (1987)
11. Schenck, H.R., in Protoplasts 1983, Poster Proceedings, Experientia Suppl. Vol. 45, Birkhäuser Basel, pp 280-281 (1983)
12. Terada, R., et al., Mol. Gen. Genet. 210: 39-43 (1987)
13. Galun, E. and Aviv, D., in Weissbach A and Weissbach H. (eds) Plant Molecular Biology. Methods in Enzymology, Vol. 118, Academic Press, Orlando (USA) pp 595-611 (1986)
14. Imamura, J. et al., Theor. Appl. Genet. 74: 445-450 (1987)
15. Spangenberg, G. and Schweiger, G., Eur. J. Cell Biol. 41: 51-56 (1986)
16. Caboche, M. and Deshayes, A., C.R. Acad. Sci. (Paris) 299: 663-666 (1984)
17. Paszkowski, J. et al., EMBO J. 3: 2717-2722 (1984)
18. Potrykus, I. et al., Mol. Gen. Genet. 199: 169-177 (1985)
19. Shillito, R.D. et al., Biotechnology 3: 1099-1103 (1985)
20. Krüger-Lebus, S. and Potrykus, I., Plant Mol. Biol. Rep. 5(2), 289-294 (1987)
21. Negrutiu, I. et al., Plant Mol. Biol. 8: 363-373 (1987)
22. Reich, T. et al., Biotechnology 4: 1001-1004 (1986)
23. Neuhaus, G. et al., Eur. J. Cell Biol. 43, Suppl. 17, pp 39
24. Potrykus, I. et al., Mol. Gen. Genet. 199: 183-188 (1985)
25. Lörz, H. et al., Mol. Gen. Genet. 199: 178-182 (1985)
26. Schocher, R.J. et al., Biotechnology 4: 1093-1096 (1986)
27. Mouras, A., et al., Mol. Gen. Genet. 207: 204-209 (1987)
28. Fromm, M. et al., P.N.A.S. USA 82: 5824-5828 (1985)

OBSTACLES AND PERSPECTIVES IN PROTOPLAST RESEARCH

Tage R. Eriksson
Institute of Physiological Botany, Box 540
S-751 21 UPPSALA, Sweden

Introduction

A list of obstacles in using protoplasts in basic studies of plant physiology and genetics will be long if we expected the isolated plant protoplast to behave as a unicellular micro-organism in culture. A protoplast is a component of a very complex developmental system and in addition it is, in general, cultured under conditions which induce heavy stress reactions. It is today known that different types of stress can affect gene regulation that genes are turned on or off. Therefore, it is not surprising that there are difficulties in inducing normal growth with consecutive divisions, new wall formation, and normal hormone metabolism in cultured protoplasts. It has also been reports of increased RNase activity and lower amount of mRNA available in protoplasts to mention a few problems. Some of these problems can be partly overcome by pretreatment of the doner plant or by using other tissues like meristems and stam segments or by modifying the procedure of protoplast isolation and culture. Shoot cultures seem to be an excellent material which gives high and reproducable yields of protoplasts. The use of agarose as a gelling agent or for embedded cultures seems to be an important improvement for many protoplasts which are difficult to culture (Eriksson, T.R., 1985).

Plant regeneration

Unfortunately, the new techniques of plant molecular biology have not so far been used to solve the problems concerning the protoplast isolation, culture and plant regeneration. Instead the molecular biology research seems to focus on rapid progress in genetic engineering of some model plants which are easily regenerated after a protoplast phase. The workers on protoplast growth and plant regeneration should make use of the molecular biology techniques to be able to better understand the difference between different species in respect to their ability to regenerate normal plants after a protoplast phase.

The role of genotype and explant on plant regeneration

The protoplast-derived cell clones have to form new meristems in a mass of undifferentiated cells to initiate the formation of adventitious shoots or somatic embryos which later should be rooted. The role of plant genotype on plant regeneration after a protoplast phase is under debate. Some workers claim that the choice of genotype is of crucial importance, while others believe that the relationship of genotype to morphogenetic competence *in vitro* is complex and indirect. The morphogenetic competence is strongly influenced by physiological and environ-mental factors, which in turn influence the synthesis, transport and the availability of plant growth regulators. (Vasil, J., 1986). Embryogenesis from protoplasts may occur indirectly, via callus proliferation and subsequent determination of embryogenic cells, or directly from the cells which develop from protoplasts. The physiological conditions of the doner plant and tissue, isolation procedures, culture conditions and genotype influence the development patterns. (Evans DA and Bravo JE., 1983 and Gleddie et al., 1985). The use of 2,4-dichlorophenoxyacetic acid (2,4-D) as a growth regulator has allowed the induction and long-term maintenance of embryogenic cultures in a wide variety of gramineous species (Vasil, J., 1986). The presence of mature cells or cells of more differentiated tissues appears to inhibit the expression of embryogenic competence by the meristematic cells. Results from several studies indicate that the physiological and developmental states of the explant at the time of excision and culture exercise more control on its morphogenetic competence than its genotype. An uninterrupted preservation of meristematic differentiation from the doner cell to the establishment of polarity is the most widely used method. The regeneration of plants from protoplasts - derived cell clones is on the whole poorly understood but scientific insight into the many factors involved is slowly emerging. A number of biochemical markers for somatic embryogenesis have been identified in cell cultures (Stuart et al. 1985, Sung 1985, Carlberg et al. 1986 and Carlberg et al. 1987) and are already being used to assess the value of the various manipulations being employed to obtain plant regeneration.

Advantage and disadvantage of protoplasts

In many studies protoplasts can be replaced by isolated embryos, meristem tissues, pollen embryos and pollen mother cells (Neuhaus et al. 1987). The advantage of all these tissues and cells is the ease to regenerate whole plants.

On the other hand a culturable protoplast system offers a lot of advantages in selection on cellular level, small space for a large number of individuals, suitable for direct gene transfer, membrane studies and plastid isolation.

Protoplasts are a prerequisite for somatic hybridization and for the chemically mediated DNA uptake as well as for induced DNA uptake by electroporation. Protoplast fusion has produced some interesting hybrids and cybrids. The techniques of protoplast fusion are now well developed but the identification of the hybrids has been hampering the progress. Now it seems however, that the technique of flow cytometry could overcome the problems of identification and selection of somatic hybrids. There is a need for large numbers of hybrids since there is an elimination of one genome or part of it as well as ongoing plastid elimination. Unfortunately, there is no good selection system available on cellular level for agricultural important traits. When a toxin is produced and can be isolated selection can take place on protoplast or cell level.

Protoplast populations

To some extent the plant cell and protoplast culture techniques are based on methods developed for microbiology and the culture of animal cells. These methods include work with high numbers of individuals and the use of selection markers. There is however, a clear disadvantage in working on cell populations since only direct conclusions can be drawn with respect to the individual cells. A look through a microscope at a freshly isolated protoplast population reveals a great heterogeneousity in size, subcellular structure and cell cycle stage. Careful cytological and genetical studies have shown that plant cell populations often contains a small and sometimes a large fraction of genetically different cells.

Protoplasts can be viewed as injured cells which must embark on a program of repair before they are capable of sustained division, and thus cultured plant protoplasts are in a linear rather than a cyclical developmental sequence (Gould and Daines 1985). Studies of this program of repair should preferably be done on synchronized systems, but staining methods do not allow quantitation of nuclear DNA content in live protoplasts. This is a major technical problem which present flow sorting of protoplasts in terms of cycle phase (Gould and Daines 1985). Careful cell cycle studies could contribute with knowledge of importance for genetic modifications of plant cells. In an asynchronous protoplast population, used in gene-transfer experiments some fraction of the population should be in the "competent" phase. However, cycle phase manipulation of protoplast populations still offers an opportunity to increase the efficiency with which transformation of protoplasts can be achieved. In attempts to isolate chromosomes cycle phase manipulations also play a key role.

Unfortunately, plant cell populations are difficult to fully synchronize. Partial synchrony can be induced but the methods used are rather hazardous (Constabel et al. 1977). A non-hazardous method of synchronization has however, recently been described (Weber et al. 1986). A general method of bypassing the problem of heterogeneity would be the use of individual cells or protoplast (Schweiger et al. 1987). Plant cells are unfortunately, very difficult to culture in single cell cultures in defined culture media. In most studies there is a need of large numbers of identical cells or cells in the same cell cycle stage to make analysis possible. Therefore, studies on cell cycle synchronization should have high priority in plant cell biology research.

Protoplasts and gene-transfer

When the aim is to produce transgenic plants it is possible to choose among several gene-transfer methods of which only some require protoplasts (Table 1).

Table 1 The most common gene-transfer methods and somatic fusion to obtain transgenic plants and somatic hybrids or cybrids.

I Plant viruses as biological vectors
(Need no protoplasts or cell cultures)

II DNA-transfer via *Agrobacterium tumefaciens*
A) Wound-infection methods
B) Co-cultivation techniques
a) Regenerable protoplasts needed
b) Explant material (especially useful when protoplast can not be regenerated and is more rapid than a protoplast system)

III Chemically mediated DNA uptake (PEG) and polyvinyl alcohol together with Ca^{2+} and high pH (regenerable protoplasts needed)

IV Electroporation (regenerable protoplasts needed)

V Microinjection (microspores, cells, embryos, protoplasts etc)

VI Somatic hybridization (regenerable protoplasts needed)

Proembryos or immature embryos as recipients of foreign DNA will reduce the need of protoplasts and provide an alternative were the plant regeneration problem is solved and the genetical instability in culture is greatly reduced. The method of microinjection into embryos will provide an excellent experimental tool for the study of gene expression and regulation during plant development. Furthermore, it should become an efficient means for producing genetically improved plants (Neuhaus et al. 1987). The high frequency of transformation is also encouraging. The procedure is a way to overcome the lack of success in regenerating whole plants in many protoplast systems. After microinjection into plant protoplasts, only transformed calli have been obtained (Melnikov et al. 1985; Crossway et al., 1986; Reich et al. 1986) and transgenic plants only of *Nicotiana tabacum* and *N. debneyi* (Neuhaus et al., in prep.)

DNA technology and the need of protoplast

Breeding for improved varieties is the traditional method of genetic manipulation. Plant breeding involves the selection of desirable traits from progeny derived from crosses between parents, one or both of which may also possess undesirable traits. In a typical plant breeding program, selection that favors desirable gene combination is used to eliminate undesirable traits. This is accomplished by a variety of means, taking advantage of genetic segretion and the techniques of recurrent selection and backcrossing. Recombinant DNA technology, with the theoretical capability to transfer any gene from any organism into a plant, has the potential to complement the efforts of traditional plant breeders. In principle, a gene determining a single trait can be transferred to a superior variety without concomtant transfer of undesirable traits. Multigenic traits will be more difficult to identify, isolate, clone and transfer into plants, but the technique is available. The limitation is our knowledge in plant biochemistry and plant physiology. The gene pool available for crop improvement by traditional methods is limited since gene transfer among plants is confined to those species that can be naturally mated. The genetic engineering of plants through recombinant DNA methods has the potential for introducing any gene into any plant improvement. In addition, a gene can theoretically be designed to be expressed in specific tissues at specific stages of the plant's development. Therefore, this new technology not only promises the introduction of many new desirable genes into plants but also the precise regulation of these genes. The combination of recombinant DNA, transformation and plant regeneration technologies provides the tools necessary to improve crop varieties. The genetical instability of cultured cells and tissues is one of the greatest obstacles in using protoplasts as recipients in transformation studies as well as in methods of mass propagation of plant material. The real interest is various aspects of whole plant improvement and therefore we should minimize the lenght of time cells and tissues are subjected to *in vitro* culture. In this connection it should be pointed out that research ought to be directed to elucidate the mechanism involved in the genetic instability. When we know how to reduce genetic variation we can also direct the variation and make use of it.

Conclusions

There are some obstacles in using protoplasts in research and plant engineering but most of these can be overcome. Protoplasts constitute a special type of plant cells, deprived of the wall and cultured under stress condition which must be taken into consideration when protoplasts are compared with other sources of plant material.

Today, there is a lack of fundamental knowledge in plant cell biology and especially in the regulation of growth and differentiation of plant cells. There is a lack of precision and reproducability in many of the reports on protoplasts. It is a hope that molecular biology and plant cell biology in an interplay will correct this deficiency (Cocking 1986). It is obvious that protoplast research is dependent on the interest of protoplasts for molecular biology and genetic engineering. Therefore it is urgent to work out the fundamental problems of genetic instability, plant regeneration and low reproducability for some important crop plants. On the other hand, protoplast research should have good support in future since protoplasts are an excellent source of experimental material in fundamental plant cell biology, plant physiology and genetics.

References

Cocking, E. C., (1986). Plant cell biology in the 21st Century: The needs of plant cell and tissue culture. InGreen, C.E., Somers, D.A., Hackett, W.P. and Biesboer, D.D.(eds):Plant Tissue and Cell Culture: Alan R. Liss, Inc., New York, pp 3-15.

Gleddie S, Keller WA, Setterield G (1985) Production of new hybrid plants through protoplast fusion. In: Cheremisinoff PN, Oullette RP (eds) Biotechnology: Applications and Research. Lancaster, Pennsylvania: Technomic Publishing Company. Inc, pp 231-242.

Constabel F, Kurz WGW, Chatson KB, Kirkpatrick JW (1977) Partial synchrony in soybean cell suspension cultures induced by ethylene. Exp Cell Res 105:263 -268.

Crossway A, Oakes JV, Irvine JM, Ward B, Knauf VC, Shewmaker CK (1986) Integration of foreign DNA following microinjection of tobacco mesophyll protoplasts. Mol Gen Genet 202:179-185.

Eriksson, T.R., (1985). Protoplast isolation and culture: In: Fowke, L.C. and Constabel, F. (eds): Plant Protoplasts: CRC Press Inc. Roca Raton, Florida, pp 1-20.

Evans D.A., Bravo JE (1983) Plant protoplast isolation and culture. Int Rev Cytol 316:33-53.

Melnikov P.V., Pasternak TP, Gleba YY, Sytnik KM (1985) Microinjektion von DNA in Zellen höherer Pflanzen. Dokl Akad Nauk SSSR Ser Geol, Khim Biol Nauki 10:69-71.

Neuhaus, G., Spangenberg, G. Mittelsten Scheid, O. and Schweiger. (1987). Transgenic rapeseed plants obtained by the microinjection of DNA into microspore-derived embryoids. Theor Appl Genet (in press).

Reich T.J., Iyer VN, Miki BL (1986) Efficient transformation of alfalfa protoplasts by the intranuclear microinjection of Ti plasmids. Biotechnol 4:1001-1004

Stuart D.A., Nelsen J, McCall CM, Strickland SG, Walker KA (1985). Physiology of the development of somatic embryos in cell cultures of alfalfa and celery. In Zaitlin M, Day P, Hollaender A (eds): "Biotechnology in Plant Science," Academic Press, pp 35-47.

Sung Z.R., (1985). Developmental Biology of plant embryogenesis. In Freeling, M (ed): "Plant Genetics", New York: Alan R Liss, 35, pp 115-128.

Schweiger, H.G., Koop, H.K., Kranz, E., Neuhaus, G., Spangenberg, G. and Wolff, D. (1987). Individual selection, culture and manipulation of higher plant cells.

Weber G, de Groot E, Schweiger HG (1986) Synchronization of protoplasts from Glycine max (L.) Merr. and Brassica napus (L.). Planta 168:273-280.

Vasil, J.K., (1987) Developing cell and tissue culture systems for the improvement of cereal and grass crops. J. Plant. Physiol. Vol. 128, pp 193-218.

ASPECTS OF PROTOPLAST CULTURE AND PLANT REGENERATION

M.R. DAVEY and J.B. POWER
Plant Genetic Manipulation Group, Department of Botany, University of Nottingham, Nottingham NG7 2RD, UK.

1. INTRODUCTION

Protoplasts are now used as source material for a range of genetic manipulations. Indeed, protoplast fusion has already produced somatic hybrid and cybrid plants, whilst techniques in molecular biology have advanced so rapidly that transgenic plants have now been reported. In addition to stable transformation, there is increased use of protoplasts in transient gene expression studies designed to provide information on promoters, vectors and methods of delivering genetic materials to plant cells.

Even with these major technological advances, successful culture and regeneration of protoplasts following various manipulations remain fundamental to this science. Although well established culture methods can be applied when initiating new programmes, much of the work remains empirical in nature. At the last Protoplast Symposium, general principles of culture were discussed(1). The literature in this area is now so extensive that only specific examples will be cited, reflecting the wide range of species under investigation in the Plant Genetic Manipulation Group, University of Nottingham. Such crops range from cereals, legumes and woody species to medicinal plants and ornamentals. Recent technological advances include improvements in protoplast fusion, particularly in relation to flow cytometry, and the use of electromanipulation to improve protoplast division and plant regeneration.

2. CEREALS

2.1. Plant regeneration

Since cereals and grasses are the most important group of crop plants, it is to be expected that there is intense effort worldwide to reproducibly regenerate plants from protoplasts/cultured tissues. As discussed recently (2), plants have been recovered from protoplast-derived tissues of *Pennisetum americanum*, *P. purpureum*, *Panicum maximum*, *P.miliaceum*, *Saccharum officinarum*, and *Oryza sativa*. However, plants of *Panicum miliaceum* were albino.

The age and physiological status of the explants used to establish the cultures for protoplast isolation had a major influence on the success of the regeneration protocols. Similarly, in developing an efficient plant regeneration system for rice protoplasts of the two japonica varieties Taipei 309 and Fujisaka 5, Abdullah et al. (3) stressed the need to pay attention to the source of cell suspensions for protoplast isolation. In the case of Taipei 309, a dry, compact, nodular callus was established from the basal 1-2 cm of seedling leaves or from embryos; mature embryo scutellum was used for Fujisaka 5. Callus transferred to amino-acid based culture medium, produced fast-growing suspensions composed of fine clusters of small, cytoplasmic cells which were maintained in exponential growth by weekly sub-culture. It was necessary for such suspensions to be maintained for at least 6-8 months to ensure satisfactory protoplast isolation, division and plant regeneration. Heat shock (4) of the freshly isolated protoplasts followed by plating in a rich medium, made semi-solid with 1.2% Sea Plaque agarose (5), produced plating efficiencies of 20 to 40% after 14 days of culture, with 1% of plated protoplasts forming compact colonies by the fourth week of culture. Three to four somatic embryos were regenerated from 15 to 20% of the cell colonies on a suitable hormone-free medium; the nitrogen source particularly influenced this stage of the procedure. It was essential to allow the coleoptiles of the embryos to elongate in the dark, since premature transfer to the light resulted in development of atypical leafy structures. All regenerated plants were morphologically normal.

2.2. Transformation assessments

Transformation studies are an important aspect of any genetic manipulation programme. PEG-mediated uptake of plasmid into rice protoplasts resulted in kanamycin resistant tissues expressing neomycin phosphotransferase (NPTII) activity, with the presence of foreign DNA being confirmed in transformed tissues (6). In studies using pCaMVNEO (7) with protoplasts from clone LB1 of rice Taipei 309, transformed protoplast-derived colonies were also selected on kanamycin medium using PEG, electroporation and a combination of these two plasmid delivery methods (8). Although cultured cells of most gramineous species are highly resistant to kanamycin (2) and even though there was variation between experiments in the natural resistance of non-transformed rice protoplasts to this antibiotic, these results illustrated that kanamycin resistance can be used effectively as a selectable marker. Further experiments should confirm whether other selectable markers, such as resistance to hygromycin, are more widely applicable to this protoplast system. Currently, the challenge is still to combine transformation with plant regeneration to produce transgenic rice plants.

Sugarcane protoplasts have also been transformed by direct uptake of a plasmid carrying a chimaeric gene, kanamycin resistant colonies being selected following PEG-induced uptake of Sma1 linearised pABD1 (9). The level of transformation was low (8 in 10^7), and for this system electroporation was unsuccessful as a method of delivery. It was essential, as in rice, to use agarose-medium for protoplast culture. The cell suspensions used for protoplast isolation were non-morphogenic, thus eliminating the possibility of regenerating transgenic plants. Since this report (9), additional information has been obtained concerning the maintenance of morphogenesis in long-term cultured sugarcane callus. Interestingly, three tissue types can be recognised in such cultures; a white compact callus capable of plant regeneration, a friable non-morphogenic callus and a mucilaginous nodular callus which can revert to the other types. Alternating sub-culture on a high-low 2,4-D régime, combined with visual selection, are essential to maintain morphogenesis for at least 30 months after callus initiation (10). This knowledge now enables plants to be recovered from non-transformed sugarcane protoplasts isolated from morphogenic cell suspensions (11). Reports of plants from protoplast-derived sugarcane calli by others (12) now make the production of transgenic plants a distinct reality for this important monocotyledonous crop.

3. LEGUMES

3.1. Studies with soybean

Soybean (*Glycine max*) continues to present a challenge to the tissue culturist. Indeed, progress with this crop species now falls behind that of cereals. Although plants have been regenerated from zygotic embryos (13,14) and complex explants (15) and soybean seedling hypocotyl protoplasts divide rapidly to form callus (16), reproducible plant regeneration from protoplasts has yet to be reported.

Since wild species are important germplasm reserves, considerable effort has been directed towards developing regeneration systems for perennial *Glycine* species. In a study of 5 accessions of *G. clandestina*, two accessions, G1231 and G1145, exhibited morphogenesis from seedling explants (17). An extension of this work with 30 accessions showed 12 within the species *G. canescens*, *G. falcata*, *G. latrobeana* and *G. tomentella* to be morphogenic, with G1171 (*G. canescens*) superior to the rest (18). Although it does not necessarily follow that accessions capable of plant regeneration from explants will also regenerate from protoplast-derived callus, it has been possible, using this basic information, to develop a successful protocol for plant regeneration from seedling cotyledon protoplasts of *G. canescens* G1171 and *G. clandestina* G1231 (19,20).

In somatic hybridisation studies, electrofusion of FDA-labelled hypocotyl protoplasts from dark-grown seedlings of *G. max* with chlorophyll-containing cotyledon protoplasts of *G. canescens* produced heterokaryons, which, following sorting by flow cytometry, produced callus which showed hybridity based on isoenzyme analysis (21). Future work should clarify whether buds on the hybrid callus will develop into somatic hybrid plants.

3.2. Culture of Medicago and Trifolium protoplasts

Since seedlings are a readily available source of protoplasts, attention has been focussed on the culture of cotyledon protoplasts of wild *Medicago* species. In a comparison of three culture methods, the technique of mixing protoplasts of the wild species with those of albino *M. sativa* protoplasts, as nurse, resulted in plating efficiencies in excess of 60% for 5 species tested (22). Plants were regenerated from free-floating somatic embryos of *M. falcata*, *M. glutinosa* and *M. hemicycla*, but in these three species and in *M. difalcata* and *M. varia*, optimum plant regeneration occurred via a callus phase.

Somatic hybridisation (see section 6) normally requires a selection procedure in order to encourage the throughput of putative hybrid or cybrid tissues prior to plant regeneration. Although automated flow cytometry has applications in this respect, this technique is, and may remain, restricted, because of its expense. Consequently, the ability to manually select small numbers of cells and to culture these successfully is an attractive alternative for many laboratories. Assessments of the use of feeder layers, conditioned medium, and membrane culture chambers, employed protoplasts of the forage legumes *M. falcata*, *M. sativa*, *Onobrychis viciifolia* and *Trifolium arvense* as model systems (23). Conditioned medium permitted growth of cell suspension protoplasts of *M. falcata* and *M. sativa* at densities of 10^3/ml and 10^2/ml respectively. When the protoplasts of interest were enclosed in a cylindrical membrane chamber surrounded by nurse protoplasts in a liquid medium over an agarose layer, fewer than 50 protoplasts from cell suspensions of *M. sativa*, *O. viciifolia* and *T. arvense* and seedling cotyledon protoplasts of *M. sativa* could be induced to form callus. Division frequencies within the membrane chambers were comparable to those of protoplasts plated as nurse in the liquid layer outside the chambers. The optimum pore size of filter membranes used to construct such chambers was found to be 5 to 12 µm.

Heterokaryons selected manually with a micromanipulator have been cultured to hybrid callus (24). Following PEG-high pH/Ca^{2+} induced fusion of chloroplast-containing M. sativa seedling cotyledon protoplasts with FDA-labelled M. falcata cell suspension protoplasts, and FDA-labelled M. sativa cotyledon protoplasts with chloroplast-containing M. quasifalcata cotyledon protoplasts, heterokaryons were identified by their dual fluorescence under UV illumination. Heterokaryons, selected with a micro-manipulator, were introduced into an albino M. sativa nurse protoplast culture. Following initial culture in the dark, green, putative hybrid colonies were selected from the albino nurse cells and cultured to callus. Chromosome and isoenzyme analyses confirmed the hybrid nature of the heterokaryon-derived tisues, but the latter have, to date, failed to produce plants.

Studies with Trifoliuim pratense have shown that plant regeneration is possible from callus (25,26) or cell suspensions (27). It is surprising that plant regeneration from protoplast-derived cells has not been described, despite a report of callus production from mesophyll protoplasts (28). This lack of a method to regenerate plants from protoplasts has limited hybridisation with other clovers (e.g. T. sarosiense) to conventional crossing procedures followed by embryo rescue (29). However, a protocol has now been developed at Nottingham to regenerate plants through somatic embryogenesis from mesophyll and cotyledon protoplast-derived callus. Such plants have been established in soil (unpublished data) for the cultivars Norseman, Merkur, Lakeland and Astra; each had a specific requirement in relation to the embryogenesis-induction medium. Although the level of NAA, 2,4-D, and kinetin was kept constant, the thiamine concentration required for embryogenesis was cultivar specific. Subsequent culture on the same medium, but with increased thiamine, induced globular stage somatic embryos. Later development occurred only when the globular structures were transferred to medium with low levels of GA_3. Thus, 0.05 mg/l GA_3 allowed globular embryos to mature only to the heart stage, while 1.0 mg/l GA_3 forced a rapid development through all of the comparable zygotic embryo stages of maturation but produced fasciated shoots which failed to root. An intermediate concentration of GA permitted a more controlled development of globular embryos to the torpedo stage. Further embryogenic development of these somatic embryos was induced on a medium with NAA and thiamine, whereupon embryos matured and germinated on a hormone-free medium. In this method, primary embryogenesis occurred on agar-solidified medium, while subsequent steps were on filter paper placed over the agar surface. Visual selection of morphologically normal embryos was important at each stage of transfer. Surprisingly, the material used to seal the Petri dishes also influenced the growth of embryos, their germination being stimulated when Nescofilm was replaced with Clingfilm during the final stage on hormone-free medium.

3.3. Investigations with legume seedling root hairs

Recent studies have shown that cell walls at the apices of root hairs can be enzymatically digested (30). Whilst many protoplasts are released, others remain in situ within the undigested tube-like side walls of the root hairs. This enzyme treatment removes the barrier to Rhizobium-host plant specificity (31). Normally, seedlings of Trifolium repens are nodulated only by Rhizobium trifolii, but following enzyme-PEG treatment and inoculation, seedlings develop nitrogen fixing root nodules with Rhizobium loti and Rhizobium 101/U, normally specific to Lotus corniculatus. Sequential enzyme-PEG treatments and inoculation on days 2,6 and 10 following seed germination stimulate nodulation. This system, in which the protoplasts are cultured in situ within the partially digested root hairs, could remove the barrier to Rhizobium-host specificity in a number of legumes and may, in the future, extend the Rhizobium association to non-legumes.

4. WOODY PLANT SPECIES

Success in regenerating plants from protoplasts of woody species, has, until recently, been slow. This has been due to a suppression of totipotency in tissues of species that exhibit distinct juvenile or mature phases of growth, and the inability to prepare protoplasts. However, a general method is now available for isolating protoplasts from axenic shoot cultures of woody species of several families including the Betulaceae, Juglandaceae, Rosaceae and Rutaceae (32). Axenic shoot cultures can, in this context, be considered as juvenile tissues, with young, rapidly expanding leaves providing an ideal source of mesophyll protoplasts capable of division for a range of deciduous fruit and nut species of the genera Corylus, Juglans, Malus, Prunus, Pyrus, Poncirus and Actinidia (32).

For some of these species, plant regeneration from protoplasts has now been established, e.g. with the cherry rootstock Colt (Prunus avium x pseudocerasus) (33) using protoplasts from cell suspensions or leaf mesophyll. In such an example, the conventional pathway to regeneration applies, whereas, for other woody species an intermediate state is required involving root formation. Protoplast-derived roots, in turn, can be induced to give rooted shoots. This pathway, frequently ignored in terms of protoplast regeneration, has been used for the regeneration of plants from protoplasts of sour cherry (Prunus cerasus) and at a low frequency Conference pear (Pyrus communis) and will, no doubt, be extended to several other plants of commercial value.

The efficiency of regeneration is central to the realistic utilisation of genetic engineering technology for woody plant improvement. It has been shown that, for Colt cherry protoplasts, there is a marked carry-over effect of electroporating cell suspension protoplasts (see section 7.2) leading to a more pronounced growth of protoplast-derived callus, an elevated frequency of shoot regeneration from such calli, and the development of a more prolific root system on the regenerated shoots (34). It would seem, therefore, that woody species protoplasts should no longer be regarded as recalcitrant, since recent modifications to standard protoplast isolation and handling procedures have shown that they could soon be on a par with the herbaceous species.

For some woody species, very simple modifications to the cultural procedures can be effective. Such an example is Solanum dulcamara (woody nightshade, bittersweet), a woody medicinal plant valued for its steroidal alkaloids (35). Although leaves and axenic shoot cultures have been used for protoplast isolation, it has been difficult to maintain cell suspensions suitable for protoplast isolation (36). Recent work has confirmed that, as in the case of rice, the age and quality of the suspensions determine the type of protoplasts released and their plant regeneration capability. Leaf callus transferred to UM liquid medium (37), produced cell clumps of various sizes from which a 64 μm fraction could be harvested. This fraction on return to liquid medium produced granular clumps of isodiametric, highly cytoplasmic cells. Cell suspensions, used 4 to 5 days after transfer and 5 to 7 months after initiation, could be digested with a Biogel P6 purified-mixture of 2% Rhozyme HP150, 2% Meicelase and 0.03% Macerozyme R10 to release protoplasts capable of rapid division in agarose-solidified medium. The division frequency was approximately seven fold higher in agarose than in liquid culture. It was essential to reduce the osmotic pressure of the culture medium during the third week of culture; earlier or later reductions lead to necrosis. Protoplasts released from cells harvested 8-9 days after sub-culture underwent extensive budding rather than division, while cell suspensions older than 12 months were unsuitable as a source of protoplasts since they contained a high proportion of elongated or vacuolated cells. Protoplast-derived tissues regenerated shoots on zeatin containing medium.

5. ORNAMENTALS

The successful regeneration of plants from protoplasts has been extended to species of prime ornamental value such as Pelargonium aridum, P. x hortorum and P. peltatum (38), and to species within several genera of the Compositae such as Dimorphotheca, Rudbeckia (39) and Chrysanthemum.

Plant regeneration from protoplasts of Chrysanthemum, arguably the most important ornamental in the Compositae, has proved difficult (40). An alternative approach to regeneration can involve a triggering of totipotency through protoplast fusion with an actively dividing but unrelated protoplast system. Plants have been recovered from mesophyll protoplasts of C. morifolium following their fusion with protoplasts of an albino cell suspension of Salpiglossis sinuata. Although the underlying mechanism is unclear, some form of cross-feeding in the heterokaryons, followed by unilateral loss, in this case, of the Salpiglossis genome, may prevail. Plants regenerated using this strategy can be compared in terms of somaclonal variation with those obtained directly from explants or callus. In this context, ornamentals such as Chrysanthemum, display considerable variation in regenerant populations in terms of flower colour and morphology.

6. SOMATIC AND GAMETOSOMATIC HYBRIDS

An alternative strategy to somaclonal variation, is, of course, somatic hybridisation. In addition to some of the developments mentioned earlier in relation to legumes, an intergeneric somatic hybrid between Nicotiana tabacum and Petunia inflata has been produced. "Nicotunia", produced following the electrofusion of protoplasts, is a fertile plant. The ability to produce such fertile intergeneric somatic hybrids may facilitate gene flow, in this particular case, leading to improvements in ornamentals such as Petunia. There is a long history of attempts to produce "Nicotunia" (41), the main factor determining success on this occasion being the substitution of the traditionally used Petunia genus representative P. hybrida with P. inflata.

Success in somatic hybridisation does not necessarily depend upon the establishment of plant regeneration from protoplasts of both parental species. Indeed, in this context, it can be a distinct disadvantage. Provided regeneration is possible for one parent in the cross, a half-selection strategy can be developed. If, as in the case of tomato, this is coupled with inherent morphological differences between the two parental species, a selection based on limited growth of one species together with visual identification of the somatic hybrid can be effective in terms of recovering somatic hybrid plants (42).

Taking this concept a stage further has formed the basis of

gametosomatic hybridisation in plants. Two such hybrids now exist for Nicotiana (43) and Petunia (44); in both systems the inability of pollen tetrad protoplasts to grow in culture formed part of a selection scheme. Gametosomatic hybridisation, the fusion of haploid (or diploid from a tetraploid parent) tetrad protoplasts with, usually, diploid somatic protoplasts, leads to the production of triploid gametosomatic hybrids. This approach, applicable to any male fertile flowering plant species used as a haploid partner in fusion, offers the potential of fertility in the hybrid product, ironically, more likely in hybrids of more distantly related parental species. Fertility in the gametosomatic hybrid is provided by the diploid genome, and, during the process of meiosis, fosters an elimination, to varying degrees, of the haploid genome (chromosomes) through selfing or backcrossing. From a plant breeding angle, this allows for the incorporation, to varying extents, of the haploid genome (through mitotic or meiotic recombination) into the basic diploid (recipient) species, against a background of more or less guaranteed fertility. The gametosomatic hybrid may prove far more useful in an applied sense, since the problem of infertility in intergeneric and some intrageneric somatic hybrids can be eliminated. The question of likely fertility in these triploid gametosomatic hybrids, as yet, is not fully proven experimentally, since only two examples of these novel hybrids exist at present. However, one could predict fertility at the interspecific level, and, because of a high degree of chromosomal homology infertility at the intraspecific level. This would seem to be the case based on the two examples in the genera Nicotiana (43) and Petunia (44).

Gametosomatic hybridisation offers a clear advantage, not only in a plant breeding context, but also at the culture level since it is not dependent on any prior knowledge of the culture requirements of the haploid (pollen tetrad protoplast) partner. In this sense, recalcitrant species can now be seen as possible sources for novel characteristics, with an in-built half- selection capability. Possibly, gametosomatic hybridisation may supercede somatic hybridisation; the latter, to date, has produced few, if any, novel hybrids of actual or potential agronomic value.

7. IMPROVED METHODOLOGIES FOR CULTURE, FUSION AND PLANT REGENERATION FROM PROTOPLASTS

7.1. PEG fusion

PEG has been used extensively as a fusogen, but certain protoplast systems, particularly those from leaf mesophyll and young seedling cotyledons, suffer extensive damage with this agent. It has been shown (45) that PEG 1540 purchased from Polysciences Inc., contains acidic impurities which can be removed by treatment with a mixed ion exchange resin. In various combinations involving Nicotiana glauca and Medicago sativa leaf protoplasts and those from cotyledons of young Glycine max seeds, deionised PEG as compared to commercial grade PEG promoted viable heterokaryon formation, while high pH/Ca^{2+} combined with purified or commercial grade PEG was preferable to PEG alone. Low PEG concentrations (17% w/v) for short periods (3 min) were preferable to an exposure to higher levels of longer duration. Deionised PEG also stimulated protoplast division. The Ca^{2+} concentration was important for specific combinations, with, for example, a need to increase the Ca^{2+} to 2200 mg/l in the fusion of Glycine with Medicago. In the case of Glycine with Nicotiana, a hypotonic salt solution was preferable to a Ca^{2+} washing solution.

In an extensive study (16), PEG from several sources was assessed for its contamination with carbonyl compounds. PEG from Aldrich (MW. 1500; 8,000), Böehringer (MW. 1500) and Koch-Light (MW. 1500; 6,000) had the lowest contamination and produced viable heterokaryons at high frequency (8-11%) in a range of protoplast combinations. Such preparations were free of protoplast clumps after treatment, and therefore ideal for flow cytometry. Buffering with 75 mM HEPES, filter sterilisation rather than autoclaving, and storage at 4°C in dark bottles, ensured protoplast division after fusion. The amount of carbonyl compounds increased with storage of the commercially available products. Samples older than three years from manufacture could be purified by distillation in the presence of carbonyl-free methanol.

7.2. Electroporation of plant protoplasts

Currently, there is interest in the effects of electromanipulating eukaryotic cells. Since it was reported (47) that short duration high voltage electric pulses modify the structure of the plasma membrane of animal cells by the formation of pore-like openings, this "electroporation" treatment has been shown to increase the permeability of the plasma membrane to molecules of various sizes, including DNA. Interestingly, this technology, with both commercially available and home-built instruments, has been used to introduce RNA, DNA and intact TMV particles into plant protoplasts.

Although information on the effects of electrical treatments on cell growth and development is scarce, electromagnetic fields have been reported to stimulate DNA synthesis in cultured mammalian cells (48) and somatic embryogenesis in Medicago protoplasts (49). Long duration, low current electric fields stimulated shoot formation and regeneration in tobacco callus (50,51), by promoting the polar transport of auxins (52).

In studies to determine the effects of electroporation on protoplast culture (53), protoplasts of Glycine canescens, Prunus avium x pseudocerasus, Pyrus communis, Solanum dulcamara and S. viarum were electroporated. In all cases, three successive exponential pulses at 10 sec intervals between 250 and 1000V caused a small decrease in protoplast viability, lower voltages with longer pulses being more detrimental to protoplast viability than high voltages of short duration. Protoplast size influenced their survival, and those less than 20 µm in diameter (e.g. Prunus) were less sensitive than larger protoplasts (e.g. Pyrus). An electroporation-treatment stimulated wall regeneration and cell division. This was particularly noticeable in Prunus, Pyrus and Solanum viarum, with division within 5 days as compared to 15, 9 and 11 days respectively in untreated preparations. In all cases, the onset of division was preceeded by a marked increase in cell diameter which could not be accounted for by electrofusion.

Electroporation also increased the plating efficiency, with treated protoplasts forming cell colonies within 10 days of culture. The plating efficiency was inversely related to pulse duration; longer pulses giving less stimulation than shorter pulses at the same voltage. The number of cells in each colony, indicative of the rate of cell division, was also greatest at lower voltages and longer pulses. The enhancement of cell division was sustained throughout culture, with more cell colonies from electroporated protoplasts compared to the controls.

In an extension of these studies (34), enhanced shoot regeneration in tissues from electroporated protoplasts of Colt cherry (Prunus avium x pseudocerasus) was noted. A similar effect has also been observed in Solanum dulcamara. In Colt cherry, pulses of 250 to 500V increased callus growth, with all these tissues showing increased shoot formation.

Such shoots were larger and more leafy than shoots from tissues derived from untreated protoplasts, facilitating the rooting of such shoots without the need for an intermediate growth phase in order to encourage internode elongation. All shoots derived from treated protoplasts rooted readily to produce branched root systems. In contrast, rooting on shoots from untreated protoplasts was restricted to 75% of the regenerants. Roots were also unbranched in the controls. No differences were observed between rooted plants of electroporated or untreated protoplasts in terms of their survival on transfer to the glasshouse.

Electroporation also increased the shoot regeneration capacity in S. dulcamara with rooting being stimulated on plants from protoplasts electropulsed at 500 to 750V for 10 or 20 µsec. The time required for plant regeneration was 84 days as compared to 105 to 119 days for the untreated preparations. More morphological variation of regenerants was observed than for Colt cherry; for example, shoots from protoplasts electroporated at high voltages (1250V, 10 to 30 µsec) were dwarf. Root development was also inhibited in such dwarf shoots. However, these plantlets developed to phenotypically normal plants after transfer to hormone-free medium. It will be interesting, in future work, to determine whether electroporation exerts its effects on the internal hormonal balance of protoplast-derived callus in a similar way to that proposed for electro-treated tobacco callus (52).

The stimilatory effect of electroporation on protoplast growth does not imply that this technique can be employed to induce cell division in hitherto non-dividing protoplasts. Indeed, the general applicability of this technology still needs to be fully evaluated. Collectively, the results indicate that electroporation of isolated protoplasts has practical application not only in transient and stable transformation studies, but also for enhancing cell growth and plant regeneration.

8. CONCLUSION

In this very brief asessment of protoplast isolation, culture and regeneration, it is apparent that several exciting technological developments have taken place recently, with clear implications in relation to somatic hybridisation, gametosomatic hybridisation, and transformation. The recalcitrance in culture of several plants has now become a problem of the past such that more crop species can be realistically incorporated into improvement programmes based on their genetic manipulation.

9. REFERENCES

1. Davey MR: in: Protoplasts 1983, Lecture Proc. 6th Internatl. Protoplast Sympos., Basel. Experientia Supplementum 46 (eds.) I Potrykus et al., Birkhäuser Verlag, Basel, Boston, Stuttgart: 19-29, 1983.
2. Vasil I: J. Plant Physiol. 128: 193-218, 1987.
3. Abdullah R, Cocking EC, Thompson JA: Bio/Technology 4: 1087-1090, 1986.
4. Thompson JA, Abdullah R, Chen WH, Gartland KMA: J. Plant Physiol. 127: 367-370, 1987.
5. Thompson JA, Abdullah R, Cocking EC: Plant Science 47: 123-133, 1986.
6. Uchimiya H, Fushimi T, Hashimoto H, Harada H, Syono K, Sugawara Y: Mol. Gen. Genet. 204: 204-207, 1986.
7. Fromm ME, Taylor LP, Walbot V: Nature 319: 791-793, 1986.

8. Yang H, Zhang H, Davey MR, Mulligan BJ, Cocking EC: Proc. 7th Internatl. Protoplast Sympos., 1987.
9. Chen WH, Gartland KMA, Davey MR, Sotak R, Gartland JS, Mulligan BJ, Power JB, Cocking EC: Plant Cell Rep. 6: 297-301, 1987.
10. Chen WH, Davey MR, Power JB, Cocking EC: J. Exp. Bot. (in press).
11. Chen WH, Genetic Manipulation of Sugarcane. Ph.D. Thesis, Univ. of Nottingham, 1987.
12. Srinivasan C, Vasil I: J. Plant Physiol. 126: 41-48, 1986.
13. Ranch JP, Oglesby L, Zielinski AC: In Vitro Cellular and Developmental Biol. 21: 653-658, 1985.
14. Hammatt N, Davey MR: J. Plant Physiol. 128: 219-226, 1987.
15. Wright MS, Ward DV, Hinchee MA, Carnes MG, Kaufman RJ: Plant Cell Rep. 6: 83-89, 1987.
16. Hammatt N, Davey MR: In Vitro Cellular and Developmental Biol. (submitted).
17. Hammatt N, Davey MR, Nelson RS: Physiol. Plant.68: 125-128, 1986.
18. Hammatt N, Nelson RS, Davey MR: Plant Cell, Tissue and Organ Culture (in press).
19. Hammatt N, Kim H-I, Davey MR, Nelson RS, Cocking EC: Plant Science 48: 129-135, 1987.
20. Davey MR, Hammatt N: Soybean Genetics Newsletter 14: 144-147, 1987.
21. Hammatt N, Nelson RS, Lister A, Cocking EC, Davey MR: Proc. 7th Internatl. Protoplast Sympos., 1987.
22. Gilmour DM, Davey MR, Cocking EC: Plant Science 48: 107-112, 1987.
23. Gilmour DM, Davey MR, Cocking EC, Pental D: J. Plant Physiol. 126: 457-465, 1987.
24. Gilmour DM, Davey MR, Cocking EC: Plant Science (in press).
25. Phillips GC, Collins GB: Crop Sci. 19: 59-64, 1979.
26. Horvath-Beach K, Smith RR: Plant Sci. Lett. 16: 231-237, 1979.
27. Phillips GC, Collins GB: Crop Sci. 20: 323-326, 1980.
28. Oelck MM, Rao PS, Ozias-Akins P, Schieder O: Poster Proc. 6th Internatl. Protoplast Sympos., Basel (eds.) I Potrykus et al., Birkhäuser Verlag, Basel, Boston, Stuttgart: 50-51, 1983.
29. Phillips GC, Collins GB, Taylor NL: Theor. Appl. Genet. 62: 17-24, 1982.
30. Cocking EC: Bio/Technology 3: 1104-1106, 1985.
31. Al-Mallah M, Davey MR, Cocking EC: Bio/Technology (in press).
32. Revilla MA, Ochatt SJ, Doughty S, Power JB: Plant Science 50: 133-137, 1987.
33. Ochatt SJ, Cocking EC, Power JB: Plant Science 50: 139-143, 1987.
34. Ochatt SJ, Chand PK, Rech EL, Davey MR, Power JB: Plant Science (in press).
35. Emke A, Eilert U: Plant Cell Rep. 5: 31-34, 1986.
36. Kowalczyk TP: Somatic Cell Studies on Solanum viarum Dunal with Solanum dulcamara L. Ph.D. Thesis, Univ. of Nottingham, 1984.
37. Uchimiya H, Murashige T: Plant Physiol. 54, 936-944, 1974.
38. Yarrow SA, Cocking EC, Power JB: Plant Cell Rep. 6: 102-104, 1987.
39. Al-Atabee JS, Power JB: Plant Cell Rep. (in press).
40. Otsuka H, Svematsu N, Toda M: Bull. Shizuoka Agr. Exp. Stn., 30: 25-33, 1985.
41. Pental D, Hamill JD, Pirrie A, Cocking EC: Mol. Gen. Genet. 202: 342-347, 1986.
42. Kinsara A, Patnaik SN, Cocking EC, Power JB: J. Plant Physiol. 125: 225-234, 1986.
43. Pirrie A, Power JB: Theor. Appl. Genet. 72: 48-52, 1986.
44. Lee CH, Power JB: Plant Cell Rep. (in press).

45. Kao KN, Saleem M: J. Plant Physiol. 122: 217-225, 1986.
46. Chand PK, Lister A, Davey MR, Power JB, Cocking EC: Proc. 7th Internatl. Protoplast Sympos., 1987.
47. Neumann E, Rosenheck K: J. Membrane Biol. 10: 279-290, 1972.
48. Takahashi K, Kaneko I, Date M, Fukuda E: Experientia 42: 185-186, 1986.
49. Dijak M, Smith DL, Wilson TJ, Brown DCW: Plant Cell Rep. 5: 468-470, 1986.
50. Rathore KS, Goldsworthy A: Bio/Technology 3: 1107-1109, 1985.
51. Rathore KS, Goldsworthy A: Bio/Technology 3: 253-254, 1985.
52. Goldsworthy A, Rathore KS: J. Exp. Bot. 36: 1134-1141, 1985.
53. Rech EL, Ochatt SJ, Chand PK, Power JB, Davey MR: Protoplasma (in press).

RECENT DEVELOPMENTS IN THE REGENERATION OF AGRONOMICALLY IMPORTANT CROPS FROM PROTOPLASTS

K. Judith WEBB
AFRC Institute for Grassland and Animal Production,
Welsh Plant Breeding Station, Aberystwyth, Dyfed, Wales, UK

INTRODUCTION

Of more than 350,000 species of plant known to science, only 100-200 are of major economic importance. The bulk of the world's food comes from just 15 species (1), which are mainly cereals, legumes and root crops. Improvements of many of these species by the powerful tools of genetic manipulation presents difficulties, mainly because tissue culturists have yet to identify successful protocols for _in vitro_ regeneration in such plants.

Potato is a notable exception in that plants can now be regenerated from protoplasts with a high degree of reliability. This paper examines the refinement of culture techniques in potato and surveys progress made in leguminous and gramineous species since the last Protoplast Symposium (2).

POTATO

Just a decade ago, protoplasts of potato were first persuaded to regenerate into plants. Now, the situation has improved to the point where these protoplast-derived somaclones and somatic hybrids are being evaluated for agronomic purposes.

Exactly what steps were involved in this process? In 1977, Shepard and Totten (3) induced mesophyll protoplasts of the commercial tetraploid cultivar Russet Burbank, to regenerate into complete plants. The described procedure was complicated, involving specific growing conditions for both donor plants and resulting colonies. Even so, the number of responsive cultures bearing one or two fully developed shoots was low.

Other workers (4,5) used a different approach: they isolated protoplasts from shoots grown _in vitro_. Wenzel and his co-workers (4) used dihaploid lines of potato, whereas Thomas (5) used commercial tetraploid varieties. In the latter experiments, protoplasts isolated from micropropagated shoots gave low but reliable plating efficiencies when cultured on a relatively simple medium. Initially colonies from cv. Maris Bard alone produced plants. The number of plants produced was still low with about 10% of colonies responding.

The identification and use of both of these convenient and competent tissues as sources of protoplasts has resulted in substantial and rapid progress. Other genotypes of potato, including tetraploid cultivars (6,7,8,9) and dihaploid lines (10) and also related _Solanum_ species (11,12) can now be cultured successfully. However, the efficiency of plant regeneration via protoplast cultures remains dependent on genotype. In cultivars Bintje, Majestic, King Edward and Desiree (6,8,9) about 70% of protoplast-derived callus cultures produced shoots. In others, the efficiency remains lower; for example, shoots regenerated from about 12% of cultures in cv. Maris Piper (8). Similarly, other species of potato showed differences in their responsiveness; for example, 60% of cultures of S.etuberosum (11), about 40% of S.brevidens (12) and a maximum of 5% of S.fernandezianum (11) regenerated shoots.

Manipulation of the culture media has, in its turn, improved plating efficiencies of protoplasts and regeneration of shoots (3,8). Modifications of the media may also bring additional benefits, such as greater cytological stability of the regenerants (6).

Protoclones have been carefully analysed for both morphological and cytological variation (6,13,14). Somatic hybrid plants produced between wild species of Solanum and both tetraploid and dihaploid potatoes (10,15,16) are also under scrutiny. These attempts to transfer useful genes into potato have met with some success. For example, Austin et al. (15) found expression of resistance to leaf roll virus in somatic hybrids produced between S.brevidens and a dihaploid line of S.tuberosum.

The development and application of protoplast technology in potato clearly illustrates the speed with which sporadic regeneration of plants can evolve into reliable protocols. These techniques can then be extended to other, often less amenable, material which is relevant to the plant breeder's needs.

Three main approaches - use of appropriate tissues, use of responsive genotypes and manipulation of media - were responsible for success in potato. These basic approaches remain important for progress in other crop species. In both leguminous and gramineous species, significant steps have been made towards their successful culture. Within both groups, herbage crops are generally more amenable in culture than their economically more important counterparts, grain legumes and cereals. Developments in related herbage species can highlight alternative routes to the desired goal.

LEGUMINOSAE

The forage species, lucerne (Medicago sativa) and bird's-foot trefoil (Lotus corniculatus) (24), are both particularly responsive in culture. The efficiency of these two systems permits the production of hybrid plants by somatic hybridization: interspecific hybrids have been produced between M.sativa and M.falcata (17) and between L.corniculatus and L.coimbrensis (18). M.arborea (19) is another potential fusion partner for lucerne and hybrid callus has been produced following electrofusion (F.Damiani, pers. comm.).

The clovers are more difficult to culture. Three Trifolium species regenerate plants from protoplasts: T.rubens (21), T.hybridum (alsike clover) (20) and T.repens (white clover) (22,23,24). In the latter species, the use of a selected genotype (22,24) improved the efficiency of shoot regeneration but failed to overcome the difficulties of inducing roots on regenerants (24). In both white and alsike clover, analysis of those plants with roots revealed gross cytological and morphological abnormalities (20,24). By contrast, protoplast-derived plants regenerated from T.rubens (21) were either diploid or tetraploid depending on the source of protoplasts. Mesophyll tissues yielded only diploid plants whereas plants from suspension cultures were all tetraploid, raising the possibility, that in the latter cells, the chromosome number had already doubled during culture (21).

Successful reports of protoplasts developing into plants in two grain legumes were previously recorded (2). Of the three most important grain legumes, bean, peanut and soybean (1), reports of the latter species and its wild relatives are the most encouraging. In soybean, protoplasts have been isolated directly from various tissues, for example, cotyledons (25) and primary leaves (26), and also from suspension cultures (27). These approaches have met with some success. Protoplasts divided to form callus (25,26,27) but shoot formation has not been reported.

Few reports exist of analysis of protoplast-derived cultures. Cultures resulting from experiments using a photoautorophic suspension culture retained the characteristics of the parental cell line and showed little cytological change, with 99% of the analysed cells remaining diploid (27). However, these cultures of soybean only yielded colonies. By contrast, protoplasts isolated from hypocotyl tissues of a wild relative of soybean, Glycine canescens (28) yielded shoots at low frequencies. The plants established from these shoots appeared morphologically normal. Higher frequencies of plant regeneration (0.2-1.2% protoplasts regenerate into whole plants) have since been achieved using cotyledonary protoplasts of G.canescens and G.clandestina (29). Protoplasts of G.canescens and G.max are currently being used to establish conditions required for selection of fusion products and regeneration of hybrid plants (N. Hammatt, pers. comm.).

Plant regeneration in legumes, particularly grain legumes, generally occurs directly from explants, not via callus or suspension cultures. The use of morphogenic suspension cultures has, however, proved to be critical in the regeneration of plants from protoplasts for the Gramineae.

GRAMINEAE

Protoplasts isolated from morphogenic suspension cultures of several species of grass and one species of cereal are capable of regeneration of green plants. Tall fescue and Italian ryegrass reliably and routinely yield plants which can be established in soil (30, Dalton, pers.comm.). Similarly, protoplasts isolated from suspension cultures of two other gramineous species - sugarcane (34) and rice (31,32,33) - also produce green plants capable of growing in soil. In rice, plant regeneration was enhanced by embedding protoplasts in agarose (32) and subjecting them to heat shock prior to culture (35).

Plants produced from tall fescue and Italian ryegrass (Dalton, pers. comm.) and rice (32,33) are under scrutiny to determine the degree of somaclonal variation. In the latter species, about 80% plants showed normal characters but variation was found in the remaining plants. These variations included changes in chromosome number, reductions in fertility and kernel weight but an overall increase in yields of grain - possibly resulting from the larger number of stems with grain-bearing panicles.

By contrast, cereals such as wheat, barley, maize and triticale remain difficult. But protoplasts isolated from some of these most difficult species can now be cultured. For example, protoplasts isolated from suspension cultures of triticale (36), maize (37), bread wheat (38) and barley (39) divided to form colonies. Of these, protoplast-derived colonies of maize (40,41) and bread wheat (MGK Jones, pers. comm.) showed limited morphogenesis, while those from barley produced albino plantlets (42). The induction of albino plantlets from protoplasts has also been reported in perennial ryegrass (30).

Most successful protocols use suspension cultures as a source of protoplasts. However, there is always a conflict in handling such cultures: the faster a culture grows, the easier it is to isolate protoplasts but the more quickly it loses its morphogenic potential. Thus, cultures must be re-initiated at intervals or parental stocks must be stored at low temperatures.

Protoplast cultures of many Gramineae are now being used in investigations of gene uptake and expression (43,44,45) and in establishing techniques for somatic hybridization (46,47,48). These techniques and the use of species which allow regular and reliable production of green plants will undoubtedly result in successful production of hybrid and genetically transformed plants.

CONCLUDING COMMENTS

Cereals, legumes and root crops are all represented by responsive species, but plant regeneration is often sporadic and many species have not yet fully succumbed to techniques of tissue culture. But, in potato, a precedent has been set: prior to 1977, no reports claimed plant regeneration from protoplasts and few reports existed of plant regeneration even directly from explants. Today, just 10 years on, breeders are poised to use techniques of tissue culture in potato breeding programmes. Gramineous species are already showing some conspicuous successes. With an equivalent input of time and effort we can expect similarly rapid developments in our remaining difficult crops.

ACKNOWLEDGEMENTS

I wish to thank many scientists for providing reprints and pre-prints for inclusion in this article and my colleagues for helpful discussions.

REFERENCES

1. Janick J, RW Schery, FW Woods and VW Ruttan: Plant Science. An introduction to world crops. WH Freeman & Co., 1981.
2. Dale PJ: 6th Int. Protoplast Symposium, Basel (eds. I. Potrykus et al.), pp. 31-42, 1983.
3. Shepard JF, RE Totten: Plant Physiol. 60, 313-316, 1977.
4. Wenzel G, O Schieder, T Przewozny, SK Sopory, G Melchers: Theor. Appl. Genet. 55, 49-55, 1979.
5. Thomas E: Plant Science Letters 23, 81-88, 1981.
6. Creissen GP, A Karp: Plant Cell Tissue and Organ Culture 4, 171-182, 1985.
7. Gunn RE, JF Shepard: Plant Science Letters 22, 97-101, 1981.
8. Foulger D, MGK Jones: Plant Cell Reports 5: 72-76, 1986.
9. Bokelmann GS, S Roest: Z. Pflanzenphysiol. 109, 259-265, 1983.
10. Fish N, A Karp, MGK Jones: In vitro (in press).
11. Barsby T, JF Shepard: Plant Science Letters 31, 101-105, 1983.
12. Nelson RS, GP Creissen, SWJ Bright: Plant Science Letters 30, 355-362, 1983.
13. Secor GA, JF Shepard: Crop Sci. 21, 102-105, 1981.
14. Sree-Ramulu K, P Dijkhuis, S Roest: Theor. Appl. Genet. 65, 329-338, 1983.
15. Austin S, MA Baer, JP Helgeson: Plant Sci. 39, 75-82, 1985.
16. Austin S, MK Ehlenfeldt, MA Baer, JP Helgeson: Theor. Appl. Genet. 71, 682-690, 1986.
17. Teoule E: 6th Int. Protoplast Symposium, Basel (eds. I. Potrykus et al.), p. 279, 1983.

18. Wright RL, DA Somers, RL McGraw: 6th Intl Congress Plant Cell and Tissue Culture, Minnesota, (eds. D.A. Somers et al.) Abstr. No. 108, 1986.
19. Mariotti D, S Arcioni, M Pezzotti: Plant Science Letters 37, 149-156, 1984.
20. Webb KJ, PJ Dale: In: Biomolecular Engineering in the European Community (ed. E. Magnien), p. 793, 1986.
21. Grosser, JW, GB Collins: Plant Science Letters 37, 165-170, 1984.
22. Bhojwani SS, DM White: Plant Science Letters 26, 265-271, 1982.
23. Ahuja PS, Lu DY, EC Cocking, MR Davey: Plant Cell Reports 2: 269-272, 1983.
24. Webb KJ, S Woodcock, DA Chamberlain: Plant Breeding 98, 111-118, 1987.
25. Lu DY, S Cooper-Bland, D Pental, EC Cocking, MR Davey: Z. Pflanzenphysiol. 111, 389-394, 1983.
26. Tricoli DM, MB Hein, MG Carnes: Plant Cell Reports 5, 334-337, 1986.
27. Chowhury VK, JM Widholm: Plant Cell Reports 4, 289-292, 1985.
28. Newell CA, HT Luu: Plant Cell, Tissue and Organ Culture 4, 145-149, 1985.
29. Hammatt N, H-I Kim, MR Davey, RS Nelson, EC Cocking: Plant Science 48, 129-135, 1987.
30. Dalton SJ: J. Plant Physiol. (in press).
31. Abdullah R, EC Cocking, JA Thompson: Biotechnology 4, 1087-1090, 1986.
32. Kyozuka J, Y Hayashi, K Shimamoto: Mol. Gen. Genet. 26, 408-413, 1987.
33. Ogura H, J Kyozuka, Y Hayashi, T Koba and K Shimamoto: Theor. Appl. Genet. 74, 670-676, 1987.
34. Srinivasan C, IK Vasil: J. Plant Physiol. 126, 41-48, 1986.
35. Thompson JA, R Abdullah, W-H Chen, KMA Gartland: J. Plant Physiol. 127, 367-370, 1987.
36. Stolarz A, H Lorz: In: Genetic Manipulation and Plant Breeding (eds. W. Horn et al.), pp. 499-501, 1986.
37. Ludwig SR, DA Somers, WL Peterson, RF Pohlman, MA Zarowitz, BG Gegenbach, J Messing: Theor. Appl. Genet. 71, 344-350, 1985.
38. Maddock SE: Plant Cell Reports 6, 23-26, 1987.
39. Luhrs R, H Lorz: In: Genetic Manipulation in Plant Breeding (eds. W. Horn et al.), pp. 483-486, 1986.
40. Vasil V, IK Vasil: Theor. Appl. Genet. 73, 793-798, 1987.
41. Kamo, KK, KL Chang, ME Lynn, TK Hodges: Planta 172, 245-251, 1987.
42. Luhrs R, PA Lazzeri, H Lorz: 7th Int. Protoplast Symposium, Wageningen (in press).
43. Chen, WH, KMA Gartland, MR Davey, R Sotak, JS Gartland, BJ Mulligan, JB Power, EC Cocking: Plant Cell Reports 6, 297-301, 1987.
44. Lorz H, B Junker, J Scell, A de la Pena. In: Plant Tissue and Cell Culture, pp. 303-316, Alan R. Liss, Inc. 1987.
45. Junker B, J Zimny, R Luhrs, H Lorz: Plant Cell Reports 6, 329-332, 1987.
46. Ozias-Akins R, RJ Ferl, IK Vasil: Mol. Gen. Genet. 203, 365-370, 1986.
47. Tabaeizadeh Z, RJ Ferl, IK Vasil: Proc. Natl Acad. Sci. USA 83, 5616-5619, 1986.
48. Terada R, J Kyozuka, S Nishibayashi, K Shimamoto: Mol. Gen. Genet. (in press).

SELECTION, EXPERIMENTAL MANIPULATION AND CULTURE OF SINGLE PROTOPLASTS.

H.-U. Koop*[1], and G. Spangenberg[2]

[1] Botanical Institute, Menzinger Straße 67, D-8000 München 19, F.R.G.

[2] Max Planck Institute for Cell Biology, Rosenhof, D-6802 Ladenburg, F.R.G.

Heterogeneity of plant cell cultures with respect of response to culture conditions is one of the major obstacles in experimental manipulation of higher plant cells or protoplasts. One way to overcome this problem would be by manipulating single selected cells or protoplasts individually.

An instrumental setup is presented, which allows microscopical selection of single protoplasts and their individual culture in a microculture chamber with regeneration of cells, microcolonies and, subsequently, plants at high frequencies. The system, assembled in a flow cabinet, consists of an inverted microscope with a stepmotor-driven programmable stage, a computer-controled micropump, allowing for the delivery and withdrawal of nanoliter amounts of liquids through a hydraulically coupled microcapillary pipet, and another programmable stepmotor, which vertically drives a holding device for different tools to be used under microscopical control. These tools may be a microcapillary pipet for preparing microculture chambers (Koop and Schweiger 1985a) and for transfering single cells or a pair of microelectrodes for the controled electrofusion of individually selected protoplasts (Koop and Schweiger 1985b). The experimental system has successfully been used with a variety of plant species for individual culture of single cells, controled fusion of protoplasts and cell reconstitution experiments including microinjected subprotoplasts (Spangenberg et al. 1986) and has recently been summarized by Schweiger et al. (1987).

REFERENCES

1. H.-U. Koop and H.G. Schweiger, J. Plant. Physiol. 121 (1985a) 245-257.
2. H.-U. Koop and H.G. Schweiger, Europ. J. Cell. Biol. 39 (1985b) 46-49.
3. G. Spangenberg, G. Neuhaus and H.-G. Schweiger, Europ. J. Cell. Biol. 42 (1986) 236-238.
4. H.G. Schweiger, J. Dirk, H.-U. Koop, E. Kranz, G. Neuhaus, G. Spangenberg and D. Wolff, Theor. Appl. Genet. 73 (1987) 769-783.

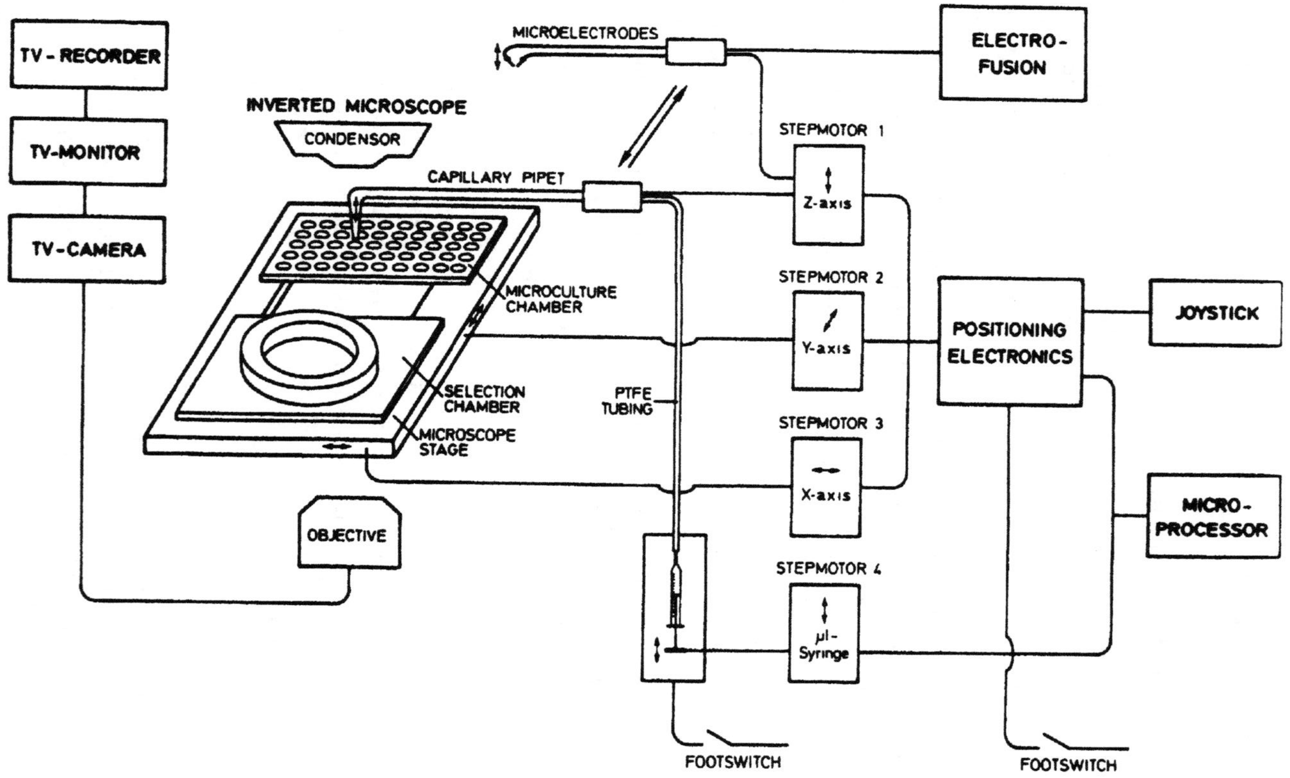

SETUP FOR HANDLING SINGLE CELLS INDIVIDUALLY

UNIPARENTAL, FUSANT, AND CHIMERIC PLANTS REGENERATED FROM PROTOPLASTS AFTER STREAK PLATING IN AGAROSE GELS

H.BINDING, D.BÜNNING, E.GÖRSCHEN, J.JÖRGENSEN, R.KOLLMANN, G.KRUMBIEGEL-SCHROEREN, LING H.Q., J.MONZER, G.MORDHORST, J.RUDNICK, A.SAUER, D.WITT, M.ZUBA

Botanisches Institut, Christian-Albrechts-Universität, Olshausenstraße 40-60, D-2300 Kiel 1, FRG

1.INTRODUCTION

Some results are discussed,here, which were obtained through plating of protoplasts in streaks of locally high density embedded in 10 µl agarose media droplets (culture lenses) covered with liquid media (FIGURE 1; Binding and Kollmann 1985). This technique is a modification of the common agarose plating method first introduced by Shepard 1980. Advantages of the streaky plating technique appeared in three areas namely in the regeneration of protoplasts, the fusion of protoplasts and the production of cell graft chimeras (Binding et al., in press).

2.RESULTS

2.1.Protoplast Regeneration

Protoplast regeneration was investigated in 19 species of the Liliatae class and 60 species of the Mangnoliatae class (in Linum: Ling and Binding 1987; in potato: Zuba and Binding, in press; other species: Binding et al., in press and prep.). Regeneration was promoted in the majority of the dicots and limitedly in some of the monocots tested. The advantages were expressed in the following features: Mitosis was induced in some of monocot genera which were recalcitrant in other culture systems, namely in 10 genera such as *Anthericum*, *Anthurium*, *Chlorophytum*, *Gladiolus*, *Orchis*, *Tacca* and *Vanilla*. As a second advantage cell clones of several dicots were fast stabilized to survive transfer to low osmotic media. Cultures of *Petunia hybrida* and *Solanum nigrum* could even be transferred after three days. Growth of the protoplasts in streaky cultures was essential for subsequent plant regeneration in an albino genotype of *Asparagus*.Shoot regeneration after protoplast culture in streaky lenses was obtained in the following species: *Arabidopsis thaliana*, *Asparagus officinalis*, *Biscutella laevigata*, *Brassica napus*, *Cichorium intybus*, *Clianthus formosus*, *Hypericum montbretii*, *H. perforatum*, *Linum alpinum*, *L. amurense*, *L. hologynum*, *L. perenne*, *L. salsoloides*, *L. usitatissimum*, *Nicotiana tabacum*, *Petunia hybrida*, *Reseda alba*, *R. lutea*, *R. luteola*, *R. odorata*, *R. phyteuma*, *Solanum nigrum* and *S. tuberosum*. The organogenesis was accellerated in nearly every tested species that was susceptible to organogenesis. Less than 20d were needed in the genera: *Reseda*, *Nicotiana*, *Petunia* and *Solanum*.

These results show up some progress in protoplast regeneration. However, it must be stated that no essential break-through in plant regeneration of recalcitrant species was achieved neither with agarose nor in particular with streaky cultures.

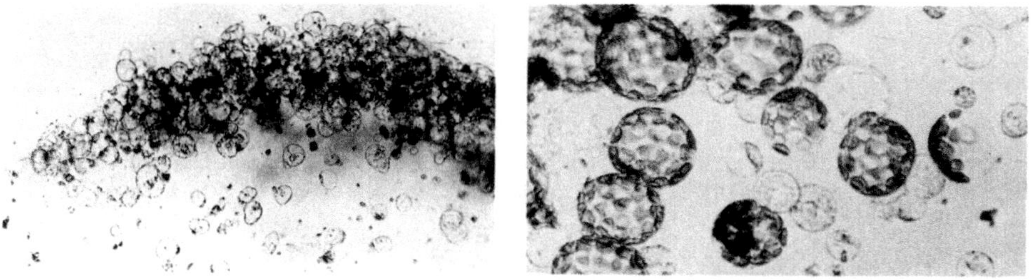

FIGURE 1: Parts of streaky culture lenses: a) albino potato clone St-W1; b) St-W1 and tomato cv..

2.2. Protoplast Fusion

Protoplast fusion was mentioned as another field of application of the streak plating technique. The protoplasts are packed tightly together. Therefore it was near at hand to make use of this for protoplast fusion which was induced with high calcium and high pH (Keller and Melchers 1973). The method offers two benefits: It enables one to watch the fusion and to follow up the early development of particular fusants in fixed orientation. Moreover the stabilization of the original shape of the protoplasts by the gel forces may increase the probability of cybridization. Good viability and sufficient yields of hybrid and cybrid plants were observed in experiments with Solanum tuberosum and Solanum nigrum. In one well passed off experiment 10^4 protoplasts of each partner were submitted to fusion treatment, a total of 545 callus lines formed shoots, five lines were hybrids, one line segregated to a hybrid and cybrid protoclones and one line produced only cybrid shoots with S. nigrum nuclei and S. tuberosum plastids (FIGURE 2). The technique was also applied in the genetic analysis of an albino potato line. Hybridization with two types of Solanum nigrum revealed evidence for a potato plastid mutation.

2.3. Cell Grafting

Streaky culture lenses furthermore offer a means of studying cell to cell interactions. Chimeric plants were obtained through the co-culture of heterospecific protoplasts. Adjacent protoclones were able to react as regenerative units resulting in the formation of chimeric plants, which were called cell graft chimeras and symbolized by (+). A single sectorial chimera was found in the combination albino Solanum nigrum (+) Nicotiana tabaccum. A number of 22 chimeral lines were selected composed of Solanum nigrum (+) tuberosum. A dihaploid potato clone was associated to a green and to an albino nightshade. The constitution of the plantlets was proven by the segregation of uniparental lines through adventitous shoot formation and protoplast regeneration. The chimeral nature was stabilized in periclinal arrangement of a potato epiderm covering a nightshade core. The heterospecific cells were interconnected with numerous well organized plasmodesmata. The variegated chimeras were investigated under tissue culture conditions.

Plants composed of the green partners were grown to maturity in the greenhouse. Some observations with the periclinal chimeras revealed informations on the contribution of the layers to the development and differentiation of the plants. The leaf morphology was determined mainly by the core tissue. However the leaf edges showed some influence of the potato layer 1 (FIGURE 2). It is selfevident that no tubers were formed. The chimeras flowered independently of the daylength - a character of nightshade located in the core. Shape and pigmentation of the crown are due to the epiderm which is the main constituent of this organ. The size of the flowers is intermediate. Seedset was rare. Limited fertility may depend on incompatibility between the L 2 derived nightshade pollen and L 1 derived potato epiderm. All seeds germinated to form pure nightshade seedlings. Besides small malformations of variegated leaves reduced fertility was the only indication of incompatibility.

FIGURE 2: Solanum tuberosum (St) / S. nigrum protoclones. From left: St, Sn, St(+)Sn, St(x)Sn hybrid, St-nucleus(x)Sn-plastid cybrid.

Other combinations are under investigation to study further interspecific interactions and to detect taxonomic limits of cell grafting and allelopathic factors.

Acknowledgements.
The investigations were supported by the DFG and BMFT.

REFERENCES

1. Binding H, R Kollmann. In: Schäfer-Menuhr A (ed) In vitro techniques. Propagation and long term storage. Nijhoff/Junk Dordrecht Boston Lancaster, 1985, pp 93-99.
2. Binding H, D Witt, J Monzer, G Mordhorst, R Kollmann: Protoplasma, in press, 1987.
3. Keller WA, G Melchers: Z Naturforschg 28c, 1973, pp 737-741.
4. Ling HQ, H Binding: Plant Breeding 98, 1987, pp 312-317.
5. Shepard JF: Plant Sci Lett 18, 1980, pp 327-333.
6. Zuba M, H Binding. In: Bajaj YPS (ed) Potato, Biotechnology in Agriculture and Forestry 3, Springer Berlin, Heidelberg, New York, Tokyo, in press, 1988 .

ELECTROPORATION-MEDIATED ENHANCEMENT OF PLANT PROTOPLAST DIVISION AND PLANT REGENERATION

E.L. RECH, S.J. OCHATT, P.K. CHAND, J.B. POWER and M.R. DAVEY
Plant Genetic Manipulation Group, Department of Botany, University of Nottingham, Nottingham NG7 2RD, U.K.

1. INTRODUCTION

Electroporation has been shown to increase the permeability of the plasma membrane, presumably by the formation of pore-like openings (1), facilitating the uptake of molecules of various sizes, including DNA (2; 3; 4). Attempts have been made to explain the possible biophysical influence of electrical fields on biological systems (5), and although knowledge of these effects on cells is scarce, pulsing electromagnetic fields stimulate DNA synthesis in cultured mammalian cells (6). In plants, electric fields have been shown to stimulate shoot regeneration from tobacco callus (7), and somatic embryogenesis of alfalfa protoplasts (8). Electroporation also promotes division and colony development from protoplasts (9) and increases plant regeneration (10). This report summarises these studies with protoplasts of *Glycine canescens*, *Prunus avium* x *pseudocerasus*, *Pyrus communis*, *Solanum dulcamara* and *S. viarum*.

2. MATERIALS AND METHODS

Protoplasts were isolated enzymatically from seedling cotyledons (*G. canescens* and *S. viarum*), callus (*P. communis*) and cell suspensions (*P. avium* x *pseudocerasus* and *S. dulcamara*), and resuspended in buffer solution (5mM 2-N-morpholinoethane sulfonic acid, 6mM $MgCl_2$ and 0.5-0.7M mannitol, pH 5.8). Samples (400 µl) of each protoplast suspension were pulsed at different voltages (250-2000V) and capacitances (10-50nF) in an electroporator (DIA-LOG, G.m.b.H.). After electro-pulsing, protoplasts were resuspended in culture medium (9), viability measured with FDA after 24 h and cell wall regeneration assessed after 1-10 days using calcofluor white. The plating efficiency, number of cells per colony, growth response of callus and frequency of regeneration were recorded.

3. RESULTS AND DISCUSSION

The percentage of viable protoplasts decreased with increasing voltage and capacitance. The size of protoplasts influenced their ability to survive the electric pulses. Assessments of cell wall regeneration showed that protoplasts electroporated with pulses up to 1250V, regenerated a wall and entered division after 5 days of culture, whereas wall regeneration was less advanced in untreated protoplasts. This response was particularly noticeable for protoplasts of *Prunus*, *Pyrus* and *S. viarum*, which inherently have long lag phases of 15, 9 and 11 days respectively prior to division. Voltages between 250V and 1500V stimulated division in *Prunus*, *Pyrus* and *S. dulcamara*. The voltage range could be extended up to 1750V for *S. viarum*, but the range was less (250-750V) for *Glycine*. In addition to stimulating protoplast division, electroporation also increased the plating efficiency. This enhancement of cell division was sustained throughout culture; after 30 days significantly more colonies were recovered from electropulsed protoplasts compared to untreated protoplasts.

In Prunus and S. dulcamara, electroenhancement of growth was sustained to the callus stage, with pulses of 250V or 500V, inducing the largest fresh weight increases. Furthermore, the electrical treatment of protoplasts influenced plant regeneration from protoplast-derived tissues with a significantly great number of shoots being produced from tissues derived from electroporated protoplasts. Additionally, shoots regenerated from callus of electro-treated protoplasts at 250V or 500V were longer and more leafy compared to controls. This made it possible to transfer such shoots directly for rooting without the need for an intermediate growth phase to encourage internode elongation.

The stimulatory effect of electroporation on growth of protoplasts used in this investigation does not imply that electromanipulation can be employed to induce cell division in hitherto non-dividing protoplasts. Indeed, the general applicability of this technique to a range of systems remains to be evaluated. These observations show that electromanipulation of isolated protoplasts has considerable practical application not only in the context of transformation, but also in the enhancement of growth and plant regeneration. This additional benefit should facilitate the recovery of genetically engineered plants.

REFERENCES

1. Neumann E, Schaefer-Rider M, Wang Y, Hofschneider PH: EMBO J. 1: 841-845, 1982.
2. Nishigushi M, Langridge WHR, Szalay AA, Zaitlin M: Plant Cell Rep. 5: 57-60, 1986.
3. Fromm ME, Taylor LP, Walbot V: Nature 319: 791-793, 1986.
4. Shillito RD, Saul MS, Paszkowski J, Muller M, Potrykus I: Bio/Technology 3: 1099-1103, 1985.
5. Sugar IP, Neumann E: Biophys. Chem. 19: 211-226, 1984.
6. Takahashi K, Kaneko I, Date M, Fukuda E: Experientia 42: 185-186, 1986.
7. Rathore KS, Goldsworthy A: Bio/Technology 3: 1107-1109, 1985.
8. Dijak M, Smith DL, Wilson TJ, Brown DCW: Plant Cell Rep. 5: 468-470, 1986.
9. Rech EL, Ochatt SJ, Chand PK, Power JB, Davey MR: Protoplasma, 1987 (in press).
10. Ochatt SJ, Chand PK, Rech EL, Davey MR, Power JB: Plant Sci., 1987 (in press).

SELECTION OF POTATO CELL SUSPENSION PROTOPLAST SUBPOPULATION BY RHODAMINE 123

S. LUCRETTI[1]*, R. TAVAZZA[1], F. MORETTI[1], A. TABASSUM[2], P. IANNI PALARCHIO[1]
[1] ENEA, TECAB IGEN ; C.P. 2400 00100 ROMA, Italy
[2] NIAB, Plant Tissue Cult. Lab., Faisalabad Pakistan.

1. INTRODUCTION

Protoplasts derived from cell suspension cultures show a wide range of variations according to different parameters such as metabolic activity or DNA content (Lucretti et al., 1987; Carlberg et al.,1984). Flow-cytometry (FCM) has been applied for assessing the homogeneity of a potato cell suspension, concerning viability and trans-membrane potential as related to cell membrane integrity and functional mitochondria; for this purpose we used the cationic fluorescent dye rhodamine 123 (R123) specifically taken up by mitochondria of living cells (Johnson et al.,1981). The toxicity of the fluorochrome to the protoplast culture has been tested to establish the more convenient concentration to employ for FCM.

2.0. Materials and methods

2.1. Protoplast isolation and culture. Potato cell suspension protoplasts were isolated from exponential growing cultures obtained from 6 month old ones of cv. Desire. The protoplasts were plated at a density of 7.10^4/ml and incubated at 26°C under continuous light (1000 lux). The cultures were diluted 1:3 after 4-5 days; then the small cell colonies were subcultured on a medium softened with agar (0.2%). After 2-3 weeks, the microcalli were transfered according to the regeneration procedure established by Roest et al. The plating efficiency has been evaluated after 3 months as the number of microcalli on the initial number of plated protoplasts; the regeneration percentage was evaluated after 7 months.

2.2. FCM measurements. R123 from SIGMA (EXT 430nm; EMS 535 nm) was dissolved in absolute ethanol at a concentration of 5 mg/ml and the solution was stored for 3-4 weeks at -20°C.The protoplast suspensions, stained with R123 at 0.1, 1.0, 5.0, 10.0, and 30.0 µg/ml, were incubated with the fluorochrome for 10', 20', 40', 80', 180' and 360'. After centrifugation (100gx3') the samples were resuspended into fresh culture medium and nylon filtered (53 µm), just before the FCM analysis. The ionophore 2,4 dinitrophenol (DNP) and the electron transport inhibitor sodium azide (NaAz) were given at 10^{-3} M and 10^{-2} M respectively for 80' or 180' to 30 µg/ml R123 stained protoplasts. Flow cytometry analyses were carried out by a BD FACS 400 equipped with an argon laser tuned at 488 nm and run at 200 mW power output; the filter set included a BP 535/15 nm for the green channel, and a 570 nm dicroic mirror. 30000 protoplasts per point were analysed at a flow rate of 500/sec, during three repetitions. Data analysed were performed with an Ahrens " FDAS III " dedicated computer.

3.0. Results and discussion. The potato cell suspension protoplasts showed a low toxicity to R123 in the range of concentration of 0.1 and 10.0 µg/ml for all the incubation periods; at 10.0 µg/ml and 180' the dye did not affect the growth of the treated protoplasts in respect of the control. After 7 months of culturing the 40% of the 5 µg/ml R123 treated calli regenerated shoots (Fig.1) while at 30.0 µg/ml the protoplasts were unable to undergo to first mitotic division (data not showed); this high concentration allowed the green fluorescence of R 123 to diffuse from the mitochondria to the cytoplasm also. To confirm that the mitochondrial acculation of the cationic dye in intact cells is infact caused by the mitochondrial trans-membrane potential (Johnson et al., 1981), we tested DNP and NaAz in order to disrupt the metabolism of the organelles. The FCM analyses showed a decreasing in fluorescence values to 65% and 55% for DNP and NaAz in respect of the control. The rhodamine 123 uptake kinetics were determined at different concentrations; the protoplasts were able to reach the fluorescence maximum during 20' -40' and they kept this level up to 360' (Derzynkiewicz et al. 1982. Fig.2). At the concn. of 10.0 μg/ml and 30.0 μg/ml and incubation time from 80' to 360', the flow cytometric analyses showed two distinct populations caracterized by two different levels of fluorescence (Bernal et al., 1982), connected with the region R_1, where the main emission peak was located and the region R_2, with lower values of fluorescence.The ratio R_1/R_2, according the number of protoplasts detected per region of interest, indicated different behaviours of the protoplast population (tab.1).The literature on R123 uptake kinetics and on the effects of the inhibitors on the mitochondria, mainly available for animal cells, is in good agreement with our results. R123 is reported to exhibit an higher staining for exponentially growing cells in comparison to quiescent or differentiated ones. At higher R123 concentrations (> 10 μg/ml) and for long exposure (> 6h) it has a cytostatic effect, blocking the cells at G_{1A}phase (DNA-presynthesis ; Derzynkiewick et al.).Therefore R123 can be utilized in FCM multiparametric analysis by allowing the sorting out and consequently the isolation of plant protoplast subpopulations having different metabolic conditions or occurring at different phases of the cell-cycle.

Tab.1.Protoplast distribution kinetics(R123 30ug/ml)

TIME	PEAK	%R1	%R2	R1/R2
10'	78	23	77	0.29
20'	91	40	60	0.66
40'	93	42	59	0.71
80'	99	44	56	0.79
180'	98	43	57	0.74
360'	94	36	64	0.56

Fig.1 Regeneration from potato cell sospension protoplasts

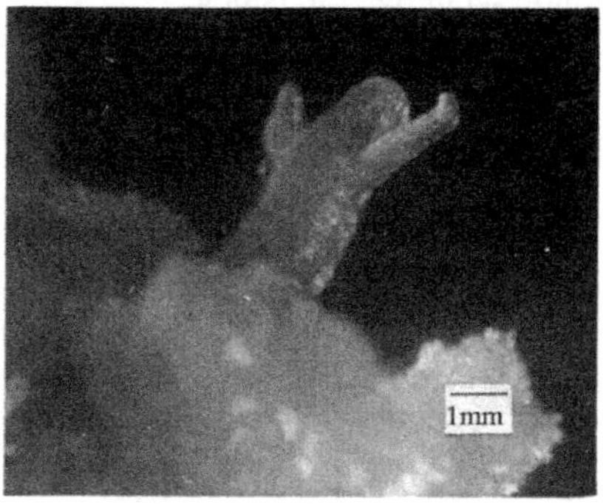

Fig.2 R123 protoplast uptake kinetics

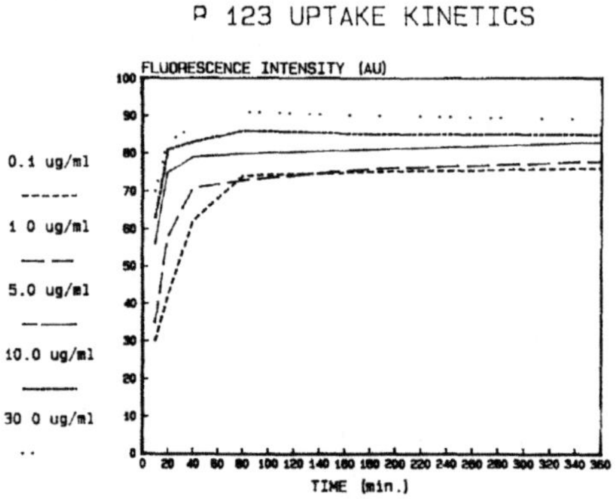

REFERENCES

1)Bernal S.D.,Shapio H.M. and Chen L.B.: Int.J.Cancer 30 (1982) 219-224
2)Carlberg I.,Glimelius K. and Tage E.: Plant Sci.Lett. 35 (1984) 225-230
3)Darzinkiewicz Z.,Traganos F.,Staiano-Coico L. and Melamed M.R.: Cancer Res. 42 (1982) 799-806
4)Johnson L.V.,Walsh M.L.,Chen L.B.: PNAS USA 77 (1980) 990-994
5)Lucretti S.,Moretti F.,Tavazza R. and Dolezel J.: Cytometry Suppl.1 (1987) 95
6)Roest S. and Bokelmann G.S.: Protoplast 1983 Post.Proc. ed. Potricus I.,Harms C.T.,Hinnen A.,Hutter R.,King P.J. and Shillito R.D. Birkhauser Verlag Basel 282-283

PROTOPLAST CULTURE OF DIOSCOREA OPPOSITA THUNB. (D. Batatas DECNE)

A. SAUER and F. WALTHER
Federal Research Centre for Horticultural Plant Breeding
D-2070 Ahrensburg, Fed. Rep. Germany

1. INTRODUCTION

Within the scope of an in vitro mutation induction program the treatment of a single cell culture with ionizing radiation provides the opportunity to avoid the development of chimeras. Protoplast cultures meet this condition. Dioscorea opposita -Chinese yam- can be propagated by in vitro procedures without any problems and this was the reason to initiate protoplast cultures.

2. PROCEDURE

2.1. Materials and methods

Axillary buds as well as shoot tips propagated on MS-medium (1) containing 2 mg/l IAA and 0.2 mg/l BAP were used as explants for protoplast isolation. They were sliced into small strips and infiltrated in enzyme solution containing macerozyme R-10 and cellulase 'Onozuka' R-10. After digestion of cell walls protoplasts were separated by sieving through a 0.4 mm stainless steel screen. The filtrate was diluted by an equal volume of the 'washing' medium V-47 supplied with 0.3 M NaCl (2). The enzyme solution was washed out by centrifugation. Later the protoplasts were diluted with liquid V-KM-medium (3, 4, 5). An illumination (1.000 lx) during a 16 hs period was best for cell division.

3. RESULTS

The highest shoot production was found on media without any release of phenols by the explants. B 5 medium (6) among others could not be used for in vitro propagation, because of high production of phenols. Protoplasts placed on this culture medium reacted with death. Using liquid V-KM-medium resulted in a higher rate of cell division than protoplast cultivation on solidified droplets of this medium (2.5 % agarose) flooded by liquid medium. Another increase of cell division was achieved by a 16 hs illumination compared to propagation under complete darkness. The culture medium was changed every second day providing high numbers of single cells, cell clusters and cell colonies. During several weeks numerous subcultures could be produced, but there was no development of dense callus. Transfer of cell clusters to soft V-KM-medium (1 % agarose) was without any improvement and the protoplasts died some days later. About three months after starting protoplast cultivation a contamination by endogenous bacteria was detected. Up to now, i.e. 6 months after initiation of in vitro propagation, no contamination of any in vitro derived explant could be observed. Possibly the propagation of bacteria was favoured by an increase of temperature up to 35° C - 38° C by a defect of the cooling system of the growth chamber. The use of cuttings from internodal segments of in vitro derived plantlets for in vitro propagation resulted in the detection of contamination by endogenous bacteria after 3 days.

4. CONCLUSIONS

From the results mentioned above the following conclusions can be drawn:

- in vitro mass propagation without any difficulties is no guarantee for a successful protoplast culture,
- the kind of culture medium not usable for shoot production seems to be also not applicable in protoplast culture,
- relative high cell division rate of protoplasts under optimal culture conditions seems to mask latent contamination by endogenous bacteria, but may negatively influence the growth of callus and the regeneration of plants,
- to avoid high inputs for single cell cultures on the base of protoplasts it seems to be reasonable to apply different kinds of tests to detect any latent contamination by very slowly propagating endogenous bacteria - first of all in vegetatively propagated species.

5. REFERENCES

1. Murashige T and Skoog F: A revised medium for rapid growth and bioassays with tobacco tissue cultures. Physiol. Plant. 15: 473-497, 1962.
2. Binding H: Regeneration von haploiden und diploiden Pflanzen aus Protoplasten von Petunia hybrida L.. Z. Pflanzenphysiol. 74: 327-356, 1974.
3. Kao K N and Michayluk M R: Nutritional requirements for growth of Vicia hajastana cells and protoplasts at a very low population density in liquid media. Planta (Berl) 126: 105-110, 1975.
4. Bokelmann G S and Roest S: Plant regeneration from protoplasts of potato (Solanum tuberosum cv. Bintje). Z. Pflanzenphysiol. 109: 259-265, 1983.
5. Binding H and Kollmann R: Regeneration of protoplasts. In Schäfer-Menuhr A(ed): In vitro techniques. Nijhoff, Junk, Dordrecht, Boston, Lancaster, pp. 93-99, 1985.
6. Gamborg O L Miller R A and Ojima K: Nutrient requirements of suspension cultures of soybean root cells. Exp. Cell Res. 50: 151-158, 1968.

ISOLATION AND CULTURE OF BUCKWHEAT (FAGOPYRUM ESCULENTUM MOENCH.) CALLUS PROTOPLASTS

N.I. RUMYANZEVA, V.V. LOZOVAYA

Kazan Institute of Biology, Postbox 30, 420084 Kazan, USSR

I. INTRODUCTION

In oder to develop new breeding strategies new opportunities must now be used to buckwheat, that is an agriculturally important plant. At present time there are data on regeneration of plantlets from buckwheat tissue culture (Yamane I974) and vegetative propagation of buckwheat in vitro (Srejović, Nešković I98I, I983).

Isolation and culture of protoplasts has become an important and integrate component of plant biotechnology. However key information on isolation and culture of buckwheat protoplasts permitting the possibilities to use cell engineering approaches to overcome incompatibility between distant buckwheat species has been lacking.

2. PROCEDURE

2.I. Materials and methods

For protoplast isolation we used callus culture of buckwheat (cv. Kazanskaya Krupnozernaya). Callus was induced from buckwheat seeds germinated on Gamborg's B5 medium (I968) with (mg/l) 2,4-D - 2, IAA - 0.5, NAA - 0.5, kinetin - 0.2. The white friable callus was obtained by further transfer of primary nodular callus on B5 medium supplemented with (mg/l) casein hydrolysate - 2 000, sucrose - 30 000, ascorbic acid - 25, succinic acid - 25, 2,4-D - 0.5, NAA - 0.I, kinetin - 0.4. Protoplasts were isolated from callus cells 5-6 days after subculture. Approximately I g fresh weight of callus was added to I0 ml of enzymatic digestion medium (I% Onozuka R-I0, 0.6% Macerozyme R-I0, 0.25% Driselase, 0.55 M sucrose) and incubated at 26 °C for I4-I6 h. After incubation I/3 volume of 0.5 M sucrose was added and digest filtered through nylon mesh of I00 m. Protoplasts were centrifugated at 75 g for 5 min. The floating protoplasts were withdrawn with a pasteur pipette and the protoplasts suspension was diluted and washed with salt solution W-5 (Menczel et al., I98I). The protoplasts were centrifugated at 50 g for 2-3 min and then the pellet was washed once in the culture medium. Protoplasts were cultured at a density $I0^4$-$I0^5$ per ml in microplates in a thin layer of medium in the dark at 26 °C. The culture medium contained Gamborg's B5 mineral salts, 0.27 M mannitol, 0.04 M sucrose, 0.I4 M glucose and the following addenda (mg/l): glutamine - 300, serine - I00, casein hydrolisate - I5o, yeast extract - I50, nicotinic acid - 0.5, pyridoxine-HCl - 0.5, thiamine - 0.5, maleic acid - 40, citric acid - 40, succinic acid - 40, fumaric acid - 40, ascorbic acid - 25, sodium pyruvate - 20.

2.2. Results

2.2.I. Since wild plant species are the donors of economically valuable traits, the analysis of the possibility of cross-pollination of various wild species and cultivars of buckwheat is made. It was established that in the case of F. tataricum pollination by the pollen of common buckwheat fertilization and embryo development do not take place. The results testify to the expendiency of using the possibilities of somatic hybridization to overcome incompatibility barrier.

2.2.2. Callus cultures were initiated from germinating buckwheat seeds, completely differentiated embryos and immature embryos of 0.5-2 mm length of F. esculentum and F. tataricum. II callus lines of common buckwheat and 2 callus lines of wild buckwheat with high morphogenic potentialities were obtained.

2.2.3. Protoplasts isolated from callus initiated from cotyledones of germinating seeds (Fig. I) formed the cell wall on the 2nd day of culture (Fig. 2). The budding of protoplasts was rare. The first divisions were detected on 6-7 days of protoplast culture.

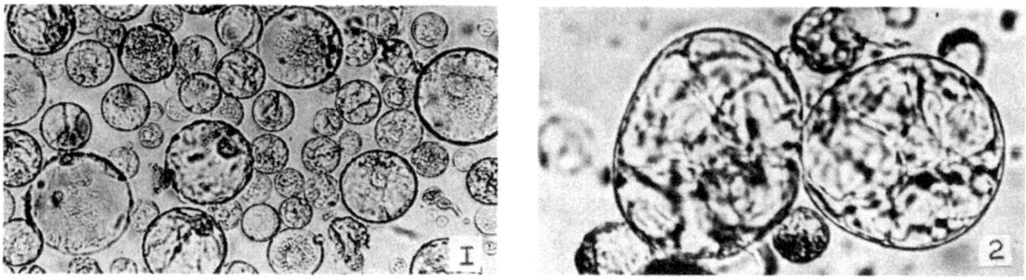

FIGURE I. Freshly isolated buckwheat protoplasts from callus tissue.
FIGURE 2. Formation of cell wall by callus protoplasts.

REFERENCES

I. Gamborg OL, Miller RA, Ojima K: Exp. Cell Res. 50: I5I, I968.
2. Srejović V, Nešković M: Zeitschrift für Pflanzenphysiol. I04: 37, I98I.
3. Srejović V, Nešković M: Buckwheat newsletter. 3: 5, I983.
4. Kreft J: Zbornik Biotechniske fakultete Universe E. K. v Ljubljani. 4I: 7, I983.
5. Yamane Y: Japan. J. Genetics. 49: I39, I974.

PARTIAL PURIFICATION OF A CELLULASE USED FOR PROTOPLAST ISOLATION FROM BARLEY (*Hordeum vulgare*)

HELLE KROGAARD

The Biotechnology Group, Danish Research Service for Plant and Soil Science. Lottenborgvej 2, DK-2800 Lyngby, Denmark

1. INTRODUCTION

Isolation of viable protoplasts requires treatment of plant tissue with cell wall degrading enzymes. Commercially produced enzymes may contain impurities, which may damage the protoplasts and thereby adding variability to the experimental system. The aim of the present experiments was to investigate the effect of partial purification of a commercial cellulase preparation (R-10, Onozuka, Japan) on barley protoplast isolation and viability.

2. PROCEDURE

2.1. Materials and methods

2.1.1. Plant material. Six-day-old barley plants (*Hordeum vulgare* L cv. Perth).

2.1.2. Enzyme preparation. Cellulase (R-10) from Onozuka, Japan.

2.1.3. Chromatography material for enzyme purification. Sephadex G-75 column.

3. RESULTS

3.1. It was found that only a small part of the total cellulase preparation was necessary for protoplast isolation from barley leaves (Fig. 1). The unpurified cellulase contained compounds toxic to the protoplasts (Fig. 1). The protoplast viability was improved when the purified cellulase (fraction 32 to 35) was used for protoplast isolation (Fig. 2).

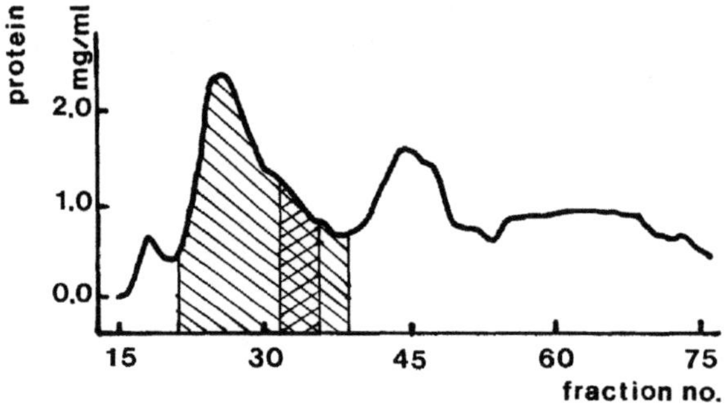

FIGURE 1. Protein elution profile for the cellulase preparation. ▨ The fractions which were able to liberate protoplasts. ▧ The fractions which were toxic to the protoplasts.

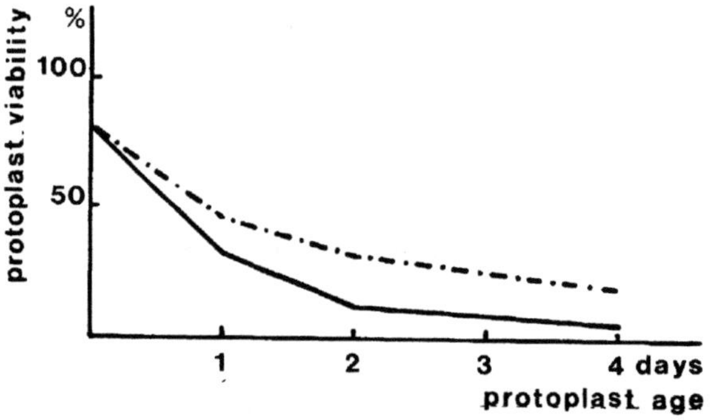

FIGURE 2. Viability of protoplasts isolated with commercial cellulase preparation (——) and with the gelfiltrated cellulase (-··-·-).

REFERENCES

1. Beldman, G., Searle-van Leeuwen, M.F., Rombouts, F.M. & Voragen, F.G.J. 1985. The cellulase of Trichoderma viride. Purification, characterization and comparison of all detectable endoglucanases, exoglucanases and β-glucosidases. - Eur. J. Biochem. 146: 301-308.
2. Fitzsimons, P.J. & Weyers, J.D.B. 1985. Properties of some enzymes used for protoplast isolation. - In The Physiological Properties of Plant Protoplasts. (P.E. Pilet, ed.), pp. 12-23. Springer Verlag, Berlin. ISBN 0-387-15017.
3. Ishii, S. & Mogi, Y. 1983. Identification of enzymes that are effective for isolating protoplasts from grass leaves. - Plant Physiol. 72: 641-644.
4. Kikuta, Y., Fujino, K., Saito, W., Masuda, K. & Okazawa, Y. 1986. Protoplast culture of potato: An improved procedure for isolating viable protoplasts. - J. Fac. Agric. Hokkaido University. 62: 429-439.
5. Oleski, N.A., Saxton, M.J., Lyons, J.M. & Breidenbach, R.W. 1983. Affinity purification of enzymes from cellulysin for protoplast production. - Plant Physiol. 72: 176.
6. Van der Valk, H.C.P.M. (1984): Determination of proteases in cell wall degrading enzyme mixtures used in protoplast isolation. - Plant Sci. Lett. 36: 201-204.

PLANT REGENERATION FROM CELL SUSPENSION PROTOPLASTS OF FESTUCA ARUNDINACEA SCHREB., LOLIUM PERENNE L. AND L.MULTIFLORUM LAM.

S. J. Dalton
AFRC Institute for Grassland and Animal Production,
Welsh Plant Breeding Station, Aberystwyth, Dyfed, Wales, UK

1. INTRODUCTION

A prerequisite for the use of many genetic manipulation techniques is the ability to regenerate plants from protoplasts, but in the cereals and grasses, plants have been regenerated in this way in only six species (1). Of these, only Oryza (2,3,4,5,6) and Saccharum (7) have been successfully transferred to soil. In vitro plants of three forage grasses: Festuca arundinacea (tall fescue), Lolium perenne (perennial ryegrass) and L.multiflorum (Italian ryegrass), have now been regenerated from protoplast-derived cell colonies and green Festuca and L.multiflorum plants established in soil (8).

2. CELL SUSPENSION AND PROTOPLAST CULTURE

2.1. Cell suspension initiation and culture

The regenerating protoplasts were isolated from young embryogenic cell suspensions initiated directly from groups of chopped mature embryos. The three species used were the most responsive of seven Lolium, Festuca, Dactylis and Phleum grass species tested with this method of suspension initiation. The original reasons for establishing cell suspensions of outbreeding species from a number of mature embryos were a) to use a convenient source of meristematic tissue, b) to save time by eliminating the callus phase and c) to obtain a 'typical' suspension of each species by culturing a mixture of genotypes. However, within each species, the cell suspension lines were as morphologically variable as those begun from calli of single genotypes. Iso-enzyme analysis of some of the plants regenerated from chopped embryo derived cell suspensions indicated that heterogeneity within individual cell suspensions (or at least of the regenerating tissues) was low.

Cell suspensions were initiated by culturing up to 30 mature embryos (chopped up with a scalpel) in liquid medium containing a high concentration of 2,4-D (10 mg l^{-1}). Fresh culture medium containing the normal cell suspension concentration of 2,4-D was added after 10 days (8). Once established the cell suspensions were subcultured weekly by transferring small, dense colonies to 80% fresh culture medium.

2.2. Protoplast isolation and culture

Protoplasts were isolated four days after subculture and could be isolated as soon as four weeks after suspension initiation. However, the yields were generally low (less than 3 x 10^6/g) until about 20 weeks after suspension initiation. A higher proportion of the protoplasts isolated from the young cell suspensions were small and dense compared with those isolated from older, friable suspensions.

Protoplasts initially isolated from young (13-20 week old) Festuca and L.perenne suspensions did not survive in liquid culture media found most suitable for callus colony formation from protoplasts isolated from older cell suspensions containing 3% sucrose and 10% mannitol (835-870 mOsm). However, colonies were formed when the protoplasts were cultured in a similar medium (PC4) of higher osmotic potential (895 mOsm), achieved by the addition of 6.84% glucose and 5% mannitol (8). Protoplasts of the three species isolated from most of the cell suspension lines tested were capable of producing cell colonies when cultured in medium PC4. Inconsistent results were obtained when protoplasts were cultured in medium solidified with seaplaque agarose.

2.3. Protoplast plating efficiency and colony culture

Festuca protoplasts were the most responsive of the three species tested, with protoplasts isolated from 15 out of 20 cell suspension lines achieving plating efficiencies of 0.1-1.0% (400-4000 colonies per plate) (Table 1). Protoplasts from only one out of nine L.perenne suspension lines and two out of six L.multiflorum suspension lines were capable of the same efficiency. Strands of single cells, friable and compact cell colonies were formed. However, less than 50 colonies per plate were compact colonies, depending on the cell suspension line and its age.

TABLE 1. Plating efficiency of cell suspension protoplasts

Species	No. of protoplast isolations per cell suspension line (total)	No. of protoplasts ($x10^6$) cultured per cell suspension line (total)	No. of lines with 0.1-1.0% plating efficiency
F.arundinacea (20)	1-15 (78)	0.1-91 (460)	15/20
L.perenne (9)	1-9 (22)	2-98 (230)	1/9
L.multiflorum (6)	2-4 (17)	8-31.5 (107)	2/6

The protoplast-derived colonies were plated onto solidified medium of the same composition as medium PC4 but without mannitol (8). After several weeks culture, any embryogenic and nodular calli developing from the compact colonies were transferred to regeneration medium. Green and white shoots emerged from embryoids in Festuca and L.perenne, but embryogenesis was less evident in L.multiflorum.

3. REGENERATION FROM CELL SUSPENSION AND PROTOPLAST-DERIVED COLONIES

3.1 Comparison between cell suspension and protoplast-derived colonies

Plated protoplast-derived cell colonies were generally capable of the same morphogenesis as colonies derived from the respective cell suspension lines. Most cell suspensions were initially able to regenerate green shoots when plated, but the regenerative ability declined with age (Table 2). The number of shoots produced declined and the proportion of shoots which were albino and the proportion of friable tissue increased. However, for a majority of the cell suspension lines which produced responsive protoplast-derived colonies, shoots could be obtained from isolations performed after the suspensions had apparently stopped regenerating. These shoots were usually albino. Protoplast-derived colonies regenerating green shoots tended to be formed in isolations performed within the regenerative life of the cell suspension as determined by plating.

3.2. Plant regeneration from F.arundinacea cultures

Protoplasts were isolated from 20 cell suspension lines 6-122 weeks from suspension initiation. Suspension colonies of 12 of the lines were able to regenerate green shoots when plated, but this response ceased

TABLE 2. The effect of cell suspension age on shoot formation from plated cell suspension and protoplast-derived colonies

	Age range in weeks of cell suspension at the time: Protoplasts were isolated which subsequently developed shoot forming cell colonies		Shoot regeneration from plated cell suspension colonies ceased	
	Green	White	Green	White
F.arundinacea (20)	18-34 (6)	11-63+ (12)	12-36 (12)	11-48 (11)
L.perenne (9)	24-30 (1)	21-63+ (5)	15-29 (6)	19-63 (7)
L.multiflorum (6)	25-33 (1)	21-33 (1)	12-43 (5)	18-43 (3)

() Number of cell suspension lines

after a maximum of 36 weeks (Table 2). Eleven lines produced albino shoots and this response generally continued for a longer period, ceasing after a maxium of 48 weeks. Two lines produced only embryogenic tissue and this response ceased after 14 and 15 weeks. However, both of these lines produced shoots from protoplast-derived colonies (Table 3). A green shoot was also produced from protoplast-derived colonies of another suspension which was apparently capable of albino shoot regeneration only. Altogether, shoots regenerated from protoplast derived cell colonies of 12 cell suspension lines. Nine of these responsive lines regenerated shoots from protoplast isolations performed after the cell suspension had apparantly ceased regenerating shoots. The great majority of shoot forming colonies were derived from only two lines (261 green and 510 white). Over 120 plants were established in soil, but require vernalisation before the ability to set seed can be determined.

3.3. Plant regeneration from L.perenne cultures

Protoplasts were isolated from nine cell suspension lines 11-63 weeks from suspension initiation. Suspension colonies of six of the lines regenerated green shoots when plated, but this response ceased after a maximum of 29 weeks, while seven lines which regenerated white shoots when plated, continued this response until a maximum of 63 weeks (Table 2). One line which did not regenerate shoots and did not appear to be embryogenic, nevertheless produced three protoplast-derived colonies with albino shoots. In the most responsive line, white shoots formed from 371 protoplast-derived colonies obtained in isolations performed 22-63 weeks after suspension initiation (Table 3). This was also the only line to have a plating efficiency above 0.1%. Two green plantlets formed from colonies obtained from protoplasts isolated 24-30 weeks after suspension initiation did not survive in soil. Three more responding lines produced 15 colonies with white shoots from protoplasts isolated 21-34 weeks after suspension initiation; in two of these this was after the respective cell suspensions had ceased regenerating shoots. Some protoplast-derived colonies had green areas but these did not develop further.

3.4. Plant regeneration from L.multiflorum cultures

Protoplasts were isolated from six cell suspension lines 12-33 weeks after suspension initiation. Suspension colonies of five of the lines retained the capacity for green shoot regeneration for up to 43 weeks while three of the lines regenerated white shoots for up to 43 weeks (Table 2). One suspension produced only embryogenic tissue up to 18 weeks. Protoplasts of five suspensions formed cell colonies. However,

TABLE 3. Shoot regeneration from cell suspension protoplast-derived colonies

Species	No. of protoplast derived colonies with shoots Green	White	No. of green plants established in soil
F.arundinacea (20)	159* (1)	379* (1)	77* (1)
	122 (5)	118 (11)	47 (3)
L.perenne (9)	2* (1)	371* (1)	0
	0	18 (4)	0
L.multiflorum (6)	24* (1)	4 (1)	5* (1)

() Number of cell suspension lines; * Most responsive line.

colony formation was generally low (less than 0.1%) and only the two lines with the highest plating efficiencies produced protoplast derived colonies capable of shoot formation (Table 3). Colonies from one line formed albino shoots although these were from isolations performed after the cell suspension had ceased regenerating. Colonies of the other line produced green shoots and five green plants were established in soil.

4. DISCUSSION

Further work is required to improve the rate of plant regeneration and the applicability of this technique to a wider range of genotypes. However, these results indicate that green plants can be regenerated from gramineous protoplasts if they are isolated from fast growing, young, embryogenic cell suspensions, before the ability of cell suspension colonies to regenerate green plants has ceased (typically up to 30 weeks after suspension initiation). The embryogenic protoplasts isolated from such cell suspensions respond when cultured in high osmotic potential medium achieved by the addition of glucose. Green and albino shoots have been obtained from protoplast-derived colonies of all three species although only plants of Festuca and L.multiflorum have been established in soil. L.multiflorum cultures may be inherently less prone to produce albino shoots than L.perenne although the L.multiflorum material used was selected for embryogenic response while the Festuca and L.perenne material were ordinary cultivars (WPBS cultivars S.170 and Aurora respectively). The tendency for gramineous cultures and L.perenne in particular, to rapidly become albino may now be one of the main problems slowing further progress in this family.

5. REFERENCES

1. Vasil IK: J. Plant Physiol. 128, 193-218, 1987.
2. Abdullah R, EC Cocking, JA Thompson: Biotechnology 4, 1087-1090, 1986.
3. Coulibaly MY, Y Demarly: Z. Pflanzenzüchtg. 96, 79-81, 1986.
4. Kyozuka K, Y Hayashi, K Shimamoto: Mol. Gen. Genet. 206, 408-413, 1987.
5. Toriyama K, K Hinata, T Sasaki: Theor. appl. Genet. 73, 16-19, 1986.
6. Yamada Y, Y-Q Yang, D-T Tang: Plant Cell Reports 5, 85-88, 1986.
7. Srinivasan C, IK Vasil: J. Plant Physiol. 126, 41-48, 1986.
8. Dalton, SJ: J. Plant Physiol. (in press).

ISOLATION, CULTURE AND REGENERATION OF *LOLIUM PERENNE* AND *LOLIUM MULTIFLORUM* PROTOPLASTS.

T. Creemers- Molenaar*, J.P.M. Loeffen* and M.A.C.M. Zaal**

* Barenbrug Holland BV, Duitse Kampweg 60,6874 BX Wolfheze,The Netherlands.
**Foundation for Agricultural Plant Breeding, Droevendaalsesteeg 1, 6700 AC Wageningen, The Netherlands

INTRODUCTION

For the application of most genetic manipulation techniques a reliable system for plant regeneration from single cells and protoplasts is a prerequisite. In the Gramineae plant regeneration from protoplasts has been achieved only when morphogenically competent cell suspensions were used for protoplast isolation. So far, protoplast-derived callus without competence for morphogenesis was obtained in only one *Lolium multiflorum* variety(4). The aim of this study was to establish a rapid and reproducible method for regeneration of plants from cell suspension-derived protoplasts of *Lolium perenne* and *Lolium multiflorum*.

MATERIALS AND METHODS

Segments of immature inflorescences were cultured on MS medium (5) supplemented with 5 mg/l 2,4-D (MS5) to induce callus. Incubation was in the dark at 25°C. For regeneration compact callus was successively subcultured on MS0.2 in the dark and MS0 in the light. Cell suspension cultures were established from compact callus or directly from mature embryos in liquid MS5 medium and were maintained by weekly subculture of compact cell groups. Protoplasts were isolated by incubation of 1 to 3 g fresh weight suspension cells in 15 ml CPW-13M medium (3) supplemented with 1% Cellulase RS, 0.1% Pectolyase Y23 and 5mM MES buffer (1) at pH 5.6 for 3 to 4 hours. The enzyme-protoplast mixture was passed successively through 88, 53 and 28 um nylon sieves. The protoplasts were washed three times by centrifugation (50xg) for 3 min. Protoplasts were cultured in liquid or agarose-solidified RY-2 medium (6) supplemented with 0.8% foetal calf serum. After 5 weeks microcalli grown in liquid RY-2 medium were plated in N6 medium (2) without hormones, supplemented with 3% sucrose and 0.8% agarose. Microcalli grown in agarose-solidified RY-2 medium were subcultured by transferring agarose blocks to N6 medium.

RESULTS AND DISCUSSION

Green plants were regenerated from inflorescence derived compact callus of 20 *L.perenne* and *L.multiflorum* varieties. Embryogenic cell suspension cultures (Fig.1) were established from several cultivars and up to 40 green shoots per gram fresh weight were regenerated from *L.perenne* on MS medium without 2,4-D. The capacity of cell suspension cultures to regenerate green shoots decreased rapidly when the cultures were older than 4 months. The yield of protoplasts was 1 to 4 x 10^6 per gram fresh weight of cell suspension. Within one week cell-wall formation and the first divisions were observed. The plating efficiency was determined after 4 weeks and was up to 0.5% in liquid medium and up to 1% in agarose-solidified RY-2 medium. On N6 medium without 2,4-D fast growing morphogenic callus was regenerated from microcalli in 4 *L.perenne* varieties. Within 9 to 12 weeks after protoplast isolation albino plantlets (Fig.2) regenerated from this callus in two L.perenne varieties.

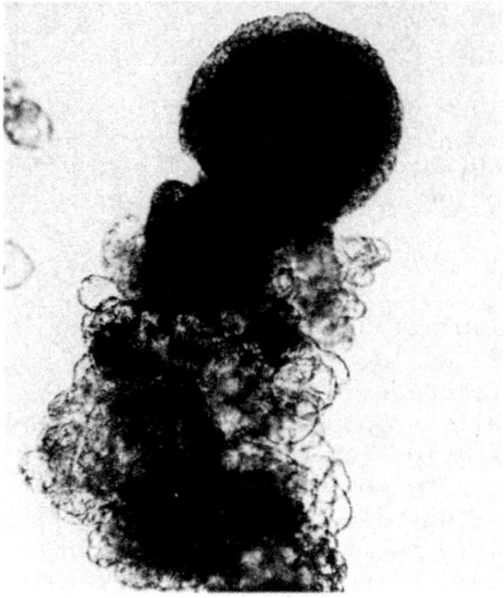

Fig. 1. Globular somatic embryo in a cell suspension culture of Lolium perenne.

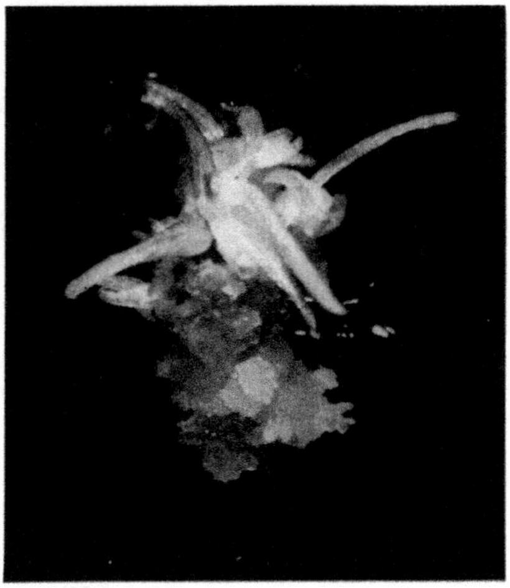

Fig. 2. Shoots regenerated from protoplast-derived microcallus of Lolium perenne.

DISCUSSION

Plant regeneration from callus and cell suspension cultures in L.perenne and L.multiflorum was established. Protoplast-derived microcalli were obtained from different L.perenne varieties in liquid and agarose-solidified media. Albino plantlets were regenerated from these microcalli. The capacity to regenerate green shoots from cell suspension cultures decreased rapidly with prolonged culture period. This suggests that in order to regenerate green shoots from protoplasts it is best to use freshly initiated cell suspension cultures for the isolation of protoplasts.

ACKWOWLEDGEMENTS

This work is part of a research cooperation between Barenbrug, DSM Research (Geleen, The Netherlands) and the Foundation for Agricultural Plant Breeding.

REFERENCES

1 R.Abdullah, E.C. Cocking and J.A. Thompson, Bio/Technology 4 (1986) 1087-1090.
2 C.C. Chu, C.S. Sun, C. Hsu, K.C. Yin, C.Y.Chu and F.Y. Bi, Sci. Sinica 18 (1975) 659-668.
3 E.M. Frearson, J.B. Power and E.C. Cocking, Dev. Biol. 33 (1973) 130-137.
4 M.G.K. Jones and P.J. Dale, Z. Pflanzenphysiol. 105 (1982) 267-274.
5 T. Murashige and F. Skoog, Physiol. Plant. 15 (1962) 473-497.
6.Y. Yamada, Y. Zhi-Qi and T. Ding-Tai, Plant Cell Reports 5 (1986) 85-88.

MATURE PLANT REGENERATION FROM CULTIVATED PROTOPLASTS OF RICE (*ORYZA SATIVA* L.)

G.Y,WANG, C.A.HSIA (ZHEN-AO XIA)
Shanghai Institute of Plant Physiology, Academia Sinica,
300 Fonglin Road, Shanghai, China, 200032.

1. INTRODUCTION

Plant regeneration from rice protoplasts has been reported recently (Fujimura et al,1985,Abdullah et al,1986,Coulibaly & Demarly 1986,Lei et al,1986,Toriyama 1986,Yamada et al,1986, Wang & Hsia 1986,Kyozuka et al,1987). In this paper the results of rice protoplast culture in our laboratory will be described.

2. MATERIALS AND METHODS

2.1.The calli were induced from mature embryos of *Oryza sativa* L.subsp.japonica cv Longhu No.6 cultured in MS medium(2,4-D 2 mg/l). After 3 month subculture pale yellowish calli with nodular appearence were obtained, which were incubated in enzyme solution (EA3-867 cellulase 1%, hemicellulase 2%, macerozyme 0.3%), 28°C, dark, 18 hr.

2.2.The protoplasts isolated from these calli were cultured in N6 medium(2,4-D 2, 6BA 0.2mg/l,mannitol 0.6 mol/l,pH 5.8) with liquid/solid double layer method in 28°C,dark,density $3x10^5$/ml.

3. RESULTS AND DISCUSSION

3.1.After 3 days culture the protoplasts were enlarged. After 6 days culture the first cell divisions were observed.The cell colonies were appeared after 30 days culture,then the thick and nodular calli were selected after 50 days culture (Fig.1-6).

3.2.These calli(dia. about 2 mm) were transferred to MS medium (KT 2, ZT 0.2 mg/l) and cultured in scattered light (3000 lux), 25°C to induce cell differentiation.Buds were initiated 1 month later,then formed roots and young seedlings(Fig.8).After transplanting them to pots, mature plants were obtained (Fig.9).

3.3.The seed fertility is 4% in G_1 and 20% in G_2.

3.4.From our results several points are mentioned.

(1)The nodular calli derived from scutellum of seed are suitable material for protoplast isolation and culture.

(2)To supplement fresh medium after 6 days and fresh medium with lower osmotic pressure after 30 days are necessary.

(3)Development of somatic embryos in protoplast culture has been observed(Fig.7).

(4)The morphogenic ability of calli is decreased with subculture time. It is better to use the protoplasts prepared from calli subcultured within 3 months.

REFERENCES

1.Wang GY,Hsia CA,Acta Biolog.Experimen.Sinica 20,252-257,1987.

calli induced
from embryos

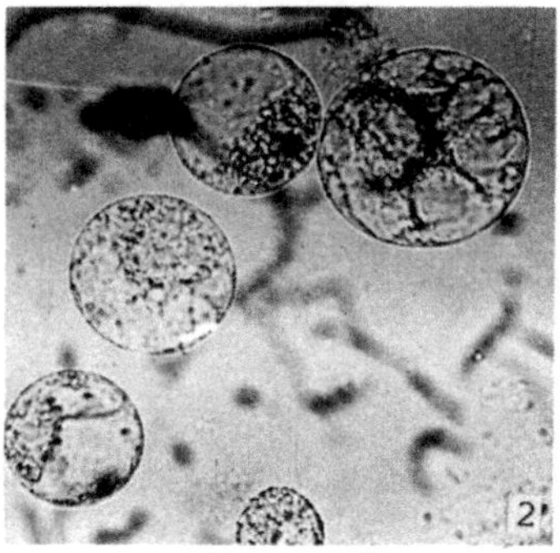

Protoplasts

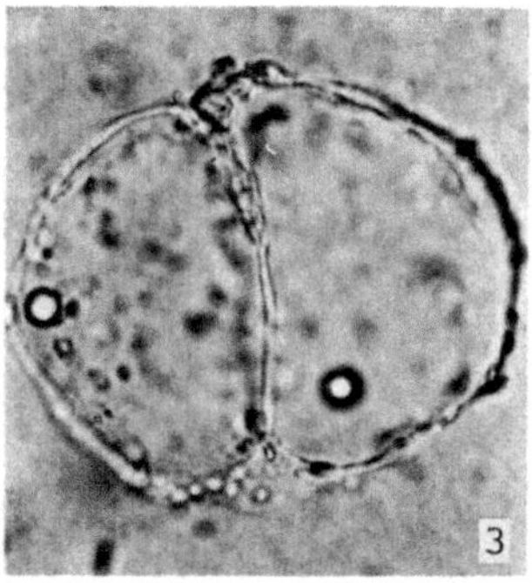

First
cell division

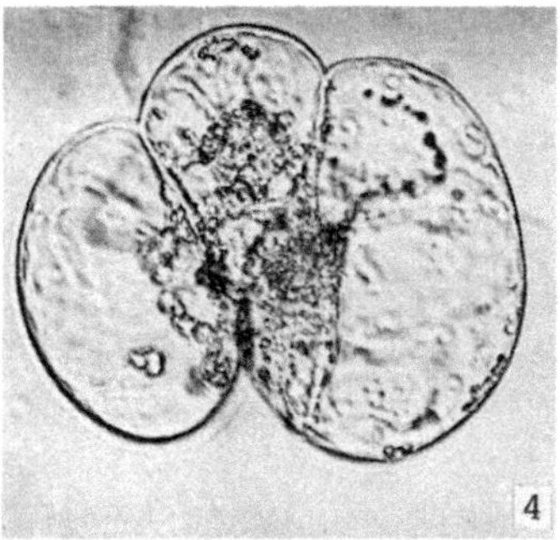

second
cell division

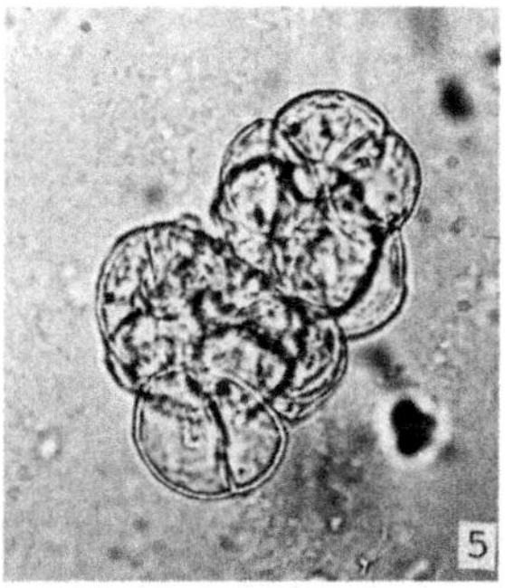

Cell clusters
after 20 days culture

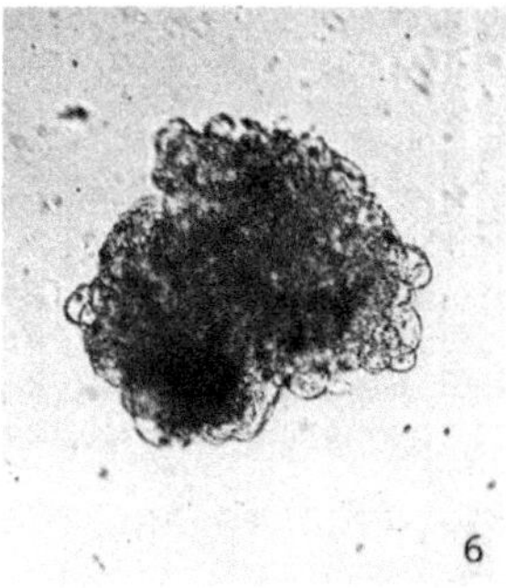

Cell clusters
after 30 days culture

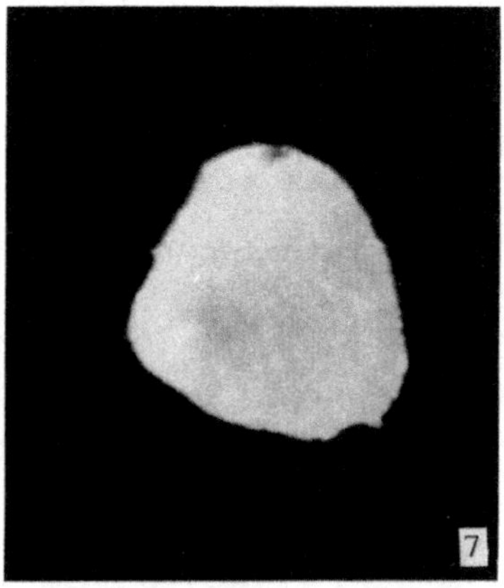

Somatic embryo

Regenerated
plantlets

Regenerated
plants

THE PLANTS REGENERATED FROM PROTOPLASTS OF SUSPENSION CELLS OF RICE(ORYZA SATIVA L)

Li Xiang-hui, Lei Ming, Sun Yong-ru and She Jiang-ming

Institute of Genetics, Academia Sinica, Beijing, China

1. INTRODUCTION

Recent success in plant regeneration from protoplasts of rice (Fujimura et al.1985, Coubibaly et al.1986, Yamada et al. 1986. Toriyama et al, 1986, Abdullah et al.1986, Wang et al.1986, Kyozuka et al.1987) made a breakthrough at the barrier of regenerating plants from protoplasts of important crops. However only a few of genotypes belonging to Japanica subspecies were responsive, and the frequency of normal reproducibility was lower in some reports. The feasibility of applying genetic manipulation in improvement of rece remains yet to be studied.

2. MATERIAL AND METHODS

2.1. Material. The calli used to initiate cell suspension were derived from young inflorescence of rice (Oryza sativa L.) F-77-170 line and cultured on MS medium with 2mg/l 2,4-D for about 2 months. The suspension cultures were established from these calli with differentiating competence in A4 medium (modified from AA medium) on rotary shaker at 150rpm, 20°C 12h/25°C 12h under diffuse light, and subcultured every 3-4 days.

2.2. Isolation of protoplasts. After being subcultured for 2 months, the suspension cultures composed predominantly of homogenous cell aggregates were used to isolate protoplasts in the enzyme solution (1). The released protoplasts were filtered through 50um stainless steel net and collected by centrifugation.

2.3. Protoplast culture. Protoplasts were incubated in the A2 medium (modified from KM medium) supplemented with Royal jelly (400mg/l), calfserum(8ml/l), CM(50ml/l), CH(250mg/l), 2,4-D(1mg/l), zeatin(0.5mg/l) and NAA(0.5mg/l) in 2x4 cm petri-dishes. After 15-20 days, 1-2ml of fresh A2 medium containing 0.3 mol glucose was added to each dish which was then placed in growth chamber with 12h illumination of 2000 Lx at 25°C. Thereafter the cultures were regularly diluted by adding equal volume of A2 medium with 0.2 mol glucose and liquid A3 medium once every 2 weeks. The cell aggregates from protoplasts in liquid medium were then transferred onto the solid A3 medium(2) and cultured under the same conditions as before.

2.4. Regeneration of plants. The colonies growing vigorous were transferred to the differentiating A4 medium(2). The regenerated plants were transplanted to the pots with soil and cultivated in greenhouse. The seeds from regenerated plants were separately collected.Observations of biological and morphological characteristics of these plants were carried out.The plants of second generation were propagated in greenhouse.

3.RESULTS AND DISCUSSION

After 3h of incubation in enzyme solution,a large number of protoplasts were released.About $5x10^6$ protoplasts were isolated from every 1g clusters.There were 2 types of protoplasts by size.About 10% of them were bigger in volume with specially rich cytoplasm,and about 90% were smaller.

The frequency of typical divisions was 3-5% at day 8.A part of the regenerated cells showed the characteristics similar to the state of embryogenesis.A lot of colonies developed after 2 months in culture.During this period the osmotic concentrations were gradually reduced by addition of A2 medium with 0.2 mol glucose.When these colonies were transferred on solid A3 medium for about 10 days,the small calli formed.After transferred them on the A4 medium without auxin,the calli grew slowly,but became to be tight.There were some noduls with white granules on the callus surface.The green bud spots emerged in 10 days after transfering,followed by shoots with 1-2 leaflets and roots another week later.The frequency of plant regeneration was 25%.More than 30 intact plantlets were obtained after a half of month.The albino was never found.The plants grew normally in the pots and tillerred vigorously. 1) All effective tillering plants have flowered and set seeds. The fertile frequency was only 30% in the greenhouse. No morphological differences were seen between regenerated plants and parents.

Fig.1.Second generation plants from protoplasts of rice

In 1987,more than 400 seedlings of second generation developed(Fig.1).Their morphological characteristics resembled original parents. However,the height of the seedlings(60cm) was shorter than the parents(80cm).The fertile rate was obviously recovered to 73.6%.The mature period was earlier than that of the parents.The cytological studies and isozyme analysis are going on.The plant regeneration system from rice protoplasts might be useful in improvement of rice quality by genetic manipulation.

REFERENCES

1.Lei M,Li XH,Sun YR and Huang MJ. Plant Regeneration from Protoplasts of Rice,Ke Xue Tongbao,22:1729-1730,1986.(in chinese)

2.Lei M,Li XH,Sun YR and Huang MJ.Plant Regeneration from protoplasts of Rice (Oryza sativa).Ke Xue Tongbao.32(20):1427-1430,1987.

REGENERATION OF PLANTLETS FROM PROTOPLASTS OF *POA PRATENSIS* L. (KENTUCKY BLUEGRASS)

Pieter van der Valk, Monique A.C.M. Zaal

Foundation for Agricultural Plant Breeding, P.O. Box 117,
6700 AC Wageningen, The Netherlands

Regeneration of plants from single cells, including protoplasts, is a prerequisite for the successful application of most genetic manipulation techniques.
The aim of this study was to develop a plant regeneration system from protoplasts of the apomictic grass species *Poa pratensis* L. (Kentucky bluegrass).
Here results are presented from experiments concerning

1. screening of 14 cultivars for callus induction and plant regeneration using both immature inflorescences and mature seeds as the explant material.
2. isolation and culture of protoplasts, derived from cell suspension cultures, which had been established from immature inflorescence-derived callus.

Screening of cultivars for callus induction and plant regeneration

14 cultivars of *P. pratensis* were screened for callus induction and plant regeneration, using immature inflorescences and mature seeds as the explant material. Callus was initiated on Murashige-Skoog basal medium (1), supplemented with 0.4 mg/l thiamine-HCL, 30 g/l sucrose and 2 mg/l 2,4-D (MS2). 6-7 weeks after initiation callus was subcultured on MS2, and again 3 weeks later. Then it was transferred to MS-medium with 0.2 mg/l 2,4-D (MS0.2), and 3 weeks later to MS-medium without 2,4-D (MS0). Initially, a watery, smooth, and translucent callus developed, that only produced roots. 4-8 weeks after initiation, morphogenic callus was produced. Such callus was slow-growing, rather compact, rough or nodular, and either white or light green/yellow. In some cultivars a rapidly growing, friable (F) callus developed after 5-10 weeks of culture.
On MS0.2 somatic embryos developed on the callus surface. Such embryos were prominent in cultivars producing friable callus. Strong variation was observed between cultivars and explant types for callus induction and plant regeneration capacity. Immature inflorescences proved superior to seeds for their ability to produce morphogenic callus and plants. Browning of seed-derived callus was more prominent than that of inflorescence callus. With seeds as the explant material, plants could only be recovered in cultivars producing F-callus (Table 1). Up to 40 plants were easily regenerated from callus of a single explant. Plants were generally green but albino shoots were produced at a low frequency (along with green ones) from F-callus in some cultivars. Plant regeneration appeared to take place through shoot morphogenesis, rather than by germination of somatic embryos.

Protoplast isolation and culture

Cell suspension cultures were initiated in liquid MS2 from 6-month old immature inflorescence-derived F-callus (var. Kimono). Albino plantlets

Table 1. Callus induction and plant regeneration from immature inflorescences and mature seeds.

Cultivar	Immature inflorescences[a]				Germinating seeds[b]			
	explants (N)	explants producing morphogenic callus (%)[d]	shoot-forming cultures (%)	somatic embryos[c]	explants (N)	explants producing morphogenic callus (%)[d]	shoot-forming cultures (%)	somatic embryos[c]
Ampellia	94	60	18	-	90	27	-	-
Aquila	89	43(F)[e]	21	+	170	11	-	-
Barblue	64	83	17	+	125	18(F)	-	-
Baron	77	69(F)	56	++	166	13(F)	3	+
Cynthia	69	54(F)	43	+	136	10	-	-
Delft	82	72	54	+	189	2	-	-
Entopper	56	59(F)	50	-	152	3	-	-
Fylking	70	94	79	-	70	9	-	-
Geronimo	20	90(F)	65	++	150	15(F)	1.5	+
Julia	55	78(F)	13	-	-	-	-	-
Kimono	58	69(F)	66	++	176	1(F)	0.3	+
Monopoly	65	44	32	+	165	13	-	-
Obelisk	72	71	54	+	70	1	-	-
Parade	66	52	-	-	113	19	-	-
VB 577	47	95	75	++	79	10	-	+

a) Tillers were obtained from field-grown plants and surface sterilized. Inflorescences of 4-8 mm long were used for experiments. Two basal segments (each 2 mm long) were cultured.
b) Seeds were surface sterilized; callus only developed from germinating seeds.
c) present (+), abundantly present (++), absent (-)
d) morphogenic callus was slow growing, rather compact, and white or greenish/yellowish
e) friable, fast growing callus

could be regenerated from this suspension at a low frequency on solidified MS0 via MS0.2, as described above.
Protoplasts were isolated from suspension cells using a solution containing 1% (w/v) cellulase RS (Yakult), 0.1% (w/v) pectolyase Y23 (Seishin), and 5mM MES in CPW-13M medium (2), pH 5.6. Protoplasts (yield: $(0.8\text{-}1.9) \times 10^7$/g F.W.) were purified by filtration (88, 53, and 28 µm filters, sequentially), washed in CPW-13M medium, and cultured in KM-8P medium (3), as modified by Vasil (4), at a density of 5×10^5/ml. Calcofluor White staining revealed the absence of wall remnants in these protoplast preparations. Protoplasts were able to regenerate a cell wall, divide, and form callus. Protoplast-derived compact callus was then cultured on MS0.2 (+ 2% sucrose and 3% sorbitol)(4 weeks), and then on MS0 (+ 2% sucrose and 3% sorbitol). Albino shoots appeared 15 weeks after the start of protoplast culture.

Conclusions

Regeneration of plants from callus was achieved for several cultivars of the grass species *Poa pratensis* L., using immature inflorescences and mature seeds as the explant material. Immature inflorescences proved superior to seeds for callus induction and plant regeneration.
Protoplasts, isolated from morphogenic cell suspension cultures, were able to divide and form callus. Albino shoots were regenerated from protoplast-derived compact callus.

References

1. T.Murashige and F.Skoog (1962). Physiol. Plant. 15:473-497
2. E.M.Frearson, J.B.Power, and E.C.Cocking (1973). Dev. Biol. 33:130-137
3. K.N.Kao and M.R.Michayluk (1975). Planta 126:105-110
4. V.Vasil and I.K.Vasil (1980). Theor. Appl. Genet. 56:97-99

This study was supported by INPLA grant number G 85-01

SOMATIC EMBRYOGENESIS AND PLANT REGENERATION FROM PROTOPLAST OF MAIZE (*Zea mays* L.)

Cai Qi-gui (Tsai chi-kuei), Kuo Chung-shen, Qian Ying-qian (Y.C. Chien), Jiong Rong-xi and Zhou Yun-luo

Institute of Botany, Academia Sinica, Beijing, China

INTRODUCTION

It is well known that protoplast is the ideal receptor for gene enginnering research. Strains with good traits could be selected from somaclonal vatiation in protoplast culture as well. Therefore, plant regeneration of recalcitrant cereal crops through protoplast culture would be followed consistently with great interest in the world. There were only two papers concerning the regeneration of plantlets from protoplast of maize (Cai et al., 1987; Kamo et al., 1987).

MATERIALS AND METHODS

Materials for protoplast isolation were taken from embryogenic calli derived from anther culture of hybrid line No. 4 of maize and subcultured for half and one year. The methods for isolation of protoplasts, culture and differentiation have been described elsewhere (Cai et al., 1987). The basic culture medium of rice protoplast (Cai et al.,1978) were used for culture of maize.

RESULTS

Division and Regeneration of Protoplasts; Fresh protoplasts isolated from embryogenic calli were globular in shape (Fig. 1). Most of them had dense cytoplasms in which large granular contents were present. Some protoplasts increased in volume after 1-2 days in culture and became oval in shape. At the same time, granular contents disappeared rapidly. 1st division of regenerated cells occurred after 4-6 days in culture (Fig. 2) and sustained divisions could be seen afterwards. Small calli which could be visible with naked eye developed after 3 weeks in culture. Except the developmental pathway of callus, embryogenesis was the another pathway. Both of them occurred simultaneouly. Proembryoid emerged after 1 week in culture. Globular embryoid with smooth surface and in different size could be seen after 3 weeks in culture (Fig. 3). The diameter of protoplast-derived calli could reach 1-2 mm in size after 4-6 weeks in culture (Fig. 4). During that time, addition of Z_2 medium in which mannitol and glucose were omitted and the concentration of sucrose was increased to 3%, was necessary. Addition of another about 0.5 ml fresh Z_2 medium was carried out 2 weeks later when calli were 2-3 mm in diameter.
Regeneration of Plants, Light yellow, protoplast-derived calli, 2-4 mm in size, were transferred onto solid differentiation medium Z_3 for differentiation. Calli grew rapidly on Z_3 medium and occasionally few embryoids and shoots could be induced on it after 2-3 weeks. Organogenesis occurred mainly after rapidly growing calli were transferred onto Z_4 medium. Calli grew slowly and became solid in nature. Tubular

coleoptiles, white or purple in colour, differentiated firstly and green leaves stretched out from coleptiles. Some of them accompanied with rhizogenesis (Fig. 5). The frequency of differentiation was more than 10% under our conditions. 46 green plantlets were obtained and some of them were more than 10 mm in height.

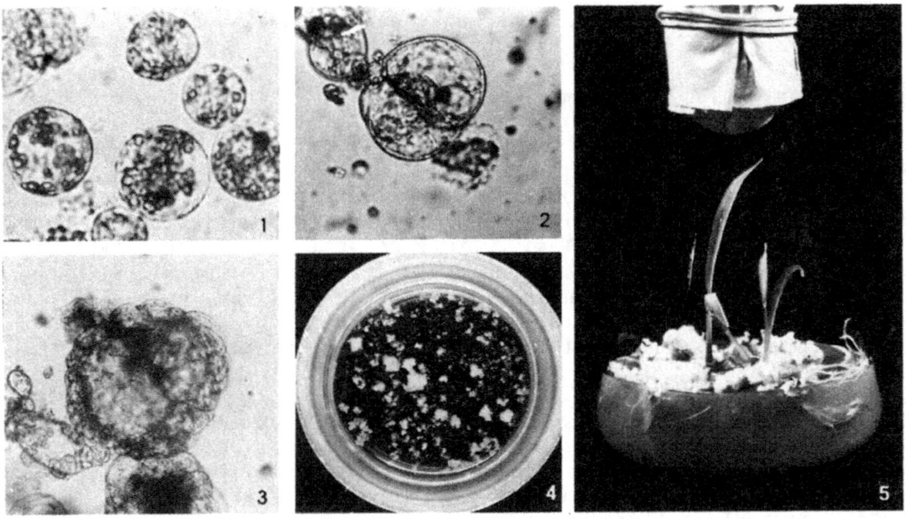

Fig. 1. Protoplast newly isolated from embryogenic calli of maize. Fig. 2. 1st division of regenerated cell. Fig. 3. Globular embryoid derived from single protoplast. Fig. 4. A lot of regenerated calli. Fig. 5. Plantlets differentiated on Z_4 differentiation medium.

DICUSSION

Concerning the plant regeneration of maize, genotype, embryogenic calli and step induction method used for differentiation probably are the 3 key factors. Step induction method was used for differentiation just as used for plant regeneration in Actinidia chinensis (Tsai, 1988). Protoplast-derived calli transferred on Z_3 medium grew rapidly. The basic medium of Z_3 is similar to Z_2 and concentration of auxin is lower; but that of cytokinin is higher. We consider that it is a transitional medium which is not only available for growth of callus, but also for preparation of organogenesis in large amounts. The basic medium of Z_4 is similar to Z_1. The basic medium and concentration of phytohormones are not suitable for callus growth but good for organogenesis.

REFERENCE

1. Cai, Qi-gui (Tsai chi-kuei), Qian Ying-qian (Y.C. Chien), Zhou Yun-luo and Wu Su-hsuen, Acta Bot. Sinica, 20 (1978) 97-103.
2. Cai, Qi-gui (Tsai chi-kuei), Kuo Chung-shen, Qian Ying-qian (Y.C. Chien), Jiong Rong-xi and Zhou Yun-luo, Acta Bot. Sinica, 29 (1987) 453-458.
3. Kamo, K.K., Chang, K.L., Lunn, M.E. and Hodges, T.K., Planta, 172 (1987) 245-251.
4. Tsai Chi-kuei (Cai Qi-gui), Plant Sci., (1988) (In press)

PROTOPLASTS FROM *ACTINIDIA DELICIOSA* (KIWI): OBTENTION AND CULTURE

M.C. PEDROSO, M.M. OLIVEIRA AND M.S. PAIS
Departamento de Biologia Vegetal, Faculdade de Ciências de Lisboa, R. da Escola Politécnica, 1294 Lisboa Codex Portugal

1. INTRODUCTION

Actinidia deliciosa (*A. chinensis*) is a fruiting species of great economic interest. Experiments on genetic manipulation for its improvement include, as for other crops, the obtention and culture of protoplasts. Protoplast isolation from *A. chinensis* leaves and calli was reported (1,4). In this study we describe the suitable conditions for the isolation and culture of protoplasts from calli and suspension cultured cells of kiwi.

2. MATERIAL AND METHODS

2.1. Establishment of calli and suspension cultures

Aseptic explants from stem and petiole of plantlets of *A. deliciosa* were used for calli induction in a solid Schenk and Hildebrandt modified basal medium (SH3) (5). Cultures were maintained at 24±2°C and 16h light / 8h dark (1700 lux). Cell suspensions were initiated by transfer of calli (3 g, F.W.) to 60 ml of SH3 liquid medium. Cultures were maintained in a rotary shaker (120 rpm), in the dark at 23±2°C, and subcultured every 7 days.

2.2. Protoplast isolation

Protoplasts were isolated from calli and suspension cultured cells at the log growth phase. Thin slices of calli and the pellet of cell suspension cultures (centrifuged at 300 x G for 5 min) were incubated in a mixture containing cellulase Onozuka R-10 and pectinase (Sigma) (Table 1) prepared in CPW, with 0.5 M mannitol, at pH 5.8. Mannitol 0.45 M and Sorbitol (0.45, 0.5 M) were also tested for E6. Incubation was performed in the dark, at 40 rpm and 22°C, for 3 hours.

ENZYMES	E1	E2	E3	E4	E5	E6	E7
Cellulase	2.0	2.0	2.0	2.0	1.5	1.5	1.5
Pectinase	2.0	1.5	1.0	0.5	1.5	1.0	0.5

TABLE 1. Enzyme concentrations tested

The protoplast suspension was then filtered through a 50 µm nylon mesh and washed twice by centrifugation at 50 x G, for 5 min, in culture medium with 0.3 M mannitol. Protoplast density was determined using a hemocytometer. Viability and presence of cell wall were checked by staining with fluorescein diacetate and calcofluor white, respectively.

2.3. Protoplast culture

Protoplasts were cultured either in suspended droplets (50µl) or in agarose (0.4%, w/v), in SH3 medium with 0.3 M mannitol at $(2,5-5)x10^4$ prot./ml. Different combinations of growth regulators (2,4-D, NAA, BAP,

Zeatin, Kin) were assayed. Cultures were performed at 24°C, in the dark for 24 h, and then under reduced light, and observed daily.

3. RESULTS AND DISCUSSION

From the enzyme mixtures assayed, E6 proved to be the best for isolation of protoplasts. The best results were achieved using mannitol 0.5 M, which agrees with those of Cassio and Marino (1983) (1). Viability of freshly isolated protoplasts was of 97% (fig.1). The ability of protoplasts to produce calli and regenerate plantlets has been reported for several species (for reviews see, (2, 6). The success in differentiating protoplasts of woody species is scarce (5). The studies on the culture of *A. deliciosa* protoplasts (fig.2) revealled that among the different hormonal combinations tested for protoplast regeneration, 2,4-D and Kin, used in SH3 medium, was found the best. After 24 h the first stages of cell wall regeneration were observed. Total regeneration was achieved within 48 h (fig.3). After the 3rd day of culture about 30 - 40% of the cells started dividing. Cell clusters were obtained within 10 days being the dilution of the osmoticum performed on the 7th day. Small colonies were obtained after 4 weeks of culture (fig.4). Experiments are still being developed in order to obtain calli and, ultimately, to regenerate plantlets.

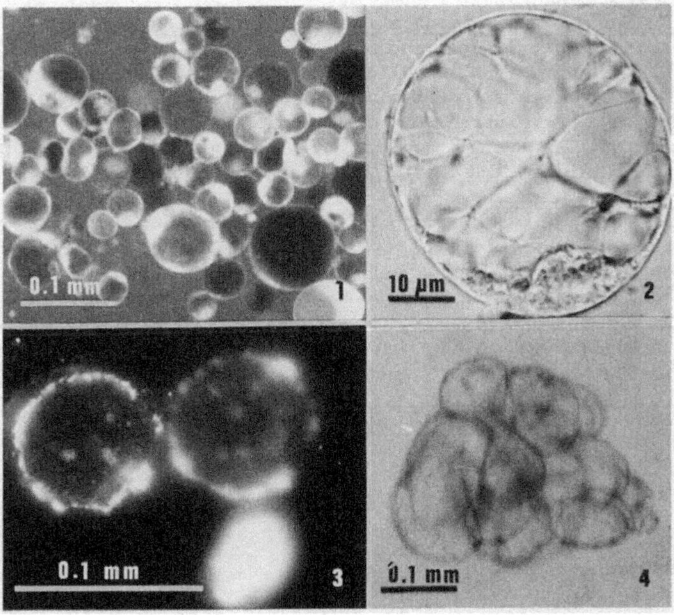

REFERENCES

1.Cassio F & Marino G: Ortoflorofrutt. Ital. 67: 455-464, 1983.

2.Evans DA & Bravo JE: Int. Rev. Cytol. Suppl. 16: 33-53, 1983.

3.Pais MS, Oliveira MM & Barroso J: Acta Horticulturae 212: 687-690, 1987.

4.Pedroso MC, Oliveira MM & Pais MS (Abstr.) Ciência Biol. (in press).

5.Vardy A. & Spiegel-Roy P: Theor. Appl. Genet. 62: 171-176, 1982.

6.Vasil IK: Int. Rev. Cytol. Suppl. 16: 79-88, 1983.

COLONY FORMATION FROM CELL SUSPENSION CULTURE PROTOPLASTS OF KIWIFRUIT

M.A. Revilla[1] and J.B. Power
[1]Departamento de Fisiologia Vegetal. Facultad de Biologia. Universidad de Oviedo, Oviedo, Spain.
Plant Genetic Manipulation Group, Department of Botany, University of Nottingham, Nottingham NG7 2RD, UK.

INTRODUCTION

Actinidia deliciosa Liang and Ferguson, commonly known as kiwifruit is a dioecious climbing shrub of vigorous growth, recently introduced in Europe. Kiwifruit production mainly utilises one cultivar, Hayward; effectively clones of a genetically uniform nature. This practice could render the kiwifruit crop vulnerable to disease (1). For this reason, it is important to exploit new technologies that could lead to a broadening of the genetic base for this crop. Protoplasts provide such a starting point for many techniques aimed at the genetic modification of plant cells and ultimately whole plants. Currently little information is available concerning kiwifruit protoplast culture, with only two reports describing protoplast isolation (2,3). We report here on the first successful culture of kiwifruit protoplasts isolated from cell suspension cultures.

MATERIALS AND METHODS

Callus establishment and culture

Petiole-half sections taken from field-grown plants of *A. deliciosa* cv. Hayward were surface sterilised with 7.5% (v/v) Domestos solution (15 min.), rinsed several times with sterile water and transferred to a medium which consisted of Cheng's mineral salts (4) supplemented with 0.25 mg/l thiamine HCl, 0.25 g/l m-inositol, 20 g/l sucrose, 8 g/l Bacto-agar, 0.5 mg/l 2,4-D and 0.05 mg/l NAA (pH 5.6). After 10 days, callus was separated from the petiole and maintained on the same medium, with monthly subcultures. All cultures were kept at 25±2°C with a continuous illumination of 2,000 lux provided by cool white day-light fluorescent tubes.

Cell suspension cultures

Callus portions (5 g f.wt. approx.) were dispersed in 70 ml of Linsmaier and Skoog's (L.S.) medium (5) containing 2 mg/l of 2,4-D, 3% (w/v) sucrose and 1.0 mg/l thiamine-HCl (pH 5.6) in 250 ml Erlenmeyer flasks. Incubation was on a gyratory shaker (100 rpm) and under growth conditions as used for callus initiation. Suspension cultures were diluted 1:1 after seven days with fresh medium until a rapid growth rate was established and thereafter subcultured (1:4) every 18-20 days.

Protoplast isolation and culture

Cell suspension cultures in their exponential growth phase (10 to 18 days after subculture) were harvested by centrifugation (150 x g) and plasmolysed (1h) in CPW13M medium (6). Cells (1.0 g f.wt.) were incubated in 10 ml of enzyme solution which consisted of Cellulase R-10, Hemicellulase, Driselase and Macerozyme R-10, all at concentrations of

0.5% (w/v) in CPW13M (pH 5.8). Incubation was overnight on a shaker (40 rpm) with 500 lux illumination (25°C). Protoplasts were separated from cell debris by passage through a 80 µm nylon sieve and collected by centrifugation (150 x g; 5 min.). The protoplast pellet was washed once in CPW13M and resuspended on a Percoll (Sigma, U.K.) gradient made up with CPW13M and then centrifuged (150 x g; 10 min.). Protoplasts, floating on 10% Percoll were collected, and washed once with CPW13M medium. Protoplast yields were determined and viability was assessed using fluorescein diacetate (FDA).

For the culture of protoplasts, several media based on that of MS (7) were used along with K8P medium (8). Protoplasts (2.0 or 4.0 ml aliquots) were plated in disposable 3 or 5 cm Petri dishes as either liquid or agarose (0.6% w/v) solidified cultures and at a density range of 10^4 to 10^6 protoplasts/ml. Cultures were maintained at 25°C with a continuous illumination (500 lux) provided by cool white day-light fluorescent tubes.

RESULTS AND DISCUSSION

A soft and friable callus was produced from the leaf-petiole of A. deliciosa. Fast-growing cell suspensions were obtained from this callus when the cells were cultured on LS medium where the organic components were replaced by 1.0 mg/l thiamine HCl alone.

Freshly isolated protoplasts (10-80 µm diam.) had a high viability (97%) with yields of 3.2 x 10^6/g f.wt. cell suspension.

The optimum plating density was 2 x 10^5 protoplasts/ml with K8P solidified medium; the best in terms of % protoplast division, cell wall regeneration as detected by changes in protoplast shape (after 48h), and first division (day 3). The plating efficiency (day 7) was 29%. In the liquid media the onset of cell wall regeneration was delayed with a much reduced plating efficiency, comparable to that obtained previously by Marino (3) for a similar kiwifruit protoplast system. After 15 days, colonies (16-25 cells) were observed, but to date, sustained growth of these has not been established.

Although further studies are needed, these results, coupled with the fact that an efficient system for plant regeneration from callus of A. deliciosa has been established (9), represent an important step towards establishing the basis for manipulation of kiwifruit via somatic hybridisation or somaclonal variation.

ACKNOWLEDGEMENT

M.A.R. was in receipt of an E.M.B.O. Fellowship.

REFERENCES

1. Young S: New Scientist 1469: 36-39, 1985.
2. Cossio F, Marino G: Riv. Ortoflorafruitt. It. 67: 455-464, 1983.
3. Marino G: in: Proc. Int. Symp. of Genetic Manipulation in Plant Breeding (eds.) W Horn, CJ Jensen, W Odenbach, O Schiedez, Walter de Gruyter, Berlin, 1986.
4. Cheng TY: Plant Sci. Lett. 5: 97-102, 1975.
5. Linsmaier EM, Skoog F: Physiol. Plant. 18: 100-127, 1965.
6. Power JB, Chapman JV, Wilson D: Laboratory Manual: Plant Tissue Culture, Univ. of Nottingham, U.K.
7. Murashige T, Skoog F: Physiol. Plant. 15: 473-497, 1962.
8. Kao KN: Mol. Gen. Genet. 150: 225-230, 1977.
9. Revilla MA, Power JB: J. Hort. Sci. (submitted for publication).

ISOLATION AND CULTURE OF GRAPEVINE PROTOPLASTS (VITIS VINIFERA L.)

G. Alleweldt, G. Reustle

Bundesforschungsanstalt für Rebenzüchtung Geilweilerhof
6741 Siebeldingen, BRD

1. INTRODUCTION

The isolation and regeneration of protoplasts deriving from perennial woody plants is very difficult and so far, regeneration experiments were occasionally successful in very few species only. Therefore, intensive efforts were undertaken to develop a system for regeneration of grapevine protoplasts in order to improve breeding programs, e.g. somaclonal variation, gene transfer and protoplast fusion. A very interesting aspect is the fusion to produce somatic hybrids between Euvitis species (n = 19) and Muscadinia species (n = 20) as a high resource for fungus resistant genes. In preliminary experiments, the isolation procedure (cellulase concentration, duration of digestion, purification) was optimized. Following this, culture conditions, different donor material as well as preconditioned leaf material were tested, to increase viability and regeneration capacity of grapevine protoplasts.

2. MATERIAL AND METHOD

2.1. Isolation

2.1.1. Starting material (Vitis vinifera L. cv. Optima).
- leaves of in-vitro plants
- shoot tips of in-vitro plants, 0.5 to 1cm in size
- suspension callus (derived from internodes)

2.1.2. Digestion. A mixture of different cellulases, described by De Filippis (1985) was used. The enzyme solution for leaves and shoot tips contained 0.4% cellulase from Aspergillus niger, 0.4% cellulase from Penicillium funiculosum, 0.8% cellulase from Trichoderma viride, 0.2% macerozyme R-10, 0.5% bovine serum albumin, 0.65M mannitol, 20mM MES-KOH, and 1mM $CaCl_2$. For suspension callus, a higher concentration of the cellulases was necessary (0.5% cell. A.niger, 0.5% cell. P.fun., 1% cell. T.viride, 1% macerozyme R-10). The pH was adjusted to 5.6. The incubation time lasts 3h in case of leaf material, for shoot tips 3h and 5h were tested. The digestion duration of callus depends on the fineness of the suspension. 8h to 12h were found to be suitable for good yield of viable protoplasts.

2.1.3. Purification. To separate intact protoplasts from cell debris, a centrifugation step (50g) with 0.75M sucrose for 10min was used.

2.1.4. Yield and viability. Protoplast yield were quantified with a Thoma Hemacytometer. Viability was confirmed with fluorescein diacetate (FDA).

2.1.5. Preconditioning. Whole leaves were incubated in different media for different durations. Pretreatment 1: 2d in medium A in darkness at 21°C, thereafter 1d in medium B in darkness at 4°C. Pretreatment 2: 7d in medium A in darkness at 21°C, thereafter 5d in medium B in darkness at 4°C. Medium A: 147mM $CaCl_2$, 80mM NH_4NO_3, 1ppm NAA, 0.5ppm BAP, pH 5.8. Medium B: 1/10 strength MS salts, without NH_4NO_3, 1/10 strength NN69 organics, 1ppm NAA, 0.5ppm BAP, pH 5.8.

2.1.6.Culture. The achieved protoplasts were cultured in darkness at 26°C, with a density of 1 to 3 x 10^5 cells per ml medium. Different modifications of liquid V/KM-medium were tested.

3.RESULTS

3.1.Isolation

TABLE 1. Yield and Viability of protoplasts derived from different starting material.

donor material	leaves 3h digestion	shoot tips 3h digestion	shoot tips 5h digestion	callus 10h digestion
yield per gram tissue (10^6)	4.25	2.53	5.25	0.28
viability (%)	87,5	92,2	83,3	90,3

3.2.Culture

Glucose was found to be the best osmoticum compared with sorbitol and mannitol. The osmotic value was adjusted at 650mosmol/kg. Two different hormone treatments were tested (1ppm 2,4-D / 0.5ppm BAP resp. 1ppm NAA / 0.5ppm BAP), but no clear differences could be found on the development of protoplasts. BSA (0.25%) has turned out to be a useful supplement for extending viability of protoplasts. Sometimes, the addition of PVP (0.5%) or PVPP (1%) was necessary, to prevent the cultures from browning in an early stage. Although these chemicals improved the viability of the protoplasts, they did not promote further development as cell division and colony formation. Preconditioning the starting material did not clearly improve viability and cell wall formation. Protoplasts derived from suspension callus showed best viability and development in culture, compared with mesophyll protoplasts from leaves and shoot tips. So far, protoplasts can be kept alive for several weeks, cell wall formation can be induced and in some cases initial cell division occurs, but no further development.

REFERENCES

De Filippis L.F., Ziegler H. The physiology of grapevine (Vitis vinifera L.) protoplasts isolated from green and senescing leaves. Biochem. Physiol. Pflanzen 180, 645-653 (1985).

STUDIES TOWARDS A PROTOPLAST ISOLATION AND CULTURE PROCEDURE IN SUGARBEET (BETA VULGARIS L.) APPLICABLE TO A VARIETY OF GENOTYPES

F.A. KRENS & D. JAMAR
FOUNDATION FOR AGRICULTURAL PLANT BREEDING, P.O. BOX 117, 6700 AC WAGENINGEN

INTRODUCTION

For the application of genetic manipulation techniques it is a prerequisite to have a good protoplast system followed by regeneration into plants. Sugarbeet is recalcitrant in some of these aspects, e.g. protoplast culture leading to microcolony formation. While some results have been obtained starting from suspension cells, success has been poor with mesophyll protoplasts (1). Our aim is to develop procedures for protoplast culture in sugarbeet applicable to a variety of genotypes.

PLANT MATERIAL

A triploid, commercial cultivar (Monohil, Hilleshög, Sweden) and a diploid, normal fertile line (SVP no.31-188) were chosen at random and routinely used in experiments unless stated otherwise. Seedlings were grown aseptically at 26^{0}C on MS medium at half strength, 3% (w/v) sucrose and solidified with 0.8% (w/v) agar and taken at an age of 4-6 weeks. Suspension cultures were maintained at 25^{0}C on PGo medium (2) supplemented with 1 mg/l 2,4-D and 0.5 mg/l kinetin (3) and subcultured every 7 days by 1:5 dilution.

RESULTS

Isolation and culture of mesophyll protoplasts

Source. As starting material for the isolation of mesophyll protoplasts the leaves of in vitro grown seedlings turned out to be the best. Leaves from plants grown in the greenhouse at a relative humidity (R.H.) of 70% developed a thick cuticle, which did not allow penetration of the enzymes used for protoplast induction. In a growth cabinet the R.H. was increased to approx. 85% and plantlets grown here gave high yields of protoplasts but without any further development like e.g. cell wall formation. Using the in vitro grown material many parameters have been varied in an effort to optimise yields, viability, cell wall formation, cell division and growth.

Preplasmolysis. Yields could be improved dramatically by preplasmolysis in a medium containing 250 mM $CaCl_2.2H_2O$ (Osmotic value: 1322 mOsm/kg H_2O). At lower osmolalities yields were less. The optimal time for preplasmolysis was 6 hours (see Table 1).

TABLE 1 The effect of preplasmolysis on protoplast yield

Time	0	1	2	6	12
Yield (as %-age of t=0)	100	133	211	357	23

Enzyme mixtures. Different enzymes in different combinations were tested as well as different media in which the enzymes were dissolved. For complete digestion a mixture of 2% (w/v) Cellulase Onozuka R-10 and 3% (w/v) Macerozyme Onozuka R-10 dissolved in culture medium proved to give the best results with regard to yield. In later experiments lower concentrations of enzymes gave lower yields but protoplast viability increased. The use of unpurified Driselase was detrimental to sugarbeet protoplasts in our hands.

Growth media. Five basal media were initially tested i.e. MS, B5, PGo, K3 and K8p. Ms resulted in lower viability, while K8p gave rise to early browning of cells. The other media showed no differences among them, but since PGo was developed for sugarbeet, further experiments used this as the basal medium. Initially protoplasts were very unstable after isolation. Filtersterilization of all media instead of autoclaving proved beneficial, as well as the addition of 0.1 M glycin to the growth medium. However, in later experiments with further optimised conditions glycin appeared not to be essential any longer. In order to increase membrane stability and cell wall formation numerous compounds were subsequently added to the growth media like CaCl2, myo-inositol, rhamnose, ribose, fructose, mannose, arabinose, galactose, galacturonic acid, cellobiose and xylose. Other additives were tested for their influence on cell division e.g. edamin, fetal calf serum. No significant positive effects were observed. Best results were obtained with PG1B, a derivative of PGo with lower inorganic nitrogen (KNO_3 and $(NH_4)_2SO_4$) and increased organic N (4 amino acids: serine, glutamine, proline, asparagine), 10 mM $CaCl_2$, 500 mg/l myo-inositol, 250 mg/l cellobiose and xylose, 0.4 M sucrose, 0.05 M glycin. As growth regulators 1 mg/l NAA, 0.2 mg/l 2,4-D and 0.5 mg/l BAP were added.

Environment. Growth at 25°C was superior to growth at 20°C. Protoplasts preferred culturing in the dark to culturing at 1200 lux for 16 hours/day. Extra stimuli came from the use of 10% conditioned medium and the presence of actively dividing suspension cells. Embedding in agarose was favourable using PG1B, however not with other media.

Protoplast isolation from cell suspension cultures

Isolation. Protoplasts were isolated as described by Bhat et al.(4).

Culture. For growth media K8p was compared to PG1B at first and K8p appeared to give higher plating efficiencies of approx. 10-20%. Also many cells turned brown in K8p and died. To avoid this K8p was adapted by eliminating certain groups of constituents (e.g. coconut water, organic acids, almost all vitamins, etc.) Highest plating efficiency combined with the least browning was achieved in a medium containing K8p macro salts, PGo micro salts, 0.25 M glucose, 3.5% mannitol, 1 mg/l nicotinic acid, pyridoxine.HCl and thiamin.HCl, 100 mg/l myo-inositol, NaFeEDTA as normal, 1 mg/l NAA, 0.2 mg/l 2,4-D and 0.5 mg/l BAP, pH6.0. Embedding the protoplasts in agarose did not have a negative effect, but gave no higher plating efficiency either. Routinely calli could be obtained, unfortunately without any regeneration ability.

CONCLUSIONS

A procedure has been developed which allowed the isolation of mesophyll protoplasts at high yields. Viable, cell wall regenerating cells can be obtained with occasionally cell division. However, no sustained growth into microcolonies occurred sofar. Several lines of sugar and fodder beet have been tried with identical results. Preliminary experiments have shown that the mesophyll protoplasts can be induced to fuse by electrofusion. Research is now focussed on protocols with lower yield, but prolonged vitality and sustained cell divisions. Suspension cultures have proven to be a good source of protoplasts with sustained growth and plating efficiencies which allow the use of these protoplasts in genetic manipulation studies.

REFERENCES

1. B.V. Ford-Lloyd & S. Bhat. In: Genetic Manipulation in Plant Breeding; pp 437-440. Walter de Gruyter & Co., Berlin, 1986.
2. W. De Greef & M. Jacobs. Plant Science Lett. vol. 17: 55-61, 1979.
3. L. Szabados & C. Gaggero. Plant Cell Reports vol. 4: 195-198, 1985.
4. S.R. Bhat, B.V. Ford-Lloyd & J.A. Callow. Plant Cell Reports vol. 4: 348-350, 1985.

IMPROVEMENT OF PROTOPLAST CULTURE IN Helianthus annuus (L)

M. BURRUS, C. CHANABE, M.H. MONTANE and G. ALIBERT

Laboratoire de Biotechnologie et amélioration des plantes, INP-ENSAT, 145, Avenue de Muret, 31076 Toulouse cedex France

1. INTRODUCTION

Breeding improvement in Helianthus annuus involves the hybridization with wild (Helianthus) species. However H. annuus is sexually incompatible with most of the wild species. Somatic hybridization by protoplast fusion represents an alternative to overcome this difficulty but is still limited by plant regeneration.

This work deals with the isolation and the purification of protoplasts from two genotypes of sunflower in order to obtain a high plating efficiency.

2. MATERIAL AND METHODS

. Seven day old hypocotyles of the RHA 274 and PHA 018 were grown on solid MS medium. Sliced hypocotyles were digested in different enzyme mixtures at 25°C in" S" medium (1). After an overnight incubation, protoplast suspensions were pelleted by low speed centrifugation and washed twice.

Purification was performed by floatation on 10% Ficoll gradient. The floated protoplasts were then washed twice. In all cases, protoplasts were resuspended in "S" medium .

The diameter of 400 protoplasts was measured for each population (purified or not), by observation on a video-monitoring system (JVC Japan).

The protoplasts were cultured in" L4" medium (1) as a test medium.

3. RESULTS

The yield of protoplasts produced after digestion with different enzyme mixtures is given in table I.

TABLE I

	Protoplast yield (10 6 / g fresh weight)		
Enzyme mixtures	I	II	III
RHA 274	1.8	2.1	2.7
PHA 018	1.5	2.3	3

I :Macerozyme R10: 0.2%-Cellulase R10:0.1%-Driselase 0.5%. II:Pectolyase Y23 0.004%-Caylase 345 0.05%- Macerozyme R10 0.05%. III: Caylase M2: 0.075%- Caylase T:0.075%

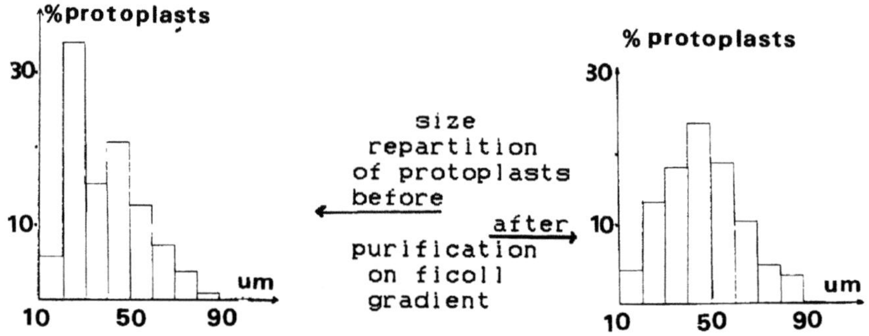

The purified fraction is devoid of cellular debris and more homogeneous in size. When cultured on L4 medium, these fractions show an improved plating efficiency and microcolony yield compared to non purified protoplasts (Table II)

TABLE II

genotype	enzyme mixture	protoplast yield (%) NP	protoplast yield (%) P	PE after7days NP	PE after7days P
RHA 274	I	49	42	20	23
	II	65	50	23	45
	III	60	45	14	30
PHA 018	I	75	33	nd	28
	II	85	53	30	37
	III	89	60	32	38

NP= non purified, P= purified

4. CONCLUSION

The results presented in this publication clairly show that the conditions for protoplast isolation in view of subsequent culture are to be adapted to each genotype. Moreover, the conditions favouring high yields are not necessarily the best for optimal plating efficiency.

The purification of the protoplasts on Ficoll gradient greatly improves the plating efficiency. In addition the fractions are more homogeneous in size and thus more suitable for protoplast electrofusion (2) an efficient technique for somatic hybride production (3, 4).

REFERENCES

1.Lenée P. and Chupeau Y.: 1986, Plant Science, 43, 69-75.
2.Montané M.H., Teissié J. and Alibert G: 1987, This issue.
3.Chapel M., Montané M.H., Ranty B., Teissié J. and Alibert G. : FEBS lett., 196, 79-85.
4.Montané M.H., Alibert G. and Teissié J.: 1987, Studia Biophysica, 119, 1-3.

ACKNOWLEDGMENT

This work was supported by Pioneer France.

FACTORS INFLUENCING PROTOPLASTS PLATING EFFICIENCY AND DIRECT SOMATIC EMBRYOGENESIS INDUCTION IN SUNFLOWER (HELIANTHUS ANNUUS L.)

J.M. DUPUIS*, P. CHAGVARDIEFF

Laboratoire Cellules Végétales Photosynthétiques, A.R.B.S.,
C.E.N., Cadarache, Bât. 161, 13108 Saint-Paul-lez-Durance Cedex, France.

A rather limited number of studies on Helianthus annuus protoplasts culture has been developed. In culture, protoplasts divisions usually lead to the formation of micro-calli (Lenée and Chupeau, 1986 ; Bohorova et al., 1986).

In our conditions, protoplasts give rise either directly to early somatic embryogenic forms, or to micro-calli in more or less important proportions. For the first time we describe direct somatic embryogenesis from protoplasts culture in sunflower (see Fig. 1).

The incubation of hypocotyl fragments and the medium used for protoplasts culture arise from a modified Caboche protocol (1980) using the plating Shillito et al. method (1983).

We measured 2 parameters after 10 days of culture : i) the plating efficiency which is the percentage of protoplasts showing more than 2 divisions (see Fig. 2 a) and ii) the somatic embryogenesis which is the percentage of somatic embryos to the whole dividing protoplasts population (see Fig. 2 b).

Four factors are studied :

. The density of protoplasts culture : 25.10^3 to $10.^5$ proto/ml.

. The type of gelling agent : Difco bacto-agar ; seaplaque LMT agarose ; Sigma agarose.

. The varietal effect among INRA F1 hybrids : 76 D ; 260 E ; 60 E ; cv "Bolero" ; cv "Primasol".

. The concentration of α-naphthaleneacetic acid (NAA : 0 to 1 mg/l) at the first subculture on day 7.

These four factors influence the plating efficiency of protoplasts. Among the combinations tested, the maximum plating efficiency of cv "Bolero" protoplasts was obtained using an initial density higher than 25.10^3 proto/ml in Difco bacto-agar with less than 1 mg/l NAA. In these conditions cv "Primasol" and 76 D gave higher efficiencies than the other hybrids.

On the other hand, somatic embryogenesis was only affected by NAA concentrations and varieties tested.

For cv "Bolero", rising NAA concentration from 0.1 to 0.5 mg/l enhanced embryogenesis from about 35 % to about 80 %. With 0.1 mg/l NAA, the best results were obtained with cv "Primasol" and 76 D.

Further evolutions of embryos and micro-calli are presented.

REFERENCES
1. N.E. Borohova, E.C. Cocking, J.B. Power, Plant Cell Rep. 5 (1986) 256-258.
2. P. Caboche, Planta 149 (1980) 7-18.
3. P. Lenée, Y. Chupeau, Plant Sci. 43 (1986) 69-75.
4. R.D. Shillito, J. Paszkowski, I. Potrykus. In : Potrykus, I., *et al.* (eds) Protoplasts 1983, Poster Proc., 6th Intl. Protoplast Symp. Birkhaüser Verlag Basel (1983) pp. 266-267.

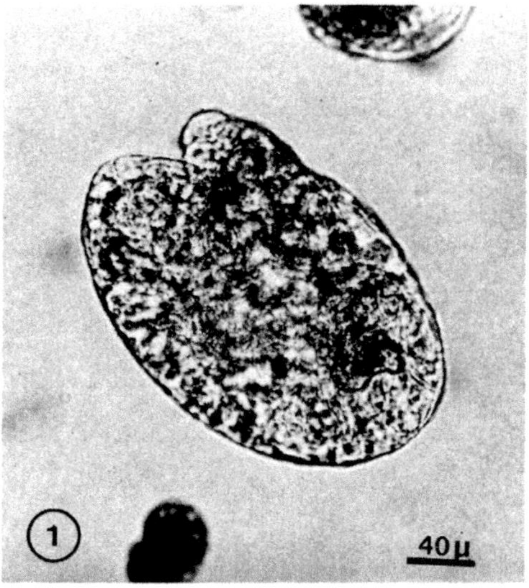

Fig. 1 : Somatic embryogenic form derived from a sunflower protoplast at day 20 of culture.

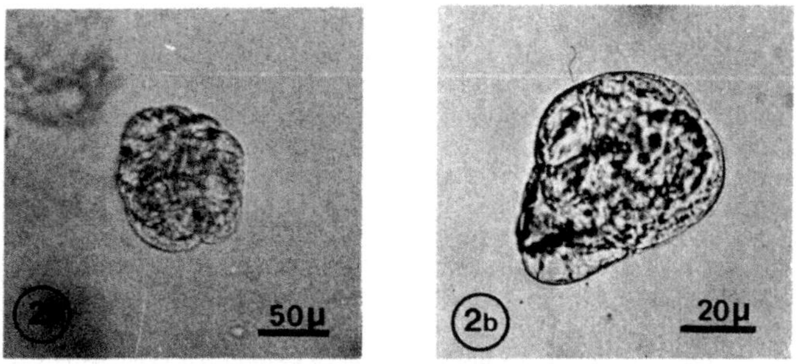

Fig. 2 : Evolution of sunflower protoplast at day 10 of culture

2a : dividing protoplast (further evolution will be a microcallus)
2b : early somatic embryogenic form.

COMPARISON OF SUNFLOWER PROTOPLAST DEVELOPMENT IN LIQUID OR SOLIDIFIED MEDIA

*V. Thor, A-L. Moyne, C. Sarda, C. Bergounioux, and P. Gadal
Physiologie végétale moléculaire, bât. 430, Faculté des Sciences, UA 1128 CNRS
Université Paris-Sud, 91405 Orsay Cedex, France.

1.INTRODUCTION

Different procedures have been reported for the culture of plant cell protoplasts in solidified media. Agar was the first gelling agent used for protoplast cultures (1). Plating efficiency was markedly improved in a number of plant species by culturing in media solidified with agarose (2, 3). Protoplast viability was not affected by immobilization in various matrices such as agarose, alginate or kappa-carrageenan (4).

The aim of this report is to compare cell division and cell structure in both suspension and solidified media. Using two complementary techniques (flow cytometry and image analysis), we followed DNA quantity and cell wall regeneration per cell during 48 hours of culture.

2.MATERIALS AND METHODS

Protoplasts of Sunflower hypocotyls (var. Rodéo) were obtained according to Lénée and Chupeau (5) and plated at a density of 6.10^4 proto./ml in 3 ml of liquid or solidified (1% agarose) culture media Mc. Cultures were maintained at 25°/22°C with a 16 hours photoperiod at 13 $\mu E.m^{-2}.s^{-1}$. For DNA measurements by image analysis, protoplast-derived cells cultured in liquid medium were entrapped in agarose beads before fixation and staining (Feulgen technique) in order to be in the same conditions than agarose-cultured ones. Protoplasts and cells were laid on 0.45 µm filters after melting by incubation at 80°C for approximately 10 mn. For the cell wall staining, Calcofluor White was added to the plated non-fixed protoplasts at a concentration of 5mg/l 24 hours before image analysis. Image analysis of nuclei was performed by means of the SAMBA 200 (TITn and Co.). Flow analysis was conducted on an EPICS V cytofluorimeter (Coulter, Florida).

3.RESULTS AND DISCUSSION

Plating protoplasts in agarose stimulated division compared to culture in liquid medium and calli were obtained only when protoplasts were entrapped in agarose solidified medium. After 24 h culture, the histograms representing the amount of DNA per nuclei were similar whatever the culture medium was. Marked differences appeared after 48h (Fig. 1): G_2 phase was characterized by a shifted sharp peak representing an increase in nuclei when protoplasts are cultured in agarose medium.

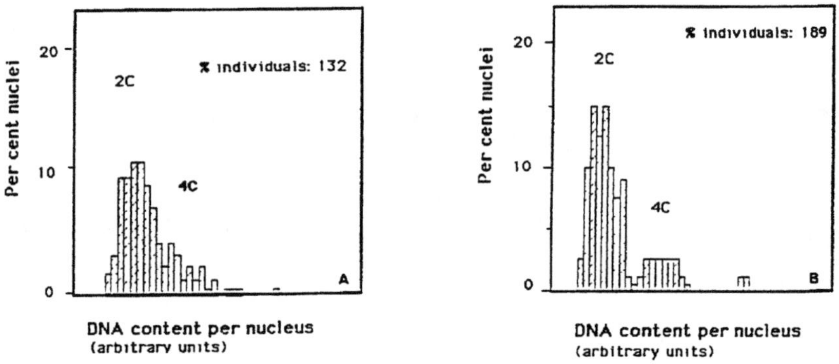

Fig 1: Optical density of nuclei after 48 hours in liquid (A) and agarose (B) media.

During the first 24 hours cell wall synthesis was more intensive in liquid medium than in solidified medium but the phenomenon reversed after 48 h.(Fig. 2) This result had been confirmed by flow cytometry analysis.

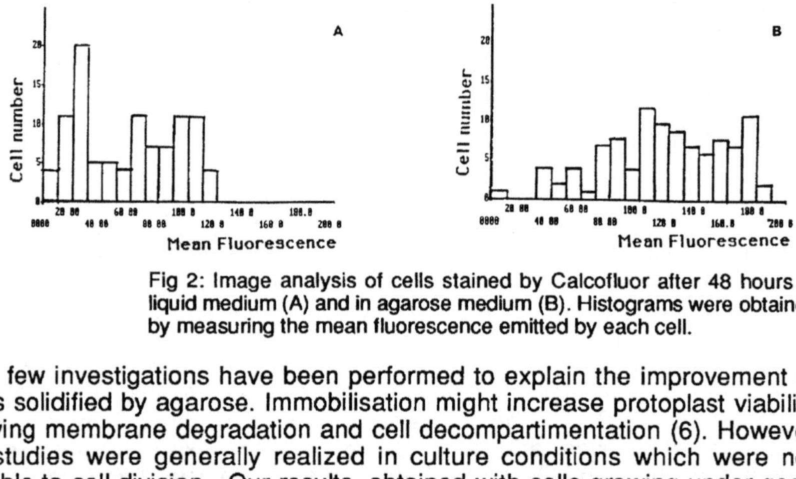

Fig 2: Image analysis of cells stained by Calcofluor after 48 hours in liquid medium (A) and in agarose medium (B). Histograms were obtained by measuring the mean fluorescence emitted by each cell.

Very few investigations have been performed to explain the improvement of cultures solidified by agarose. Immobilisation might increase protoplast viability by slowing membrane degradation and cell decompartimentation (6). However these studies were generally realized in culture conditions which were not favourable to cell division. Our results, obtained with cells growing under good culture conditions, showed differences in DNA synthesis and cell wall regeneration during the first 48 h in culture.

REFERENCES

1. Nagata T., Takebe I. (1971). Planta 99:12-20.
2.Shillito R.D., Paszkowski J. and Potrykus I. (1983). Plant Cell Rep. 2:244-247.
3. Lörz H., Larkin P.J., Thomson J. and Scowcroft W.R. (1983). Plant Cell Tiss. Org. Cult. 2:217-226.
4. Linsefors L. and Brodelius P. (1985). Plant Cell Rep. 4:23-27.
5.Lénée P. and Chupeau Y. (1986). Plant Sci.43:69-75.
6. Schnabl H. and Youngman R.J. (1985). Plant Sci.40:65-69.

This work has been supported by a Rhône-Poulenc grant. The authors wish to acknowledge TITn and Dr.MARTY-BATTLE for allowing us to work with SAMBA., Dr.S. BROWN for his assistance on EPICS V.

PLANT REGENERATION FROM PROTOPLASTS OF RUDBECKIA AND DIMORPHOTHECA SPECIES (COMPOSITAE)

J. AL-ATABEE and J.B. POWER[1]
Scientific Research Council, P.O. Box 2441, Baghdad, Iraq.
[1]Plant Genetic Manipulation Group, Department of Botany, University of Nottingham, Nottingham NG7 2RD, UK.

1. INTRODUCTION

There are only few reports describing plant regeneration from protoplasts of ornamental species (Compositae), Gaillardia (1) and Chrysanthemum (2). Regenerating protoplast systems have now been established for Rudbeckia hirta, R. laciniata and R. purpurea (Cone Flower), and Dimorphotheca aurantiaca (Cape Marigold) (3).

2. MATERIALS AND METHODS

2.1. Plant material

Plants of D. aurantiaca (cvs. Giant Orange and Glistening White), R. hirta cv. Marmalade, R. laciniata cv. Irish Eyes and R. purpurea cv. Brilliant Star were grown in the greenhouse. For all species leaves (3rd/4th leaves of 25-40 day old plants), stem and ray florets were surface-sterilised in 10% Domestos (15 min.) and used directly as a source of protoplasts. Leaf callus was established for Dimorphotheca using MSD_3 (4) and using MS medium (5) with 1.0 mg/l NAA, 2.0 mg/l BAP, 3% sucrose, 0.8% agar (pH 5.8) for Rudbeckia. Cotyledons from in vitro cultured seedlings were also used for protoplast isolation.

2.2. Rudbeckia protoplast isolation and culture.

The lower epidermis of leaves, cotyledons and ray florets was removed by peeling and then tissues plasmolysed, for 1h in CPW13M medium (6) or CPW8M medium (ray florets). They were transferred to an enzyme mixture which consisted of 0.8% Cellulase R10, 0.4% Driselase, 0.8% Rhozyme HP150, 0.2% Macerozyme R10, CPW salts, 8% mannitol (or 13% mannitol for ray florets) (pH 5.8). All tissues were incubated (3-4h) on a rotary shaker (30 cycles/min) followed by a stationary incubation (8-10h), 23°C (dark). Protoplasts were released from Rudbeckia leaf callus using the same enzyme mixture and protocol but with 0.4% Macerozyme R10 (12h on a rotary shaker). Protoplasts were maintained as liquid or agarose-bead cultures (7) at a density of 2.5 x 10^4/ml, 23°C (dark or light). The protoplast liquid culture medium was replaced every 5-7 days with the same medium before a reduction of the osmotic pressure was initiated after 16 days for callus protoplasts and 30 days for the other systems. The following régimes were used to reduce the osmotic pressure by the addition of the appropriate medium of a lower osmotic pressure. The three media used, $MSP_1$9M (4), K8P (8) and KM8P (9) were progressively diluted with $MSP_1$0M, K8 and KM8 respectively and in the ratios (ml) 3:1, 2:2, 1:3, 0:4 at 5 day intervals. Protoplast-derived calli (1.5 - 2.0mm diam.) were transferred to MS medium with 2.0 mg/l BAP, 1.0 mg/l IAA, 3% sucrose, 0.8% agar (pH 5.8) for 3 weeks for further growth and then to plant regeneration medium (MS medium with 0.1 mg/l IAA, 0.4 mg/l kinetin, 0.8% agar, 3% sucrose (pH 5.8)). Shoots were rooted in BGS medium (4) whereupon they were transferred to the open greenhouse.

2.3. Dimorphotheca protoplast isolation and culture.

Leaves, cotyledons and ray florets (lower epidermis removed) plus stem segments were plasmolysed for 1h in CPW8M medium (leaves/cotyledons) or CPW13M medium (stem/ray florets). Protoplasts were released from leaf/cotyledon tissues using an enzyme mixture which consisted of 2.0% Meicelase, 2.0% Rhozyme HP 150, 0.03% Macerozyme R10, CPW salts, 8% mannitol (pH 5.8). For Dimorphotheca leaf callus/stem/ray floret tissues, the enzyme mixture was as used for Rudbeckia callus. A static incubation (dark, 23°C) was adopted for leaves (5-6h), cotyledons (10-12h) or ray florets (10h), whilst for stem or callus tissues incubation was on a rotary shaker (30 cycles/min) (4-5h) followed by static conditions (8-10h). Protoplast culture was as for Rudbeckia.

Protoplast-derived calli of Dimorphotheca (1.5-2.0mm diam.) were transferred to MSD_3 medium (3-5 weeks) for proliferation and then to MS medium with 1.0 mg/l IAA, 0.5 mg/l BAP, 0.3 mg/l GA_3, 3% sucrose, 0.8% agar (pH 5.8) (3-7 weeks) in order to induce shoots. Shoots, detached from the callus, were grown on BGS medium (3-5 weeks) and then rooted following transfer firstly to MS medium (with 0.5 mg/l NAA, 3-5 days) and then to MS medium lacking hormones.

3. RESULTS AND DISCUSSION

First division of D. aurantiaca protoplasts was after 2-3 days with the formation of 8-celled colonies by day 12. Ray floret protoplasts failed to divide in all media, and $MSP_1$9M medium did not support division of any of the Dimorphotheca protoplast systems. Upon transfer to regeneration medium, the efficiency of plant regeneration of the leaf protoplast derived-calli was 26% and 20% respectively for the varieties Glistening White and Giant Orange. This efficiency of plant regeneration for Dimorphotheca leaf protoplasts was superior to that described for the related species Chrysanthemum morifolium (2) where only 1.5% of leaf protoplast-derived calli underwent organogenensis.

The pattern of protoplast development for the Rudbeckia species was basically similar to Dimorphotheca. Although callus tissues were superior in terms of initial protoplast viability, it was only the leaf protoplasts of R. hirta cv. Marmalade that gave callus capable of plant regeneration. In this case, regeneration efficiency was 100% after two subcultures on the regeneration medium. This study now forms a basis for somatic hybridisation programmes designed, ultimately to foster their improvement in relation to their floral and marketable qualities. For example, in preliminary experiments leaf protoplasts of R. laciniata, electrofused with non-dividing ray floret protoplasts of D. aurantiaca, gave heterokaryons capable of sustained division in agarose-solidified KM8P medium.

REFERENCES

1. Binding H, Nehls R, Kock R, Finger J, Mordhorst G: Z. Pflanzenphysiol. 101: 119-130, 1981.
2. Otsuka M: Agric. Hort. 41: 50-51, 1986.
3. Al-Atabee JS, Power JB: Plant Cell Rep. (in press).
4. Power JB, Chapman JV, Wilson D: Laboratory Manual: Plant Tissue Culture, Nottingham University, U.K.
5. Murashige T, Skoog F: Physiol. Plant. 15: 473-497, 1962.
6. Frearson EM, Power JB, Cocking EC: Dev. Biol. 33: 130-137, 1973.
7. Shillito RD, Paszkowski J, Potrykus I: Plant Cell Rep. 2: 244-247, 1983.
8. Kao KN: Mol. Gen. Genet. 150: 225-230, 1977.
9. Kao KN, Michayluk MR: Planta 126: 105-110, 1975.

Optimization of lettuce calli regeneration from protoplasts.

ROGUET Y.

Laboratoire d'amelioration des plantes - Bt 360- Faculté d'Orsay, Paris XI- 91405 Orsay cedex- FRANCE.

Lettuce, a crop of major economic importance in the world, is placed fourth behind tomatoes, orange and potatoes in the rating scheme of stapple food.

Despite the progress of the breeding programs, lettuce is continually beset with problems that reduce quality and yield. Numerous laboratories in the world explored the potential of crop improvements through the use of protoplasts.

Technics for regeneration of lettuce plants from leaf protoplasts (4-5) and calli (1-3) have been reported. But the rate of regeneration of protoplasts are too low to allow the use of transformation technology. A method with high regeneration rates from protoplasts must be developed.

This abstract describe the different steps of an optimization of lettuce protoplasts regeneration.

1- The protoplasts are isolated following Sihachachr method (6) for the Solanaceae. The cotyledons were cut and incubated over-night into an enzymatic solution. The yields of protoplasts per gram of fresh material were :

- 36 10^6 for the cotyledons ;
- 16 10^6 for the first leaf.

The protoplasts are diluted at 70 000 per milliliters in the medium A wich induced cell division. The first division appears between 3 to 6 days after protoplast isolation according to the medium used. The cell clusters to divided until 32 cells need a maximum of 21 days.

2- The microcalli were platted into a solid medium B for the growth of the calli. 3 mediums (4,5,6) were tested, all the mediums allowed the growth of the calli and further plant regeneration, but the medium n°6 is the best.

3- Four weeks after, the calli were transfered on a medium C for shoot induction. Shoot buds are visible after a month at a frequency of about 0.46. Nine-agar medium differring by hormonal composition were tested :

- 4 mediums do not allowed plant regeneration ;
- 3 mediums allowed low rates of plant regeneration ;
- 1 medium allow a good rate of plant regeneration ;
- 1 medium allow a very high rate of plant regeneration.

4- A fourth medium D was necessary for root initiation and shoot growth. An average of 53 plants per calli was obtained.

The calli produced plants during five months.

The plants were transfered into a green-house. 75 % of the regenerated plants survived transfering into soil, flowering and seed production by self-pollination were normal. Only 11 % of the regenerated plants were sterile.

Although all regenerated plants have a normal chromosome number (2n = 18). On the other hand, they showed a great phenotypic variability.

We have a four fold increase the yield of protoplasts. The number of plants regenerated per callus has been enhanced.

REFERENCES :

1- KOEVARY K. Hort Science 13 (1978) 39 ;

2- DOERSCHUNG M.R., MILLER C.O. Am. J. Bot. 54 (1967) 410 ;

3- SIBI M. Ann. Am. Pl. 26 (1976) 523 ;

4- BERRY S.F. Z Pflansenphysiol. 108 (1982) 31 ;

5- ENGLER D.E., GROGAN R.G. Plant Sci. Letters 28 (1982/83) 223 ;

6- SIHACHAKR D., DUCREUX G. Plant Science 52 (1987) 117.

OPTIMAL DENSITY AND OSMOLALITY OF PROTOPLASTS FOR IMPROVED CULTURAL PERFORMANCE

G. B. Poulsen
The Biotechnology Group
Danish Research Service for Plant and Soil Science
Lottenborgvej 2, DK-2800 Lyngby, Denmark

1. INTRODUCTION

Osmolality is known to affect the replication and translation processes in protoplasts (1,2).

In *Brassica napus* was found that small unvacuolated protoplasts were the first to begin cell division (3). Microcallus-formation was encouraged at an osmolality of 0.600 osmol/kg, while the viability of protoplasts were unchanged over a wide osmolality range 0.400-0.650 osmol/kg (4).

The aim of the present experiment was:

- determination of the optimal conditions for protoplast growth,
- definition of optimal buoyant density of the purification medium in order to select a fraction of protoplasts with high viability and ability to grow, and
- optimization of the osmolalities and densities in the media.

2. MATERIALS AND METHODS

Four day old etiolated hypocotyls of *Brassica napus* cv. Topas were isolated and cultivated as described by Glimelius (3).

The varying osmotic treatments were created by addition of different amounts of glucose to the medium. The osmolalities of the media were determined on a Vescor vapourpressure Osmometer 5500. Viability was determined by fluoresceindiacetet (FDA)-stain on the 2nd and 5th day after protoplast isolation.

The densities of the protoplasts were determined by density centrifugation on a colloid silica sol - Percoll® (5) adjusted to appropriate osmolality by addition of 0.21 M NaCl. A continuous gradient was generated by high speed centrifugation of 25 ml Percoll/NaCl in 25 x 120 mm centrifuge tubes. Sample and density markers were added on top of the gradients and centrifuged (100 x g, 7 min.).

A discontinuous gradient with density steps at 1.01 g/ml, 1.02 g/ml, 1.03 g/ml, 1.04 g/ml, 1.05 g/ml and 1.06 g/ml were composed according to the manufacturer's instructions (5) and stacked in 2 ml steps in 16 x 120 mm centrifuge tubes.

Individual fractions were pipetted, diluted with 0.21 M NaCl in water, and precipitated by low speed centrifugation. The precipitate was resuspended in culture medium, and viability of protoplasts were determined 2 and 5 days after isolation.

3. RESULTS AND DISCUSSION

Osmolalities of purification and cultivation media varied from 0.440 osmol/kg (TVL) to 0.630 osmol/kg (W5).

Viability tests of protoplasts cultivated at varying osmolalities proved that protoplasts cultivated in media with 0.500-0.600 osmol/kg had the highest rate of survival. The plating efficiencies showed no differences between the treatments, indicating that protoplasts which have divided and formed microcalli, are able to survive the following cultivation.

Centrifugation on a continous density gradient resulted in a protoplast buoyant density of 1.028-1.031 g/ml at 0.4 osmol/kg. The discontinuous gradient showed that the highest number of protoplasts was found in the fraction of 1.03 g/ml and 1.04 g/ml, which is a little more dense than the average value found with the continuous gradient. Viability of protoplasts was highest in the 1.04 g/ml and 1.05 g/ml fractions.

Microscopic examination of the density fractions revealed that the lighter fractions (1.01-1.02 g/ml) consisted of small protoplasts and debris. The 1.03 g/ml fraction contained amorphous protoplasts and highly vacuolated protoplasts. The dense fraction (1.05 g/ml) had a high number of sphaerical unvacuolated protoplasts, and was practically free of impurities. However, the most dense fraction investigated (1.06 g/ml) had collected many chloroplasts, residues from the explant tissue and broken protoplasts.

Density centrifugation is thus an efficient method for selecting the most competent protoplasts for growth, which is a necessary condition for succesful regeneration of plants after protoplast fusion and introduction of foreign genes into protoplasts.

REFERENCES

1. Cress, D.E.: Plant Cell Reports 1, 186-188, 1982.
2. Zelcer, A.; Galun, E.: Plant Science Letters 7: 331-336 (1976).
3. Glimelius, K.: Physiol. Plant. 61, 38-44, 1984.
4. Spangenberg, G; Koop, H.-U.; Lichter, R.; Schweiger, H.G.: Physiol. Plant. 66, 1-8.
5. Anonymous. Percoll® Methodolygy and applications. Pharmacia Fine Chemicals, 1982.

DEVELOPMENT OF PROTOPLAST CULTURE PROCEDURES IN BRASSICA CROPS SUITABLE FOR IN VITRO SELECTION HERBICIDE RESISTANCE MUTANTS AND CYBRIDS

C. O'NEILL[1], P.J. DIX[1], G. DOUGLAS[2]
[1]Dept. of Biology, St. Patrick's College, Maynooth, Co. Kildare, Ireland.
[2]Agricultural Research Institute, Kinsealy, Malahide Road, Dublin 17, Ireland.

1. INTRODUCTION

The present report is related to our project on improvement of resistance to a triazine herbicide, atrazine, in *B. napus*, *B. oleracea* and *B. campestris*. Previous reports on selection for herbicide resistance involve the use of partially photosynthetic cultures (Cseplo *et al.*, 1985). Sensitive cells bleach in the presence of atrazine whilst resistant biotypes do not. It should be possible to use the same selection system to detect resistant p-calli after fusion of protoplasts of atrazine resistant rape with susceptible cultivars. The current report describes the development of an efficient protoplast culture and regeneration system for *Brassica* species, including bright green p-calli, in which it should be possible to obtain the primary symptom of atrazine sensitivity (bleaching) and therefore to select resistant lines.

2. PROCEDURE

The protocol used is based on that outlined by Pelletier *et al.*, 1983. A number of changes have been employed of which the most important are a) a slow reduction of the osmoticum of the protoplast culture medium by addition of small amounts of fresh medium with no sucrose or mannitol, and b) zeatin and IAA are used in *B. oleracea* shoot-regeneration media and the cytokinin 4Pu for shoot regeneration from *B. napus* p-calli. Using this method a high percentage of the p-calli survive the transfer to agarose regenerating medium. It has also been found that darker green *B. oleracea* p-calli develop on zeatin and IAA medium than on medium with 2,4-D and B.A.P. Fig. 1 shows some *B. oleracea* var Jura p-calli regenerating shoots on 0.2 mg l^{-1} IAA and 3.0 mg l^{-1} zeatin. On this medium 18.3% of the calli produce shoots. Table 1 shows the effects of a range of 4Pu concentrations and silver nitrate addition on *B. napus* var Tower and *B. oleracea* var Jura regeneration frequencies.

3. CONCLUSION

A good reliable system for culture of *Brassica* protoplasts has been developed. It satisfies the requirements for selection of atrazine resistance *in vitro*. This improves the prospects for *in vitro* mutagenesis, and the selection of resistant calli, as well as successful recovery of resistant cybrids after transfer of chloroplasts from resistant to susceptible cultivars.

Figure 1.

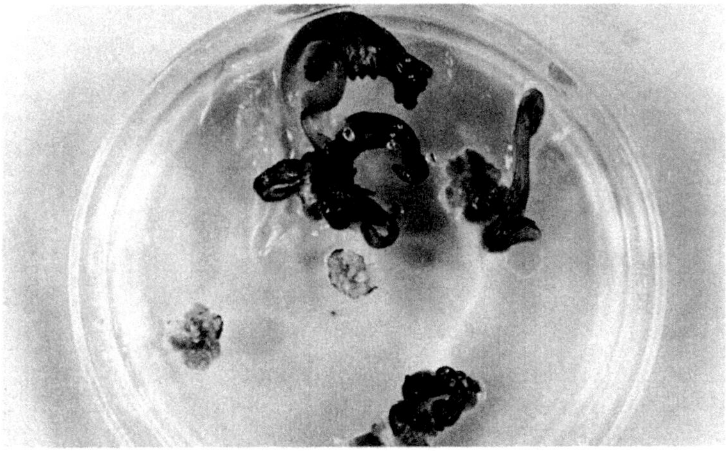

Table 1. Regeneration frequencies (% of p-calli giving shoots) of 13 month old Tower p-calli and 4 month old Jura p-calli on media containing various levels of the cytokinin 4Pu (in one case in combination with $AgNO_3$) are recorded.

4Pu (mg l^{-1})	B. napus var. Tower	B. oleracea var Jura
0	3.8	11.5
0.1	3.3	4.3
0.5	10.7	4.0
1.0	0	3.7
10	8.0	9.5
0.1 + 15 mgl^{-1} $AgNO_3$	43.7	12.2

REFERENCES

1. Pelletier G, Primard G, Vedel F, Chetrit P, Remy R, Rousselle and Renard M: Mol. Gen. Genet. 191, 244-250, 1983.
2. Cseplo A, Medegysey P, Hideg E, Demeter S, Marton L, and Maliga P: Mol. Gen. Genet. 200, 508-510, 1985.

Abbreviation:

4Pu: 2Chloro-4-pyridyl-n-phenyl urea.

SOMATIC EMBRYOGENESIS IN STATIONARY PHASE SUSPENSION CULTURES DERIVED FROM HYPOCOTYL PROTOPLASTS OF BRASSICA NAPUS L.

E. KRANZ
Max-Planck-Institut für Zellbiologie, D-6802 Ladenburg, Rosenhof, Federal Republic of Germany
Present address: Max-Planck-Institut für Züchtungsforschung, D-5000 Köln 30, Federal Republic of Germany

1. INTRODUCTION

To get more efficient regeneration of plants from hypocotyl protoplast-derived suspension cultures of *Brassica napus*, investigations regarding the time course of cultivation for the formation of somatic embryos were carried out. Although regeneration of plants from hypocotyl protoplasts via callus is sometimes efficient (Glimelius 1984, Chuong et al. 1985, 1987, Barsby et al. 1986), shoot morphogenesis is often not reproducible, when other genotypes are used in regeneration experiments. Reproducible plant regeneration procedures from protoplasts of Brassica species are still needed (Chuong et al. 1987). Most useful may be the regeneration of plants via somatic embryogenesis. Li and Kohlenbach (1982) obtained somatic embryogenesis from protoplast-derived cultures by testing various media, especially the hormone compositions. Our data demonstrate the importance of the choice of stationary phase aggregates as inoculum for embryoid formation in hypocotyl derived suspension cultures of *Brassica napus* L. These findings may be of general importance, since embryoid formation, greening and plant regeneration could be achieved with an embryogenic line of *Digitalis lanata*, using a similar sequence of passages. Stationary and late stationary phase cells showed the highest degree of embryoid formation followed by greening and plantlet formation.

2. MATERIALS AND METHODS

Plant material

Seeds of winter oilseed rape, *Brassica napus* L., line EGK 10, were a gift from Dr. Böhm, Kartoffelzucht Böhm, Langquaid, FRG. Preparation of hypocotyl protoplasts from *Brassica napus* has been described elsewhere (Spangenberg et al. 1986). The embryogenic line of *Digitalis lanata* Ehrh., derived from flower bud filament, was established by Tewes et al. (1982).

Culture procedure for Brassica napus suspensions

The protoplast culture of *Brassica* was stepwise switched from 0.4M glucose to 0.1M sucrose. The established suspension line was transferred weekly on medium I and cultured in the dark at 25°C on a reciprocal shaker (50 strokes/min) in 100 ml Erlenmeyer flasks with 25 ml medium. The inoculum was 3 ml (1,5-2,0 mg dry weight/ml). For further development embryoids were cultivated for 2 passages on liquid medium IV and then transferred to medium IV, solidified with 8 g/1 agar for 3 to

4 passages. For shooting structures were placed onto solid modified K_3 medium, modified MS medium with various concentrations of NAA, BA and in some cases additional GA_3. Light-grown cultures were grown in a 12h light / 12h dark cycle (4000 lx, fluorescent lamps, "Warmton", Philips TL65W/32).

Culture procedure for Digitalis lanata suspensions

The maintenance suspension was cultivated in the dark in 500 ml flasks containing 100 ml medium I on a rotary shaker (125 rpm) at 25°C. The inoculum was 20 ml (2-3 mg dry weight/ml). The culture was transferred after 3 1/2 days. Light-grown cultures were cultivated in a 12h light/12h dark cycle (5000 lx, fluorescent tubes "Day light white" and "lumoflor", VEB Narva, Brand-Erbisdorf).

Suspension media

Modified MS media (Murashige and Skoog, 1962): with 30 g sucrose/l

I : 1,0 mg 2,4D/l; 0,02 mg Kinetin/l (Garve et al. 1980), Digitalis line was cultivated in MS with 680 mg KH_2PO_4/l)

IV : 0,1 mg NAA/l; 2,0 mg BA/l; 30 g /l sucrose

O : hormone free, half salt concentration of MS

Shooting media for Brassica

Modified K_3medium, macronutrients (Nagy and Maliga, 1976), micronutrients and vitamins (Gamborg et al.1986) and $FeSO_4$/Na_2-EDTA from Nagata and Takebe (1971), 0,1 mg IAA/l and 0,5 mg BA/l, 10g sucrose/l.

Modified MS media with 1 mg NAA/l, 2,0 mg BA/l and 20g sucrose, Modified MS media with 0,1 mg NAA/l, without BA or 0,5; 1,0 or 3,0 mg BA/l. Additional to these combinations variants with 0,1 and 1,0 mg GA_3/l, respectively, 30 g sucrose/l.

3. RESULTS AND DISCUSSION

Using cell suspensions of Brassica napus, it became evident that only aggregates derived from the stationary or late stationary phase of the growth cycle were a suitable inoculum for embryoid formation. To get organized structures it was important to culture the suspension, which was routinely transferred weekly, 20 to 40 days without any medium exchange. By reducing drastically the inoculum from 1,5 - 2,0 to 0,15 - 0,2 mg dry weight/ml in the differentiation passage, globular-, heart- and torpedo-shaped structures, anomalous structures and embryos with fused cotyledons were formed over about 9 to 16 days in culture (Fig. 1; 2 b, c, d).

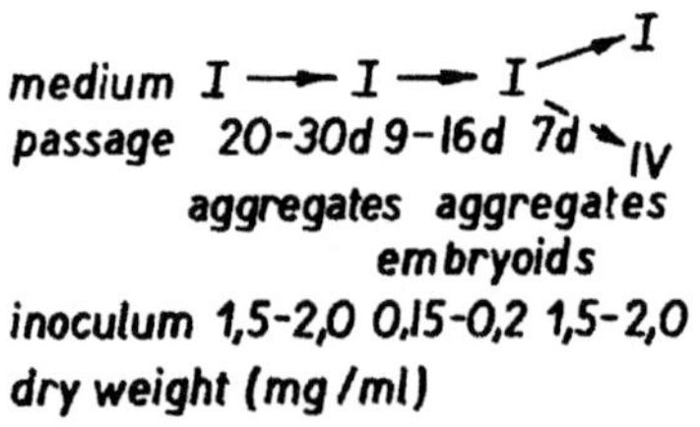

FIGURE 1. Time course of cultivation of Brassica napus suspension leading to the formation of embryoids.

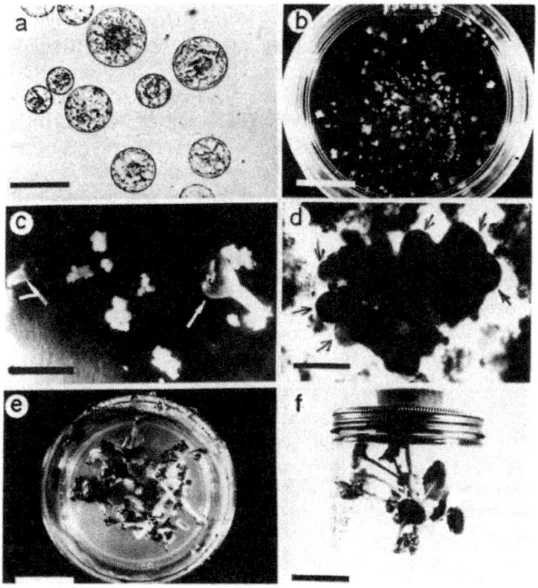

FIGURE 2. Plant regeneration from hypocotyl protoplast-derived embryogenic suspension of Brassica napus.

a) Freshly-isolated protoplasts. Bar: 50 µm.
b) Embryogenic suspension. Bar: 15 mm.
c) Embryos with fused cotyledons. Bar: 5 mm.
d) Globular structures. Bar: 500 µm.
e) Shoot morphogenesis. Bar: 20 mm.
f) Regenerated plant. Bar: 30 mm.

During further passages of 7 days such structures were formed continuously with a frequency up to 40% of inoculated aggregates, which have characteristic features of dividing meristematic cells: dense cytoplasm, small vacuoles and rich in starch. With all tested hormone concentrations and combinations shoot regeneration occurred from the compact green structures after 6 to 8 weeks (Fig. 2 e, f). The frequency of shoot regeneration was 20 - 36% (average 26%) on modified K_3 medium. On the modified MS medium with different NAA and BA concentrations shoot formation frequency was 2 - 40% (average 18%). The addition of GA_3 to regeneration medium had no detectable effect.

The importance of the preculture for indirect embryogenesis found in Brassica napus may be a general finding, since embryoid formation, greening and plant regeneration could be observed also with an embryogenic strain of Digitalis lanata (Tewes et al., 1982), applying a similar sequence of passages. Inocula, derived from several stages of the growth cycle of the suspension on medium I with and without sucrose were tested in the following passage on medium IV in the light (Fig. 3 a). Resting cells, grown only 2 days without sugar, were suitable as inoculum, whereas cells cultivated with sucrose needed 9 days or more to differentiate into green embryoids. Inocula taken from the growth phase showed no response in the following differentiation medium. Several inocula, grown one passage on medium IV, were used for the examination of the capability to differentiate in a second passage on medium IV in the light (Fig. 3 b).

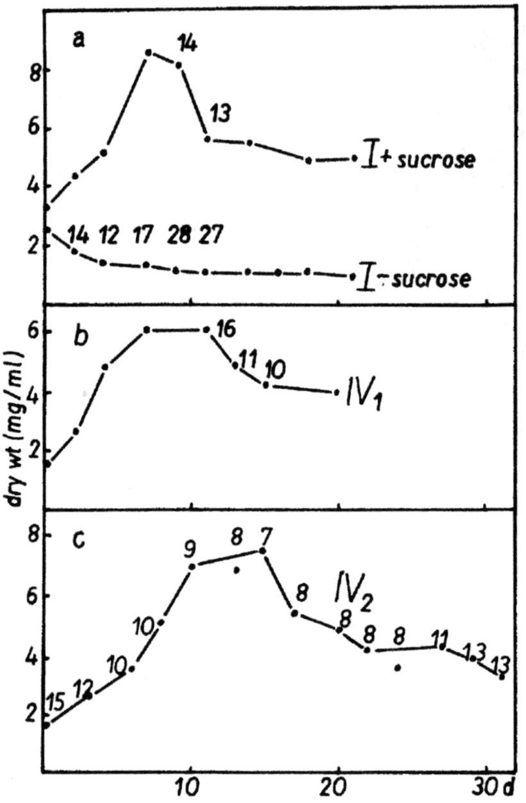

FIGURE 3. Growth cycles of Digitalis lanata suspensions. The numbers indicate the duration of days, necessary for the formation of green globular structures in the following passage. Inocula were used after time intervals of 2 and 3 days respectively.

a) Medium I with and without sugar
b) Medium IV 1. passage
c) Medium IV 2. passage

The experiment clearly showed, that only material from the late stationary phase was suitable. The later the stationary phase, the shorter was the time for the formation of green embryoids. Preculture for two passages on medium IV in the dark was more suitable (Fig. 3 c). All inocula transferred from the growth cycle of the second passage on medium IV developed to a homogeneous suspension of green globular structures. Also in this experiment it was observed that inocula, taken from the stationary phase, more quickly turned to green embryoids in the following passage, than the aggregates, derived from the growth phase, did. We finally developed the following standard procedure, which gave the best results in respect of fast formation of homogeneous green embryoids (Fig. 4).

	← dark			light →	
medium	I →	IV →	IV →	IV →	0 →
passage	9d	9d	17d	10d	30d
	aggregates		globular structures	green embryoids	green leaflets shoots
inoculum dry weight (mg/ml)	2-3	2-3	2-3	0,2-0,5	0,5

FIGURE 4. Time course of cultivation of Digitalis lanata suspension leading to the formation of embryoids.

After 3 passages on medium IV the green globular structures of about 2 - 3 mm diameter were transferred to a medium without hormones and half of the concentration of mineral salts of MS medium and given a weekly medium exchange to form green leaflets. With the same procedure leaflets were also formed in the dark.

In a batch culture system the cell division phase is followed by stationary phases. These phases are characterized by depletion of nutrients of the medium, by cessation of cell division and by changes in cell metabolism. During this growth phase a change in differentiation stages may occur and specialized cells, leading to regeneration and production of secondary products, may develop. From the above experiments with *Brassica* and *Digitalis* it was shown that the suspensions have to pass a stationary phase, an intermediate time of cessation of growth for switching to a new developmental course. This growth cessation was achieved 1. by using aggregates from the stationary phases and resting cells as inoculum and 2. by inoculating cell aggregates at a level near the critical cell density. Furthermore a certain degree of cell aggregation was necessary for these developmental processes. As pointed out by White (1967) and Lindsey and Yeoman (1985), the cells within multicellular structures have a high cell-to-cell contact and this microenvironment can provide chemical and physical gradients, essential for the induction of structural organization. Aggregation may be of importance also for creating nutrient stores for differentiating cells maintained in the stationary phase.

3. ACKNOWLEDGEMENTS

The skilful technical assistance of Barbara Widera and the careful reading of the manuscript by Dr. P. Lazzeri are gratefully appreciated.
Experiments with *Digitalis* were carried out at the Institut für Biochemie der Pflanzen der Akademie der Wissenschaft der DDR.

4. REFERENCES

1. Barsby TL, Yarrow SA, and Shepard JF: Plant Cell Reports 5, 101-103, 1986.
2. Chuong PV, Pauls KP, and Beversdorf WD: Plant Cell Reports 4, 4-6, 1985.
3. Chuong PV, Pauls KP, and Beversdorf WD: Plant Cell Reports 6, 67-69, 1987.
4. Gamborg OL, Miller FA, and Ojima K: Exp. Cell Res. 50, 151-158, 1968.
5. Garve R, Luckner M, Vogel E, Tewes A, and Nover L: Planta med. 40, 92-103, 1980.
6. Glimelius K: Physiol. Plant. 61, 38-44, 1984.
7. Li L and Kohlenbach HW: Plant Cell Reports 1, 209-211,1982
8. Lindsey K and Yeoman MM: In: Vasil IK (ed): Cell Culture and Cell Genetics of Plants, Vol. 2, pp. 61-102, Cell Growth, Nutrition, Cytodifferentiation and Cryopreserva-

tion, Harcourt Brace Jovanovich, Publishers, Academic Press, Inc. 1985.
9. Murashige T and Skoog F: Physiol. Plant. 15, 473-497, 1962.
10. Nagata T and Takebe I: Planta 99, 12-20, 1971.
11. Nagy JI and Maliga P: Z. Pflanzenphysiol. 78, 453-455, 1976.
12. Spangenberg G, Koop HU, Lichter R, and Schweiger HG: Physiol. Plant. 66, 1-8, 1986.
13. Tewes A, Wappler A, Peschke E-M, Garve R, and Nover L: Z. Pflanzenphysiol. 106, 311-324, 1982.
14. White PR: In: Plant Cell, Tissue and Organ Cultures, pp. 12-19, U.G.C. Centre of Advanced Study in Botany, Delhi, India, 1967.

PLANT REGENERATION FROM CUCUMBER PROTOPLASTS

C.M. Colijn-Hooymans, R. Bouwer and J.J.M. Dons
Institute for Horticultural Plant Breeding, P.O. Box 16,
6700 AA Wageningen, The Netherlands.

INTRODUCTION

Cucumber (Cucumis sativus) is a major glasshouse crop in the Netherlands. Although breeding has resulted in cultivars resistant to many diseases, the crop is still threatened by a number of diseases such as Cucumber Green Mottle Mosaic Virus and powdery mildew (Sphaerotheca fuliginea). Certain wild Cucumis species harbour dominant monogenic resistances against these diseases. Crossing of these wild species with cucumber, however, is inhibited by strong crossing barriers [1]. Transfer of the genes responsible for the resistances may be provided for by somatic hybridization. To realise this, methods for protoplast culture and plant regeneration must be developed.
In this paper we describe the effects of different plating procedures on protoplast division of Cucumis sativus. Furthermore, we report upon the differences between leaves and cotyledons as source for protoplast isolation and plant regeneration.

MATERIAL AND METHODS

Plant material

Seeds of Cucumis sativus cv. Hokus were sterilized for 30 min. in 2% (w/v) NaOCl and washed three times with sterile water. The seeds were germinated on MS medium [2] supplemented with 0.8% (w/v) agar and 0.2% (w/v) sucrose in a growth chamber at 26°C and 16 h of light (26.6 W/m^2). Ten days after sowing, the cotyledons were used for protoplast isolation. Shoot cuttings including the cotyledons were excised from seedlings eight days after germination and subcultured on fresh MS medium with 3% (w/v) sucrose for production of leaves. The first fully developed leaves (21 days after germination) were used for protoplast isolation.

Protoplast isolation

The lower epidermis of the leaves and cotyledons was damaged by brushing and after removal of the midrib 1 g of tissue was digested for 16 h in the dark at 25°C and 30 rpm in 10 ml enzyme solution. For the digestion of leaves 0.6% (w/v) Cellulase R-10 and 0.6% (w/v) Macerozyme R-10 were used. In the case of cotyledons 1.5% (w/v) Cellulase R-10 and 0.3% (w/v) Macerozyme R-10 were added. Enzymes were dissolved in a washing solution consisting of 0.1M glycine, 0.25M mannitol, CPW salts [3], 3mM 2-morpholino-ethanesulphonic acid (MES), pH 5.6. After incubation the protoplast suspension was filtered through nylon sieves (297 and 53 um) and centrifuged for 10 min. at 20 g. The protoplast pellet was washed once with the same medium without enzymes.

Protoplast culture

Cotyledon and leaf protoplasts were plated in different culture media (CMC and CML, Table 1). They were cultured in the dark at 25°C at a concentration of $1x10^5$ protoplasts per ml. Three different culture procedures were compared: agarose bead culture [4], agarose disc culture [5] and liquid culture. A concentration of 0.6% (w/v) SeaPlague agarose (FMC corporation, Rockland, USA) was used.

Media used in cucumber protoplast culture.

Compound in basal medium	CML leaf	CMC cot.	CM 1	CM 2	CM 3	CM 4	CM 5
Sucrose (g/l)	20	20	20	20	30	30	20
Mannitol (M)	0.25	0.25	0.2	0.15	--	--	--
NAA (uM)	25	5	5	--	--	--	--
IPA (uM)	15	15	5	4	--	--	--
2,4-D (uM)	--	--	--	4	4	--	--
BA (uM)	--	--	--	--	4	--	--
Kinetin (uM)	--	--	--	--	--	--	0.5
Agar pur. (% w/v)	--	--	--	--	0.55	0.6	0.65
Osmol (mOsmol/kg)	460	460	345	280	108	108	108

Basal medium consists of 1/2 MS macro elements, MS micro elements, MS vitamins, 250 mg/l edamine (Serva Heidelberg, Germany), pH 5.6.
Agar: Oxoid purified agar, Oxoid Ltd., Hampshire, England. For the agarose-bead or agarose-disc cultures Seaplaque agarose (0.6% w/v) was used.

Plant Regeneration

Yellow calli, which had developed after several transfers on CM-3 medium, were placed on medium without phytohormones (CM-4). Embryoids that developed on this medium were transferred to the same medium but with 2% (w/v) sucrose and 0.5 uM kinetin (CM-5), placed in the light (2.9 W/m^2).

RESULTS AND DISCUSSION

Comparison of plant material used for protoplast isolation

To be informed about the condition and the variation of cells from cotyledons and leaves, the ploidy level of the genome was determined by flow-cytometry. The frequency distribution of the nuclear DNA content of leaves from in vitro grown plants showed a slight tendency to shift towards the tetraploid level. In cotyledons, only about 20% of the nuclei established the 2C DNA content and even octaploid cells were present. Leaves from greenhouse grown plants, used as reference showed the typical histogram expected for diploid tissue. Therefore, in respect to ploidy level of the cells, differences exist between cotyledons and leaf tissue used for protoplast isolation. This heterogenity in ploidy level is important in somatic hybridization where cotyledon protoplasts are used as fusion partner, assuming that protoplast isolation and division is efficient for all cell types. Both from cotyledons of 10 days old and from leaves of 21 days a yield of $5\text{-}10x10^6$ protoplasts per gram of tissue was obtained.

Comparison of plating procedures on protoplast division

Differences were observed between growth of protoplasts when cultured at a density of $1x10^5$ pps/ml in liquid medium, in bead [4] or in disc culture [5]. In all the experiments performed, protoplasts only occassionally

divided in liquid medium (PE 2%), whereas in agarose bead and disc culture high plating efficiencies were obtained (33 to 42% for protoplasts of leaves and 39 to 42% of protoplasts of cotyledons). The first cell divisions in these cultures were observed within 5 days and after 10 days microcalli were already visible.
(Fig. 1).

Fig. 1. Development of microcalli in agarose droplets.

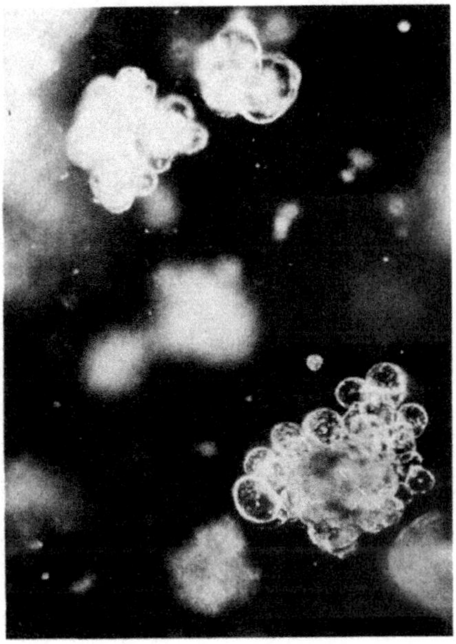

Fig. 2. Shoot regeneration.

Regeneration of plantlets

On CM-3 medium the calli looked white and soft. They grew rapidly but no embryogenesis was observed, not even after several transfers on CM-3 medium or after transfer of these calli to hormone free medium. These yellow calli were very compact and consisted of small meristematic cells. Transfer of these yellow calli to hormone free medium (CM-4), regularly resulted in the development of numerous embryoids. Approximately 5% of these embryoids developed to small shoots in the light (2.9 W/m^2) on CM-5 medium (Fig. 2). Embryo development was obtained with both cotyledon and leaf protoplasts. However, only in the case of leaf protoplasts morphologically normal plants were obtained. From the nuclear measurements it is obvious that the ploidy level has increased in cotyledons compared to leaves.

Although there is no evidence that this influenced the plant regeneration one should be aware of the fact that approximately 80% of the protoplasts isolated from cotyledons may be tetra or octaploid.
With the procedure established so far, shoots were obtained three months after protoplast isolation. Calli which had been maintained on CM-3 medium for one year, were still able to regenerate into plants.
The described protocol was also tested for other cultivars, viz. Salty, GY-3, and Borszczagowski. These cultivars reached a similar plating efficiency as Hokus both for cotyledon and leaf protoplasts. Induction of somatic embryos and shoot development were also observed, except for GY-3. In conclusion it can be stated that we developed a procedure for the regeneration of protoplasts from different Cucumis sativus cultivars.

REFERENCES

1. A.P.M. den Nijs and J.B.M. Custers. Introducing resistances into the cucumber by interspecific hybridization, in: D.M. Bates (Ed.). The Biology and Chemistry of the Cucurbitaceae, Cornell University Press, Itaca, New York, 1987, in press.
2. T. Murashige and F. Skoog. Physiol. Plant. 15 (1962) 473.
3. A.M. Frearson, J.B. Power and E.C. Cocking. Develop.Biol. 33 (1973) 130.
4. R.D. Shillito, J. Paszkowski and I. Potrykus. Plant Cell Reports 2 (1983) 244.
5. J.J.M. Dons and R. Bouwer. Improving the culture of cucumber protoplasts by using an agarose-disc procedure, in: Nuclear techniques and in vitro culture for plant improvement, International atomic energy agency, Vienna, 1985, p. 497.

CALLUS FORMATION FROM MESOPHYLL PROTOPLASTS OF FORAGE LEGUME ASTRAGALUS HUANGHEENSIS FU ET LIN

Y.C. Chien (Qian Ying-qian)1* and Zou Ji-tao2
1Bureau of Bio-Sciences & Biotechnology, Academia Sinica, Beijing, China
2Institue of Botany, Academia Sinica, Beijing, China.

INTRODUCTION

Astragalus huangheensis Fu et Lin having origin in areas of old course of the Huanghe River (Yellow River), China is a kind of perennial legume with values of forage grass, green manure crop and water and soil conservation. This herb has characteristics of different stress resistances, high yield and high protein contents, and is a good germplasm for promoting agriculture and animal husbandry in the areas of drought and windy-sandy. But several agronomic characters, such as long growth period, low maturity of seeds etc, should be improved. Somatic variation may occur by protoplast culture and clones with good traits may be selected. These have been certificated by Larkin & Scowcroft (1981). The aim of this experiment is seeking for obtaining new cultivars with good traits through protoplast culture.

MATERIALS AND METHODS

The first and second pinnate leaves of seedling were used for protoplast isolation. Leaves, the lower epidermis of which was stripped, were incubated in the enzymatic mixture at 28°C for 6 hours. The enzymatic mixture consisted of 0.8% Onozuka R-10, 0.3% Macerozyme R-10, 1,000 mg/l $CaCl_2.2H_2O$, 100 mg/l $NaH_2PO_4.H_2O$, 585 mg/l MES and 13% mannitol, pH 6.5. Protoplasts collected and washed were suspended in B_5 culture medium (Gamborg et al., 1968) supplemented with 2 mg/l glycine, 0.4 mg/l folic acid, 0.05 mg/l biotin, 0.05 mg/l vitamin A, 0.05 mg/l vitamin D, 100 mg/l L-glutamin, 100 mg/l L-asparatate, 125 mg/l D-xylose, 125 mg/l cellobiose, 100 mg/l casein hydrolysate, 100 ml/l coconut milk, 60 g/l mannitol, 30 g/l sucrose and with phytohormones in different kinds and concentrations, pH 5.7. Protoplast suspension in the density of 1 x 10 protoplasts/ml was cultured onto petri dish which is 0.8 cm in diameter.

RESULTS AND DISCUSSION

A. huangheensis protoplasts changed in shape obviously after 2 days in culture. First division took place in 4-5 days. Aggregation occurred frequently. Only first or second division could occur to the cells in aggregates. The more high density of protoplasts was used, the more aggregation occured. 0.8 ml liquid medium was added onto petri dish. In this condition, some areas with no medium were appeared. No aggregation occurred on edges of the medium, in which sustained division could be developed to small callus formation. It was suggested that maybe lower density in culture is available for protoplast of *A. hyhangheensis*. The suitable ratio of phytohormones in culture medium is 0.2 mg/l 2,4-D, 1 mg/l NAA and 0.5 mg/l zeatin. After one month in culture, the diameter of protoplast-derived calli could reach 1-2 mm

Fig.1. Protoplasts newly isolated.

Fig.2. 1st division of regenerated cell.

Fig.3. Small cluster consisted of few cells.

Fig.4. Callus derived from single protoplast.

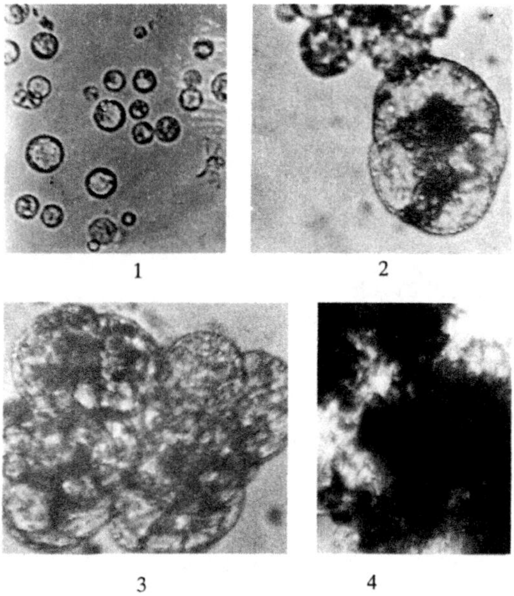

in size. They were transferred to Kao's cell culture medium (Kao et al., 1980) gradually. In this medium, some of these grew vigorously and some became browning and died eventually. Differentiation is in progress.

Plant regeneration of some forage legume, such as Medicago sativa (Xu et al., 1982) and Trifolium repens (Gresshoff, 1980) was obtained, but protoplast culture of legume plants still is a difficult problem to be solved. The conditions used in this experiment was not the best, these have to be improved further.

REFERENCE

1. O.L. Gamborg, R.A. Miller and K. Ojima, Expt. Cell Res. 50 (1968) 151-158.
2. P.M. Gresshoff, Bot. Gaz. 141 (1980) 157-164.
3. K.N. Kao and M.R. Michayluk, Z. Pflanzenphysiol. 96 (1980) 135-141.
4. P.J. Larkin and W.R. Scowcroft, Theor. Appl. Genet. 60 (1981) 197-214.
5. Z.H. Xu, M.R. Davey and E.C. Cocking, Z. Pflanzenphysiol., 107 (1982) 231-236.

PRELIMINARY STUDIES OF LUPINUS PROTOPLASTS

J.J. Rybczynski[1], J.A. Tarkowska[2], E. Podyma[1]
[1]The Botanical Garden of the Polish Academy of Sciences, Warsaw, Poland.
[2]Department of General Botany, Warsaw University, Poland.

INTRODUCTION

It has been described that successful interspecies sexual hybridization of Lupinus taxa is caused by numerous pre- and postzygotic barriers. However, in the case of numerous successful attempts of sexual hybridization the embryo collapsed because of disappearance and under-development of the endosperm (Jaranowski, 1962). The most important program of sexual hybridization includes L. albus and L. luteus. Using "bridge species" like L. hispanicus the above mentioned problems could be overcome, however, this method of hybridization requires a long-term crossing program.

Protoplast isolation, culture and fusion is an alternative system to obtain a new plant which would carry agronomically important characters from both somatic hybridization partners, however, there is no reliability of regenerant fertility.

MATERIAL AND METHODS

Our experiments have been carried out on various lupin species. Special attention was paid to L. albus cv. Hetman, L. luteus cv. Topaz, L. hispanicus and L. polyphyllus. Seeds of studied lupinus were scarified and then surface sterilized and imbibed for 2-4 h in sterile distilled water. Three explants of 1, 3, 5, 10 and 13 day old seedlings were the source of protoplasts with special attention paid to cotyledons. Cytological analysis of protein storage bodies were conducted for 1, 3 and 5 day old seedlings. Different culture conditions of seedlings on protoplast release were tested.

Fifteen plasmolytic solutions based on CPW and B_5 medium for plasmolysis and protoplasts release and various enzymes in numerous combinations were checked. Purification of already released protoplasts in CPW 13S + 0,1G was made.

RESULTS AND DISCUSSION

Considerable progress in plant tissue culture (Sator, 1985) of lupin recently reached in our laboratory (Rybczyński et al, 1986) gave the basis for protoplast studies. Because of almost only protein storage material of lupin cotyledon cells the studies of protoplasts required also initial investigation of protein degradation and protein body conversion and beginning of cell division of cotyledon parenchyma cells in the culture conditions.

During the first five days of seedling culture the storage proteins undergo continuous degradation. After three days the number of protein structures was reduced and moreover these structures could be present in cells of cotyledons. Different pathways of storage protein body degradation were observed during 5 days of culture, however, in the fifth day of culture only small numerous protein bodies in cells of one cotyledon were found.

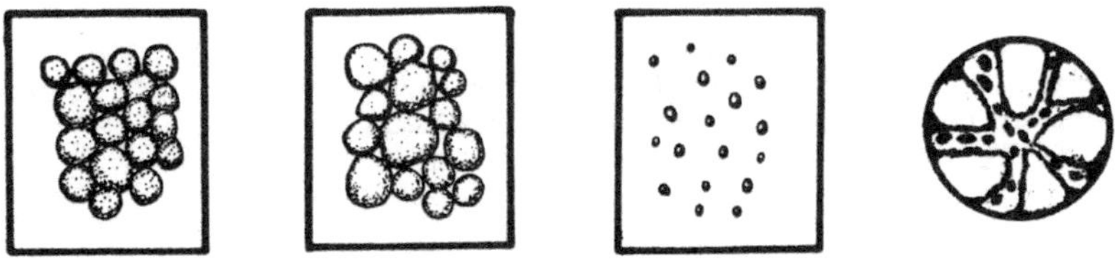

1 day 3 day 5 day protoplast

In cotyledons cultured on solidified medium cell division after the fifth day of culture was observed too. This physiological stage was the most effective for protoplast isolation. Among various seedling culture conditions for protoplast release the best turned out to be as follows: 3 days of culture in 16/8 h of day/night period at 22 ± 1 °C in culture chamber and the next 2 days of seedling culture in 16 °C in diffuse light. From among fifteen plasmolytic solutions the best was CPW salt solution supplemented with 13% sorbitol and 0,1% glucose and higher concentration of $CaCl_2$. Preplasmolysis required low temperature at 4 °C treatment during 2-4 h. Table 1 presents different enzyme mixtures.

TABLE 1. Enzyme combination used for Lupinus cotyledon protoplasts release (%)

Meicelase	95	0,25	0,25	0,25	025	0,25	0,25
Cellulase YC	0,5	0,25	0,25	0,25	0,25	0,25	0,25
Pectolyase T-23	0,5	0,25	0,25	0,25	0,25	0,25	0,25
Macerozyme	2,0	1,0	1,0	2,0	2,0	2,0	3,0
Cellulase R-10	2,0	1,0	2,0	1,0	2,0	3,0	2,0
Pectinase	1,0	0,5	0,5	0,5	0,5	0,5	0,5
Cellulase R-S				1,0			
Driselase				0,25			

The best enzyme mixture appeared to be the one including: Meicelase, Cellulase YC, Pectolyase Y-23 all in 0,25%. Macerozyme and Cellulase R-10 1% and Pectinase 0,5%. Next steps of protoplast isolation and culture are under investigation.

REFERENCES

1. Jaranowski H., 1962. Fertilization and embryo development in the genus Lupinus Tourn. Part II. Fertilization and embryo development following reciprocal species hybridization. Genetics Polonica, 3, 333-367.
2. Rybczyński J., Badzian T., Podyma E., 1986. Attempts of plant regeneration from various explants of some economically important Legumes. In: Abstracts 4th Symposium of the Socialist countries on Biotechnology, p. 27.
3. Sator Ch., 1985. Studies on shoot regeneration on lupines (Lupinus ssp.). Plant Cell Report, 4, 126-131.

We are grateful to J. Nowakowska MSc for excellent cytochemical analysis of Lupinus cotyledons.

REGENERATION AND DIFFERENTIATION OF PROTOPLASTS IN GRAIN LEGUMES

R.Gill, S.Eapen and P.S.Rao
Bio-Organic Division, Bhabha Atomic Research Centre
Bombay-400085, India

INTRODUCTION

Somatic hybridization following protoplast fusion is a reliable tool to combine genotypes which are difficult to cross sexually Since the first report of sustained division and regeneration of plants from protoplasts in Nicotiana tabacum, success has been achieved in many plant families. Unfortunately, unlike other plants, regeneration of protoplasts of chief food crops have proved difficult though sporadic reports are available (Hammatt et al. 1986). Regeneration of plants from protoplasts of a few forage legumes have been achieved, but grain legumes are relatively retractable to available culture stategies. We have investigated regeneration potential of different species of the genus Vigna in order to use them for our future somatic hybridization programme.

MATERIALS AND METHODS

Plant Material: Mature seeds of Vigna aconitifolia Jacq Marechal, V. mungo (L) Hepper, or hypocotyl-derived callus of V. sublobata (Phaseolus sublobata Roxb.) and V. unguiculata(L) Walp (Syn. V. sinensis(L) Saviex Hassk) served the source material. Seeds were aseptically raised on Murashige and Skoog's (1962) basal medium, and callus was raised on the same basal medium supplemented with 2,4-D (2mg/l) and coconut water (15%).

Isolation and Culture of Protoplasts: The hypocotyls or hypocotyl-derived calli were incubated overnight at 25±1°c in enzyme mixture comprising 2.5% Cellulase (Onozuka R-10) and 0.5% each of macerozyme,(Onozuka R-10) Hemicellulase (Sigma Chemical Co.) and Driselase (Kyowa Hakko Koggo) in 0.5M sorbitol at pH 5.8. Protoplasts were recovered by repeated centrifugation followed with purification by floatation over sucrose solution (25%) and cultured in MS medium supplemented with BA, NAA and 2,4-D each at 1.0 mg/l level. Cultures were incubated at 25±1 °c under continuous light. After a month, protoclones were cultured on top of agar medium of the same composition but with reduced level of sucrose (7%). For organogenesis the colonies were transferred to basal medium randomly supplemented with 2,4-D, IAA, BA or Zeatin.

Bacterial Strains and Co-cultivation: Cultures were co-cultivated with Agrobacterium tumefaciens wild type strain B6S3 and 'shooter' mutant pG V2215 and pGV2298 for 2 days for transfomation studies.

RESULTS AND DISCUSSION

Protoplasts either isolated from hypocotyl pieces (*V. mungo* and and *V. aconitifolia*) or hypocotyl-derived callus (*V. sublobata* and *V. unguiculata*) divided within 3-4 days and developed into colonies. In mothbean and cowpea, division frequency of 60-70% could be achieved whereas in *V. mungo* and *V. sublobata* it was restricted to 10-20%. Even regenerated calli started turning brown despite repeated subculture and addition of charcoal. Protoclones of *V. aconitifolia* showed distinct dark green patches which differentiated rapidly to shoot buds after transfer to basal medium. These shoot buds developed roots on transfer to basal medium supplemented with NAA (0.5 mg/l). Regenerated plants could be transplanted to field.

However,protoclones of cowpea did not differentiate shoot buds and plants despite using several auxin/cytokinin combinations. Cultures exhibited root formation predominantly. The protoclones which were co-cultivated with 'shooter' strains also develop roots. These strains have been reported to produce shoots in protoclones of *Physalis minima* and *Nicotiana paniculata* where otherwise shoot regeneration was never achieved by conventional regeneration methods (Steffen et al. 1986).

Presently, attempts are being made to introduce selectable kanamycin and hygromycin marker genes in order to use transformed lines for somatic hybridization programme.

REFERENCES

Hammatt N, Ghose TK and Davey MR. (1986) In: Cell Culture and Somatic Cell Genetics of Plants, Vasil IK (ed).Academic Press Inc. pp 67-95.

Murashige T and Skoog F (1962) Physiol. Plant.15: 473-497.

Steffen A, Eriksson T and Schieder O. (1986) Theor. Appl. Genet. 72: 135-140.

PLANT REGENERATION FROM STRAWBERRY(FRAGARIA ANANASSA) MESOPHYLL PROTOPLASTS

MARIE NYMAN AND ANITA WALLIN
Institute of Physiological botany of the university of Uppsala
Box 540, S-75121 Uppsala, Sweden

INTRODUCTION

In vegetatively propagated species with strong heterozygoty and high ploidy levels, plant breeding through somatic cell genetics is of special interest. In strawberry breeding, where plant breeders wish to introduce specific traits such as earliness, winterhardiness and disease resistance into already established cultivars these techniques might be of great value. A prerequisite for somatic hybridization and direct gene transfer is an efficient procedure for isolation of protoplasts and plant regeneration.
In this paper we present a method for protoplast isolation and plant regeneration in strawberry, a species which we intend to use in a coming fusion project.

MATERIAL AND METHODS

Plant material

The leaf material used for protoplast isolation was taken from shoots of two in vitro cultured strawberry clones, Sengana and Canoga. The shoots were grown on shoot-inducing medium, a modified MS-medium(1) with 4 % glucose, 1.5 mg/l BA and 0.75 mg/l IBA. Young shoots were transferred to hormone-free media with different sugar concentrations(Fig.1) 2-3 weeks before protoplast isolation.

Protoplast isolation and purification

Leaves were cut into pieces, preplasmolyzed in TVL(2) and exposed to an enzyme solution containing 1.0 % cellulysin and 0.05 % pectolyase. Incubation took place overnight(15-18 h) in darkness at 25°C. The released protoplasts were purified by repeated centrifugation in a sucrose gradient.

Protoplast culture and plant regeneration

Freshly isolated protoplasts were cultured at a density of 50 000 protoplasts/ml in a modified 8p-medium(3) containing 0.4 M glucose, 1.0 mg/l 2,4-D and 0.5 mg/l BA. After three days in culture, the auxin concentration and protoplast density was lowered by diluting the culture medium 3:1 with the same medium as above but lacking 2,4-D. When cell colonies had formed after 3-4 weeks, the liquid culture was poured onto semi-solid medium, a modified 8p-medium containing 1 % sucrose, 0.2 M mannitol, 1.0 mg/l BA 0.1 mg/l NAA and solified with 0.2 % agarose. As the cell colonies reached a size of about 2-3 mm in diameter, those micro-calli were transferred to plant regeneration medium, a MS-medium with 2 % sucrose, 5 mg/l BA, 0.2 mg/l NAA and 0.6 % agarose.

RESULTS AND DISCUSSION

Culturing strawberry shoots on medium containing BA resulted in rapid shoot production. This has previously been reported by Boxus(4). Shoots grown on shoot-inducing medium had short petioles, more or less folded leaves and no root growth. Under these culture conditions, the plant material only sporadically yielded protoplasts after enzyme treatment. When the shoots were transferred to hormone-free medium they developed into plants with unfolded leaves, extended growth of the petioles and root growth, but still the protoplast yield was very low. However , by reducing the sugar concentration in the culture medium, the protoplast yield was greatly increased. In Sengana, no difference in protoplast yield was noticed when sucrose and glucose were compared, while results with Canoga indicated a slight advantage of using glucose(Fig.1). To obtain leaves suitable for protoplast preparation, young shoots growing on shoot-inducing medium were transferred to fresh hormone-free medium with low sugar concentration 2-3 weeks before isolation. A reproducible high yield of protoplasts was obtained only when pectolyase was included in the enzyme solution. In many species, plant regeneration is the limiting step in protoplast techniques. This seems not to be the case with strawberries, which in our experiments have shown a great tendency to regenerate plants from callus. After four weeks on regeneration medium, the first shoots were visible on the surface of the calli. These shoots were then easily rooted by simply transferring them to hormone-free medium. Preliminary experiments indicate that shoot formation occurs only on medium containing high levels of BA(5-10 mg/l).

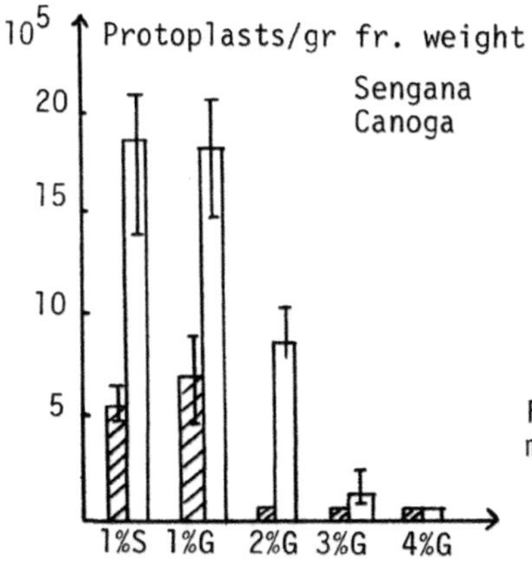

Fig.1 Protoplast yield from leaves cultured on medium with different sugar concentrations. G=glucose, S=sucrose

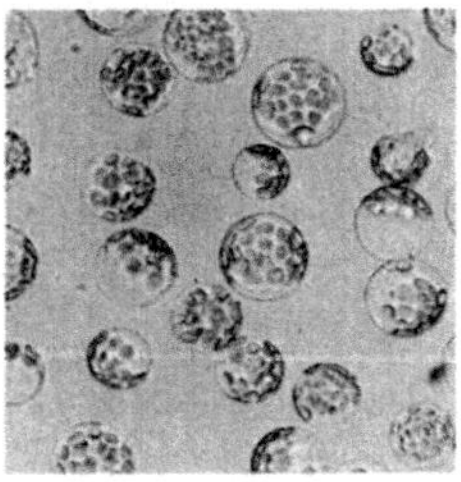

Fig.2 Freshly isolated mesophyll protoplasts

Fig.3 Protoplast derived strawberry plant

References

1. Murashige, T. and Skoog,F.(1962) Physiol Plant. 15:473-497

2. Carlberg,I. et.al.(1983) Plant. Sci. Lett. 35:225-230

3. Glimelius,K. et.al.(1986) Plant Science 45:133-141

4. Boxus,P.(1974) J. Hort. Sci. 49:209-210

ISOLATION AND CULTURE OF APPLE MESOPHYLL PROTOPLASTS

A. Wallin

Dept. of Physiological Botany, University of Uppsala, Box 540
S-751 21 Uppsala, Sweden

1. INTRODUCTION

Most commercial apple cultivars are diploid or triploid. Efficient apple breeding programs are obstructed by the long generation time and by the strong heterozygosity, which decreases the chances of preserving desired characters during sexual crossing. Somatic cell genetics may be an alternative method for introducing genetic variation and selecting new plants. Somaclonal variation might offer the opportunity of making "onestep" improvements in established cultivars and somatic hybridization could be used to combine interesting genomes by fusion. In order to realize these possibilities, reliable methods for regenerating plants from protoplasts have to be developed. The Rosaceae family is known to be recalcitrant to protoplast handling. However, shoots have been regenerated from protoplasts of wild pear (1) and sweet cherry (2) and calli from apple protoplasts derived from callus or suspension cultures (3,4).

2. PROCEDURE

2.1 Plant material

Leaves from shootcultures of an apple cultivar with a dominant gene for columnar habit (A1583(Cox x Court Pendu Plat) x Wijcik McIntosh, East Malling) were used. The shoots were cultured for about 2 weeks at 50 uE $m^{-2}s^{-1}$ in LS medium containing 1 mg/l BA and gelled with 0.7 % agar. The release of protoplasts increased significantly if the shoots were placed at low light intensity (1 uE $m^{-2}s^{-1}$) 6 days before enzyme treatment. Shoots with welldeveloped leaves and few axillary shoots were preferable to bushy shoots.

2.2 Protoplast isolation and purification

A yield of $2 \cdot 10^6$ protoplasts/g fresh weight was obtained if leaves from shoots grown at low light intensity were preplasmolyzed in W5(5) and digested for 18 h in 1 % cellulysin, 0.1 % macerase and 0.1 % pectolyase in K3 medium (5) containing 0.4M sucrose. Survival after 1 day in medium A (Table 1) was 80 - 85 %. The addition of pectolyase not only increased the number of protoplasts but also raised the proportion of protoplasts containing chloroplasts. Protoplasts were purified according to Carlberg et al. (1983).

2.3 Culture of apple protoplasts

Protoplasts were cultured at 27°C at 1 uE $m^{-2}s^{-1}$ in a modified protoplast plating medium PPP (7) containing 2 mg/l 2,4D and 0.5 mg/l BA and gelled with agarose (Sea Plaque, 1%). Sorbitol was excluded, the concentrations of mannitol, sucrose and glucose were modified as indicated in Table 1 and xylitol was replaced by xylose. No or very few divisions were obtained in liquid medium in a thin layer or fixed to a net (7), in bead cultures or in culture media such as 8p or K3 which are richer in constituents than PPP. The highest frequency of cell divisions, measured after 8 days of culture, was observed in modified PPP medium gelled with 1% Sea Plaque agarose containing 0.5M sucrose and 0.1M glucose (medium A). In media B - E (Table 1), the protoplasts were more enlarged and budding was more frequent than in medium A or F with sucrose. Initial divisions, which started after 5-7 days, were stimulated by a high protoplast density ($1 - 2 \cdot 10^6$ protoplasts/ml). After 3 weeks, small aggregates were formed in medium A (Fig 1). So far no plants have been regenerated.

3. DISCUSSION

Due to strong phenol formation, a different cell wall structure and a low nutrient requirement, the basic methods developed for isolation and culture of protoplasts from herbaceous plants cannot be applied directly to apple. Apple protoplasts go through the same stages of development as fast growing herbaceous protoplasts but the development is slower. The prolonged time could result in genetic changes. Therefore, it is important to find optimal culture conditions.

Table 1

Division frequency of apple mesophyll protoplasts cultured for 8 days in a modified PPP medium with different carbohydrates

		Division frequency %
A.	0.5 M sucrose 0.1 M glucose	11
B.	0.5 M glucose	3
C.	0.6 M glucose	3
D.	0.4 M mannitol 0.1 M glucose	3
E.	0.5 M mannitol 0.1 M glucose	4
F.	0.5 M sucrose	5

Fig. 1 Cell aggregates of apple protoplasts formed after 3 weeks in medium A

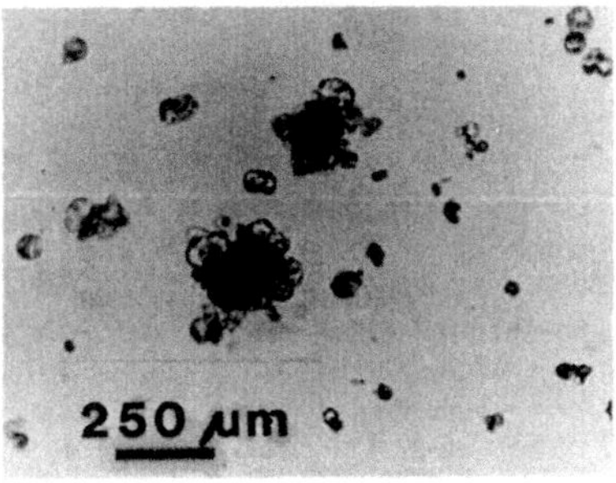

4. REFERENCES

1. S.J. Ochatt and O.H. Caso, Plant Physiol. 122 (1986) 243-249
2. S.J. Ochatt, E.C. Cocking and J.B. Power, Plant Science 50 (1987)139-143
3. M. Koudier, R. Hauptmann, J.M. Widholm, R.M. Skirvin and S.S. Korban, Plant Cell Rep. 3 (1984) 142-145
4. D.J. James, A.J. Passey, K.A.D. Mackenzie and E.C. Mehinick In: Withers, L.A., Alderson, P.G. (eds) Plant tissue culture and its agricultural applications. Butterworths London (1986) pp 349-357
5. L.Menczel, F. Nagy, Z. Kiss and P. Maliga, Theor. Appl. Genet. 59 (1981) 191-195
6. I.Carlberg, K.Glimelius and T.Eriksson, Plant Cell Rep.2 (1983) 223-225
7. J.A. Russel and B.H. McCown, Plant Science 46 (1986) 133-142

RHIZOGENESIS AS AN INTERMEDIATE STEP TOWARDS PLANT REGENERATION OF PROTOPLASTS OF SOUR CHERRY (PRUNUS CERASUS L.), ROSACEAE

S.J. OCHATT AND J.B. POWER
Plant Genetic Manipulation Group, University of Nottingham, Nottingham NG7 2RD, UK.

1. INTRODUCTION

The list of woody species, where sustained protoplast growth has been reported, has lengthened but several problems still need to be resolved before genetic manipulation techniques, using protoplasts, can be routinely applied to this economically important group of plants. The reproducible regeneration of plants, from isolated protoplasts is a prerequisite and, in this respect, woody species are considered by many to be recalcitrant (1) with only two examples of successful protoplast regeneration for members of the Rosaceae (2,3). We report here on firstly, a differential cultural requirement for isolated mesophyll protoplasts of three clones of sour cherry and secondly, on the pattern of plant regeneration, in one case, via rhizogenesis as an intermediate step.

2. MATERIALS AND METHODS

Mesophyll protoplasts were isolated from the leaves of sour cherry (*Prunus cerasus* L.) clones CAB4D, CAB5H and CAB11E as described before (4), and were suspended at a range of densities ($0.01\text{-}5.0 \times 10^5$/ml) in MS salts medium (5) supplemented with combinations of NAA, BAP and/or Z (at a range of 0.0 to 2.0 mg/l) plus 9% mannitol (pH 5.8). Protoplast suspensions were transferred to Petri dishes in liquid media, media solidified with 0.625% agarose as semisolid layers or as agarose drops submerged in a liquid mannitol-free culture medium of the same composition. All cultures were kept at 25°C and in the dark. The mannitol concentration of the original medium (e.g. as liquid or semisolid layers) was reduced every 10 days from the 20th day to the 70th day of culture by dilution with a protoplast-mannitol-free medium (2:1 ratio). Plating efficiency was calculated at day 15 (up to first division: IPE) and at day 75 (up to microcallus stage: FPE).

Microcalli (1.0 mm diam.) were transferred for callus proliferation to a range of agar-solidified (0.8%) media based on MS medium with 2.0 mg/l NAA plus a range (0.25 to 1.0 mg/l) of BAP. For plant regeneration assessments callus pieces were transferred to MS media with a range of NAA (0.0 to 0.1 mg/l), BAP (1.0 to 5.0 mg/l) and/or Z (0.0 to 1.0 mg/l) concentrations. For indirect plant regeneration, via rhizogenesis, protoplast-derived roots (ca. 1.5 cm long) or their decapitated basal sections (ca. 1.0 cm long) were detached from the appropriate callus and transferred to an agar-solidified (0.8%) MS medium with 0.01 mg/l NAA and 2.0 mg/l BAP. All cultures were kept at 25°C with a continuous illumination of 1000 lux (fluorescent, cool white).

3. RESULTS AND DISCUSSION

Comparable yields (2.3-2.9 x 10^6 protoplasts/g f.wt.) were obtained for all three sour cherry clones with viabilities of 68%, 69% and 52% respectively for clones CAB4D, CAB11E and CAB5H. Protoplasts, of all clones, required an initial plating density of 1 x 10^5/ml for maximum survival, but their requirements differed thereafter.

For clone CAB4D, the best IPE (22%) and where most colonies comprised of 6 or more cells, was obtained when protoplasts were cultured, as agarose beads, in a medium with Z (1.0 mg/l) as the sole growth regulator. Growth was usually arrested at the 10-cell stage, but by replacing (after 20 days) the initial liquid medium with a mannitol-free MS medium with 3.0 mg/l IAA and 1.0 mg/l Z this problem was resolved. After 75 days, colonies (1.0mm diam.) were recovered (FPE = 1.73%) and were transferred to MS medium with 2.0 mg/l NAA and 0.25 mg/l BAP, and taken to the callus stage. When maintained on MS medium with 0.025 mg/l NAA, 1.0 BAP and 0.1 mg/l Z, 70% of these protoplast-derived calli developed one or more thick roots. Shoot buds were regenerated from 40% of these roots, when their tips were removed and the remainder of the root placed on a medium with 0.01 mg/l NAA and 2.0 mg/l BAP. Shoots were detached and transferred, for multiplication and subsequent rooting, to the same cultural conditions/media as used for the shoot cultures that provided the protoplasts.

For clone CAB5H, MS medium with 1.0 mg/l NAA, 0.5 mg/l BAP, 1.0 m//l Z and 9% mannitol provided the best results (IPE = 13.8%; FPE = 1.04%) either as liquid, semisolid layers or agarose drops. Protoplast-derived colonies were converted to callus on a medium with 2.0 mg/l NAA and 1.0 mg/l BAP, and handled thereafter as described above. Rhizogenesis was an intermediate step, for plant regeneration. In addition to this approach, shoots could be recovered, directly from the protoplast-derived callus, on MS medium with 0.05 mg/l NAA, 5.0 mg/l BAP and 0.5 mg/l Z. The number of shoots could be doubled every 4 weeks. Shoots were rooted and transferred to the greenhouse.

For clone CAB11E, the best medium was with 0.05 mg/l NAA and 0.5 mg/l Z, using semisolid layers (IPE = 25.6%; FPE = 1.45%). This method of culture was more efficient than the use of agarose drops (IPE = 17.5%; FPE = 0.87%). Colonies were taken to the callus stage on MS medium with 2.0 mg/l NAA and 0.5 mg/l BAP but, in contrast to the other clones, clone CAB11E did not exhibit rhizogenesis or caulogenesis.

The regeneration of shoots and whole plants from the protoplast-derived roots of sour cherry has shown that, for this species, root regeneration from protoplast-callus is not inconsistent with plant regeneration. Woody plants should therefore be no longer regarded as inherently recalcitrant when it comes to plant regeneration from their protoplasts (1). It is likely that this approach to regeneration may be applicable to other woody or non-woody protoplast systems.

REFERENCES

1. Bajaj YPS: in: Biotechnology in Agriculture and Forestry 1: Trees I (ed) YPS Bajaj, Springer-Verlag, New York: 1-24, 1986.
2. Ochatt SJ, Caso OH: J. Plant Physiol. 122: 243-249, 1986.
3. Ochatt SJ, Cocking EC, Power JB: Plant Sci. 50: 139-143, 1987.
4. Revilla MA, Ochatt SJ, Doughty S, Power JB: Plant Sci. 50: 133-137, 1987.
5. Murashige T, Skoog F: Physiol. Plant. 15: 473-497, 1962.

ISOLATION AND CULTURE OF PROTOPLASTS FROM TOTIPOTENT CELL CULTURES OF POPULUS CILIATA

G.S. CHEEMA

TATA ENERGY RESEARCH INSTITUTE
90 Jor Bagh
New Delhi - 110 003 INDIA

INTRODUCTION

Attempts at isolation and culture of mesophyll protoplasts of tree species has revealed problems of low yields (1,2) coupled with poor digestibility which has necessitated lengthy exposure times to enzymes. There have also been problems of leaf leachates (3). Only recently, a case of successful mesophyll protoplast culture has been reported (4). Some researchers have overcome these problems by isolating protoplasts from callus cultures derived from mature trees (5,6) and have reported the complete protoplast systems utilizing such tissues for Santalum and Ulmus respectively. It has now been possible to induce embryogenic cell suspension of Populus ciliata. The somatic embryos produced by this cell suspension developed into complete plants which have been transferred to field conditions (7,8). This study was undertaken to determine the suitability of totipotent cell suspensions for isolation and culture of protoplasts.

MATERIALS AND METHODS

Development of embryogenic cell suspension cultures

The cell suspension culture used for isolation and culture of protoplasts were derived from embryogenic callus obtained from long-term 'leaf' cultures of mature trees of Himalayan Poplar (P. ciliata wall ex.Royle) [For details see Refs. 7,8). The callus tissue gave rise to finely dispersed and highly friable suspension cultures, which almost consisted entirely of unorganised small groups of cells. The cells were small, richly cytoplasmic and contained many starch grains. The suspensions were subcultured at 1 wk intervals with a dilution of 1:2 or 1:3 with fresh medium. The cell suspensions were maintained on gyrotory shaker at 120 rpm at 25±2°C in diffuse light. Seven to 14 days old cell suspensions have been used for isolation and culture of protoplasts.

Protoplast isolation and culture

The requisite cell suspensions were centrifuged at 500 rpm for 5-7 min. The pellet of cells was plasmolysed in CPW salts with 13%(w/v) mannitol(CPW 13M). The plasmolysed cells were digested in CPW 13M or 0.35M mannitol solution containing enzymes-1%(w/v) Cellulase R-10, 0.5%(w/v) Macerozyme R-10 and 0.5%(w/v) Drieslase, pH 5.8. The cells were incubated overnight in the enzyme mixture at 25°C on a gyrotory shaker set at 50 rpm. After the completion of digestion the mixture was filtered through 50µ stainless steel mesh, followed by two washes in CPW 13M solution. The pellet was resuspended in CPW 13M solution and layered on a CPW salt solution containing 18% (w/v) sucrose(CPW 18S)and centrifuged at 500 rpm for 10 min. The band of protoplasts, floating at the top, was removed and resuspended in fresh CPW 18S solution for further purification and concentration. This procedure resulted in debris free protoplasts. Protoplasts were counted and diluted to approximately 10^6 protoplasts per ml. Protoplasts were cultured in 5ml of liquid medium at densities of 1×10^4 protoplasts/ml; 1×10^5 protoplasts/ml and 8×10^5 protoplasts/ml, in each sterile dish. Murashige and Skoog (1962) (MS)medium with the following hormone concentration was used for protoplast culture. MS+2,4-D (0.2, 0.25 or 0.5mg/l)+0.4M sucrose, MS+2,4-D (0.2 or 0.5mg/l)+6% (w/v)mannitol,MS+NAA (1.5mg/l)+BAP(0.5mg/l)+2,4-D(0.5mg/l)+6% or 9%(w/v) mannitol. Protoplasts were incubated in dark at 25±2°C. Cultures were observed regularly for development. Fresh culture medium was added to each dish at 1 wk intervals. Cultures were brought to diffuse light conditions after 10 days.

RESULTS & DISCUSSION

Cell suspensions grown for 7 days after subculture yielded good, stable protoplasts capable of division. Older cell suspensions(10-15 days after subculture) generally yielded highly vacuolated protoplasts which though remained viable for considerably long time, did not divide. The concentration of Drieslase affected the health and time taken for the release of protoplasts. Incubation for more than 3-4 hr at 25°C in enzyme mixture resulted in over digestion of cells resulting in total loss of protoplasts. Pretreatment of cells in MS+NAA(1.5 mg/l)+BAP(0.5mg/l)+2,4-D (0.5mg/l)+6% mannitol medium for 12 hr prior to protoplast isolation had a positive effect on stability and viability of protoplasts as manifested in early division. Preliminary experiments revealed several factors having important, inhibitory or beneficial effects. The accumulation of exudates in the culture medium is aggravated by using aged cultures, high light intensity and agar base of the culture medium. Protoplasts cultured in liquid medium in dark exhibited cell wall formation and division within 3-4 days culture. Fresh culture medium had to be added at weekly intervals to produce multicellular colonies. However, for yet unknown reasons, cell growth became stagnant and signalled senescence after 6-8 wks. The problem of browning and division arrests could be sorted out to some extent by embedding the protoplast in 0.7% agarose.

This study showed that protoplasts isolated from rapidly dividing cell suspension usually divided within 3-4 days of isolation. The plating efficiency calculated for dividing protoplast was 13-14%. This contrasts with very low (1-2%) plating efficiency and slow growth of mesophyll protoplasts reported for poplar (4). The important parameters that need to be taken care for successful protoplast system are (a) the nature and conditions of suspension used, (b) the duration of enzyme treatment and (c) culture conditions. As callus induced from P. ciliata can regenerate at high frequencies on MS+BAP (2mg/l) medium (9), it should be possible to regenerate plants from the protoplast-derived colonies.

REFERENCES

1.Ahuja MR: Silvae Genetica, 32: 225-227, 1983.
2.Smith MAL and BH McCown: Plant Sci.Lett., 28:149-156, 1982/83.
3.Butt AD: Plant Sci., 42: 55-59, 1985.
4.Russell JA and BH McCown: Plant Sci., 46: 133-142, 1986.
5.Rao PS and P Ozias-Akins: Protoplasma, 124: 80-86, 1985.
6.Sticklen MB, RD Lineberger and SC Domir: Hort. Sci., 20: 571, 1985.
7.Cheema GS: In: IUFRO International Workshop on Agroforestry for Rural Needs. Vigyan Bhawan, New Delhi, India, p.39-40, Feb. 22-26, 1987.
8.Cheema GS: In: Symposium "Genetic Manipulation of Woody Plants" Kellogg Centre, Michigan State Univ., East Lansing, Michigan. Abstract # 9. June 21-25, 1987
9.Mehra PN and GS Cheema: Phytomorphology, 30: 336-343, 1980.

A PROTOPLAST SYSTEM FOR POPULUS WHICH WILL SUPPORT GENETIC ENGINEERING RESEARCH OF TREE SPECIES

Julie A. Russell and Brent H. McCown
Dept. of Horticulture, University of Wisconsin, Madison, WI 53706 USA

The use of genetic engineering is an appropriate technology for the improvement of tree crops because breeding by classical methods has been difficult. Of the 2 main avenues for plant transformation, the *Agrobacterium* vector system has been successfully used to insert an economically important gene into a commercial forest tree (*Populus*) [1]. The protoplast-based transformation systems (eg. fusion, microinjection, direct gene transfer), however, have not been tested in woody plants because of the lack of suitable methods to recover whole plants from protoplasts of most tree genotypes.

Our laboratory has initiated a project to study the use of direct gene transfer methods for the genetic improvement of *Populus*. As a prerequisite for such research, we launched a program to develop a protoplast system with the following characteristics:

1) One should be able to use superior genotypes selected in the mature phases of growth as the starting material for protoplast isolation. However, this presents a major problem because tissues from adult trees are not readily manipulated *in vitro*.
2) One should be able to isolate high yields of highly viable protoplasts.
3) The protoplasts must progress through the stages of cell wall formation and cell division. The incidence of cell division should be high.
4) The division of the protoplast-derived cells must continue such that multicelled colonies are formed.
5) Whole plants must be recovered from a high percentage of the protoplast-derived calli.

A protoplast system which meets the five desired characteristics was developed as follows:

*Characteristic 1: Shoot cultures are used as the protoplast source because they establish tissue from trees selected in the mature phases of growth in a form that is responsive to *in vitro* culture and is available year-around. The genotypes used for this research were field-grown clonal cultivars of the poplar hybrid NC-5339 (*P. alba* x *P. grandidentata* 'Crandon'), the poplar hybrid NC-5331 (*P. nigra* 'Betulifolia' x *P. trichocarpa*) and the Upright European Aspen (*P. tremula* 'Erecta').

*Characteristic 2: High yields (3 to 11 million protoplasts per g fresh weight) of viable protoplasts can be isolated when the following techniques are used:

a) The leaf-grinding technique [2] which involves processing the leaves in an Omni-mixer for 3 to 5 s prior to the digestion period. This procedure serves to strip away the epidermis and provides uniform exposure of the cells to the cell wall degrading enzymes.

b) The leaf-debris washing technique [2] which involves washing the digested leaf tissue with medium using a hypodermic syringe and needle. This procedure aids in releasing protoplasts that are trapped in the leaf-debris.
c) Collecting and washing the protoplasts by sucrose flotation. A total of 3 sucrose flotations are used [4].

*Characteristic 3: The protoplasts develop in a manner that is typical of that reported for other species. Development of the protoplasts is aided by culturing them in a monolayer together with a polyester screen disc floated at the surface of a liquid medium (the floating disc culture system) [3]. The incidence of protoplast division is typically 10 to 30% depending on the plating density and whether or not the growth medium is supplemented with coconut water and casein hydrolysate [4].

*Characteristic 4: Continued division of the protoplast-derived cells is stimulated by replenishing the growth medium on a weekly basis.

*Characteristic 5: Shoots can be regenerated from the protoplast-derived calli by transferring the screen discs with the attached calli to a semi-solid regeneration medium (genotype dependent). The resultant protoclones can be maintained as shoot cultures and can be transferred to the greenhouse and field using micropropagation methods.

The protoplast system described above has provided the framework for research on the direct gene transfer of *Populus* in our lab. Future research will include:

1) Increasing the protoplast plating efficiency (>50% desired) in order to improve our chances of recovering transformants. Our approach will be to select subpopulations of competent protoplasts before transformation is performed.
2) Definition of parameters for electroporation.
3) Increasing the level of gene insertion.

REFERENCES

1. Fillatti, J.J., J. Sellmer, B. McCown, B. Haissig, and L. Comai. 1987. *Agrobacterium* mediated transformation and regeneration of *Populus*. Mol. Gen. Genet. 206:192-199.
2. Russell, J.A. and B.H. McCown. 1986. Techniques for enhanced release of leaf protoplasts in *Populus*. Plant Cell Rep. 5:284-287.
3. Russell, J.A. and B.H. McCown. 1986. Culture and regeneration of *Populus* leaf protoplasts isolated from non-seedling tissue. Plant Sci. 46:133-142.
4. Russell, J.A. and B.H. McCown. Recovery of plants from leaf protoplasts of hybrid-poplar and aspen clones. Plant Cell Rep. (In Press).

FACTORS INFLUENCING DIGITALIS OBSCURA MESOPHYLL PROTOPLAST DEVELOPMENT

R. JOVER, M.C. BRISA and J. SEGURA

Department of Plant Physiology,Faculty of Pharmacy,University of Valencia, Avda. Blasco Ibañez, 13. 46010 Valencia (Spain)

INTRODUCTION

In a previous work (1), we found that a combination of liquid-over-agar-solidified MS (2) modified medium was superior to either all liquid or all solid medium for culture of Digitalis obscura L. mesophyll protoplasts. In order to achieve optimum culture conditions we now report on the influence of a different gelling agent (agarose) as well as the effect of several nutrient media on D. obscura mesophyll protoplasts development.

MATERIALS AND METHODS

Mesophyll protoplasts of D. obscura were isolated as previously described (1). In a first experiment, protoplasts, at different culture densities, were either plated in agarose or cultured in liquid-over-agarose bed. Medium culture was MS1 (1), solidified with 0.6% agarose and supplemented with 5.7 µM IAA, 4.4µM BA, 0.01% casein hydrolysate (CH), 3% sucrose and 0.5M mannitol. The pH was adjusted to 5.8. On the basis of this experiment, a culture density of $5x10^4$ prot/ml was selected to define the most suitable medium for protoplast regeneration. Protoplasts were cultured onto liquid-over-agarose bed. Five major salt formulations were tested: B_5 (3), V-47 (4), MS1 (1), KM (5) and NT (6). In all cases, media included MS1 minor salts, MS vitamins, 5.7µM IAA, 4.4uM BA, 3% sucrose and 0.5M mannitol, pH 5.8. Media were also tested with 0.01% CH, without NH_4NO_3 or replacing NH_4NO_3 by L-Glu. Protoplasts were cultured in plastic Petri dishes and incubated in darkness at 27±1ºC. Cell wall regeneration and cell division percentages were determined after 10 days of culture.

RESULTS AND DISCUSSION

In both culture techniques employed, the percentage of cell wall regeneration and cell division was influenced by protoplast culture density (Figs. 1 and 2). Best results were obtained with liquid-over-agarosa bed, which is in agreement with our previous results with agar (1). It confirms that for D. obscura mesophyll protoplasts, plating is not a suitable technique. As a

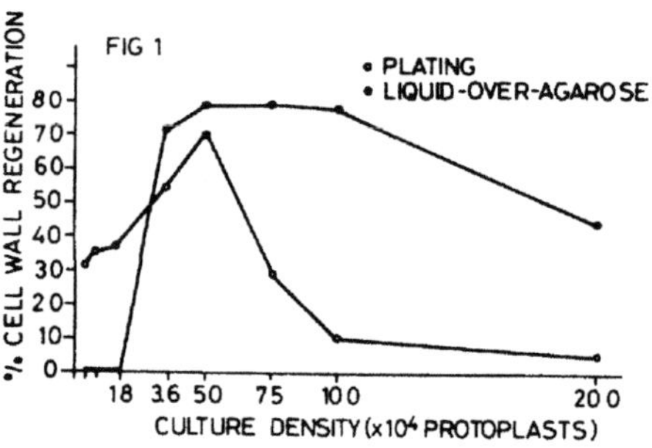

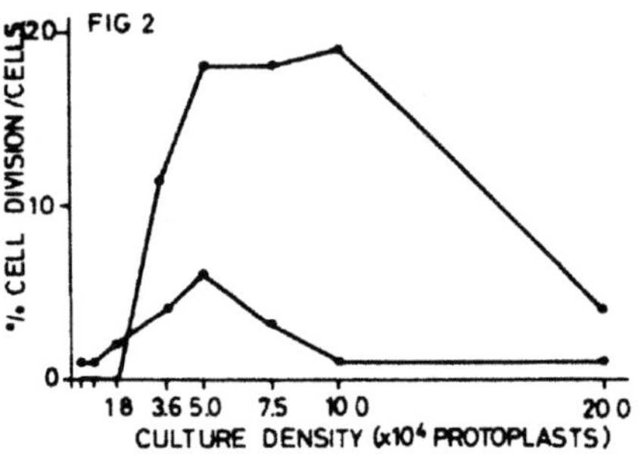

result of this information, protoplasts were subsequently cultured in liquid-over-solidified medium. Five different nutrient media were tested, being the culture density $5x10^4$ prot/ml. The influence of these media on cell wall regeneration and cell division is shown in Fig. 3. In all media, the percentage of cell wall regeneration increased with the incubation period, and after 10 days was greater than 65% in V-47, MS1 and NT media. Except B_5, all media were effective in promoting cell division, best results being obtained with MS1 medium; furthermore, in this medium protoplasts underwent sustained mitotic division and produced colonies. Even though it is difficult to generalize on the influence of mineral components on protoplast regeneration, the poor results obtained with B_5 could be attributed to the low Ca^{2+}, Mg^{2+} and NH_4^+ concentrations in this medium (Table 1).

The effect of CH on cell wall regeneration and cell division was closely related to NH_4^+ concentration in the original media (Fig. 3 and Table 1): improved both parameters in B_5 medium, strongly inhibited in KM and NT media and had a little effect in V-47 and MS1 media.

The culture of D. obscura mesophyll protoplasts required the presence of reduced nitrogen, since the omission of NH_4^+ provoked a drastic inhibition of the investigated parameters. However, the source of reduced nitrogen seems to be not specific since L-Glu can replaces ammonium ion (Fig. 3).

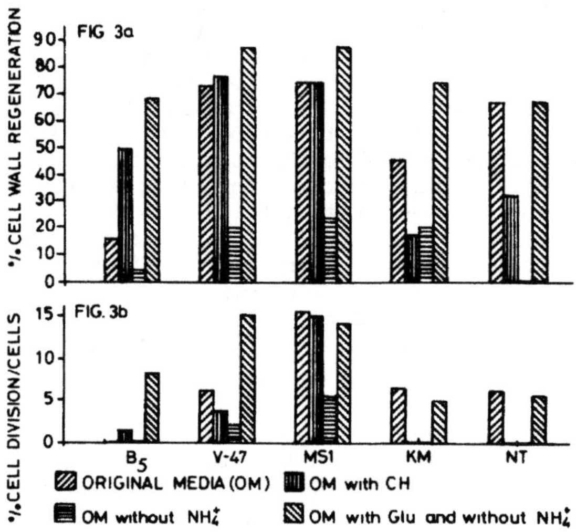

TABLE 1. Concentration (mN) of NH_4^+ and Ca^{2+} plus Mg^{2+} of the media used in this experiment.

Media	NH_4^+	Ca^{2+}plus Mg^{2+}
B_5	2.03	4.08
V-47	3.37	18.01
MS1	5.16	18.02
KM	7.50	10.40
NT	10.65	13.00

Acknowledgements. The financial support of CAICYT, Spain, project Nº 83/2026 is gratefully acknowledged.

REFERENCES

1. Brisa MC, Segura J: Physiol. Plant. 69: 680-686, 1987.
2. Murashige T, Skoog F: Physiol. Plant. 15: 473-497, 1962.
3. Gamborg OL, Miller RA, Ojima K: Exp. Cell. Res. 50: 151-158, 1968.
4. Binding H: Z. Pflanzenphysiol. 74: 327-356, 1974.
5. Kao KN, Michayluk MR: Planta 126: 105-110, 1975.
6. Nagata T, Takebe I: Planta 99: 12-20, 1971.

COTYLEDON PROTOPLASTS FROM SOLANUM MELONGENA L. REGENERATION TO FERTILE PLANTS

G.G. Li and L.Y. Zhang
South China Institute of Botany, Academia Sinica, Guangzhou, China.

INTRODUCTION

The genus Solanum has been investigated in detail for the development of plants from isolated protoplasts. Positive results have been achieved with *Solanum tuberosum*[2,3,5,6,9,10,12], *Solanum nigrum*[8,13], *Solanum uporo*[4], and *Solanum melongena*[7,11]. This paper presents the results of research in the regeneration from cotyledon protoplasts to maturity of this species.

MATERIAL AND METHODS

The seeds of eggplant (*Solanum melongena* L. cv. Yuanjing) were surface--sterilized by previously reported methods[4] and placed on SH agar medium containing only ½ macronutrients + 1% sucrose, and was cultured at constant 25 + 2 °C, and 16 h/day light of 4000 lux. The youngest, fully expanded cotyledons were used for protoplast isolation.

The enzyme solution contained 1.5% cellulose Onozuka R-10, 0.4% hemicellulase Rhozyme, 0.4% pectinase, 3 mM $CaH_4(PO_4)_2 \cdot H_2O$, 0.55 M sorbitol, pH adjusted to 5.7. Prior to enzymatic digestion, the enzyme was mixed in a ratio of 1:1 with culture medium of protoplasts[5]. The cotyledons were cut into strips and prepared by previously reported methods[5]. Finally, the protoplasts were suspended in culture medium of protoplast[5]. The protoplast suspensions (10^4/ml) were plated as 40 µl droplets in petri dishes in the dark at 26 °C. Adjustment of the osmolarity of the medium was carried out twice. First at day 15 by the addition of 0.35 M osmoticum and then at day 25 by the addition of 0.25 M osmoticum.

After two months growth the colonies were transferred to two media. Medium 1 consisted of Ms + 1 mg/l NAA + 2% sucrose (medium C_1) while medium 2 comprised Ms + 2 mg/l KT + 0.05 mg/l NAA + 2% sucrose (medium C_2). The cultures were kept under 25 + 2 °C, and 16 h/day light of about 2000 lux.

When differentiated, shoots developed to a length of 2-3 cm. They were transferred to agar media for rooting: (1) Ms + 0.1 mg/l NAA + 2% sucrose; (2) Ms + 0.1% mg/l IAA + 1% activated charcoal + 2% sucrose.

RESULTS AND DISCUSSION

The enzyme mixture adopted was effective in releasing a large number of protoplasts of viable from the cotyledons after 5-6 h. The majority of protoplasts of a uniform size after purification (Fig. 1). After 18 h in culture medium of protoplasts cell wall formation was observed. Culture of protoplasts led to a first division and three division phases were found at 2-3 day. They were early division (Fig. 1a), equal and unequal division (Fig. 2, 3). The protoplast divisions of a frequency of approximation 40% after 8-10 days were obtained. After 15 days colonies of many cells were observed (Fig. 4). Compared to those colonies continuously maintained in unreduced culture medium of protoplasts, where the colonies growth became slow after several days in culture and even stopped. It was found that the medium with lowered osmotic pressure faciitated continuous growth of colonies and formation of new colonies. This adjustment not only replaced nutrients but also improved the physiological environment of the colonies. It proveded a means of assisting the transition from liquid to solid media. This observation is similar to those reported for the culture from leaf protoplasts of potato[5] and Solanum uporo[4].

Once cell colonies of a diameter of 1-1.5 mm had formed they were transferred onto medium C_1 or C_2. Calli were induced from cell colonies of these media with 10 days. However, the calli appeared smooth and grew slowly, grey-white on medium C_1, and dense structure, but grew more rapidly and had a green yellow colour on medium C_2. When the protoplast-derived calli on medium C_2 developed to 5 mm in diameter they were transferred to a differentiation medium. This contained 2% sucrose and kinetin 2 mg/l combined with NAA 0, 0.001, 0.005, 0.01, 0.1, 0.5, 1 or 5 mg/l individually. Shoots were observed in 70 days on MS medium containing kinetin 2 mg/l, NAA 0.0005 mg/l and 2% sucrose.

When the shoots grew to a height of 3-4 cm, and had formed 3-5 fully expanded leaves, they were excised from callus and transferred on MS media containing IAA at 0.1 or 0.1 mg/l in combination with activated charcoal at 1%. The intact plants bearing well-developed roots were gradually produced (Fig. 6). In both medium, root growth was better on MS medium with IAA 0.1 mg/l, activated charcoal 1% and sucrose 2%.

Plantlets which had grown completely were hardened to under natural light. The plantlets were then transplanted to a pot of soil mixture, and kept under damp soil, air humidity 95%, and 25 + 2 °C. New growth commenced after 10 days and the plants flowered and set fruit after three months (Fig. 7).

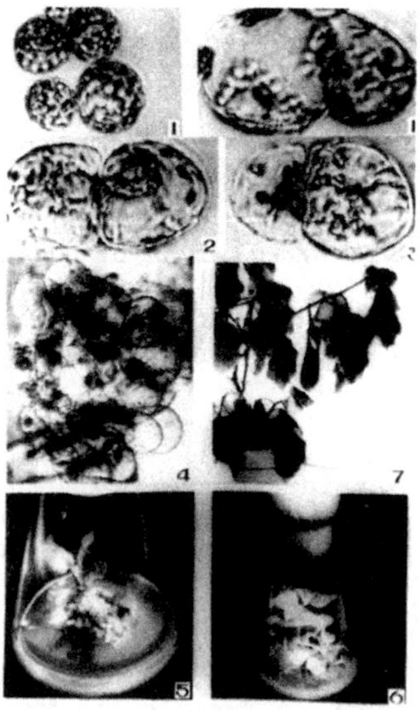

Fig. 1. Freshly isolated protoplasts from egg-plant cotyledon. 1a. Early division. 2. Equal division. 3. Unequal division. 4. Protoplast-derived colonies. 5. Shoot differentiation on callus. 6. Regenerated shoot in rooting medium 7. Regenerated fertile plant.

REFERENCES

1. Binding H: Z. Pflanzenphysiol., 74: 327-356, 1974. 2. Binding H & R Nehls: Z. Pflanzenphysiol., 85: 279-289, 1977. 3. Binding H et al.: Physiol. Plant., 43: 52-54, 1978. 4. Li GG & F Constabel: J. Plant Physiol., 117: 137-142, 1984. 5. Li GG & Zhang LY: Acta Botanica Sinica, 30, 1988. In press. 6. Gunn RE & JF Shepard: Plant Sci. Lett., 22: 97-101, 1981. 7. Jia JF & I Potrykus: Plant Cell Rep., 1: 71-72, 1981. 8. Nehls R: Plant Sci. Lett., 12: 183-187, 1978. 9. Schumann U & H Koblitz: Biologia Plantarum, 25: 180-186, 1983. 10. Shepard JK & RE Totten: Plant Physiol., 60: 313-316, 1977. 11. Saxena PK et al.: Protoplasma, 108: 355-356, 1981a. 12. Tavazza R & G Ancora: Plant Cell Rep., 5: 243-246, 1986. 13. Wang GY & ZA Xia: Acta Botanica Sinica, 25: 111-114, 1983.

ISOLATION, CULTURE AND REGENERATION OF LEAF PROTOPLASTS IN THE GENUS CAPSICUM

I. DIAZ, *J.B. POWER, J.R. DIAZ-RUIZ
UEI Fitopatología. Centro de Investigaciones Biológicas. CSIC. 28006-Madrid. Spain.
* Dpt. of Botany. University of Nottingham. Nottingham NG7 2RD U.K.

INTRODUCTION

Capsicum is a well known genus and one of the most economically important vegetable crops all over the world.

The capacity to regenerate pepper plants from tissue culture has been described (1,2,3) but there is only one report on the regeneration of plants from protoplasts, this was restricted to C. annuum v. California Wonder (4). This suggests that Capsicum species in general are recalcitrant at the protoplast level.

It has also been shown that varieties of a given species differ markedly in their cultural requirements, particularly at the protoplast stage (5,6).

This summary describes conditions for the efficient and reproducible culture and division of leaf protoplasts of several varieties of C. annuum and the wild species C. chinense.

PROCEDURE

Axenic cultures of Spanish varieties of C. annuum (Americano, Dulce Italiano, Florida Gyant and Nigrum) and the wild species C. chinense were used as the source for mesophyll protoplast isolation.

Shoots were cut into small pieces, plasmolysed for 1 h inCPW13M (7) and then incubated (14 h, dark, 25°C) in an enzyme mixture (Cellulase Onozuka R10 and Macerozyme R10). Protoplasts were pelleted by centrifugation and freeded of cell debries by resuspension. In order to iniciate and maintain protoplast division, different media with a wide range of auxin and cytokinin concentration were tried..

RESULTS

For both Capsicum species and for three of the four varieties KM8P/KM8 medium (8) and agarose bead culture (9) was found to be the only reliable approach which enable protoplast division to be iniciated and sustained for Capsicum

Protoplast yields and plating efficiencies are given in Table 1.

During the first few days of culture Capsicum protoplasts expanded as a prelude to first division. After 4-5 weeks colonies transferred to MS medium (10) with zeatin developed into green callus and C. annuum v. Dulce Italiano protoplast-derived callus developed shoot primordia after a further two weeks following final transfer to MS medium with 6-BAP. By contrast, under the same cultural regime callus of the other varieties of C. annuum and C. chinense produced meristematic-like structures, but no shoots.

Regenerated shoots of C. annuum v. Dulce Italiano were detached and rooted after 2-3 weeks of placing on MS medium which lacked growth regulators.

Species/variety	Yield per g/f.wt.	% Plating efficiency (KM8P/KM8 medium)
C. annuum		
Americano	4.0×10^5	26
Dulce Italiano	2.0×10^5	22
Florida Gyant	3.5×10^5	19
Nigrum	2.0×10^5	0
C. chinense	6.0×10^5	24

Table 1.- Leaf protoplast yields and plating efficiencies (14 days) of C. annuum (4 varieties) and C. chinense.

REFERENCES

1. G.C. Phillips and J.F. Hubstenberger (1985). Plant Cell Tiss Org Cult 4: 261-269.
2. M. Fari and M. Czako (1981). Scientia Hortic 15: 207-213.
3. A.L. Gunay and P.S. Rao (1978) Plant Sci Lett 11: 365-372.
4. P.K. Saxena, R. Gill, A. Roshid and S.C. Maheshwari (1981). Protoplasma 108: 357-360.
5. N.E. Bohorova, E.C. Cocking and J.B. Power (1986) Plant Cell Rep 5: 256-258.
6. J.B. Power, E.M. Frearson, D. George, P.K. Evans, S.F. Berry, C.Hayward and E.C. Cocking (1976). Plant Sci Lett 7: 51-55.
7. J.B. Power and J.V. Chapman (1985). In: R.A. Dixon (ed) Plant Cell Culture. IRL Press, Oxfod Washington DC, pp 37-66.
8. K.N. Kao and M.R. Michayluk (1975) Planta 126: 105-110.
9. R.D. Shillito, J. Paszkowski, I. Potrykus (1983) Plant Cell Rep 2: 244-247.
10. T. Murashige and F. Skoog (1962). Physiol. Plant 15: 473-497.

ISOLATION AND PROPERTIES OF PROTOPLASTS OF *THEOBROMA CACAO* L.

H.A.COLLIN, R.B.MUSE, S.ISAAC, K.HARDWICK
Department of Botany, University of Liverpool, P.O.Box 147, Liverpool, L69 3BX, U.K.

1. INTRODUCTION

Protoplasts provide an important technique for the study of plant:fungal interaction in disease. For cocoa (*Theobroma cacao*) the major disease is Witches' Broom, caused by the fungus *Crinipellis perniciosa*. The primary infective stage of the fungus has a biotrophic mode of nutrition whereas the succeeding secondary stage has a necrotrophic mode of nutrition which causes tissue death (Evans, 1980). Preliminary evidence has shown the existence of phytotoxic activity in culture filtrates from the secondary stage. The mechanism of action of these phytotoxins may be through an effect on membrane permeability, which can be measured most easily using protoplasts of cocoa. The aim of this project was to:-

a)Develop a method for protoplast isolation from cocoa leaves.
b)Test the effect of culture filtrate from secondary stage mycelium on mesophyll protoplasts.

2. PROCEDURE

2.1. Production of culture filtrate

C. perniciosa was isolated from infected cocoa tissue and grown in liquid MS medium with 3% sucrose, pH 5.6 at 30°C and shaken at 100 rpm.. The culture filtrate was harvested 15 days after inoculation, during late stationary phase, and concentrated five-fold.

2.2 Protoplast isolation

Protoplasts were isolated from cocoa leaves about 11 cm long (area 20 cm^2) by first homogenising for 15 sec in water. Material was then incubated in water for 30 min prior to transfer into lytic mixture containing Cellulase 1%, Macerozyme 1%, and Driselase 0.5% in the osmoticum of Wright (1985). After 4h protoplasts were filtered through a nylon net (63 um), centrifuged and resuspended in the same osmoticum, containing culture filtrate both undiluted and 5x concentrated. Viability was assessed after 6h using Evans' Blue stain. Controls included undiluted and 5x concentrated nutrient medium made up in osmoticum.

3. RESULTS

Viability counts of protoplasts, recorded in Table 1, show that in the controls there was little decline in viability with time (10%), and no significant effect of high concentrations (x5) of nutrient medium. However, there was a large

TABLE 1. Loss in viability of cocoa protoplasts maintained in undiluted (x1) and concentrated (x5) nutrient medium (MS) or culture filtrate (CF) in the presence of an osmoticum solution.

Treatment	Viable protoplasts x 10^5 g.fr.wt^{-1}		Loss in viability (%)
	0 h	6 h	
x1 MS	78.8	70.4	10.7
x5 MS	78.0	70.0	10.3
x1 CF	74.8	37.2	50.3
x5 CF	74.4	17.6	76.3

reduction in viability of protoplasts in the presence of both diluted (50%) and concentrated (76%) culture filtrates. The culture filtrate caused lysis as well as loss in viability.

4. CONCLUSIONS

a)A protocol was developed for isolating mesophyll protoplasts from young cocoa leaves.

b)Culture filtrate from the secondary stage mycelium of *C. perniciosa* caused loss of viability in cocoa protoplasts.

REFERENCES

1. Evans HC:Pleomorphism in *Crinipellis perniciosa*, causal agent of Witches' Broom Disease of cocoa. Transactions of the British Mycological Society 74, 515-523, 1980.
2. Wright DC:Factors affecting isolation of protoplasts from leaves of grape (*Vitis vinifera*). Plant Cell Tissue and Organ Culture 4, 95-100, 1985.

PLANTLET REGENERATION FROM PROTOPLASTS OF LEVISTICUM OFFICINALE KOCH. AND ITS CYTOLOGICAL INVESTIGATION

JING-FEN JIA., SHI-YU ZHANG., KUO-CHANG CHENG

Department of Biology, Lanzhou University, Lanzhou, China.

INTRODUCTION

Levisticum officinale Koch, a species of Umbelliferae, is a medical hers used as a blood tonic. Somatic embryogenesis has been obtained from tissue culture(1). This study deals with embryogenic callus induction and plantlet regeneration through protoplast culture. A comparison of cytological variations between source materials and protoplast-derived plantlets was also carried out.

MATERIALS AND METHODS

Long-term subcultured calli of L. officinale Koch used for protoplast isolation were maintained on MS medium supplemented with 1mg/l 2,4-D and subcultured at one month intervals.

To isolate protoplasts, 7-day-old calli were placed in enzyme solution containing 1% each of cellulase R-10, pectinase(Serva) 0.5% macerozyme R-10 and 0.2% hemicellulase(Sigma) in 0.38M glucose(pH 5.8). After overnight incubation at 25°C, the released protoplasts were collected by centri-fugation and purified by floating with 12% sucrose, then washed twice with 0.16M $CaCl_2 \cdot 2H_2O$. Protoplasts were suspended at a density of $5 \cdot 10^4$/ml in liquid C81V medium plus 1mg/l 2,4-D and 0.2mg/l kinetin and incubated at 25°C. After two weeks, an equal volume of fresh medium with 0.19M glucose was added to the cultures. When colonies formed, they were transferred on N6 medium(3) with 0.2mg/l 6BA and 0.1mg/l IBA. About two months leter, the embryogenic tissues were transferred to MS medium with 0.1mg/l NAA and 0.2 mg/l IBA to induce morphogenesis.

For cytological observations, the squash method and staining with modified carbol fuchsin were adopted.

RESULTS AND DISCUSSION

A large amount of survival protoplasts could be isolated from fast growing calli of L. officinale(Fig.1a). C81V medium (2)with 2,4-D(1mg/l) and Kinetin(0.2mg/l) was quite effective in supporting cell sustained divisions. The first divisions often occurred at day 3-5 (Fig.1b). The division frequency of 50% was achieved in the best case. Subsequent divisions and colony formation successfully carried on in liquid C81V medium(Fig.1c). Diluting the cultures with fresh medium reducing the osmoticum could promote this process. When colonies grew up to 1-2 mm size, they were transferred on the agar solidified N6 medium with 0.2mg/l 6BA and 0.1mg/l IBA, and cultured in the growth cabinet with a 12/12hr day/night cycle at 26/15°C,

3000lux. Under these conditions,the yellowish embryogenic tissues were produced on the surface of the callus in two months.Such hormone combination played a decisive role in inducing embryogenic calli and embryoids(Figs.1d-1g).After being transplanted to the MS medium with 0.1mg/l NAA and 0.2mg/l IBA,these tissues could differentiate further and give rise to plantlets(Fig.1h).

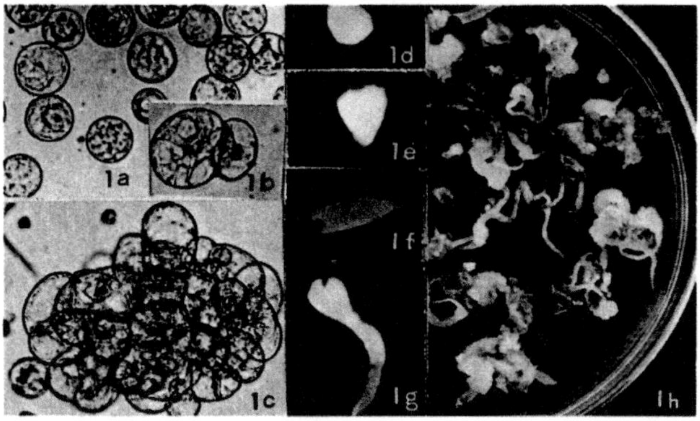

Fig.1a-1h.From protoplasts to plantlets.1a.Freshly isolated protoplasts. 1b.First division at day 3 in culture.1c. A cell cluster after two weeks of culture. 1d-1g.Protoplast derived embryoids,showing globular,heart-shaped,torpedo-shaped and a plantlet. 1h.Regenerated plantlets from L.officinale protoplasts.

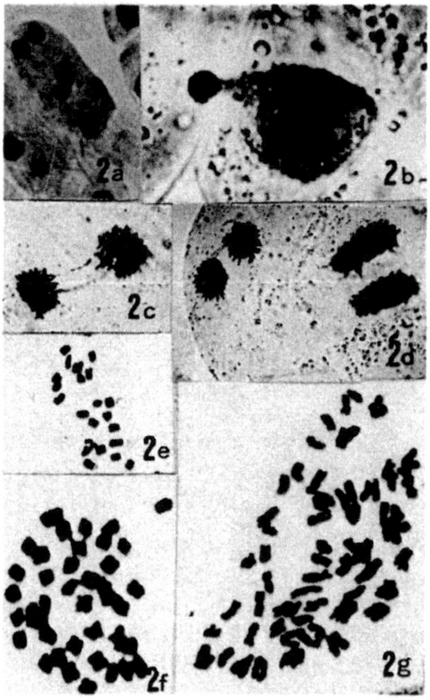

Fig.2a-2g.Abnormalities of nuclei and mitoses observed in donor callus and protoplast derived plantlets.2a.Multinuclei.2b.Chromatin migration between cells.2c.Chromatin bridge.2d.Multipolar mitosis. 2e.2n=22.2f.3n=38.2g.Polyploid.

Cytological observations revealed that the donor calli used for protoplast isolation exhibited obviously nucleus abnormalities,such as multinuclei(Fig.2a),chromatin migration from one cell to another(Fig 2b),chromatin bridge(Fig.2c),multipolar mitosis(Fig.2d),etc.The chromosome numbers in this kind of calli were also variable(Fig.2e-2g). The rate of 2n=22 cells was less than 20%,whereas the aneuploid cells nearly 40%.However,such variations found in donor materials were relatively few in protoplast regenerated plantlets.

REFERENCES

1.Zhang,S-Y and Cheng,K-C.,Act.Bot.Bor.-Occ.Sinica.in press.
2.Dudits,D.,et al.,Theor.Appl.Genet.51(1977) 127-132.
3.Chu,C-C.,et al.,Sci.Sinica 5 (1975) 484-490.

Differences in developmental morphology of protoplasts from two cell lines of white spruce (*Picea glauca*)

Attree SM[1]; Bekkaoui F[2], Saxena PK[1], Dunstan DI[2] and Fowke LC[1]

[1]Department of Biology, University of Saskatchewan, and [2]Plant Biotechnology Institute, National Research Council, Saskatoon, Saskatchewan, Canada

1. INTRODUCTION

Until recently cultured protoplasts of conifers remained recalcitrant (see Dunstan and Thorpe, 1987). The establishment of embryogenic suspension cultures of both white spruce (*Picea glauca*) (Hakman and Fowke, 1987) and (*Pinus taeda*) (Gupta and Durzan, 1986) has created a new source of protoplasts capable of somatic embryo regeneration (Attree *et al*, 1987; Gupta and Durzan; 1987). However, not all embryogenic suspension lines of white spruce have yielded protoplasts capable of regenerating embryos; protoplasts from one line gave rise to green callus but no somatic embryos (Bekkaoui *et al*, 1987). The ability to regenerate embryos from protoplasts of all cell lines may be crucial to creating improved forest trees using protoplast technology. In this communication a comparison is made between two white spruce embryogenic suspension lines.

2.1 Materials and Methods

The embryogenic suspensions of white spruce (culture lines WS1 and WS3) were derived and maintained according to previously published methods (Hakman and Fowke, 1987). Protoplasts were isolated and cultured following methods of Bekkaoui *et al*, (1987) and Attree *et al*, (1987).

2.2 Results and Discussion

The comparison of the two embryogenic suspension lines and their protoplasts is made in table 1. The major difference was that protoplasts of WS3 regenerated to form somatic embryos (fig 1); those of WS1 produced a green callus (fig 2) at reduced efficiency. When WS3 protoplasts were cultured in conditions that favored WS1 callus formation, all protoplasts died within 14 days. The culture of WS1 protoplasts in conditions that favoured WS3 embryo regeneration led to reduced callus formation and did not lead to regeneration of embryos.

Although protoplasts of WS1 regenerated only callus the initial suspension culture was highly embryogenic and possessed a higher frequency of well-defined somatic embryos compared to the WS3 culture. The calli that developed in the WS1 protoplast cultures may have originated from non-embryogenic cells in the initial suspension culture. These are considered to arise by continuous conversion of embryogenic to nonembryogenic cells (see Vasil, 1987). Their slow growth compared with embryogenic cells, as observed in the protoplast cultures, would limit them to occupying a smaller proportion of the total suspension culture cell population. Thus, in addition to yielding protoplasts from non-embryogenic cells, the WS1 suspension culture must have yielded many embryo derived protoplasts. The reason why the latter failed to regenerate embryos when cultured in the same conditons as WS3 protoplasts may reflect genetic differences between the cell lines which were initially derived from individual zygotic embryos. Perhaps by identifying more optimal media and using suspension cultures maintained only for a short term, these difficulties may be overcome yielding morphogenic protoplasts from all cell lines.

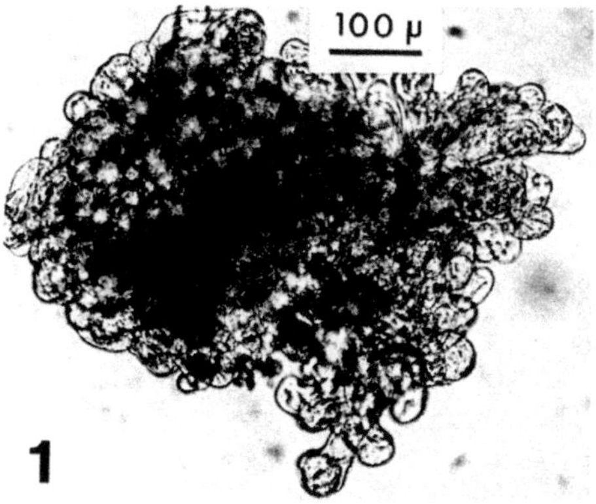

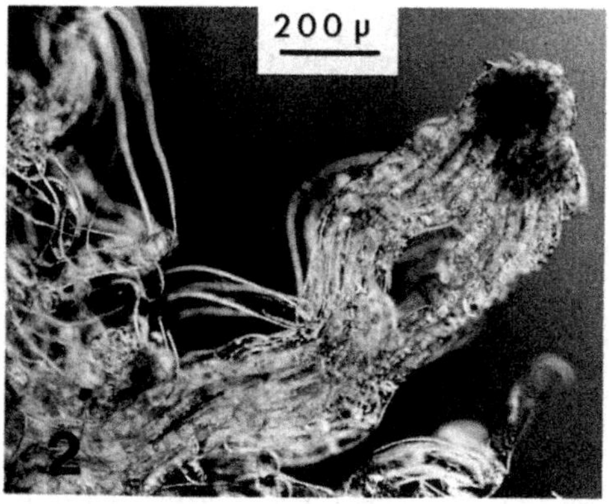

Figs 1 and 2. Cultures derived from protoplasts of two white spruce cell lines after 4-6 weeks. Fig 1. Callus derived from WS1. Fig. 2 Embryogenic colony derived from WS3.

Table 1. A comparison of two embryogenic lines (WS1 and WS3) of white spruce and their cultured protoplasts.

variable	Cell line	
	WS1	WS3
suspension culture age	< 14 months	< 14 months
suspension culture composition	many embryos Also isodiametric cell clusters and free cells.	mostly cell aggregates with few embryos.
mean protoplast yields (g^{-1})	4.5×10^6	5.2×10^6
lag to 1st division	3 days	2 days
time to 2nd division	10 days	4 days
cell types present after 1st division	isodiametric	isodiametric and embryonal suspensor
optimal plating density	$1\text{-}2 \times 10^5$	$1\text{-}2 \times 10^5$
optimal medium form	liquid	agarose solidified
optimal growth regulators	1 mgl^{-1} 2,4-D	2 mgl^{-1} 2,4-D, 1 mgl^{-1} BAP
viability after 7 days	30-40%	30-40%
additional medium supplement	5mM glutamine	5mM glutamine
plating efficiency (4 weeks)	< 1%	5-10%
final morphology	green callus	white somatic embryos

2. REFERENCES

Attree SM, Bekkaoui F, Dunstan DI, Fowke LC (1987) Plant Cell Reports (in press).

Bekkaoui F, Saxena PK, Attree SM, Fowke LC, Dunstan DI (1987) Plant Cell Reports (in press).

Dunstan DI, Thorpe TA (1987) In: Vasil IK (ed.), Cell Culture and somatic cell genetics of plants, Academic Press, NY, 223-241.

Gupta PK, Durzan DJ (1986) Bio/Technology 4:643-645.

Gupta PK, Durzan DJ (1987) Bio/Technology 5:710-712.

Hakman I, Fowke LC (1987) Plant Cell Reports 6:20-22.

PROTOPLAST DEVELOPMENT FROM PORPHYRA LINEARIS - AN EDIBLE MARINE RED ALGA*

CHEN, L.C-M., M.F. HONG AND J.S. CRAIGIE
National Research Council, Atlantic Research Laboratory
1411 Oxford Street, Halifax, Nova Scotia, Canada, B3M 3R3

INTRODUCTION

Porphyra linearis is a gastronomically acceptable alga (1) but unlike the commercially important species, *P. tenera* and *P. yezoensis* it does not produce neutral spores (monospores) (2). Our work shows how this disadvantage can be overcome by regeneration of vegetative cells from protoplasts.

MATERIALS AND METHODS

The mid sections of young thalli of *P. linearis* (Figs. 1 & 2) were sterilized by immersion in 1% Triton X-100 and 0.5% Betadine solution for 1-2 min, respectively (3) prior to being brushed to remove epibionts. The thalli were rinsed 6x with sterile seawater and left in antibiotic solution overnight (4). Clean small pieces (≤1mm²), were plasmolysed in sorbitol (1.1 M in seawater) for 30-60 min and then incubated in a mixture of 2% (w/v) of crude enzyme from sea-snails (4, 5) and 3% cellulase (R-10) (6) in 0.5 M sorbitol-seawater for 2-3 h at 15°C with gentle magnetic stirring. The released protoplasts (Fig. 3) were filtered through a nylon sieve of 30 µm pore size, pelleted at 100 x *g* for 2-5 min and washed with sorbitol (0.5, 0.3, 0.1 and 0 M) in seawater at 24 h intervals for each concentration. Protoplasts were then cultured in TC-11 medium (4) at 10°, 15°, and 20°C, 10-100 $\mu E m^{-2} s^{-1}$ with photoperiods of 8:$\overline{16}$, 12:$\overline{12}$ and 16:$\overline{8}$. The medium was changed weekly.

RESULTS AND DISCUSSION

Thallus-sheath degradation occurred within 30 min and individual cells separated. These then became spherical and slightly larger than the cells embedded in the original leafy thallus. A number of undisintegrated thalli remained despite prolonged stirring suggesting that the cell wall composition was not uniform. After two to four days in culture, cell wall formation was observed and some cells began to divide. Three weeks later most protoplasts divided and quickly differentiated into rhizoid and vegetative cells forming a leafy-thallus (Fig. 4). However, some developed into a callus-like structure (Fig. 5) which released cells (Fig. 6) that later formed leafy thalli. Other callus-like structures developed leafy thalli directly (Fig. 7). In some cases, protoplasts formed cell walls, divided 10-20 times (Fig. 8) to give individual normal sized cells, each of which later formed a conchocelis under constant culture conditions. A few protoplasts after cell wall regeneration continued to grow in size without dividing and contained dense plastid granules (Fig. 9). Most of these unusual cells eventually ruptured, but in a few cases they released "cells" from which leafy thallus or conchocelis developed as reported in *P. miniata* (4).

A few protoplasts formed cell walls and divided into uniseriate fertile cell rows (Fig. 10); others developed filamentous conchocelis (Fig. 11) similar to the development of carpospores in *Porphyra*. All seven of these morphological variations ultimately followed the normal developmental sequence of *P. linearis* (Fig. 12) as demonstrated in unialgal culture (2). Whether any of the morphological developments are associated with nuclear ploidy changes is unknown.

CONCLUSIONS

Protoplasts from *P. linearis* can regenerate into vegetative cells and redifferentiate into leafy thalli directly or indirecty, through more complex intervening developmental stages.

*NRCC No. 28092

REFERENCES

1. McLachlan, J., J.S. Craigie, L.C-M. Chen and E. Ogata. Proc. Int. Seaweed Symp. 7: 473-476, 1971.
2. Bird, C.J., L.C-M. Chen and J. McLachlan. Can. J. Bot. 50:1859-1863, 1972.
3. Chen, L.C-M. Bot. Mar. 29:435-439, 1986.
4. Chen, L.C-M. and A.R.A. Taylor. Can. J. Bot. 56:886-887, 1978.
5. Liu Wan Sun, Tang Yan Lin, Liu Xue Wu and Fang Tsung (T.C. Fang). Hydrobiologia 116/117:319-320, 1984.
6. Chen, L.C-M. Bot. Mar. 30:399-403, 1987.
7. Chen, L.C-M. Biol. Bull. Nat'l. Taiwan Normal Univ. Taiwan, 17:63-67, 1982.

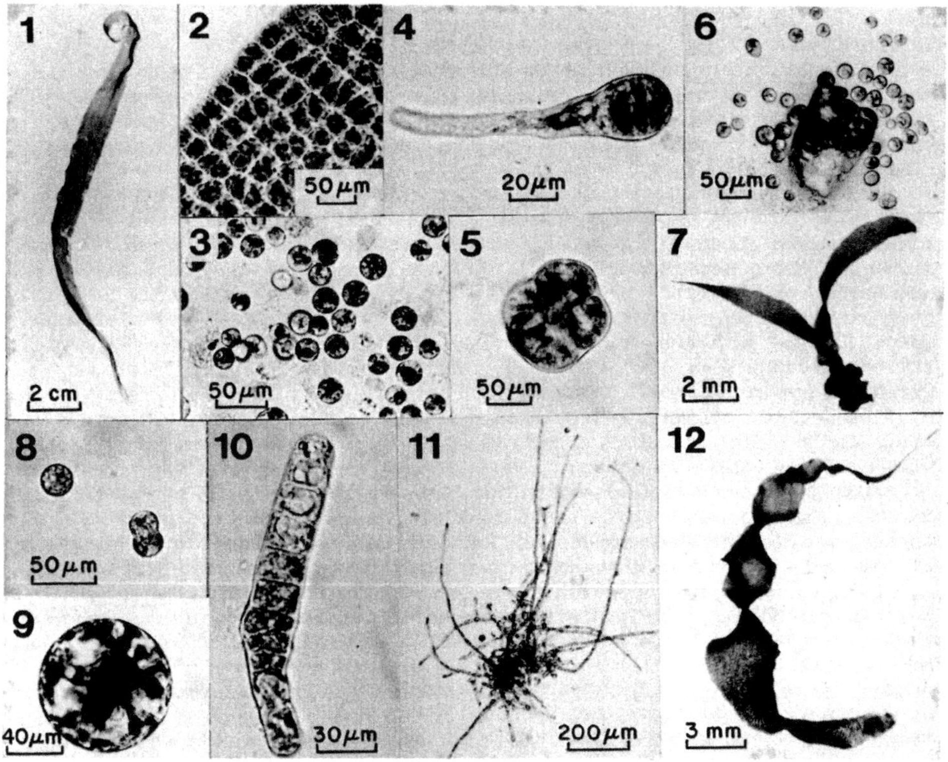

FIGURES

1. Porphyra linearis intact thallus.
2. Thallus showing vegetative cells.
3. Isolated protoplasts.
4. Regenerated leafy thallus.
5. Irregular callus-like structure.
6. Leafy thallus developed from callus-like structure.
7. Cells released from callus-like structure.
8. Cells regenerated from one protoplast.
9. Large regenerated cell.
10. Fertile cell rows.
11. Filamentous conchocelis.
12. Mature leafy thallus from protoplast.

STRUCTURE AND PHYSIOLOGY OF THE PROTOPLAST PLASMA MEMBRANE

L.C. FOWKE

Department of Biology, University of Saskatchewan, Saskatoon, Saskatchewan, Canada, S7N OWO.

The plasma membrane has a number of important functions in plant cells including the regulation of ion and metabolite transport, hormone transport and action, host pathogen interactions and cell wall biosynthesis and deposition. The plasma membrane is normally enclosed within a complex cell wall which limits experimental analysis of this important cell component. Removal of the wall to produce protoplasts permits direct access to the plasma membrane. The ready availability of protoplasts from a wide range of plant species (e.g. Vasil and Vasil 1980, Fowke and Constabel 1985) has facilitated studies of the structure, chemistry and physiology of the plasma membrane.

This chapter examines the application of protoplasts to the study of three features of the plant plasma membrane (i) plasma membrane fluidity, (ii) organelles associated with the plasma membrane and (iii) uptake across the plasma membrane.

1. PLASMA MEMBRANE FLUIDITY

The fluidity of the plasma membrane has a direct effect on membrane function. For example, a current model for the synthesis and deposition of cellulose microfibrils on plant cells involves the movement of enzyme complexes through the plasma membrane (e.g. Heath and Seagull 1982). Direct evidence for this movement is lacking, in fact until recently very little attention was paid to the fluidity of the plant plasma membrane. In contrast, numerous studies have clearly demonstrated the fluidity of animal membranes in terms of both lipids and proteins (e.g. Morré et al. 1979).

During the late 1970's a number of ultrastructural studies were undertaken to determine whether the plant plasma membrane was fluid in nature (for details see Fowke and Gamborg 1980). Briefly, lectins which bind to specific carbohydrates were used to attach electron microscope labels (e.g. hemocyanin, colloidal gold, ferritin) to the surface of plant protoplasts. Most studies utilized the lectin Concanavalin A (Con A) a metalloprotein which binds specifically to α-D-glucose and α-D-mannose sugars. Binding sites with a high affinity for Con A were present on a number of different plant protoplasts. When Con A was attached to prefixed protoplasts or live protoplasts in the cold, an even binding pattern was observed. If protoplasts labeled either in the cold or at room temperature were incubated at room temperature, clustering of Con A binding sites was observed. These results provided the first direct evidence in favour of a fluid plant plasma membrane. It was not clear from this early work whether the mobile Con A binding sites represented actual components of the plasma membrane (ie. sugars or glycoproteins) or remnants of the cell wall. These studies also did not provide any information regarding the lipid components of the plasma membrane.

During the past five years a group at Michigan State University has investigated this problem further using an entirely different experimental approach and has made a significant contribution to our understanding of the plant plasma membrane (Metcalf et al. 1983, 1986a,b). They investigated the fluidity of the plasma membrane of soybean (SB-1) protoplasts with the light microscope using a number of different fluorescent probes which detect carbohydrates, lipids and proteins. They utilized a special optical technique, fluorescence recovery after photobleaching (FRAP), which permits the direct measurement of diffusion coefficients for membrane components. Briefly, the technique involves the attachment of fluorescent probes to components in the plasma membranes of protoplasts. A laser beam is first used to photobleach a small area on the plasma membrane surface and then to scan the membrane surface in order to determine the rate of recovery as fluorescent molecules move back into the bleached area. These studies have resulted in a number of very interesting observations which are summarized in Table 1.

TABLE 1. FRAP analysis of soybean protoplast plasma membranes.

Plasma membrane Component	Fluorescent Probe	Diffusion Coefficient (cm^2/sec)
Carbohydrate	Con A	7×10^{-11}
	SBA	5×10^{-11}
	WGA	3×10^{-10}
Lipid	Lipids & P-lipids (18°C)	$5 \times 10^{-10} + 3 \times 10^{-9}$
	(37°C)	approx. 10^{-8}
Protein	MVS-1	3.2×10^{-10}
	MVS-1 (pretreat SBA)	4.1×10^{-11}
	MVS-1 (pretreat SBA + Col)	1.3×10^{-10}
	MVS-1 (pretreat SBA + CB)	1.3×10^{-10}

Abbreviations: FRAP - fluorescence recovery after photobleaching, Con A - concanavalin A, SBA - soybean agglutinin, WGA - wheat germ agglutinin, MVS-1 - monoclonal antibody against soybean plasma membrane protein, COL - colchicine, CB - cytochalasin B. (Based on Metcalf et al. 1983, 1986a,b).

Fluorescent lectins (Con A, soybean agglutinin - SBA, wheat germ agglutinin - WGA) were used to study the mobility of carbohydrates (Metcalf et al. 1983). The results with Con A confirmed the mobility of binding sites reported in the earlier study of soybean protoplasts using electron microscopy (Williamson et al. 1976). In addition, results with the three lectins indicated two distinct rates of movement of lectin-receptor complexes, slow (Con A - 7×10^{-11} cm^2/sec, SBA - 5×10^{-11} cm^2/sec) and fast (WGA - 3×10^{-10} cm^2/sec). These results provide the first suggestion of two domains within the plant plasma membrane. Again it is not clear exactly what types of molecules are binding the lectins and subsequently migrating in the plane of the plasma membrane.

Further evidence for two distinct domains within the plasma membrane was provided by studies using the FRAP analysis with fluorescent lipids and phospholipids as probes (Metcalf et al. 1986a). Experiments conducted at 18°C revealed molecules diffusing at two distinct rates (Table 1). When the temperature was raised to 37°C, a single diffusion rate was observed.

These results were interpreted to indicate the presence of two stable immiscible domains of fluid and gel-like lipids in the plasma membrane of soybean protoplasts.

Monoclonal antibodies against components of the plant plasma membrane have recently been produced (Villanueva et al. 1986; Norman et al. 1986) and are being used to explore specific proteins in the plasma membrane (Metcalf et al. 1986b). Fluorescent labeled MVS-1 antibody binds to a soybean plasma membrane protein of 400,000 molecular weight and was used to study the mobility of this protein by FRAP analysis. The results indicated that the protein moved at a fast rate (Table 1), comparable to that of the lectin WGA (see above). Pretreatment of protoplasts with SBA prior to MVS-1 slowed the movement of the protein dramatically. The binding of lectin to the membrane seemed to inhibit the motion of the protein. This inhibition could be reversed by the addition of either colchicine or cytochalasin B, suggesting that the membrane component to which the lectin is bound is attached to elements of the cytoskeleton. The interaction of membrane components with the cytoskeleton has important implications in terms of plasma membrane function. The direction of movement of enzyme complexes through the plasma membrane during cellulose deposition, for example, may be controlled by cortical microtubules which are thought to form tracks (Heath and Seagull 1982).

The results of the FRAP analysis on soybean protoplasts have confirmed and extended the results obtained earlier by electron microscopy. It is clear from this research that carbohydrates, lipids and proteins are capable of diffusing in the plane of the plant plasma membrane and that the two membrane domains influence the rate of movement.

2. ORGANELLES ASSOCIATED WITH THE PLASMA MEMBRANE

A number of organelles (e.g. microtubules, microfilaments, coated pits, coated vesicles) are known to be associated with the plasma membrane of plant cells. It is important to investigate the interaction of these organelles with the plasma membrane in order to gain a clearer understanding of plasma membrane function. Protoplasts provide a useful tool for such studies. Recent research has been concerned particularly with cytoskeletal elements and organelles involved in endocytosis.

2.1. Microtubules and Microfilaments

Most cytoskeletal studies to date have examined the relationship of the cortical microtubules to the process of cell wall formation (reviewed by Fowke et al. 1985). Immunofluorescence techniques using anti-tubulin provide a rapid method of examining the overall distribution of cortical microtubules with the light microscope and have been used to study microtubules in plant protoplasts (e.g. Lloyd et al. 1980, van der Valk et al. 1980, Simmonds 1986). Freshly isolated protoplasts are characterized by a random distribution of cortical microtubules. During cell wall regeneration and the subsequent development of cell shape the microtubules establish a transverse orientation parallel to the orientation of newly deposited cellulose microfibrils. The relationship between microtubules, cell wall formation and the establishment of cell shape has been clearly illustrated in studies of protoplasts of the alga *Mougeotia* (e.g. Galway and Hardham 1986). Regeneration of filaments from spherical protoplasts is a highly ordered process in which cortical microtubule arrays establish the axis of elongation and parallel the orientation of subsequently formed cellulose microfibrils.

Immunofluorescence has also been used to study the relationship of cortical microtubules to the process of cell division. Hahne and Hoffman (1984) provided evidence to suggest that the cortical network of microtubules is necessary for normal cell division. Protoplasts of hibiscus lacking such a network failed to divide, however, division could be induced by restoring the cortical microtubules using microtubule polymerizers.

Large fragments of plasma membrane can be prepared from protoplasts and the inner surface examined with either light or electron microscopes. Fine structure studies have provided basic information concerning the number, length, and distribution of microtubules on the plasma membrane (van der Valk et al. 1980).

Very little information is available concerning the association of actin microfilaments with the plasma membrane. Actin-like filaments have been recognized on plasma membrane fragments from protoplasts (Doohan and Palevitz 1980, van der Valk and Fowke 1981, Fowke et al. 1983) but have not been examined in detail.

2.2. Coated Membranes

Protoplasts have proved particularly useful for studies of coated membranes in plant cells. Ultrastructural studies of conventional thin sections as well as plasma membrane fragments from protoplasts have provided information concerning the structure and distribution of both coated pits and coated vesicles along the plasma membrane. Membrane fragments are particularly useful for determining the frequency of coats over large areas of the inner membrane surface (Doohan and Palevitz 1980, van der Valk and Fowke 1981). Suspension culture protoplasts which are capable of rapid regeneration of cell walls have a high frequency of coated pits and vesicles on the plasma membrane compared to leaf protoplasts which are differentiated and cannot regenerate walls rapidly (Fowke et al. 1983).

The organization of the clathrin coat of coated pits and vesicles can also be explored using stained protoplast plasma membrane fragments. Such studies reveal the lattice of hexagons and pentagons typical of animal coats (e.g. van der Valk and Fowke 1981). Large flat patches of coat have also been seen in such preparations. Recently both dry cleaving (Traas 1984) and rapid - freeze, deep - etching techniques (Hawes and Martin 1986) have also been used to study the structure of coated pits and coated vesicles associated with the plasma membrane of cultured cells.

Plant protoplasts have also proved valuable for isolating plant coated vesicles (Mersey et al. 1985). Fractions enriched in coated vesicles have been used to examine both the structure and chemistry of plant coated vesicles and results suggest that coated vesicles may have a role in the sorting and transport of proteins in plants (Griffing et al. 1986).

3. UPTAKE BY PLANT PROTOPLASTS

Plant protoplasts provide an excellent experimental system to study uptake of materials across the plasma membrane. This phenomenon is difficult to study with whole cells because of the presence of a cell wall. Protoplasts can be exposed to a wide range of particles including macromolecules, polystyrene spheres of different sizes, organelles, and other protoplasts. The interaction of each with the plasma membrane can be examined and modes of uptake determined. Uptake can occur by a number of different mechanisms depending upon the size of the objects and whether or not they are membrane bounded.

3.1. Coated Vesicles

Coated vesicles are tiny cytoplasmic vesicles (approx. 100nm in diameter) consisting of a membrane vesicle bounded by a patterned coat consisting primarily of the protein clathrin. In surface view they resemble a soccer ball. They are formed from clathrin coated pits on the plasma membrane by an invagination process. In animal cells coated vesicles have a number of important functions including receptor mediated endocytosis by which specific proteins are taken into animal cells and delivered to target organelles (e.g. Fine and Ockelford 1984).

The study of coated vesicle function in plant cells has been difficult. Attempts to introduce probes into plant cells have been hampered by the presence of the cell wall. Recent studies with plant protoplasts have overcome this difficulty and uptake studies have identified a pathway involving coated vesicles (Tanchak et al. 1984, Joachim and Robinson 1984, Hillmer et al. 1986, Tanchak and Fowke 1987). Ultrastructural work with soybean protoplasts, for example, shows binding of cationized ferritin to the plasma membrane, uptake via coated pits to coated vesicles and sequential delivery of the ferritin to the partially coated reticulum and dictyosomes, multivesicular bodies and finally to the major lysosomal component of plant cells, the vacuole (Tanchak et al. 1984, Tanchak and Fowke 1987). Uptake via coated pits and vesicles is apparently also operational in intact plant cells (Hubner et al. 1985). The significance of the internalization of externally applied markers via coated membranes is unknown. Vesicular transport of such markers to dictyosomes may replace plasma membrane which is moved to the cell surface during secretion of cell wall matrix materials. Alternatively, this pathway may play a role in host-pathogen interactions perhaps by transporting elicitors to target sites in plant cells or toxins to the vacuole for degradation. Further research is necessary to elucidate the role of coated membranes in plant cells.

3.2. Smooth Vesicles

Particles can also be transported into plant protoplasts by smooth vesicles. In general these vesicles are considerably larger than coated vesicles. Under certain experimental conditions such as hypotonic shock, numerous smooth vesicles are invaginated from the plasma membrane into the cytoplasm (Gordon-Kamm and Steponkus 1984). This mechanism has been used to introduce macromolecules into plant protoplasts (Saleem and Cutler 1986). It is conceivable that molecules small enough to cross the cell wall might enter intact plant cells by this mechanism.

Larger particles have also been taken into plant protoplasts by smooth vesicles. Suzuki et al. (1977) described the energy dependent uptake of polystyrene spheres by tobacco leaf protoplasts. More recently bacteria (e.g. *Agrobacterium*, *E. coli*) have been introduced into plant protoplasts by a similar mechanism (Hasezawa et al. 1983, Matsui et al. 1983). This type of uptake is not likely to operate in cells because of the cell wall.

3.3. Fusion

Larger membrane bounded structures such as organelles and small protoplasts apparently do not enter protoplasts via smooth vesicles. Cell organelles such as chloroplasts and nuclei are taken into protoplasts (e.g. Bonnett 1976, Saxena et al. 1986), however, the mechanism of entry is not known (see discussion in Fowke and Gamborg 1980). Studies of the interaction of small algal protoplasts (e.g. *Chlamydomonas*, *Stigeoclonium*) with carrot protoplasts clearly indicate that membrane fusion occurs (Fowke et al. 1979, 1981). In both cases fusion of adjacent plasma membranes resulted in the release of algal organelles into the carrot cytoplasm.

Recent experiments have demonstrated that particles can be taken into plant protoplasts by coated vesicles, smooth vesicles and by fusion. The evidence to data suggests that uptake via smooth vesicles could operate in intact cells under conditions of stress. Endocytosis via coated vesicles occurs in both protoplasts and cells and may be important in membrane recycling and/or host-pathogen interactions in plants.

REFERENCES

1. Bonnett HT: Planta 131, 229-233, 1976.
2. Doohan ME, Palevitz BA: Planta 149, 389-401, 1980.
3. Fine RE, Ockleford CD: Int. Rev. Cytol. 91, 1-43, 1984.
4. Fowke LC, Constabel F (eds): Plant Protoplasts, Boca Raton, CRC Press, 1985.
5. Fowke LC, Gamborg OL: Int. Rev. Cytol. 68, 9-51, 1980.
6. Fowke LC, Gresshoff PM, Marchant HJ: Planta 144, 341-347, 1979.
7. Fowke LC, Griffing LR, Mersey BG, Tanchak MA: In: Plant Protoplasts, eds. LC Fowke, F Constabel, Boca Raton, CRC Press, 1985.
8. Fowke LC, Marchant HJ, Gresshoff PM: Can. J. Bot. 59, 1021-1025, 1981.
9. Fowke LC, Rennie PJ, Constabel F: Plant Cell Reports 2, 292-295, 1983.
10. Galway ME, Hardham AR: Protoplasma 135, 130-143, 1986.
11. Gordon-Kamm WJ, Steponkus PL: Protoplasma 123, 83-94, 1984.
12. Griffing LR, Mersey BG, Fowke LC: Planta 167, 175-182, 1986.
13. Hahne G, Hoffmann F: Proc. Natl. Acad. Sci. 81, 5449-5453, 1984.
14. Hasezawa S, Matsui C, Nagata T, Syono K: Can. J. Bot. 61, 1052-1057, 1983.
15. Hawes CR, Martin B: Cell Biol. Int. Rep. 10, 985-992, 1986.
16. Heath IB, Seagull RW: In: The Cytoskeleton in Plant Growth and Development, ed. CW Lloyd, New York, Academic Press, 1982.
17. Hillmer S, Depta H, Robinson DG: Eur. J. Cell Biol. 41, 142-149, 1986.
18. Hubner R, Depta H, Robinson DG: Protoplasma 129, 214-222, 1985.
19. Joachim SA, Robinson DG: Eur. J. Cell Biol. 34, 212-216, 1984.
20. Lloyd CW, Slabas AR, Powell AJ, Lowe SB: Planta 147, 500-506, 1980.
21. Matsui C, Hasezawa S, Tanaka N, Syono K: Plant Cell Reports 2, 30-32, 1983.
22. Mersey BG, Griffing LR, Rennie PJ, Fowke LC: Planta 163, 317-327, 1985.
23. Metcalf TN, Villanueva MA, Schindler M, Wang JL: J. Cell Biol. 102, 1350-1357, 1986b.
24. Metcalf TN, Wang JL, Schindler M: Proc. Natl. Acad. Sci. 83, 95-99, 1986a.
25. Metcalf TN, Wang JL, Schubert KR, Schindler M: Biochemistry 22, 3969-3975, 1983.
26. Morré DJ, Kartenbeck J, Franke WW: Biochimica Biophysica Acta 559, 71-152, 1979.
27. Norman PM, Wingate VPM, Fitter MS, Lamb CJ: Planta 167, 452-459, 1986.
28. Saleem M, Cutler AJ: J. Plant Physiol. 124, 11-21, 1986.
29. Saxena PK, Mii M, Crosby WL, Fowke LC, King J: Planta 168, 29-35, 1986.
30. Simmonds DH: Planta 167, 469-472, 1986.
31. Suzuki M, Takebe I, Kajita S, Honda Y, Matsui C: Exptl. Cell Res. 105, 127-135, 1977.
32. Tanchak MA, Fowke LC: Protoplasma 138, 173-182, 1987.
33. Tanchak MA, Griffing LR, Mersey BG, Fowke LC: Planta 162, 481-486, 1984.

34. Traas JA: Protoplasma 119, 212-218, 1984.
35. van der Valk P, Fowke LC: Can. J. Bot. 59, 1307-1313, 1981.
36. van der Valk P, Rennie PJ, Connolly JA, Fowke LC: Protoplasma 105, 27-43, 1980.
37. Vasil IK, Vasil V: Int. Rev. Cytol. 11B, 1-19, 1980.
38. Villanueva MA, Metcalf TN, Wang JL: Planta 168, 503-511, 1986.
39. Williamson FA, Fowke LC, Constabel FC, Gamborg OL: Protoplasma 89, 305-316, 1976.

WHY DO MESOPHYLL PROTOPLASTS DEDIFFERENTIATE?

Y. MEYER, Y. CHARTIER and J. GROSSET.

Laboratoire de Physiologie Végétale, Université de Perpignan, Avenue de Villeneuve - 66025 PERPIGNAN-Cédex.

We have analysed protein synthesis and mRNA accumulation in Tobacco leaves and in freshly isolated protoplasts, during the first hours of in vitro culture of protoplasts and finally in protoplasts derived callus and cell suspension, in order to establish the timing of the shift in gene expression from mesophyll cell to callus cell.

From the beginning of culture Tobacco leaf protoplasts synthesize proteins which are different from leaf proteins. They continue to be synthesized and accumulate in cells cultivated in vitro in liquid medium or on agar as calli.

Analysis of mRNA by in vitro translation in rabbit reticulocyte followed by 2-D electrophoresis of the translation products leads to similar conclusions: during maceration of leaves poly(A)+ RNA population seems to disappear. Freshly isolated protoplasts are devoid of detectable mRNA activity. After washing, a new set of mRNA appears in cultivated protoplasts. They become detectable after half an hour and are callus specific. No leaf-specific mRNA can be detected in protoplasts. Changes in the culture conditions or medium (nutrients and hormones) modify only some mRNA species.

We have purified 3 protein groups which are synthesized by tobacco mesophyll protoplasts and prepared antibodies. These basic proteins are major callus proteins. They are synthesized at a high level in cell suspensions and protoplasts. Their synthesis is reduced by the presence of auxin in the culture medium and they accumulate in the protoplast vacuole (MEYER et al., 1987).

- Antibodies directed again protein a' and a1 (MW 33000 daltons) recognise these proteins but not spot a which as of similar molecular weight and which undergoes identical regulation and is also localised in the vacuole.

- Antibodies directed against polypeptide b' (MW 26000 daltons) recognize b' and related spot b.

- Antibodies directed against spot c (MW 16000 daltons)recognize c, as well as other spots which are also under auxinic control.

Using these antibodies we have shown that proteins a1, a', b and c are absent from aerial organs of the plant (leaf stalk, bark and pith) but accumulate in roots.

Some of these proteins are probably identical to b 1-3 glucanases and chitinases described by SHINSHI et al., 1987 and which have been shown to be implicated in different stress reactions (LEGRAND et al., 1987;KAUFFMANN et al., 1987 and MEMELINK et al., 1987).

Thus it appears that dedifferentiation, considered as a change in gene expression, takes place during maceration. From a physiological point of view ,we may envisage that dedifferentiation in mesophyll protoplasts is a cicatrisation reaction in response to the wounding occuring during isolation conditions.

It is tempting to extend these results and to postulate that all in vitro cultivated calli are initially wound calli. In fact calli appear at the level of excision in the primary explant and established calli accumulate large amounts of pathogenesis related proteins (wound induced proteins).

The chemical nature of the dedifferentiation factor(s) released by wounding or enzymatic maceration is (are) unknown. Its possible relationships with the proteinase inhibition inducing factor (PIIF) is discussed. In addition, we suggest that the inability to initiate cell division in cereal mesophyll protoplast is related to the low cicatrisation reaction of cereals. Progress in the understanding of dedifferentiation in tobacco mesophyll protoplasts may lead to a better control of the wound reaction and, possibly, to the obtention of cereal mesophyll protoplast division.

REFERENCES

Kauffmann S., Legrand M., Goeffroy P. and Fritig B. (1987) The EMBO Journal , 6, 3209-3212.
Legrand M., Kauffmann S., Geoffroy P. and Fritig B. (1987) Proc. Natl. Acad. Sci.,84, 6750-6754.
Memelink I., Hoge J.H.C. and Schilperoort R.A. (1987) The EMBO J. 6, 3579-3583.
Meyer Y. Chartier Y. and Alibert G. (1987) Plant Physiol. 83, 713-718.
Shinshi H., Mohnen D. and Meins F. Jr (1987). Proc. Natl. Acad. Sci. 84, 89-93.

RELEASE OF PHYTOTOXIC FACTORS FROM PLANT CELL WALLS DURING PROTOPLAST ISOLATION

G. HAHNE*, H. LÖRZ
MPI für Züchtungsforschung, Egelspfad, 5 Köln 30, W. Germany and
*IBMP du CNRS, 12, Rue du Général Zimmer, 67000 Strasbourg, France

INTRODUCTION

Isolation of protoplasts from plant cells involves a number of steps, the impact of which on the physiological response of the resulting protoplasts is not well understood. In particular, the usual treatment of cells with hydrolytic enzymes of fungal origin has received little attention, except for empirical selection of appropriate commercial preparations and, in some cases, their purification via gel filtration.

Studies of plant defense mechanisms suggest the existence of "endogenous elicitors" of phytoalexin synthesis, liberated from plant cell walls by the action of enzymes of fungal origin (1). In addition, these endogenous elicitors have been shown to have phytotoxic effects (2). Are similar factors released by the protoplast isolation procedure, and do they have adverse effects on protoplasts? To study this question, separately prepared cell wall digestion supernatants (CDS) were added to protoplast cultures and viability determined after various culture periods.

MATERIALS AND METHODS

Isolation of protoplasts

Protoplasts were isolated from suspension cultures of *Triticum monococcum* and *Nicotiana tabacum* cv. W38 under standard conditions, using the following enzyme solutions or their individual constituents, dissolved in C_8/IV medium (3): (I) 1% Onozuka Cellulase RS, 0.5% Onozuka Macerozyme R10, 0.05% Pectolyase Y23, pH 5.6; (II) 3% Cellulysin, 1% Rohament P, 0.5% Hemicellulase (Sigma), pH 5.6.

Cell wall digestion supernatant (CDS)

For the production of CDS, cells were incubated under identical conditions as for protoplast isolation and the supernatant collected after 2h, or 8h, when no protoplasts were formed. The supernatants were autoclaved to remove enzymatic activities, the precipitate removed by centrifugation, and CDS stored at -20°C until they were added to protoplast cultures at 50% concentration. Protoplast viability was evaluated with fluorescein diacetate.

RESULTS AND DISCUSSION

T. monococcum suspension culture cells yield viable protoplasts upon digestion under the conditions used. However, their partial digestion (10 or 30 min.) invariably resulted in death of the treated cells. Partial digestion of *N. tabacum* cells did not result in this puzzling phenomenon.

CDS prepared from *T. monococcum* cell walls with both the standard enzyme mixture (I) and an alternative mixture containing different preparations but similar activities (II) resulted in drastically reduced viability of the treated protoplasts (*T. monococcum* as well as *N. tabacum*, Fig. 1 and 2). Control experiments with inactivated enzyme preparations added instead of CDS showed that the inherent toxicity of these preparations was a minor factor. When CDS prepared from tobacco cell walls was used, it showed a similar toxicity on *T. monococcum* protoplasts, while the viability of tobacco protoplasts was not

significantly affected. When cells were used in these experiments instead of protoplasts, the same results were obtained. Thus, the presence of the cell wall is of no importance for the observed toxicity of CDS. These observations correspond well to the partial digestion experiments, where *T. monococcum* cells are killed, while tobacco cells are unaffected by their own cell wall hydrolysis products.
Desalting of enzyme preparation prior to production of CDS had no detectable effect on CDS toxicity, while purification via Sephadex G-75 removed completely the intrinsic toxicitiy of the enzyme preparation. However, CDS prepared with this purified mixture were considerably more toxic (0% viable cells after 24h). When the enzyme preparations were employed separately for CDS production, none was found capable of protoplast production, all had comparatively low intrinsic toxicity, but the toxicity of their CDS increased from cellulase, to pectinase, to hemicellulase. The active principle contained in the CDS is not released by simple autoclaving of the undigested cells, is heat- and acid stable, alcali-labile, and in ultrafiltration experiments is retained in the fraction >10 kD.
These data are consistent with the idea that during enzymatic protoplast isolation phytotoxic substances are produced. A similarity to the oligosaccharides which have been described as endogenous elicitors (4) is not excluded on the basis of their properties. Apparently, these substances may be toxic for members of different plant families, but resistance of cells to their own CDS is also possible.
In the most frequently used protoplast isolation schemes, protoplasts are in contact with their CDS for only a few hours. A kinetic study using the most active CDS showed a decrease in viability to ca. 50% already after 1h. Thus, the toxic principle acts sufficiently fast that it could provoke adverse effects on protoplasts which are later perceived by the investigator as "low quality". The release of phytotoxic substances during protoplast preparation offers one possible explanation of the "overdigestion syndrome".

REFERENCES

1. Davis FR, Lyon GD, Darvill AG, Albersheim P: Plant Physiol. 74, 52-60 (1984)
2. Yamazaki N, Fry SC, Darvill AG, Albersheim P: Plant Physiol. 72, 864-869 (1983)
3. Dudits D, Hadlaczky G, Levi E, Fejer O, Haydu Z, Lazar G: Theor. Appl. Genet. 51, 127-132 (1977)
4. Davis KR, Darvill AG, Albersheim P: Plant Mol. Biol. 6, 23-32 (1986)

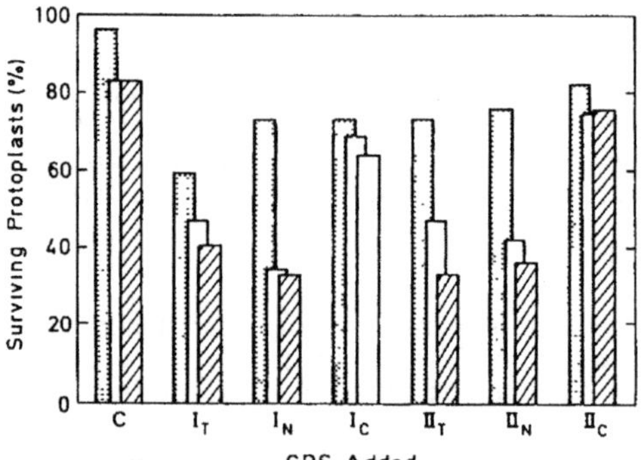

Fig.1

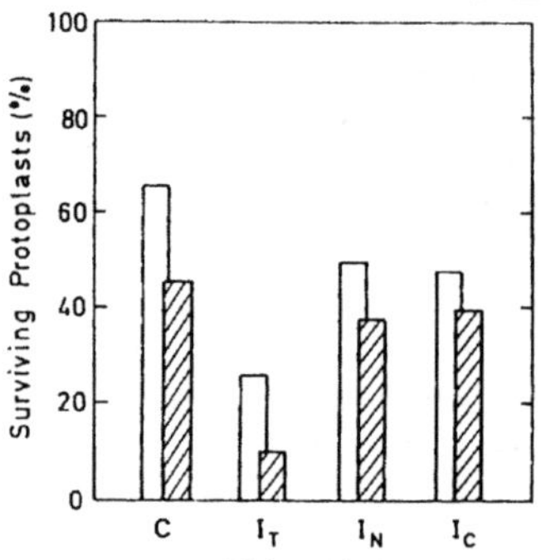

Fig.2

T. monococcum (Fig.1) and *N. tabacum* (Fig.2) protoplasts were treated with CDS obtained with enzyme mixture I and II, respectively, from *Triticum* (I_T, II_T) or tobacco cell walls (I_N, II_N). The suffix C designates control experiments with inactivated enzyme solution only.

ROLE OF PEROXIDASE IN PROTOPLAST DEVELOPMENT: THE ACTIVITY AND MOLECULAR FORMS OF PEROXIDASE IN MITOGENIC AND NON-MITOGENIC PROTOPLASTS

M. SALEEM, A.J. CUTLER, M. COFFEY and M. LOEWEN.
Plant Biotechnology Institute, National Research Council
110 Gymnasium Road, Saskatoon, Canada. S7N 0W9

INTRODUCTION

Genetic manipulation of agronomically important monocot plants, such as wheat, corn and barley remain difficult. This is primarily due to the recalcitrance of cereal protoplasts. The antithetical behaviour of cereal protoplasts in culture may be related to differences in response to protoplasting stress. Our earlier studies indicated that freshly isolated corn and wheat leaf protoplasts showed evidence of lipid peroxidation and elevated lipoxygenase activity. No such increases were detected in freshly isolated bromegrass protoplasts (1-2). There are many unresolved questions concerning the inability of cereal leaf protoplasts to divide. We have focussed our attention on peroxidase, an enzyme which has been repeatedly implicated in plant developmental processes (3-4).

MATERIALS AND METHODS

Protoplasts from corn, wheat and barley leaf tissues and cell suspension cultures of Soybean (SB-1), bromegrass-970 and Black Mexican Sweet Corn were isolated and cultured as described (2). Tissues were homogenized in 50 mM phosphate buffer (pH 7.1). High speed supernatents were assayed for peroxidase activity in the presence of 1% guaiacol. Isoperoxidases were separated by polyacrylamide gel electrophoresis and stained for activity with O-dianisidine.

RESULTS AND DISCUSSION

Comparison of peroxidase activity in mitogenic (bromegrass, Black Mexican Sweet Corn and Soybean) protoplasts showed a significantly higher activity relative to the source material (Fig. 1). There was a 10 fold increase in peroxidase activity in cultured bromegrass protoplasts, and an 18 and 3.5 fold increase in Black Mexican and soybean respectively. The specific peroxidase activity further increased during culture. In contrast, non-mitogenic (corn, barley and wheat leaf tissues) protoplasts

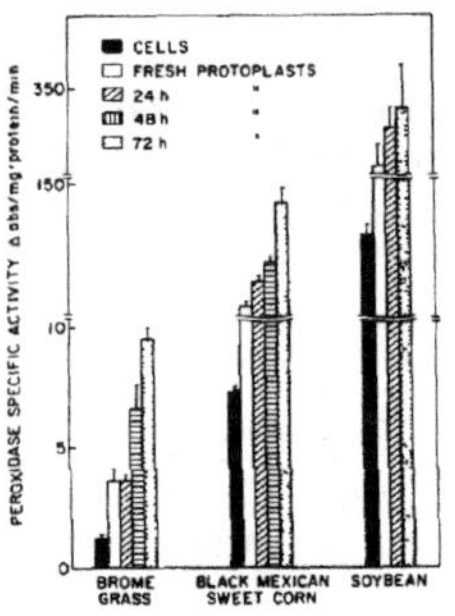

FIG. 1

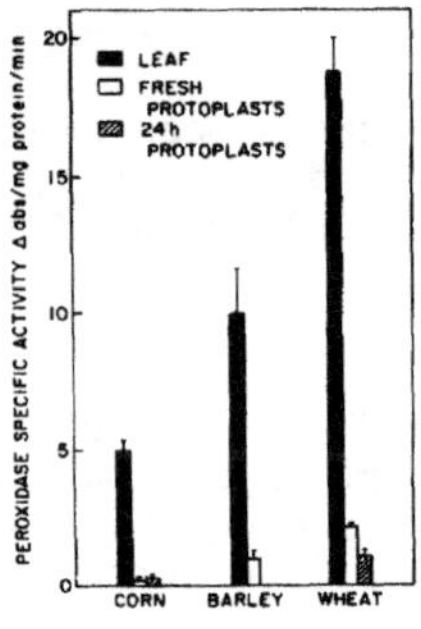

FIG. 2

FIG. 1. Specific peroxidase activity in extracts of cells and protoplasts.

FIG. 2. Specific peroxidase activity in extracts of leaves and protoplasts.

showed a sharp decrease in peroxidase activity (Fig. 2). The decrease in peroxidase activity was 5 fold for corn and 9 and 10 fold in wheat and barley respectively. Moreover the activity remained low during culture. Fig. 3 depicts the electrophoretic pattern of anodic isoperoxidases from cereal leaf tissues. Freshly isolated cereal leaf protoplasts showed a decrease in all isoperoxidases characteristic of leaf tissues. It was apparent that cultured cereal leaf protoplasts were unable to regain the original isoperoxidases characteristic of leaf tissue (Fig. 4). In contrast, mitogenic protoplasts not only retained all the isoperoxidases characteristic of the source material, but also additional isoperoxidases appeared during culture (Figs. 5-6). Since peroxidase is known to be involved in plant development processes (3-4), it is possible that loss of peroxidase activity in cereal leaf protoplasts may be a major factor in their inability to divide in culture.

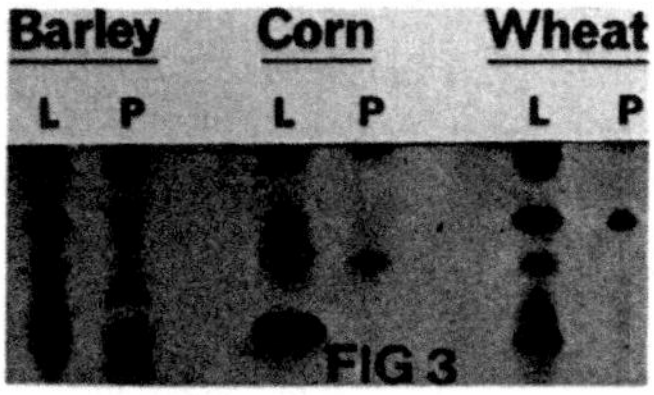

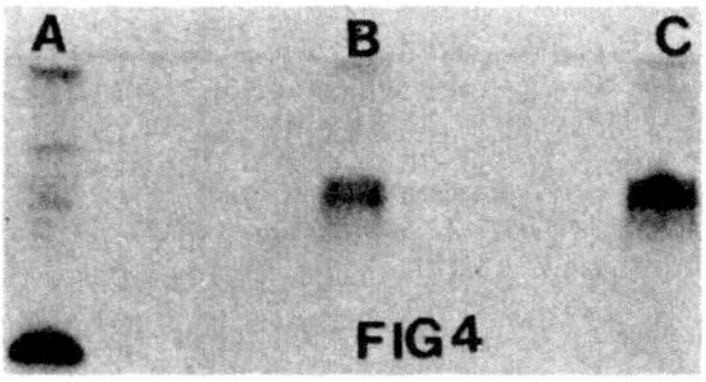

Fig. 3 Isoperoxidases from leaf (L) and freshly isolated protoplasts (P)
Fig. 4 Isoperoxidases from Corn leaf (A), fresh (B) and 8 days cultured (C) Protoplasts

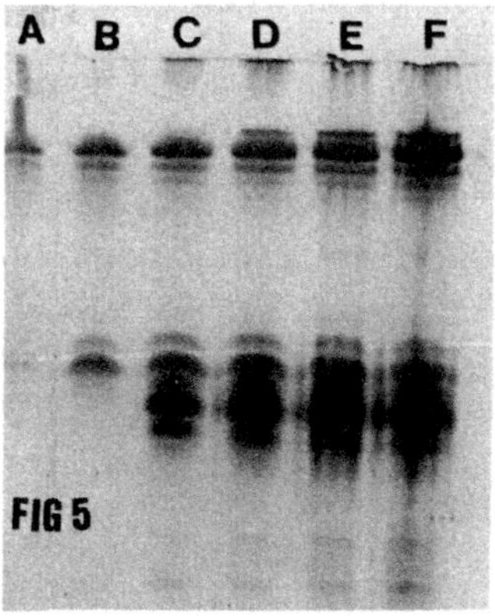

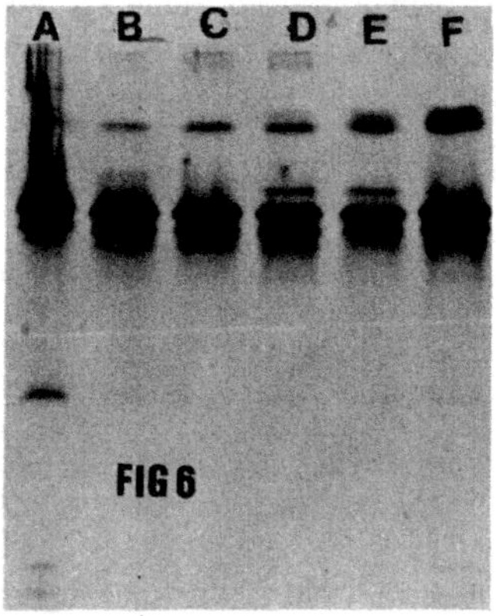

Figs. 5-6 Isoperoxidases from Black Mexican Sweet Corn (Fig. 5) and Soybean (Fig. 6) suspension cultured cells (A), fresh protoplasts (B) and 24h (C), 48h (D), 72h (E) and 80h (F) cultured protoplasts

REFERENCES

1. Cutler AJ and Saleem M: Plant Physiology 83 (1987). 324A.
2. Saleem M and Cutler AJ: J. Plant Physiology 128 (1987) 479-484.
3. Kay LE and Basile DV: Plant Physiology 34 (1987) 99-105.
4. Goldberg R, Liberman M, Mathieu C, Pierron M and Catesson AM: J. Exp. Bot. 38 (1987) 1378-1390.
This is NRCC Publication 28383

MEASUREMENT OF CYTOPLASMIC CA^{2+} AND H^+ IN BARLEY ALEURONE PROTOPLASTS

DOUGLAS S. BUSH, ASOK K. BISWAS AND RUSSELL L. JONES
Department of Botany, University of California, Berkeley, CA 94720, USA

1. INTRODUCTION

Calcium has been implicated as a regulator of a wide range of developmental responses in plants (1, 2). It is thought to be involved in phenomena as diverse as mitosis, root gravitropism, pollen tube growth, protein secretion, and photosynthesis. These observations have led to the view that Ca^{2+} plays a central role as a regulator of cell function in plant cells in a manner analogous to that which it plays in animal cells. It is also widely held that Ca^{2+} is a cytotoxic element and that its cytoplasmic concentration must be maintained at submicromolar levels. In animal cells cytoplasmic Ca^{2+} is generally in the range of 10-100 nM (3); by contrast, little is known about cytoplasmic Ca^{2+} concentrations in higher plants.

This paucity of information on cytoplasmic Ca^{2+} in plants is due largely to technical difficulties inherent in the methods used to determine Ca^{2+} levels. In animal cells Ca^{2+}-responsive proteins (3) or dyes (3) are introduced into cells using methods that have so far not been applicable to higher plant cells. For example, the protein aequorin can be introduced into animal cells by microinjection, but such an approach is difficult in walled and highly vacuolated angiosperm cells. In the case of the Ca^{2+}-sensitive dyes developed by Tsien and his colleagues (4, 5) cells that accumulate these dyes must contain esterases to facilitate their accumulation in the cytoplasm. Our experiments with barley aleurone cells (7) show that esters of fura 2, for example, do not accumulate in the cytoplasm. Rather, this dye accumulates in the vacuole, rendering it useless as an indicator of cytoplasmic Ca^{2+} (6).

The measurement of light emitted by Ca^{2+}-sensitive probes is also technically difficult in walled plant cells. Plant cell walls fluoresce brightly when irradiated with short wavelength light because of the presence of phenolic compounds such as ferulic acid (7). The cell wall would also be expected to quench light emitted by probes such as aequorin.

In this paper we report on the development of a non-intrusive method that permits loading of the Ca^{2+}-sensitive dye indo-1 into the cytoplasm of protoplasts isolated from various higher plant tissues. Using this method we have measured steady state Ca^{2+} and H^+ levels in protoplasts incubated under various conditions of incubation. Our results show that the method of dye loading may be applicable to a wide variety of higher plant protoplasts.

2. MATERIALS AND METHODS

2.1. Protoplast preparation

Barley aleurone protoplasts were prepared as previously described (8) and incubated for 32 h in Gamborg's B-5 medium modified by omitting IAA and adding 300 mM mannitol, 5 µM gibberellic acid (GA_3), and 20 mM $CaCl_2$.

2.2. Dye loading

The pentapotassium salt of indo-1 (indo-1, Molecular Probes, Junction City, OR, USA) was loaded after washing protoplasts once in an isotonic wash solution (WS) containing 60 mM sucrose, 700 mM mannitol, 100 mM glucose, 60 mM KCl and either 10 mM dimethyl glutaric acid (pH 4-5.6), 10 mM MES (pH 5.6-6.6), or 10 mM HEPES (pH 6.6-7.6). Protoplasts were loaded with dye in WS containing indo-1 at a final concentration of 1-60 µM and $CaCl_2$ at 0-10 mM for 2 h at 25°C. Loaded protoplasts were resuspended twice in WS.

The pH sensitive dye 2',7'-bis (carboxymethyl)-5 (6)-carboxyfluorescein (BCECF, Molecular Probes) was loaded into protoplasts over 2 h at a concentration of 1 µM in WS buffered at pH 4.5.

2.3. Fluorescence measurements

Fluorescence of indo-1 in individual protoplasts was measured using a modular fluorimeter (Spex Fluorolog 2, model F2C) designed and described by Tsien (5). Ca^{2+} concentration was determined using a fluorescence ratio technique described in detail elsewhere (6).

Fluorescence of BCECF was measured at 500 nm after excitation at 439 nm and 490 nm. The ratio of the fluorescence after excitation at these two wavelengths is proportional to pH (9).

3. RESULTS AND DISCUSSION

The Ca^{2+}-sensitive dyes indo-1 and fura-1 were synthesized by Tsien (3) as acetoxymethyl esters (indo-1/am and fura-1/am) to facilitate their loading into the cytoplasm of animal cells. Indo-1/am and fura-1/am are Ca^{2+}-insensitive, but when they cross the plasma membrane of animal cells, cytoplasmic esterases convert them into the Ca^{2+}-sensitive, impermeant anions. Both indo-1/am and fura-1/am have proven unsatisfactory for Ca^{2+} measurement in plant cells. Indo-1/am is accumulated in the cytoplasm but aleurone cells do not have esterases to convert it to the Ca^{2+}-sensitive indo-1 (6). On the other hand, fura-1/am is taken up by aleurone cells but accumulates in the vacuole rather than the cytoplasm.

We have adopted an alternative strategy for the loading of Ca^{2+}-sensitive dyes such as indo-1 into plant cells. This strategy takes advantage of the large pH gradient that exists across the plant cell plasma membrane to load the Ca^{2+}-sensitive form of the dye. Most plant cells acidify their exter-

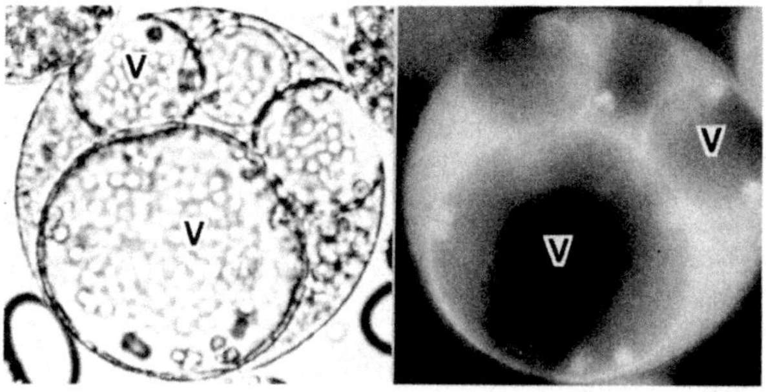

FIGURE 1. Bright-field (left) and fluorescence (right) photomicrographs of barley aleurone protoplasts loaded with the Ca^{2+} dye indo-1. Note that fluorescence is excluded from the vacuoles (V).

nal environment, and barley aleurone protoplasts maintain their incubation medium at a pH between 4 and 4.5. At this pH, aleurone protoplasts accumulate indo-1 in the cytoplasm and not in the vacuole (Fig. 1). Indo-1 loaded in this fashion is Ca^{2+}-sensitive and is not metabolized or transported to other compartments during several hours of incubation. There is almost no indo-1 accumulation in protoplasts when they are incubated at pH 7.0.

The pH-dependent loading of indo-1 into barley aleurone protoplasts can be explained by the model advanced to account for the accumulation of indole-acetic acid and other weak acids by plant cells (10). At low external pH, the carboxyl groups of these weak acids are protonated, neutralizing the charge on the carboxyl group and making the molecule membrane permeable. At a neutral cytoplasmic pH, however, the weak acid dissociates to give a charged, impermeable anion that cannot permeate either plasma membrane or tonoplast (Fig. 2).

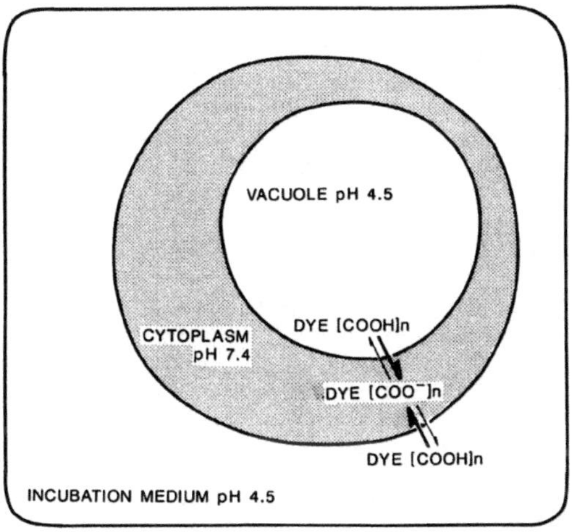

FIGURE 2. Diagram illustrating the principles of indo-1 accumulation in the cytoplasm of barley aleurone protoplasts at incubation medium pH 4.5. Indo-1 accumulates in the cytoplasm (pH 7.4) because it exists as the charged anion.

We have confirmed that at low external pH (pH 4-4.5) the cytoplasmic pH of aleurone cells is above neutrality. Using BCECF to measure cytoplasmic pH, we have shown that over a broad range of external pH (pH 4-7), the cytoplasmic pH of the aleurone cell is maintained at about 7.4. We have used the strategy of indo-1 loading at pH 4.5 to explore Ca^{2+} levels in barley aleurone protoplasts under various conditions of incubation. While cytoplasmic Ca^{2+} levels of cells incubated in a medium with GA_3 but without Ca^{2+} are maintained around 250 nm, the addition of 10 mM Ca^{2+} to the incubation medium elevates cytoplasmic Ca^{2+} levels to about 350 nm.

Protoplasts isolated from tomato leaf mesophyll (D. S. Bush, N. Sinha and R. L. Jones, unpublished observations) and barley root cortex (V. Polito and J. Lynch, personal communication) also accumulate indo-1 in the cytoplasm in a pH-dependent manner. Based on these and other observations we believe that this strategy of dye loading into plant cells will prove to be applicable to a wide range of plant protoplasts and to a range of weakly acidic dyes.

ACKNOWLEDGEMENTS

This work is supported by grants from the National Science Foundation and the Department of Energy. The assistance of Eleanor Crump in preparing this manuscript is gratefully acknowledged.

REFERENCES

1. Hepler PK, Wayne RO: Calcium and Plant Development. Ann. Rev. Plant Physiol. 36, 397-439, 1985.

2. Poovaiah BW, Reddy ASN: Calcium Messenger Systems in Plants. CRC Crit. Rev. Plant Sci. 6, 47-103, 1987.

3. Tsien RY: Intracellular Measurements of Ion Activities. Ann. Rev. Biophys. Bioeng. 12, 91-116, 1983.

4. Grynkiewicz G, Poenie M, Tsien RY: A New Generation of Ca^{2+} Indicators with Improved Fluorescent Properties. J. Biol. Chem. 260, 3440-3450, 1985.

5. Tsien RY, Rink TJ, Peonie M: Measurement of Cytosolic Free Ca^{2+} in Individual Small Cells Using Fluorescence Microscopy with Dual Excitation Wavelengths. Cell Calcium 6, 147-157, 1985.

6. Bush DS, Jones RL: Measurement of Cytoplasmic Calcium in Aleurone Protoplasts Using Indo-1 and Fura-2. Cell Calcium, in press.

7. Fulcher RG, O'Brien TP, Lee JW: Studies on the Aleurone Layer. 1. Conventional and Fluorescent Microscopy of the Cell Wall with Emphasis on Phenol-carbohydrate Complexes in Wheat. Aust. J. Biol. Sci. 25, 23-34, 1972.

8. Bush DS, Cornejo MJ, Huang CN, Jones RL: Ca^{2+}-stimulated Secretion During Development of Barley Aleurone Protoplasts. Plant Physiol. 82, 566-574, 1986.

9. Paradiso AM, Tsien RY, Demarest JR, Machen TE: Na/H and Cl/HCO_3 Exchange in Rabbit Oxyntic Cells Using Fluorescence Microscopy. Am. J. Physiol.

10. Goldsmith MHM: The Polar Transport of Auxin. Ann. Rev. Plant Physiol. 28, 439-478, 1977.

THE EFFECT OF SECRETION ON LIPIDS AND THEIR DISTRIBUTION IN BARLEY ALEURONE PROTOPLASTS

M.J. CORNEJO[1], M.E. ARNALTE[1], D.S. BUSH[2] and R.L. JONES[2]

[1]Department of Plant Biology,Faculty of Pharmacy,University of Valencia. Avda. Blasco Ibañez 13, 46010 Valencia, Spain.

[2]Department of Botany,University of California,Berkeley,CA 94720, USA.

INTRODUCTION

Barley aleurone protoplasts, when exposed to Ga_3 and Ca^{2+} synthesize and secrete the hydrolytic enzyme α-amylase (Bush et al.,1986). At the same time, protoplasts undergo important structural changes which reflect the elaboration of the membrane system involved in protein synthesis and secretion. The mobilization of lipid reserves may be a preliminar step which would provide energy as well as several components for the synthesis of membrane lipids.

We have characterized the neutral and phospholipid composition of secreting and nonsecreting aleurone protoplasts and have compared it with that of barley mesophyll protoplasts. Lipid mobilization has also been investigated in lipid body fractions. On the other hand, lipid distribution along the incubation period of protoplasts has been studied with two fluorescent lipid dyes: rhodamine b and 3,4 benzpyrene.

MATERIAL AND METHODS

Aleurone protoplasts were isolated from barley (Hordeum vulgare L. cv. Himalaya, 1985 harvest) caryopses as previously described (Bush et al., 1986) and purified on Nycodenz gradient. Mesophyll protoplasts were isolated from primary leaves of barley seedlings, using the enzimatic mixture proposed by Eriksen and Jensen (1984). Lipid body fractions were obtained from homogenized layers by two succesive centrifugations.

Lipids were extracted from lipid bodies and protoplasts by a modified Bligh and Dyer procedure described by Kates (1986). Lipid extractions were also performed with chloroform:methanol (2:1) from lyophilized protoplasts. Both extraction procedures gave similar results. Neutral and phosphoglyceride lipids were separated as described by Morrison et al. (1980), but TLC was performed on silicagel HPTLC plates with a concentrating zone. Lipids were identified by cochromatography with lipid standards.

RESULTS AND DISCUSSION

Neutral lipid determinations in aleurone and mesophyll protoplasts of barley showed the importance of the triglyceride reserve contained in aleurone protoplasts. Recently isolated aleurone protoplasts had a much higher amount of triglycerides in comparison with the remainder neutral lipids. After 48 h of incubation with GA_3 and Ca^{2+} there was a decrease in the triglyceride content while untreated protoplasts showed not noticeable changes in the pattern of neutral lipids. With respect to the phospholipid content, secreting protoplasts (GA_3-plus-Ca^{2+}-treated) showed an increase in phosphatidylethanolamine (PE) and phosphatidylcholine (PC) in comparison with freshly isolated protoplasts. However, phospholipid composition in treated and untreated aleurone protoplasts was relatively similar.

In order to investigate the relationship between GA_3-induced secretion and the mobilization of trigycerides we have studied lipid composition in lipid body fractions obtained from aleurone layers. After 24h of GA_3 and Ca^{2+} treatment there was a drastic decrease in triglycerides whereas the levels of steryl esters (SE), free fatty acids (FAA) and monoacylglycerols (MG) increased. Lipid bodies from untreated aleurone cells did not show important changes in neutral lipids along the incubation period. Our data suggest a higher lipase activity in lipid bodies of GA_3-treated aleurone cells, and therefore support the results recently reported by Fernandez and Staehelin (1987). Neither PC nor PE were detectable in the lipid body fractions of treated or untreated aleurone cells.

Lipid distribution in aleurone protoplasts was investigated with two fluorescent lipid dyes. Rhodamine b is a basic lipid dye which precipitates in the presence of negatively charged compounds such as fatty acids while 3,4 benzpyrene reacts with all lipid sites in the cell. Lipids were detected more accurately when the protoplasts presented some degree of vacuolation. Both dyes showed increasing fluorescence associated with the protein body membranes during vacuolation. Incubation of aleurone protoplasts at 35ºC for 6h prior to staining with rhodamine b, increased the level of fluorescence. Incubation at 35ºC could cause alterations in the binding of the lipids to other compounds and thus make the lipids available to the stain. On the other hand, the fluorescence intensity of GA_3-plus-Ca^{2+} treated protoplasts stained with rhodamine b was higher than that observed in untreated protoplasts. This could be the cytochemical indication of a triglyceride digestion which, according to our data, occurs in GA_3-plus-Ca^{2+}-treated protoplasts along their incubation.

REFERENCES

Busch, D.S., Cornejo, M.J., Huang, C.N., Jones, R.L.: Ca^{2+}-stimulated secretion of α-amylase during development in barley aleurone protoplasts. Plant Physiol. 82: 556-574, 1986.

Eriksen, F.D., Jensen, C.J.: An improved method for the isolation of barley protoplasts. First Nordic Symposium. Cell and Tissue Culture, Frostvallen, Sweden p. 67, 1984.

Fernandez, D.E., Staehelin, L.A.: Effect of gibberellic acid on lipid degradation in barley aleurone layers. In: Molecular Biology of Plant Growth Control. Alan R. Liss, Inc. New York, 1987.

Kates, M.: Techniques of lipidology: Laboratory Techniques in Biochemistry and Molecular Biology, 2nd ed. Elsevier, Amsterdam, 451 pp, 1986.

Morrison, W.R., Tan S.L., Hargin, K.D.: Methods for quantitative analysis of lipids in cereal grains and similar tissues. J. Sci. Food Agric. 31: 329-340, 1980.

INCIDENCE OF GROWTH REGULATORS ON RNA/ DNA LEVELS IN PROTOPLASTS AS ANALYSED BY ACRIDINE ORANGE STAINING AND FLOW CYTOMETRY .

C Bergounioux[1], C. Perennes[1], S.C. Brown[2], C. Sarda[1] and P. Gadal[1]
[1]Physiologie végétale moléculaire, bât 430, Faculté des Sciences, Université Paris-Sud, 91405 Orsay Cedex, France.
[2] Service de cytométrie, CNRS, 91190 Gif sur Yvette, France.

INTRODUCTION

Acridine orange simultaneously stains DNA and RNA. Using flow cytometry, synthesis of these nucleic acids can be followed through a series of sampling times. Darzynkiewicz (3) has thus shown the progression of cells through the phases GØ, G1A and G1B before entry into S.

In the present study we report the suitability of this technique to follow simultaneously nuclear DNA and RNA synthesis during plant protoplast culture. The nucleic acid metabolism of protoplasts cultivated in complete medium, in medium deficient in cytokinin, in auxin or both are describe. Moreover, the effect of a suboptimal supply of a growth regulator has been studied in an attempt to identify points of the cell cycle having greater or lesser dependence upon a given growth factor.

MATERIAL AND METHODE

Petunia hybrida protoplasts were isolated and cultivated from mesophyll leaves as previously described (2). They were plated at 10^5/ml, with 10^6 protoplasts in each batch.

The four basic treatments were: no hormones, both BA and 2,4-D; 2,4-D or BA alone. Triplicate cultures were analysed at 0, 18, 24 and 48 hours. After 24 hours of protoplast culture two supplementary treatments were introduced. BA 0.4mg/l was added to a replicate of dishes growing with only 2,4-D, and conversly 2,4-D 1mg/l added to replicates growing with only BA.

Nuclei isolation and staining were done as previously described (1)

After nuclear isolation, ribonuclease (RNase) treatment was done after Maniatis *et al.* with 5 units/ml for 10^6 nucléi during 30mn at 37°C.

Conversely, to prevent RNA from degradation, 25 units/ml ribonuclease inhibitor (RNasin, BRL) was added to each assay for 30mn at 30°C. Actinomycin D treatment was 40µg/ml in the initial protoplast cultures. Flow analyses were done as previously describe (1).

Abreviations: 2,4-D=2,4-dichlorophenoxyacetic acid; BA = 6-benzyladenine

RESULTS- DISCUSSION

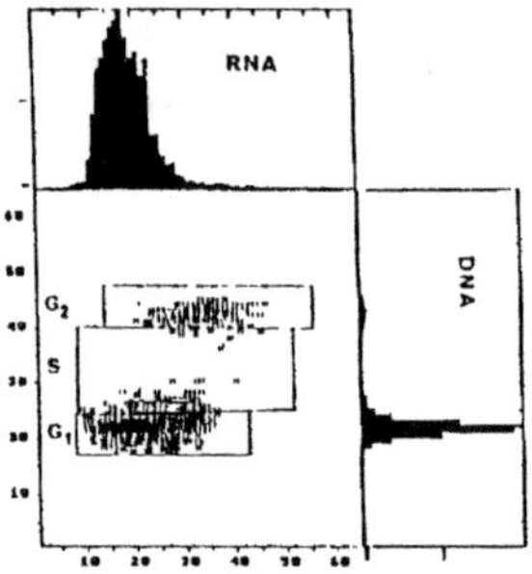

Figure 1 Bi-parametric analysis of isolated nuclei stained with acridine orange.

In presence of Triton X-100 all protoplasts burst and we obtain entire clean nuclei preparation. Howevere only one nuclear menbrane remains. From the bi-parametric analysis of the cytogramme we obtain simultaneously DNA and RNA amount as illustrated in figure 1. Rnase treatment decreased considerably RNA amount without any modification on DNA quantity. Compared to the control, actinomycin treated protoplasts did not shown any red luminescence increas even after 48 hours. Acridine orange, thus allows to quantify RNA and DNA nuclear content from isolated nuclei.

We show that neither auxin nor cytokinin are indispensable at the beginning of the culture (during 18h) for RNA increas. Nevertheless, past this previous hours, RNA increas upon hormonal control is obvoius. Only in the two hormones medium the RNA-DNA ratio revealed a characteristic pattern of changing rates of DNA replication versus transcription which were in relation to cell cycle position as represented by DNA content. Thus, when DNA was constant (2C), nuclei accumulated increasing quantities of RNA and this was reflected by the high heterogeneity of the RNA-DNA ratio before entering the S phase. A critical RNA-DNA value indicated the RNA threshold reached by G1 cells before they entered S phase. At each culture time, G2 nuclei having twice the DNA complement could be distinguished from GØ + G1 nuclei; their RNA level was also twice that of the GØ nuclei.

However we show that the two hormones are not necessary to initiate DNA synthesis. In our experimental conditions, despite low RNA synthesis in G1 nuclei from protoplasts cultivated only with 2,4-D the frequency of S phase nuclei increased after 24h and 48h. Nevertheless, these nuclei with low RNA could not attain the 4C DNA level. Addition of BA to 24h cultures with only 2,4-D reproducibly induced a slighty higher G2 frequency than the reciprocal treatment. This suggest that a restriction point near the passage S to G2 is reached in presence of 2,4-D alone, BA releasing the restriction.

REFERENCES

1- Bergounioux, C., Perennes, C., Brown, S.C., Gadal, P. Cytometry, 9, 1, in press.

2- Bergounioux, C., Perennes, C., Miège, C., Gadal, P. Protoplasma130, 138-144,1986

3- Darzynkiewicz, Z., Sharpless, T., Staiano-Coico, L., Melamed, M.R. Proc. Natl. Acad. Sci. USA, 77, 6696-6699, 1980.

IAA , ABA CONTENT AND METABOLISM IN MAIZE ROOT PROTOPLASTS

Ch. JOLLES and P.E.PILET

Institute of Plant Biology and Physiology, Biology Building, University of Lausanne, CH-1015 Lausanne, Switzerland

INTRODUCTION

Protoplasts respond to indol-3yl-acetic acid (IAA) and abscisic acid (ABA) treatments by an increase of vacuolisation until they ultimately burst (1,2,3). This disruption of the plasmalemma may depend on the levels of growth hormone in the cells used for protoplast preparation (4). Nevertheless, there are few data concerning the IAA and ABA contents of root protoplasts. Several methods have been used for protoplasts of leaves: immunoassay (5), gas chromatography-mass spectrometry (6) and gas-liquid chromatography (7,8,9). However, it was of interest to know whether, when growth hormones have penetrated across the plasmalemma, protoplasts are still able to metabolise these compounds. IAA metabolites have been detected several times in leaf protoplasts of various species (6) but these protoplasts seemed to be unable to metabolise ABA at pH 6 (9). The new approach of this work was to study the endogenous IAA and ABA levels and ABA metabolism at a lower pH for protoplasts and isolated cells prepared from root tips.

METHODS

Plant material: caryopses of Zea mays were germinated in darkness as previously described (10). After 48h , primary roots were selected and divided : cap and apex (0-1mm), elongating zone (1-4mm) and differentiating zone (4-10 mm, 10-14mm). Plant material was pretreated for 1h with sorbitol 3% followed by an enzymatic treatment with Pectolyase Y-23 at 0.075% to 0.4% and Cellulase Y-C at 2% (10). Isolated cells were obtained using only PectolyaseY-23.

IAA and ABA content : plant material was freeze-dried for 12h . Acetate buffer (pH 4) and internal standards were then added. IAA and ABA were extracted by partitioning against diethylether twice. Ether fractions were evaporated to dryness and methanol-water 1:1 (v/v) was added. ABA and IAA fractions were separated by HPLC and methylated with diazomethane. For IAA, a further treatment with HFBI was performed. Derivatised compounds were stored in hexane and analysed by GC/MS (11). Hydrolysable conjugates (esters) were obtained by treatment of the last aqueous phase for 1h in NaOH (1N) followed by acidification to pH 4 with HCl (5N) and partitioning into ether.

IAA and ABA treatments : purified protoplasts were incubated in IAA or ABA at 1 to 100 uM for 6h. After two washings, viable protoplasts were determined using the fluorescein diacetate method.

ABA metabolism: after a 6h treatment with ^{3}H-ABA at pH 5, protoplasts were freeze-dried and taken up in acetone - acetic acid 1% , 99:1 (v/v). After centrifugation, the organic layer was filtered and evaporated to dryness. The residue was dissolved in acetonitrile-acetic acid 0.3% , 88:12 (v/v) and fractions corresponding to ABA metabolites were collected after HPLC separation with an acetonitrile gradient. Radioactivity was then analysed.

RESULTS AND DISCUSSION :

IAA and ABA content : the highest levels of free IAA and ABA were found in protoplasts obtained from the cap and apex (Fig.1,2). In the case of IAA, the levels of bound compounds (hydrolysable conjugates) were two to four times greater than free levels. The ratio between free and conjugated IAA was modified in cap and apex protoplasts by comparison with those in the corresponding parts of the root. IAA conjugates found especially in the root tip might reflect the IAA levels reported in whole roots (11). ABA conjugates were localised in the elongating and differentiating zones. We suggest that these ABA conjugates could be accumulated in the vacuoles as observed in mesophyll cells (8).

IAA and ABA treatment : At high concentrations , IAA and ABA induce in both cases protoplast swelling and plasmalemma disruption (Fig.3,4). Protoplasts with the highest IAA content appear to be less sensitive to increasing concentrations of applied IAA . It seems that protoplast bursting only occurs when the concentration of applied IAA reaches the content of endogenous IAA, as recently observed with protoplasts isolated from the upper and the lower sides of gravireacting roots (4). In contrast, for ABA, protoplasts with the highest ABA level (in the cap and the apex) were less resistant to applied ABA at low concentrations (Fig.4).

Fig.1 and 2 - Content of IAA and ABA in protoplasts isolated from different parts of the root.

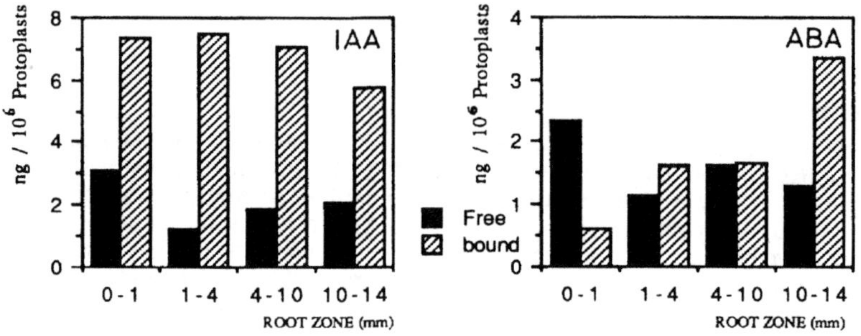

Fig.3 and 4 - Yield of protoplasts surviving in relation to IAA and ABA treatment at increasing concentrations.

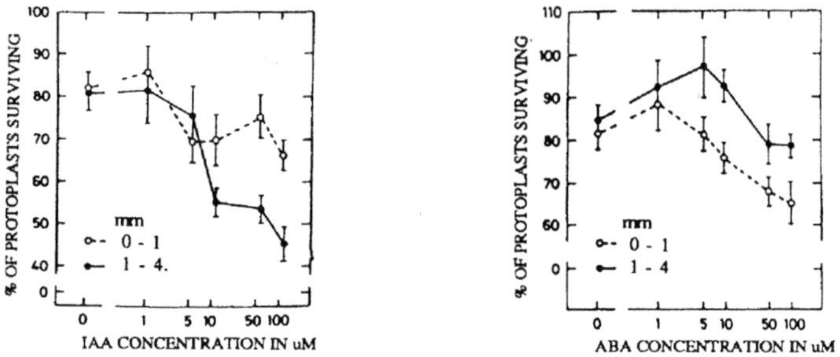

ABA metabolism: when D-L cis trans (3H)ABA was applied to protoplasts, phaseic acid (PA) and dihydrophaseic acid (DPA) and a glucoside of ABA (ABA-GE) were detected. The extent of metabolism appear to be less in protoplasts than in root segments. PA and DPA in the incubation medium represent 76% of their total amounts.

Protoplasts isolated from various zones along the root axis showed large differences in their morphological structure and also in their IAA and ABA contents which influence their reaction to applied hormones.

REFERENCES

1. Pilet P.E., C.R. Acad.Sci. (Paris),273,2253-2256 (1971)
2. Bayer M.H., Plant Physiol., 51, 898-901 (1973)
3. Jolles C. and P.E. Pilet, Abstracts of the XIV International Botanical Congress, p.131, Berlin 1987
4. Pilet P.E., Z.Pflanzenphysiol., 113, 373-376 (1984)
5. Weiler E.W., Schnabl H. and C.Hornberg., Planta, 154, 24-28 (1982)
6. Sandberg G.and A.Crozier, In: The Physiological Properties of Plant Protoplasts, Pilet P.E. ed., pp.209-218, Springer Verlag, Berlin 1985
7. Hartung W, Gimmler H., Heilmann B. and G.Kaiser, Plant Science Letters, 18, 359-364 (1980)
8. Bray E. and J.Zeevaart , Plant Physiol., 79, 719-722 (1985)
9. Loveys B.R. and S.Robinson, Plant Science, 49, 23-30 (1987)
10. Pilet P.E., Jolles C., and H.Henry, In: The Physiological Properties of Plant Protoplasts,Pilet P.E. ed., pp.24-28, Springer Verlag, Berlin 1985
11. Saugy M. and P.E. Pilet, Plant Physiol, 85:42-45 (1987)

SOME METABOLIC PECULIARITIES OF ISOLATED PROTOPLASTS

V.V.Lozovaya, T.A.Gorshkova, F.G.Karimova, I.T.Raimanov, O.I. Tarchevskaya, and M.B.Khusainov

Kazan Institute of Biology, Postbox 30, 420084 Kazan, USSR

1. INTRODUCTION

Isolated protoplasts of higher plants represent valuable and sometimes indispensable model system for a number of physiological and genetic investigations. At the same time until recently there is the lack of works giving comprehensive description of this model, its merits and limitations. The aim of the present work was to investigate some metabolic peculiarities of isolated protoplasts.

2. PROCEDURE

2.1. Materials and methods.

2.1.1. Plant material. Bean protoplasts (Vicia faba) were isolated from the leaves of 10 days seedlings and cultivated in modificated (Arnold-Erikson) medium described by Lozovaya (1987).

2.1.2. Photosynthetic metabolism analysis. The analysis of alcohol-water soluble products of Na $H^{14}CO_3$ and ^{14}C-glucose metabolism was carried out by the method of two-dimension paper chromatography followed by radioautography.

2.1.3. Structural polysaccharides analysis. Cell walls were isolated by the Talmadge et al (1973) method, the fractionation of structural polysaccharides was carried out by the Updegraff method, 1969. After protoplasts culture in ^{14}C-glucose medium, radioactivity of cell wall polysaccharides was determined.

2.1.4. cAMP content was determined according to Gilman (1970).

2.2. Results.

2.2.1. The comparison of CO_2 metabolism of protoplasts and initial tissue has been carried out (Lozovaya, 1987). By the character of changes in direction of $^{14}CO_2$ photosynthetic incorporation protoplasts after their isolation from the leaves resemble cells under extreme conditions or subjected to the effect of inhibitors or uncouplers of photophosphorylation and electron transport: the decrease of label percentage in sucrose and the increase in sugar phosphates and alanine have been revealed (Fig.1) Low label content in sucrose in isolated protoplasts may be connected with a number of reasons: ATP defficiency in protoplasts caused by the influence of unfavourable conditions of isolation procedure; by partial transport of sucrose into the environment; by the decrease of the intensity of sucrose formation due to substrates attraction for cell wall regeneration. Rapid and considerable activation of synthesis of matrix polysaccharides on the first day of culture and its subsequent maintaining practically at the unchanged level along with slow increase of rate of cellulose formation are revealed. The indirect evidence of metabolism reconstruction of cells without walls may be the alteration of cAMP endogenous level in bean protoplasts in comparison with the leaf tissue cAMP endogenous level in protoplasts has been shown to be considerably higher than in leaves, cAMP level being enhanced both in protoplasts and tissues with the increase of osmotic concentration in incubation medium (Fif.2). The more considerable cAMP content in protoplasts is connected, probably, not only with the effect of the higher osmotic concentration, but also with the other factors, because under the same incubation conditions cAMP level in protoplasts was higher than in leaves. At prolonged protoplast culture cAMP level decreased slightly, remaining higher as compared with leaves. One of the reasons of energy defficiency revealed in protoplasts may be the increase of adenylatcyclase activity in response to unfavourable conditions of isolation procedure. The results testifying to the exogenous cAMP stimulative effect on the cell wall regeneration by isolated protoplasts have been obtained (Raimanov et all.1985).

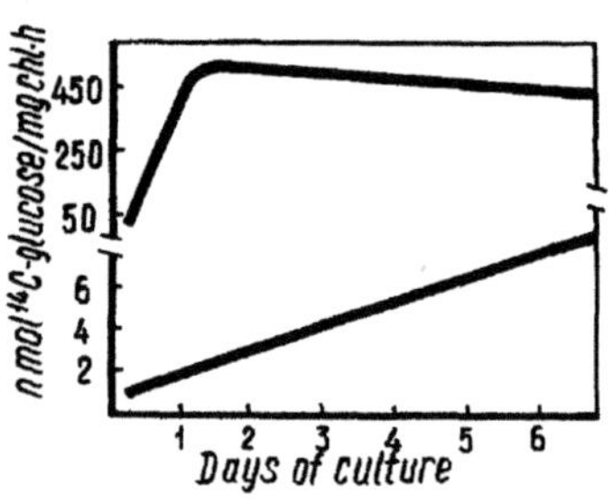

1

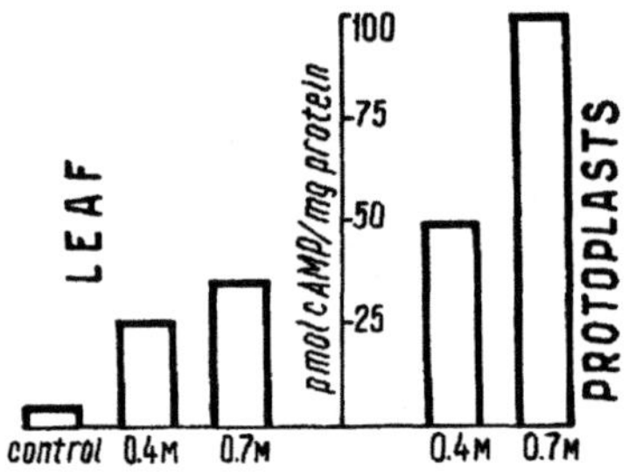

2

Fig.I

Fig.2

Fig.I-Incorporation of ^{14}C-glucose into hemicelluloses(I) and cellulose(2) of cell wall regenerated by protoplasts.
Fig.2-Effect of osmoticum concentration on cAMP level in leaves and protoplasts.

REFERENCES

1. Gilman IG: Proc.Nac.Acad.Sci. USA. v 67, 1970.
2. Losovaya VV: Peculiarities of cell wall formation by isolated protoplasts of higher plants. Kazan, 1987.
3. Raimanov IT, Karimova FG, Losovaya VV, Tarchevskaya OI, Khusainov MB. Abstracts of the 2 nd all-Union Conference on Cellulose biosynthesis. Kazan, 1985.
4. Talmadge K, Keegstra K, Bauer ED, Albersheim P. Plant Physiol., v.51, 1973.
5. Updegraff DM Analyt. Biochem., v.32, N3, 1969.

GENE TRANSFER BY INDIVIDUAL MANIPULATION OF PLANT CELLS

G. SPANGENBERG and G. NEUHAUS
Max-Planck-Institut für Zellbiologie, Abt. Schweiger, Rosenhof, D-6802 Ladenburg, FRG. Present address: Institut für Pflanzenwissenschaften, ETH Zentrum, CH-8092 Zürich, Switzerland

1. INTRODUCTION

The use of an improved microculture system originally developed for protoplasts of Nicotiana tabacum and Datura innoxia cells (Koop et al.1983a, Koop and Schweiger, 1985a) allows the individual selection of single plant cells and their transfer into microdroplets of unconditioned culture medium by using a fully automatized setup (Spangenberg 1986, Schweiger et al. 1987). This microculture system has been successfully applied for the individual culture of protoplasts of Brassica napus (Spangenberg et al. 1986a) and of the moss Funaria hygrometrica (Mejia et al. in press), allowing the establishment of defined cell clones with morphogenetic potential and the regeneration of whole plants.

Details about the assembly of the instrumental setup, as well as of the preparation of the microculture chambers are presented elsewhere (Koop and Spangenberg, this volume).

Based on this useful tool, studies on the physiology of different cell types and cell-to-cell interactions (Spangenberg et al. 1985) as well as genetic manipulations could be performed under controlled conditions at the single cell level.

Particularly, these different categories for gene transfer by individual manipulation of higher plant cells will be presented with a progressive reduction of the size of the transgenome: a) one-to-one protoplast electrofusion (the transgenome consists of a whole genome), b) transfer of partial genomes by one-to-one electrofusion of defined preselected protoplast-subprotoplast pairs (the transgenome is mainly the nuclear genome in the case of protoplast-karyoplast fusion or the cytoplasmic genophores in the case of protoplast-cytoplast fusion) and c) monofactorial transformation (the transgenome consists of plasmid DNA in the kbp-range).

2. RESULTS

2.1. One-to-one electrofusion of defined pairs of protoplasts

The feasibility of somatic hybridization of Nicotiana tabacum via controlled electric-field-induced fusion of defined preselected pairs of protoplasts was first shown by Koop and Schweiger (1985b) with the aid of an instrumental setup previously described (Koop et al. 1983b).

In this way, a perfect control on the number and type of protoplasts that are involved in the fusion event is possible.

By using this system up to 50 preselected pairs of protoplasts can be fused and the fusion products transferred into microculture within one hour (Fig. 1a-d, e-h).

Depending on the type of protoplast to be fused the efficiency of the

electrofusion varies between 45% and 60% for B. napus hypocotyl protoplasts (Spangenberg and Schweiger 1986), between 11% and 25% for protoplasts of F. hygrometrica (Mejia et al. in press) and up to 90% for N. tabacum protoplasts (Koop and Schweiger 1985b).
The fusants obtained can be microcultured until the establishment of hybrid cell clones and regeneration of whole plants.
No difference has been found in the morphogenetic capacity of microcalli obtained by individual culture of fusion products compared with non-fused microcultured control protoplasts (Koop and Schweiger 1985b, Spangenberg and Schweiger 1986).

2.2. Micro-fusion of defined protoplast-subprotoplast pairs and cell reconstitution

The transfer of alien cytoplasm by using protoplast fusion represents an important way to increase the genetic diversity of the extranuclear genomes of plants.
Particularly, transfer of partial genomes using subprotoplasts - cytoplasts and karyoplasts - as one or both of the parents in the one-to-one electrofusion process allows a priori a definition of the subcellular compartments that will participate in the fusion event.
All the techniques so far available for subprotoplast isolation yield preparations of subprotoplasts contaminated with protoplasts. In addition protoplasts contaminating cytoplast preparations are known to be more stable or efficient in the fusion process. Therefore experiments dealing with organelle transfer by mass fusion of protoplast-cytoplast populations may lead to a misinterpretation of the results.

Fig. 1

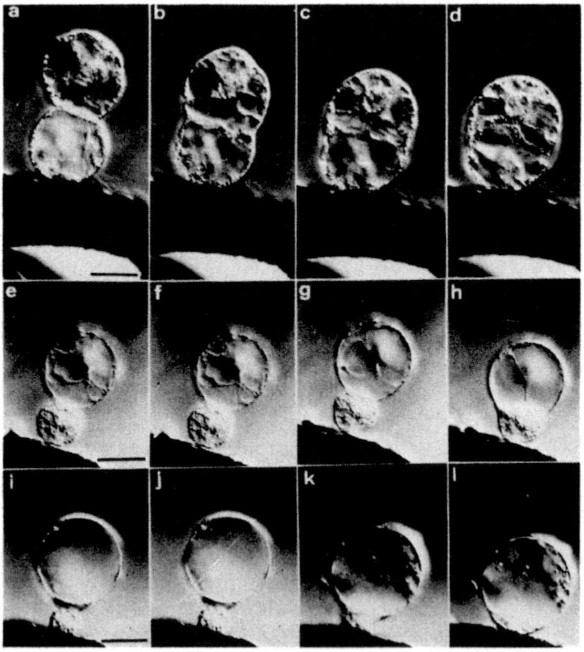

Fig. 1 Electrofusion of selected pairs of defined protoplast types and subprotoplasts of B. napus. a-d) Fusion of two type B protoplasts. e-h) Fusion of a type A and a type B protoplast. i-l) Cell reconstitution by karyoplast -cytoplast fusion (Eur. J. Cell Biol. (1986)41, 51-56)

The individual production and culture of fusants originating from a single cytoplast and protoplast, a karyoplast and a protoplast or a karyoplast and a cytoplast by using the one-to-one electrofusion approach has been reported for B. napus (Spangenberg and Schweiger 1986) (Fig. 1 i-e). Up to 30% of the defined protoplast-subprotoplast pairs could be successfully fused and transferred into microculture. As regards survival and cell division frequencies, no difference could be detected with unfused control protoplasts.
Different fusion combinations performed at the single cell level as well as the number of established callus clone after protoplast-protoplast and protoplast-subprotoplast fusion for different genotypes of B. napus are shown in Table I.

Table I

Genotype combination	N^{o} of selected pairs	N^{o} of fusion products	N^{o} of established callus clones
Ag903 + B3545	1394	946	17
Ag903 + Tower ATR	1635	1055	23
kAg903 + Tower ATR	398	161	3
Ag903 + cTower ATR	529	196	4
kAg903 + cTower ATR	394	142	3

Genotypes: Ag903: B. napus cv. Bronowski, B3545: (cms R. sativus) alloplasmatic B. napus, Tower ATR: (atrazine resistant B. campestris) alloplasmatic B. napus. Karyoplasts: k, cytoplasts: c and protoplasts: no symbol.

2.3. Transformation via microinjection of DNA into higher plant cells

Studies on transformation of defined plant cells by using micromanipulation techniques were performed while dealing with a further reduction of the transgenome size.

A) Microinjection into protoplasts and karyoplasts
The combination of a microinjection technique (Neuhaus et al. 1986) with micromanipulation methods for cell reconstruction and microculture of manipulated protoplasts (Spangenberg et al. 1986b) allowed the transformation of B. napus and N. tabacum (Neuhaus et al. 1987a).
Intranuclear microinjection of chimeric npt II gene constructions (pSV2neo and pLGV23neo) into hypocotyl protoplasts of B. napus (Fig. 2a, b) and karyoplasts (Fig. 2c) as well as cell reconstitutions with microinjected karyoplasts (Fig. 2c-e) could be performed with a rate of 100 microinjected cells per hour .
Survival rates of 80% to 90% were achieved for protoplast and karyoplast

microinjections and in the range of 20% after electrofusion mediated cell reconstitution with microinjected karyoplasts. Independently of the micro-injected cell type, expression of the transferred foreign genes could be assessed by an indirect immunofluorescence test in 60% to 70% of the cases.

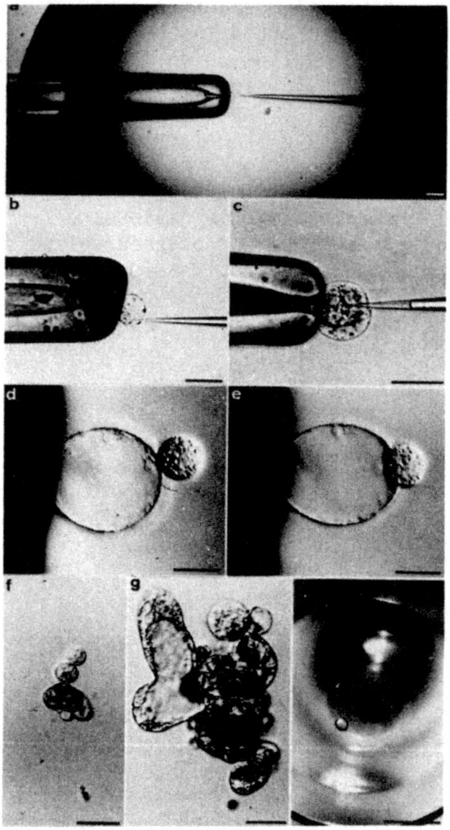

Fig. 2 Microinjection, electrofusion mediated cell reconstruction and callus formation. a,b) Intranuclear microinjection of a hypocotyl protoplast (bars: 200 and 100 um) c) Microinjection of a karyoplast (bar: 10 um). d,e) Electrofusion mediated cell reconstitution from a microinjected karyoplast and a cytoplast (bars:10 um) f-h) Development of microinjected reconstructed cells in microculture (bars: 100 um, 50 um and 1 mm respectively). (Theor. Appl. Genet (1987) 73, 769-783).

After microculture of microinjected rapeseed protoplasts until the establishment of calli, stable transformation could be detected by Southern blot analysis of genomic DNA (Fig. 4a).
The failure in plant regeneration from transformed rapeseed callus clones stimulated microinjection experiments with protoplasts of N. tabacum with partial cell walls where highly efficient transformation (up to 40%) of the regenerants could be demonstrated by detailed Southern blot analysis and npt II enzyme assay (Neuhaus et al. in preparation).

B) Microinjection into microspore-derived embryoids

In order to achieve the production of transgenic rapeseed plants a novel

method of gene transfer by microinjection into multicellular structures which have a high competence for plant regeneration through embryogenesis was used (Spangenberg et al. 1987, Neuhaus et al. 1987b).
Microspore-derived embryoids of B. napus were individually selected, microinjected with npt II gene constructions (pSV2neo and K5) and microcultured until haploid plant regeneration through embryogenesis (Fig. 3a-f).

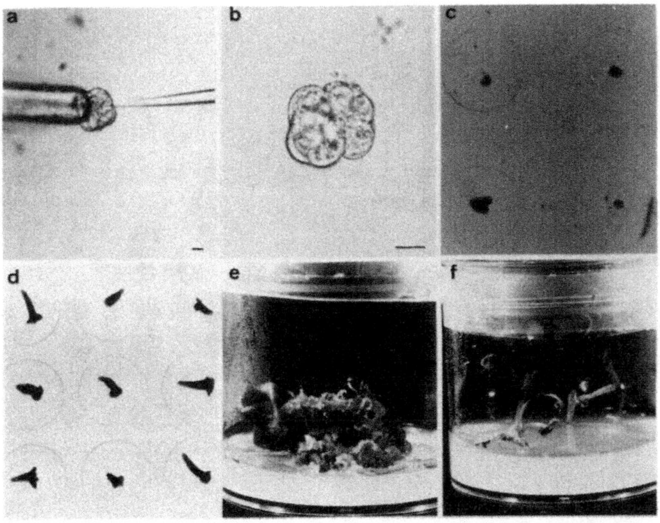

Fig. 3 a) Microinjection into individual cells of a multicellular embryoid derived from 6 day old microspore culture of B. napus. Bar: 50 um. b) Detailed view of embryoid immediately after microinjection. Bar: 50 um. c) Microculture of microinjected embryoids in 0,5 ul medium. d) Individual culture of microinjected embryoids in 5 ul medium. e) Induction of secondary embryogenesis from the primary regenerants for segregation of putative chimeras. f) Regenerated haploid plants 8 weeks after microinjection of the microspore-derived embryoids. (Theor. Appl. Genet. (1987) in press).

Transformation efficiencies in the range of 30% to 50% and up to 60% were determined by DNA dot blot analysis of primary regenerants (Fig. 4b).
Integration of the full length npt II gene as well as rearranged copies of the microinjected foreign gene was proven by Southern analysis of genomic DNA isolated from regenerated plants after induction of secondary embryogenesis from the primary microinjected microspore-derived embryoids (Fig. 4c).
The expression of the transferred gene could be assessed by the neomycin phosphotransferase II enzyme assay (Fig. 4d).
In addition, the in vitro segregation of the putative chimeras obtained after multiple injections into 8 to 20 cell-staged embryoids was achieved

by plant regeneration from secondary embryoids of a stem-embryo system. In this way, experimental evidence for the advantageous features of the use of microinjection as a gene transfer technique were presented: a) extends the range of techniques available for plant cell transformation, b) no host range limitation, c) no need for use of selectable marker genes, d) demonstrated for protoplasts with partial cell walls, for karyoplasts, for suspension cells and for microspore-derived embryoids and e) high transformation efficiency.

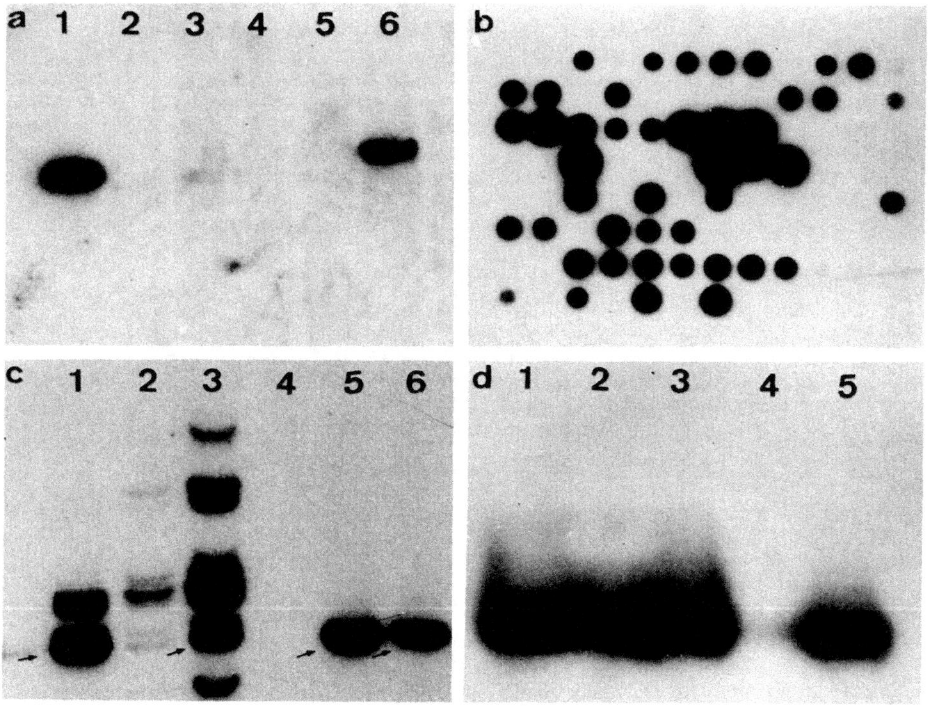

Fig. 4 a) Southern blot analysis of digested (Bam HI and Bgl I) genomic DNA isolated from calli of B. napus regenerated after microinjection of SV40 DNA into single protoplasts. Lane 1: digested SV40 DNA (0,05 ng) insert containing the T-antigen gene (2,7 kb), lanes 2-5: genomic DNA from not transformed calli, lane 6: genomic DNA from a transformed callus clone of B. napus. The Bam HI/Bgl I insert of SV40 DNA was used as a probe. b) Dot blot analysis of genomic DNA(10 ug each) isolated from primary regenerants after pSV2neo microinjection probed with the npt II sequence. c) Southern blot analysis of digested (Hind III) genomic DNA isolated from regenerants after microinjection of K5 into microspore-derived embryoids of B. napus. Lanes 1-3, 5 and 6: DNA from transformed regenerants. Lane 4: DNA from not transformed control regenerant. The plasmid insert (1 kb) and the integrated full length insert are marked by arrows. d) Assay for npt II activity in plants regenerated after K5 microinjection. Lanes 1 - 3: plants with integrated npt II sequence, lane 4: microinjected but no hybridization with npt II DNA and lane 5: bacterial enzyme.

REFERENCES

1. Koop, H.U. and Schweiger, H.G. J. Plant Physiol. 121,245-257(1985a)
2. Koop, H.U. and Schweiger, H.G. Eur. J. Cell Biol. 39,46-49(1985b)
3. Koop, H.U., Weber, G. and Schweiger, H.G. Z. Pflanzenphysiol. 112,21-34 (1983a)
4. Koop, H.U., Dirk, J., Wolff, D. and Schweiger, H.G. Cell Biol. Int. Rep. 7,1123-1128(1983b)
5. Mejia, A., Spangenberg, G., Koop, H.U. and Bopp, M. Botanica Acta (in press)
6. Neuhaus, G., Neuhaus-Url, G., de Groot, E. and Schweiger, H.G. EMBO J. 5,1437-1444(1986)
7. Neuhaus, G., Kranz, E., Spangenberg, G. and Schweiger, H.G. Eur. J. Cell Biol. 43, Suppl. 17,39(1987a)
8. Neuhaus, G., Spangenberg, G., Mittelsten Scheid, O. and Schweiger, H.G. Theor. Appl. Genet. (1987b, in press)
9. Schweiger, H.G., Dirk, J., Koop, H.U., Kranz, E., Neuhaus, G., Spangenberg, G. and Wolff, D. Theor. Appl. Genet. 73,769-783(1987)
10. Spangenberg, G. PhD thesis, University Heidelberg (1986)
11. Spangenberg, G. and Schweiger, H.G. Eur. J. Cell Biol. 41,51-56(1986)
12. Spangenberg, G., Koop, H.U. and Schweiger, H.G. Eur. J. Cell Biol. 39, 41-45(1985)
13. Spangenberg,G., Koop, H.U., Lichter, R. and Schweiger, H.G. Physiol. Plant. 66,1-8(1986a)
14. Spangenberg, G., Neuhaus, G. and Schweiger, H.G. Eur. J. Cell Biol. 42, 236-238(1986b)
15. Spangenberg, G., Neuhaus, G., Gland, A. and Schweiger, H.G. Eur. J. Cell Biol. 43, Suppl. 17,56(1987)

POTENTIALS OF PROTOPLAST FUSION IN PLANT BREEDING PROGRAMMES

K. GLIMELIUS
Department of Plant Breeding, The Swedish University of Agricultural Sciences, Uppsala, Sweden.

1. INTRODUCTION

With the development of the protoplast techniques, a new tool was obtained for the potential improvement of our crops. Since protoplast fusion enables combination of genetic material independent of species and sexual crossability, transfer of genes between unrelated species have been facilitated. This combination of new genetic material refers both to nuclear and to organellar genes. A new type of cytoplasmic composition which cannot be obtained through sexual crossings, thus being unique for protoplast fusions can be created. All these combinations allow gene flow from one species to another, which certainly is the" bottleneck" in conventional plant breeding.

2. FUSION METHODS

Efficient mass-fusions of different protoplast material from different species can be provided both with polyethyleneglycol (PEG) (1,2) in various modifications and combinations (3) and by electric stimulation (4,5). Besides these methods there are several other alternatives for inducing large scale fusions of protoplasts (for a review, see 3).

One remaining problem related to the fusion process is to be able to control fusion of only two protoplasts, one from each parent. This has been studied and analysed by Tempelaar and coworkers and the process can be manipulated to some extent using electrofusion (6,7). Although rather high heterodimeric fusion frequencies, 20-25%, can be obtained both with PEG (8) and after electrofusion (9), the frequency of the binary heterodimeric fusion products is still low.

Another area of interest to investigate more carefully is to what extent the heterogeneity of the protoplasts regarding function, developmental status and cell cycle phase at the time of fusion interferes with the hybridization process, and what the implications will be for the hybrid plant. A great deal of basic cell biological investigations are needed before these aspects can be elucidated.

An elegant and most fascinating method making it possible to investigate these features has been developed. A microculture system based on a computer controlled set-up for selection, transfer, fusion and culture of individual cells has been developed in the former Ladenburg group (10). With this system it is possible to select specific protoplast types, induce fusion with an electric pulse and culture single fusion products under controlled conditions (11). This method enables both that basic cell biological studies can be made, but also that a defined system for production of the desired hybrids can be made.

3. SELECTION METHODS

For plant breeding purposes, a selection of the fusion products is necessary. Although several methods have been developed (12,13), most of them are not general and therefore not practically usefu'. The above presented micro-culture system is one excellent and general way of selecting the desired hybrids. The idea of selecting single cells is, however, not new (14,15). The prerequisities for this type of selection procedure is that the parental material can be distinguished from the hybrids by visible markers like morphology and/or colour. The efficiency of the method was recently demonstrated by Sundberg and Glimelius (16) fusing hypocotyl protoplasts with mesophyll protoplasts from different Brassica species. The fusion products could easily be distinguished from the parental material due to differences in morphology. Isozyme analyses revealed that all (100%) presumed hybrids actually were hybrids. If natural visible markers are missing, different labelling of the parental material can be made either with fluorophores (17) or with other types of pretreatments including bleaching of the mesophyll material (18). However, in spite of the success with this single cell method, it is time consuming and laborious, requiring a lot of experience and patience from the person sitting at the micromanipulator.

An improvement and automatization of the selection of single cells has been obtained by the fluorescence activated cellsorter. With this instrument, it is possible to select and sort out thousands of cells in a very short time, making flowsorting an attractive alternative to manual isolation. The method of using the flowsorter for plant cells, primarily built for mammal cells, was initiated and developed by Galbraith and co-workers (19). They could demonstrate that the instrument was functional also for plant cells after slight modifications, necessary due to the larger size and fragility of the protoplasts compared to mammalian cells. Efficient enrichments of hybrid protoplasts after selective staining of the parental material and specific sorting conditions including positioning of the electronic sorting windows have been demonstrated by Afonso et al. (20) and by Glimelius et al (21). Flowsorting has completely replaced the microfishing of hybrid protoplasts in the hybridization program within Brassicaceae. We are routinely sorting out somatic hybrids produced between hypocotyl protoplasts from one species stained with carboxyfluorescein diacetate (CFDA) and mesophyll protoplasts, containing the intrinsinc fluorophore chlorophyll, from the parental cells. With this selection procedure, it is possible to perform mass selections in a short time with a high reproducibility and efficiency (8,22,23,).

An area which has received a great deal of interest is the production of mutants containing biochemical markers useful in the selection procedure (24). Several different types of mutants have been isolated such as auxotrophic, temperature sensitive, chlorophyll deficient and drug resistant lines. By fusing protoplasts isolated from mutants with different recessive selectable biochemical markers it is possible to select a hybrid with a restored genome through complementation (13,25). However, the usefulness of such a scheme in applied research is strongly limited, due the lack of and the problems with inducing mutants in different species.

Some years ago, however, the construction of so called universal hybridizers were introduced (26). A dominantly expressed trait, e.g. resistance, and a recessive trait, e.g. auxotrophy, is combined in one cell line. This cell line or mutant can then be fused with any wild type cell and the hybrid can be selected (27) by using a minimal medium with the inhibitory drug. The universal hybridizer, which is auxotrophic cannot grow due to the missing compund, and the wild type will be eliminated due to sensitivity to the inhibitor (28,29,30). However, irrespective of the nomenclature, "universal hybridizer", the same limitations as for the mutants having a single mutation can be argued for practical applications. One of the parents to be used in the hybridization experiments has to be selected as a mutant before utilized in the hybridization, which limits the generality of the system.

4. VERIFICATION AND CHARACTERIZATION OF SOMATIC HYBRIDS

The enrichment of somatic hybrids, in spite of using one of the efficient selection procedures, is only an indirect proof that the material actually is a hybrid. There are several reasons why the hybrid nature of the material must be proven. Parental material may slip through the selection procedure, an elimination of genetic material from the initial fusion product may take place and somaclonal variation might occur. In several laboratories hybrids are produced without any type of selection. A method is required by which genetic material from both parents directly or indirectly can be detected in order to prove the hybrid nature of a fusion product.

Traditional methods like studies of morphology and chromosome numbers are in common use (13). However, since the morphology of a plant can vary enormously due to the polygenic control of these traits and since a variation can be due to somaclonal variation and abnormalities in chromosome numbers, these markers could be inadequate. One way to confirm hybridity is to identify gene products encoded by the genomes to the parental material, such as isozymes (31). By making a systematic evaluation of several different isozymes with species specific patterns these constitute an efficient tool for determination of the hybridity. However, since only a few genes can be identified using isozymes as markers, additional ways of identifying the hybrid nature are also needed.

A powerful tool is the recombinant DNA techniques. By the restriction-fragment-length-polymorphism (RFLP) typical fingerprints can be found for each species. By choosing a species-specific DNA-clone (probe), a repetitive DNA sequence, or cloned genes from any species, all of which can give a species specific pattern after hybridization to the isolated and cut DNA from each parent (28,32), hybridity of the putative hybrid can be proved. By choosing a DNA sequence that is spread over the whole genome, i.e. a highly repetitive DNA sequence as a probe, a high probability of detecting how much of the parental genomes that is present in the hybrids will be obtained (33). This type of probe is very powerful in confirmation of hybrids, especially asymmetric hybrids, where most or even all of the chromosomes from one parent might be lost, although genetic material from this parent might have been integrated in the other parental genome (33,34). If combined with *in situ* hybridization techniques, it will also be possible to locate the genes or probes to a certain chromosome (35).

Since both the isozyme analysis and DNA analysis can be made on a very small amount of tissue, it is possible to screen the callus material for hybrid nature early in the culture procedure. Thus, it is possible to select the material of interest and only propagate that for plant production. Considerable gain in work and space will be made avoiding culture of all the parental material.

5. SOMATIC HYBRIDS OF INTEREST FOR BREEDING PURPOSES

During the last years, several reports about succesful hybridization between species of agricultural value have been published. Thus, although studies of model-materials continues, the techniques have reached some application. In spite of that, the examples found are still restricted to only a few families and species.

5.1. Somatic hybridization within Brassicaceae.

Hybridizations in the cruciferous crops have been made with the different oilcrops including *Brassica napus*, *B. campestris* and *Sinapis alba*, but also vegetables belonging to the *B. oleracea* group. Several hybridizations have been made between the species included in the triangel of U (36). Resynthesis of rapeseed fusing *B. campestris* with *B. oleracea* was the first positive result in this respect (16,37,38). In all these investigations, succesful hybridizations were performed. Of the hybrids obtained, however, about 30% got the expected chromosome number 38, while the rest either had more or less chromosomes (22,38). Thus, in spite of sexual crossability beween the species and compatibility between the genomes, the karyotype was unstable. Several of the hybrids had 56 chromosomes and could come from multiple fusions. In others, aneuploidy was registered which could be due to additions as well as eliminations of chromosomes. Most of the hybrids having 38 chromosomes were fertile although the fertility was lower than in ordinary rapeseed varieties (22).

Another hybridization made is the combination of *B. napus* with *B. nigra* (8). This hybrid results in a combination of all the 3 genomes of *Brassica*. Several hybrids with the expected chromosome number (54) were obtained and when the isozymes were analysed, the pattern typical for the two species was found. Also in this case, fertility and seed set was obtained.

Besides the combinations of the different *Brassica* species, hybridizations with other genera such as *Sinapis* and *Eruca* have been performed. These genera are closely related to *Brassica* and homologies between the chromosomes have been documented (39). In spite of these homologies, sexual hybridization is almost impossible. *B. napus* has been hybridized via protoplast fusion with *Sinapis alba* (40) and with *Eruca sativa* (23). In both cases the sum of the parental chromosome numbers was noted, although again a large variation was found. This was especially the case in the combination with *Eruca*. According to the isozyme analyses in the latter case, an elimination preferentially of the *Eruca* genome had occured. In spite of the asymmetry obtained, the hybrids were fertile.

Toriyama and coworkers (41) have made somatic hybrids between *Sinapis turgida*, having the features of a universal hybridizer, nitrate-reductase-deficiency and resistance to a drug 5MT (5-methyl-trypthophane) and either of B. cleracea or B. napus. Even here the sum of the parental chromosome numbers was found indicating a complete hybridization. No reports about fertility in these plants have been made.

Hybridizations have also been made of species belonging to different tribes. The first somatic hybrid in Brassicaceae was produced by Gleba and Hoffmann (42) combining Arabidopsis thaliana with B. campestris. After a long culture period, the callus material finally differentiated into plants, some of which still had the sum of the chromosome numbers of the two parents, while others had less. Toriyama and coworkers produced hybrids between B. oleracea and Moricandia arvensis (43). Moricandia exhibits an intermediate C_3-C_4 photosynthetic property making it very interesting as a fusion partner. Several hybrid plants have been obtained and claimed to be amphidiploid. The chloroplasts had segregated randomly, but none of the hybrids showed the intermediate photosynthetic properties found in Moricandia. Both these types of hybrids were sterile.

5.2. Somatic hybridization within Solanaceae.

In the family of Solanaceae a substantial work has been done with production of somatic hybrids. At the latest protoplast congress in Basel, mainly hybrids between different Nicotiana, Petunia and Datura species were produced with the exclusive and historical hybrids produced between tomato and potato by Melchers and coworkers (31). Today, there is a shift over to different types of crops such as potato, tomato and eggplant. In several different laboratories hybridizations have been made between S. tuberosum and wild relatives to potato like S. brevidens (9,44,45) and S. phureja (18). In all cases, complete somatic hybrids have been obtained independent of fusion or selection strategy. In most cases, the hybrids have been verified with isozymes, but also with morphological and cytological traits. Although hybrids with the expected sum of the parental chromome numbers were found, a large variation was also noted as well as genetically controlled eliminations of specific chromosomes (46). Most of the hybrid material produced have obtained traits of value for potato breeding like resistance to PLRV (potato-leaf-roll-virus) (47,48), but also to late blight (Phytophtora infestans) (47) and potato virus Y (48). The value of these somatic hybrid plants for practical breeding of potato is clear, since the wild Solanum species are important gene sources, and the introduction of these germplams into S. tubersoum is very difficult (49).

Another species where somatic hybridization is a potential tool for an improvement of the crop is tomato. Several of the wild relatives to Lycopersicon esculentum like S.lycopersicoides, S.penelii and Lycopersicon peruvianum contain important agronomic traits of interest to transfer to tomato (50). Sexual crossings between these species and tomato are limited due to bilateral or unilateral incongruity (51,52). Several attempts have been made to hybridize the wild species with tomato somatically, some of which have been successful resulting in fertile plants with the expected chromosome number (53,54,55,56).

6. SYMMETRIC VERSUS ASYMMETRIC HYBRIDS

From the examples of parasexual hybridization presented above, it is clear that distantly related species can be combined in spite of sexual incompatibility. In several cases, combinations of species have been made which is severely restricted or even impossible using conventional sexual hybridization. It is also obvious that the closer related to each other the species are, the higher is the chance for obtaining symmetric hybrids with the amphidiploid chromosome number. These types of hybrids are usually fertile making it possible to use them in further breeding. Thus somatic hybridization can be regarded as a realistic and important alternative for the breeding of some crops.

However, when combining species from different genera and tribes via protoplast fusion, hybrid plants may well be obtained, but the hybrids are usually anomalous and incapable of a subsequent sexual reproduction. Thus, although protoplast fusion permits novel genomic combinations beyond genera and taxa, the coordination of the genomes in these remote hybrids will be very limited. Yet the fusion products may survive and even differentiate into plants which usually takes place after a spontaneous, but preferential elimination of chromosomes from one parent (57). Although most or even all of the chromosomes from one partner have been eliminated, traits from this parent may still be present due to recombination of the genetic material into the partner with a more stable genome. The asymmetry obtained, however, is difficult to predict. If cultured cells constitute one of the parents, they are usually the most stable or persistent partner (31). The somatic incompatibility acting on both the nuclear and the organellar genomes, is, however, poorly understood (31,58,59).

Spontaneous asymmetrization has revealed that it is possible to transfer genes between remote species. This highlights the usefulness of the process. By enhancing and directing the elimination of the non-valuable chromosomes and by excerting a selection pressure for the trait of importance, the production of asymmetric hybrids could be of great value (8). One way of inducing elimination of the donor genomes, both nuclear and organellar, is by irradiation (60,61). There are several reports about succesful production of asymmetric hybrid plants where extra chromosomes and traits from the irradiated donor have been found in the acceptor genome (61,62,63,64,65). However, to keep and transfer the extra chromosome to the progeny will be difficult. The only way to solve this will be through introgression of the genes of interest via recombination or translocation.

7. CONCLUSION

What has happened with all the promises and potentials involved in this technique? Improvements have been made in most areas in this field and the results obtained witness about rapid progress. However, as in most research disciplines, the progress as well as the future directions differ from what was designed and proposed earlier. Instead of producing complete somatic hybrids between distantly related species, a limited gene transfer through the production of asymmetric as well as cytoplasmic hybrids (cybrids) is the new goal. Still, the production of symmetric somatic hybrids holds promise for the future, when thinking of the highly sucessful combinations spontaneously obtained in nature once, such as wheat, rapeseed and corn.

8. ACKNOWLEDGEMENTS

I wish to express my gratitude to all colleagues who provided reprints, pre-prints and slides of their work for inclusion in this article. I also want to thank all the co-workers at the Department of Plant Breeding, The Swedish University of Agricultural Sciences for informations, results and valuable discussions.

9. REFERENCES

1. Wallin, A., Glimelius, K. and Eriksson, T., The induction of aggregation and fusion of Daucus carota protoplasts by polyethylene glycol, Z. Pflanzen-physiol., 74, 64, 1974.
2. Kao, K.N. and Michayluk, M.R., A method for high frequency inter-generic fusion of plant protoplasts, Planta, 115, 355, 1974.
3. Constabel, F. and Cutler, A.J., Protoplast fusion. Plant protoplasts (Ed. L.C. Fowke, F. Constabel), 53, 1985.
4. Senda, M., Takeda, J., Abe, S. and Nakamara, T. Induction of cell fusion of plant protoplasts by electrical stimulation. Plant Cell Physiol., 20, 1441, 1979.
5. Zimmermann, U. and Scheurich, P., High frequency fusion of plant protoplasts by electric fields, Planta, 151, 26, 1981.
6. Tempelaar, M.J. and Jones, M.G.K., Fusion characteristics of plant protoplasts in electric fields, Planta, 165, 205, 1985.
7. Tempelaar, M.J., Duyst, A., De Vlas, S.Y., Krol, G., Symmonds, C. and Jones, M.G.K., Modulation and direction of the electrofusion response in plant protoplasts, Plant Science, 48, 99, 1987.
8. Sjödin, C. and Glimelius, K., Selection for resistance to Phoma lingam using pycnospore infections of plants in vivo and sirodesmin PL. as selective agent in vitro. (This volume) 1988.
9. Fish, N., Karp, A. and Jones, M.G.K., Production of somatic hybrids by electrofusion in Solanum. Theor. Appl. Gen. (submitted).
10. Schweiger, H.-G., Dirk, J., Koop, H.-U., Kranz, E., Neuhaus, G., Spangenberg, G. and Wolff, D., TAG, 73, 769, 1987.
11. Koop, H.-U. and Schweiger, H.-G., Regeneration of plants after electrofusion of selected pairs of protoplasts, Eur. J. Cell Biol., 39, 46, 1985.
12. Lazar, G.B., Recent developments in plant protoplast fusion and selection technology. Protoplasts 1983. 6th Int. Protoplast Symposium Basel, 1983 (Eds. J. Potrykus, C.T. Harms, A. Hinnen, R. Hütter, P.J. King, R.D. Shillito), 61, 1983.
13. Harms, C.T. Hybridization by somatic cell fusion. Plant protoplasts (Ed. L.C. Fowke, F. Constabel), 169, 1985.
14. Kao, K.N., Chromosomal behavior in somatic hybrids of soybean - Nicotiana glauca, Mol. Gen. Genet., 150, 225, 1977.
15. Gleba, Y.Y. and Hoffmann, F., Hybrid cell lines Arabidopsis thaliana + Brassica campestris: No evidence for specific chromosome elimination, Mol. Gen. Genet., 165, 257, 1978.
16. Sundberg, E. and Glimelius, K., A method for production of inter-specific hyibrids within Brassiceae via somatic hybridization, using resynthesis of Brassica napus as a model. Plant Science, 43, 155, 1986.
17. Galbraith, D.W., Mauch, T.J., Identification of fusion of plant protoplasts. II. Conditions for the reproducible fluorescence labelling of protoplasts derived from mesophyll tissue. Z. Pflanzenphysiol., 98, 129, 1980.
18. Puite, K.J., Roest, S. and Pijnacker, L.P., Somatic hybrid potato plants after electrofusion of diploid Solanum tuberosum and Solanum phureja, Plant Cell Reports, 5, 262, 1986.

19. Galbraith, D.W., Selection of somatic hybrid cells by fluorescence-activated cell sorting. Chapt. 50, Cell Culture and Somatic Cell Genetics of Plants, Vol. 1, Academic Press, 1984.
20. Afonso, C.L., Harkins, K.R., Thomas-Compton, M.A., Krejci, A.E., Galbraith, D.W., Selection of somatic hybrid plants in Nicotiana through fluorescence-activated sorting of protoplasts, Biotechnology, 3, 811, 1985.
21. Glimelius, K. Enrichment of somatic hybrids with flow cytometry, Newsletter, 52, 2, 1987.
22. Sundberg, W., Landgren, M. and Glimelius, K., Fertility and chromosome stability in Brassica napus resynthesised by protoplast fusion, Theor. Appl. Genet., (in press) 1987.
23. Fahleson, J., Råhlen, L. and Glimelius, K., Protoplast fusion between Eruca sativa and Brassica napus-analysis of regenerated plants, (This volume) 1988.
24. Bourgin, J.P. Protoplasts and the isolation of plant mutants, Protoplast symposium Basel, 1983 (Ed. J. Potrykus, C.T. Harms, A. Hinnen, R. Hütter, P.J. King, R.D. Shillito), 43, 1983.
25. Negrutiu, I., De Brouwer, D., Watts, J.W., Sidorov, V.I., Dirks, R. and Jacobs, M., Fusion of plant protoplasts: a study using auxotrophic mutants of Nicotiana plumbaginifolia, Viviani, Theor. Appl. Genet., 72, 279, 1986.
26. Lo Schhiavo, F., Giovinazzo, G. and Terzi, M., 8-azaquanine resistant carrot cell mutants and their use as universal hybridizers. Mol. Gen. Genet., 192, 326, 1983.
27. Hamill, J.D., Pental, D., Cocking, E.C. and Müller, A.J., Production of a nitrate reductase deficient streptomycin resistant mutant of Nicotiana tabacum for somatic hybridization studies. Heredity, 50, 197, 1983.
28. Ye, J., Hauptmann, R.M., Smith, A.G. and Widholm, J.M., Selection of a Nicotiana plumbaginifolia universal hybridizer and its use in intergeneric somatic hybrid formation, Mol. Gen. Genet., 208, 474, 1987.
29. Brunold, C., Krüger-Lebus, S., Saul, M.W., Wegmüller, S. and Potrykus, I., Combination of kanamycin resistance and nitrate reductase deficiency as selectable markers in one nuclear genome provides a universal somatic hybridizer in plants, Mol. Gen. Genet., 208, 469, 1987.
30. Toriyama, K., Kameya, T. and Hinata, K., Selection of a universal hyibridizer in Sinapis turgida Del. and regeneration of plantlets from somatic hybrids with Brassica species, Planta, 170, 308, 1987.
31. Harms, C.T. Somatic hybridization by protoplast fusion. Protoplast symposium Basel, 1983 (Ed. I. Potrykus, C.T. Harms, A. Hinnen, R. Hütter, P.J. King, R.D. Shillito), 69, 1983.
32. Saul, M.W. and Potrykus, I., Species-specific repetitive DNA used to identify interspecific somatic hybrids, Plant Cell Reports, 3, 65, 1984.
33. Imamura, J., Saul, M.W. and Potrykus, I., X-ray irradiation promoted asymmetric somatic hybridization and molecular analysis of the products, Theor. Appl. Genet., 74, 445, 1987.
34. Sala, C., Biasini, M.G., Morandi, C., Nielsen, E., Parisi, B. and Sala, F., Selection and nuclear DNA analysis of cell hybrids between Daucus carota and Oryza sativa, J. Plant Physiol., 118, 409, 1985.
35. Mouras, A., Saul, M.W. and Potrykus, I., Localization by in situ hybridization of a low copy chimaeric resistance gene introduced

into plants by direct gene transfer, Mol. Gen. Genet., 207, 204, 1987.

36. U.N. Genome-analysis in Brassica with special reference to the experimental formation of B. napus and peculiar mode of fertilization. Jap. J. Bot., 7, 389, 1935.

37. Schenck, H.R. and Röbbelen, G., Somatic hybrids by fusion of protoplasts from Brassica oleracea and B. campestris. Z. Pflanzenzüchtg., 89, 278, 1982.

38. Terada, R., Yamashita, Y., Nishibayashi, S. and Shimamoto, K., Somatic hybrids betwen Brassica oleracea and B. campestris: selection by the use of iodoacteamide inactivation and regeneration ability, Theor. Appl. Genet., 73, 379, 1987.

39. Mizushima, U., Genome analysis in Brassica and allied genera. Brassica crops and wild allies. (Ed. S.Tsunoda, K. Hinata, C. Gomez-Campo), 89, 1980.

40. Primard, C., Vedel, F., Mathieu, C., Pelletier, G. and Chèvre, A.M., Interspecific somatic hybridization between Brassica Napus and Brassica Hirta (Sinapis Alba L.), Theor. Appl. Genet., (in press) 1987.

41. Toriyama, K., Kameya, T. and Hinata, K., Selection of a universal hybridizer in Sinapis turgida Del. and regeneration of plantlets from somatic hybrids with Brassica species, Planta, 170, 308, 1987.

42. Gleba, Y.Y. and Hoffmann, F., "Arabidobrassica": A novel plant obtained by protoplast fusion, Planta, 149, 112, 1980.

43. Toriyama, K., Hinata, K. and Kameya, T., Production of somatic hybrid plants, "Brassicomoricandia", through protoplast fusion between Moricandia arvensis and Brassica oleracea, Plant Science, 48, 123, 1987.

44. Barsby, T.L., Shepard, J.F., Kemble, R.J. and Wong, R., Somatic hybridization in the genus Solanum: S. tuberosum and S. brevidens, Plant Cell Reports, 3, 165, 1981.

45. Austin, S., Ehlenfeldt, M.K., Baer, M.A. and Helgeson, J.P., Somatic hybrids produced by protoplast fusion between S. tuberosum and S. brevidens: phenotypic variation under field conditions, Theor. Appl. Genet., 71, 682, 1986.

46. Pijnacker, L.P., Ferwerda, M.A., Puite, K.J. and Roest, S., Elimination of Solanum phureja nucleolar chromosomes in S. tuberosum + S. phureja somatic hybrids, Theor. Appl. Genet., 73, 878, 1987.

47. Helgeson, J.P., Hunt, G.J., Haberlach, G.T. and Austin, S., Somatic hybrids between Solanum brevidens and Solanum tuberosum: Expression of a late blight resistance gene and potato leaf roll resistance, Plant Cell Reports, 3, 212, 1986.

48. Gibson, R.W., Jones, M.G.K. and Fish, N., Resistance to potato leaf roll virus and potato virus Y in somatic hybrids between dihaploid S. tuberosum and S. brevidens, (submitted).

49. Ehlenfeldt, M.K. and Hanneman, Jr. R.E., The use of endosperm balance number and 2n gametes to transfer exotic germplasm in potato, Theor. Appl. Genet., 68, 155, 1984.

50. Handley, L.W., Nickels, R.L., Cameron, M.W., Moore, P.P. and Sink, K.C., Somatic hybrid plants between Lycopersicon esculentum and Solanum lycopersicoides, Theor. Appl. Genet., 71, 691, 1986.

51. Rick, C.M., Hybrids between lycopersicon esculentum Mill. and Solanum lycopersicordes Dun. Proc. Natl. Acad. Sci., 37, 781, 1951.

52. Rick, C.M., Hybridization between Lycopersicon escu-lentum and Sola-num pennellii: phylogenetic and cytogenetic significance, Proc. Natl. Acad. Sci., 46, 78, 1960.

53. Handley, L.W. and Sink, K.C., Plant regeneeration of protoplasts isolated from suspension cultures of Solanum lycopersicoides. Plant Sci., 42, 201, 1985.
54. Tan, M.-L.M.C., Somatic hybridization and cybridization in some solanaceae, Vrije Universiteit te Amsterdam, Free University Press, Amsterdam, 1987.
55. Kinsara, A., Patnaik, S.N., Cocking, E.C. and Power, J.B., Somatic hybrid plants of Lycopersicon esculentum Mill. and Lycopersicon peruvianum Mill., J. Plant Physiol., 125, 225, 1986.
56. Adams, T.L. and Quiros, C.F., Somatic hybridization between Lycopersicon peruvianum and L. pennellii: Regenerating ability and antibiotic resistance as selection systems, Plant Science, 40, 209, 1985.
57. Dudits, D., Hadlaczky, G.Y., Bajszár, G.Y., Koncz, C.S., Lázár, G. and Horváth, G., Plant regeneration from intergeneric cell hybrids, Plant Sci. Lett., 15, 101, 1979.
58. Fluhr, R., The segregation of organelles and cytoplasmic traits in higher plant somatic fusion hybrids, Protoplasts 1983. 6th Int. Protoplast symposium Basel, 1983 (Eds. J. Potrykus, C.T. Harms, A. Hinnen, R. Hütter, P.J. King, R.D. Shillito), 61, 1983.
59. Pelletier, G.R., Plant organelle genetics through somatic hybridization, Oxford Surveys of Plant Molecular & Cell biology, 3, 96, 1986.
60. Menczel, L., Galiba, G., Nagy, F. and Maliga P., Effect of radiation dosage on efficiency of chloroplast transfer by protoplast fusion in Nicotiana, Genetics, 100, 487, 1982.
61. Dudits, D., Fejér, O., Hadlaczky, G., Koncz, X., Lázár, G.B. and Horváth, G., Intergeneric gene transfer mediated by plant protoplast fusion, Molec. Gen. Genet., 179, 283, 1980.
62. Gupta, P.P., Schieder, O. and Gupta, M., Intergeneric nuclear gene transfer between somatically and sexually incompatible plants through asymmetric protoplast fusion, Mol. Gen. Genet., 197, 30, 1984.
63. Dudits, D., Maroy, E., Preznovszky, T., Olah, Z., Gyorgyey, J. and Cella, R., Transfer of resistant traits from carrot into tobacco by asykmmetric somatic hyibridization: Regeneration of fertile plants, Proc. Natl. Acad. Sci., (in press) 1987.
64. Bates, G.W., Hasenkampf, C.A., Contolini, C.L. and Piastuch, W.C., Asymmetric hybridization in Nicotiana by fusion of irradiated protoplasts, Theor. Appl. Genet., 74, 718, 1987.
65. Sidorov, V.A., Zubko, M.K., Kuchko, A.A., Komarnitsky, I.K. and Gleba, Y.Y., Somatic hybridization in potato: use of γ-irradiated protoplasts of Solanum pinnatisectum in genetic reconstruction, Theor. Appl. Genet., 74, 364, 1987.

USE OF PROTOPLASTS IN PLANT BREEDING:CYTOPLASMIC ASPECTS.

G. PELLETIER(1), C. PRIMARD(1), M. FERAULT(1), F. VEDEL(2), P. CHETRIT(2), M. RENARD(3), R. DELOURME(3)

1. Laboratoire de Biologie Cellulaire, I.N.R.A, 78000 VERSAILLES, France
2. Laboratoire de Génétique Moléculaire des Plantes, C.N.R.S, 91405 ORSAY, France
3. Station d'Amélioration des Plantes, I.N.R.A, LE RHEU, France

I. INTRODUCTION

Somatic hybridization has renewed the interest for cytoplasmic genetics in higher plants by offering the possibility of manipulating chloroplast and mitochondrial genomes (reviewed by galun and Aviv, 1986; Pelletier, 1986). Moreover, these manipulations may have a direct application in plant breeding. In this perspective we shall try to discuss three main questions which could be asked by breeders to cell biologists, using examples chosen in the *Nicotiana* and *Brassica* genera. These questions are: 1) is it possible to increase the genetic diversity of cytoplasms in a given species? 2) Are cell fusion techniques the best tools for modifying organelle composition of a cytoplasm or for transferring as quickly as possible a given cytoplasm into several nuclear backgrounds? 3) If these new cytoplasmic combinations are viable, have they the required qualities to be included in the production of cultivated varieties?

2. USE OF PROTOPLASTS IN CREATING NEW SOURCES OF CYTOPLASMIC VARIABILITY

In vitro culture of mutagenized protoplasts, somaclonal variation and cell fusion allow to create new cytoplasmic characters as illustrated by the following examples.

2.1. Cséplo *et al* (1985) succeeded in selecting triazine resistant chloroplast in *Nicotiana*. Mesophyll protoplasts of *N. plumbaginifolia*, after a mutagenic treatment with N-nitro-N-ethylurea were cultured until the microcolony stage, where the herbicide was added. At the same time, the glucose concentration of the culture medium was reduced to 0,015M. These conditions allow the selection of resistant mutants by subcultures of green areas until complete sorting-out of chloroplasts. The chloroplastic location of the mutation was further confirmed through chloroplast transfer to *N. tabacum* after protoplast fusion (Menczel *et al.* 1986).

2.2. It is well known that alloplasmic associations may lead to male-sterility as a result of nucleus mitochondria interaction. These associations are generally obtained by interspecific crosses and repeated backcrosses. In *Nicotiana* for instance, the recently discovered species *N. africana* is cross incompatible with *N. tabacum*, particularly when it is used as the maternal parent. Recently, Kumashiro *et al.* (in press) succeeded in obtaining a cytoplasmic male-sterile cybrid plant by fusion between *N. tabacum* protoplasts and X-irradiated *N. africana* protoplasts.

2.3. Although mitochondrial rearrangements following *in vitro* tissue culture are not systematically observed, several reports of regenerated plants with modified mitochondrial genomes have been published (Gengenbach *et al.*, 1977; Brettel *et al.*, 1980; Kemble and Shepard, 1984). Recently Li (1987) studied the genetic variability of *N. sylvestris* plants during three successive cycles of protoplast culture and plant regeneration. Three plants were regenerated from a single callus after the second cycle, two of which were shown to be cytoplasmic male sterile, the third being fertile. Chloroplastic DNA from these three plants was found to be identical to wild type. Mitochondrial DNA variations appeared in these three lines the two c.m.s presenting deletions associated with specific changes in the in-organello protein synthesis pattern.
These three examples show that it is possible to obtain through protoplast culture or cell fusion new sources of cytoplasmic variability in a species. Atrazine resistant mutants occurring in natural populations show a decrease in photosynthetic potential and consequently in yield (Gressel 1985) except in one case recently described (Schonfeld *et al.* 1987). Therefore it is justified to try to select new triazine resistant mutants which would not affect the yield of crops. C.m.s mutants may help breeders in species when this trait is unknown or when available c.m.s types present some agronomical difficulties. The use of these methods of enrichment of cytoplasmic variability remains to be evaluated in a plant breeding perspective.

3. TRANSFER AND RECOMBINATION OF ORGANELLE GENOMES: PROTOPLAST FUSION VERSUS SEXUAL REPRODUCTION

3.1. Protoplast fusion.

Several combinations of cytoplasmic informations are theoritically possible in cybrids resulting from protoplast fusion between two parental lines.
The apparent rule concerning the chloroplast genome of cybrids can be stated as follows: chloroplastic genomes of cybrids result from the sorting out of parental genomes without any changes (Akada and Hirai, 1986). Therefore protoplast fusion allows the rapid and definitive exchange of chloroplast types between two lines (Belliard *et al.*, 1978; Aviv *et al.*, 1980; Glimelius *et al.*, 1981) and even between remote cross incompatible species like *N. tabacum* and *Petunia hybrida* (Glimelius and Bonnet, 1986) or *N. tabacum* and *Salpiglossis sinuata* (Thanh *et al.*, 1987). Interspecific chloroplast recombination has been described once in a *Nicotiana* somatic hybrid (Medgyesy *et al.*, 1985) and recently intertribal recombination between *S. tuberosum* and *N. tabacum* has been obtained (Medgyesy, personal communication).
The fate of mitochondrial (mt) genomes after protoplast fusion has been mainly studied in three genera, *Nicotiana, Petunia* and *Brassica*. In contrast with the plastid genomes, mitochondrial genomes from cybrids and somatic hybrids which have been analysed generally differ from the mitochondrial genomes in the original parental lines. The hybrid mt DNAs contains fragments characteristic of both parents as well as novel fragments not observed in parental mtDNAs (Belliard *et al.*, 1979; Nagy *et al.*, 1981; Galun *et al.*, 1982; Boeshore *et al.*, 1983; Fluhr *et al.*, 1983; Chetrit *et al.*, 1985; Matthews and Widholm, 1985; Kemble *et al.*, 1986; Ozias-Akins *et al.*, 1987). The physical proof of intergeneric recombination was given by Rothenberg *et al.*, (1985) in Petunia and Vedel *et al.*, (1986) in *Brassica*. They showed that a novel fragment in somatic

hybrid mt DNA was constituted by mt DNA restriction fragments derived from both parents.
Entire cytoplasm transfer can be efficiently achieved by protoplast fusion when the nucleus of the cytoplasm donor protoplast is destroyed by irradiation (Aviv *et al.*, 1984) and the other parent is treated with iodoacetate for inactivating its cytoplasm (Sidorov *et al.*, 1981).
Although largely investigated in *Solanaceae*, these possibilities are illustrated below (table 1) with examples chosen in the genus *Brassica*, where several groups have tried to improve cytoplasms by protoplast fusion: correction of chlorophyll deficiency and improvment of nectar production in the case of Ogura c.m.s cytoplasm (Ogura, 1968; Bannerot *et al.* 1977), combination of triazine resistant *B. campestris* chloroplasts with mitochondria bearing a c.m.s trait.

Parents of fusion	Methods of Enrichement or screening or selection	Results mitochondria (Mt) chloroplast (Cp)	References
(cms Ogura) B.napus + fertile B. napus	Screening among regenerated plants	Mt recombination cp exchange	Pelletier et al. (1983)
(cms Ogura) B. napus + AtrR B. campestris	Screening among regenerated plants	Mt exchange cp exchange	Pelletier et al. (1983)
"nap" cms B.napus + AtrR B. campestris B.napus	Micro manipulation of heterokaryons	Mt exchange cp exchange	Yarrow et al. (1986)
(cms Polima) B. napus + (AtrR B. campestris cytoplasm) B. napus	Iodoacetate on cms and γ rays on AtrR	Mt exchange cp exchange	Barsby et al. (1987)
(cms Ogura) B. napus + fertile B. napus	X irradiation of cms	Mt recombination cp exchange	Menczel et al. (1987)
(cms Ogura) B. oleracea + AtrR B. campestris	Atrazine+ Ability to regenerate	Mt recombination cp exchange	Robertson et al. (1987)

Table 1.

3.2. "Abnormal" fertilization and direct exchange of cytoplasmic genetic information.
Use of selectable markers makes it possible to screen for very rare events occurring during the fertilization process: paternal transmission of organelles or natural androgenesis. These events result, to some extend, in products similar to those obtained after somatic hybridization.

Medgyesy *et al.*, (1986) were able to obtain *Nicotiana* plants possessing paternal streptomycin resistant chloroplasts after intra and interspecific sexual hybridization. These plants carried maternal mitochondria exclusively. This experiment constitute a direct way for transferring chloroplast solely, a goal more difficult to attain by protoplast fusion.

We have used (Pelletier *et al.*, 1987) a homozygous "rootless" dominant line of *N. tabacum* as female in crosses with normal lines. Androgenetic haploids with the male nucleus into the entire cytoplasm of the female parent were directly selected by their ability to grow normally whereas hybrid plants, without root system died a few weeks after sowing.

These "sexual" methods present the advantages of escaping not desired somaclonal variations or organelle recombination and of overcoming limitations due to genotypic or specific differences in ability to regenerate plants from protoplasts. Nevertheless, the use of abnormal fertilization for direct exchange of cytoplasmic genetic information is limited by the availability of selectable markers and the knowledge of frequencies of these events in the main cultivated species.

4. AGRONOMICAL PERFORMANCE OF HYBRID CYTOPLASMS.

In *Brassica napus*, protoplast fusion experiments and interspecific crosses have provided us with plants for comparison of several combinations of nuclear, chloroplastic and mitochondrial genomes, in field conditions (table 2). Characters like restoration of male fertility, nectar secretion, flower morphology and yield were studied to evaluate the influence of these combinations in the perspective of commercial seed production.

	Nucleus	Combined genomes Chloroplasts	Mitochondria	Origin
1	B. napus	R. sativus	R. sativus	Interspecific crosses
2	B. napus	B. campestris	B. campestris	
3	B. oleracea	R. sativus	R. sativus	
4	B. napus	B. napus	R. sativus/B. napus	Protoplast fusion
5	B. napus	B. campestris	R. sativus	
6	B. napus	B. campestris	R.sativus/B. napus	
7	B. oleracea	B. oleracea	R. sativus/B. oleracea	
8	B. oleracea	B. napus	R. sativus/B. napus	Protoplast fusion Followed by interspecific crosses

Table 2: Cybrids cited in the text correspond to the combination 4 (cybrids 27 and 23) or to the combination 5 (cybrid 77).
R. sativus/B. napus means recombined mitochondrial genomes between these two parental mitochondrial genomes.

4.1. Restoration of pollen fertility of Ogura c.m.s *B. napus* and derived cybrids.

To restore male fertility of *B. napus* plants bearing Ogura cytoplasm, it is necessary to introduce specific *R. sativus* genes by interspecific crosses (Heyn, 1978). In the further generations fully male fertile plants with 38 chromosomes (*B. napus* chromosome number) are obtained only if they also possess another dominant character from radish: the white color of petals (W). This observation leads to the conclusion that at least one restorer gene is tightly linked to the W locus. Table 3 summarizes the different phenotypes obtained when Ogura c.m.s or cybrids c.m.s B. napus plants are crossed by a supposedly heterozygous genotype for restorer genes and W gene. In the case of cybrid n° 27 (see table 2) fully fertile plants are observed whatever the petal color is. Thus this c.m.s cybrid cytoplasm does not need the restorer gene(s) linked to W locus to be restored. On the contrary, cybrid n° 23 (see table 2) gives a progeny where plants with white petals are always fully fertile. For this cybrid, although not completely sterile, restorer gene(s) linked to W are sufficient for its restoration.

These observations lead to the conclusion that the Ogura c.m.s system in *Brassica napus* is in fact a composite system. Mitochondrial recombination explain the differences observed between cybrids and in some cases like cybrid 27, allows to obtain a simpler system of male sterility regarding the restoration of fertility.

Phenotype Crosses	WF	WS	YF	Y1/2F	Y1/2S	YS
OGU x Restorer	+	+	-	+	+	+
27 x Restorer	+	+	+	-	-	+
23 x Restorer	+	-	+	-	+	-

Table 3: Phenotypes observed (+) or absent (-) in crosses involving different cytoplasms (Ogu: Ogura; 27: cybrid n°27; 23: cybrid n° 23 (see table 2).

W: white petals. F: fully male fertile. S: aborted anthers-no pollen. Y: yellow petals. 1/2F: flowers with four to six narrow anthers producing pollen. 1/2S: flowers without developed anthers or at best four anthers showing only a partial development, sometimes pollen produced.

4.2. Nectar secretion and F1 hybrid seed production.
Pollination of male-sterile rapeseed plants in the field is performed by bees. Consequently, hybrid seed production of a male-sterile line depends on bee frequentation of flowers.
In a field plot system which alternated male fertile and male sterile stips we observed a strong correlation between bee frequentation and nectar production of the flowers. On one hand, *B. napus* with Ogura cytoplasm produce less than 10% the nectar of male fertile rapeseed, there was a significant decrease in bee density with distance from pollen donors and a corresponding decreasing gradient in seed production. On the other hand, Cybrid 27 produce 70% the nectar of male fertile rapeseed and bee frequentation was more homogenous over all male fertile and male sterile strips. No gradient was apparent and pollination was equivalent to that of pollen donors. The performance for nectar production of c.m.s cybrid 27 cytoplasm is due to its recombined mitochondrial genome since other c.m.s cybrids, with the same nucleus and chloroplasts but Ogura mitochondria, behaved exactly as Ogura c.m.s plants.
In this case, mitochondria recombination allowed to improve an important character for the use of cybrids in plant breeding systems.

4.3. Flower morphology and seed production.
We have compared flower and fruit morphology of Ogura c.m.s cytoplasm *B. napus* plants (*B. napus* nucleus with Ogura chloroplasts and mitochondria) and cybrid 77 (*B. napus* nucleus with *B. campestris* chloroplasts and Ogura mitochondria). (See table 2). We observed that heterospecific cytoplasm induces malformations of flower and distorsion of fruits. The seed set is very poor in this case. As a control, *B. napus* nucleus on homospecific *B. campestris* cytoplasm leads to normal flowers and fruits. These observations suggest that mitochondrial chloroplastic interactions play a role in flower and fruit morphogenesis.

4.4. Yield performance of c.m.s cybrid lines.
Different mitochondria in the same genetic background (nucleus+chloroplasts) can be compared in the field for yield. For example, when a male fertile variety is mixed with the same variety possessing cybrid 27 cytoplasm, a strong competition is observed between the two types of plants. Male sterile individual plants in these conditions can produce 50% more than fertile individuals. When the two lines are cultivated in separate plots in conditions where natural pollination of sterile plants is performed by bees, yield differences are also observed, to the advantage of the male sterile plants (table 4). As this agronomical improvement results from mitochondrial recombination, although its physiological basis remains to be elucidated, it was only possible through protoplast fusion.

These studies converge to the same conclusion: among the various cytoplasmic combinations obtained by protoplast fusion, only some of them, where recombinations had occurred mays respond positively to the agronomical demands.

Nucleus (variety)	Cytoplasm	Yield T/ha	%
Brutor	B. napus	2,41	100
Brutor	27	2,61	108
Darmor	B. napus	4,04	100
Darmor	27	4,41	109

Brutor: Spring variety
Darmor: winter variety

Table 4: Yield comparison between genotypes differing by their mitochondria.

5. CONCLUSION

We can conclude from the above examples that protoplast techniques, especially protoplast fusion, are actually interesting in a plant breeding perspective. Protoplast culture and fusion are efficient methods for increasing the cytoplasmic variability. Consequently, they can provide plant breeders with new genetic pools.

Transfer of entire cytoplasms in new nuclear backgrounds can be achieved through protoplast fusion or using natural sexual events. The choice of the methods will depend on the feasability of each technique in each particular case. When cybrids are observed in agronomical conditions, the breeder can find valuable or improved cytoplasms among different mitochondrial recombinants. Moreover these recombinants constitute a suitable material for studying c.m.s at the genetic and molecular levels. These conclusions are encouraging to develop protoplast techniques in other species than the model genera *Nicotiana* or *Brassica*, especially agronomically important plants.

REFERENCES

1. Akada S, Hiraï A: In: Mantell S and Chapman G, Eds, The chondriome, Pitman, London, 290-298, 1986.
2. Aviv D, Arzee-Gonen P, Bleichman S, Galun E: Mol. Gen. Genet. 196, 244-253, 1984.
3. Aviv D, Fluhr F, Edelman M, Galun E: Theor. Appl. Genet. 56, 145-150, 1980.
4. Bannerot H, Boulidard L, Chupeau Y: Eucarpia Cruciferae, Newsl. 2, 16, 1977.
5. Belliard G, Pelletier G, Vedel F, Quétier F: Mol. Gen. Genet. 165, 231-237, 1978.
6. Belliard G, Vedel F, Pelletier G: Nature 281, 401-403, 1979.
7. Boeshore M, Lifshitz I, Hanson M, Izhar S: Mol. Gen. Genet. 190, 459-467, 1983.
8. Brettel R, Thomas E, Ingram D: Theor. Appl. Genet. 58, 55-58, 1980.

9. Chetrit P, Mathieu C, Vedel F, Pelletier G, Primard C: Theor. Appl. Genet. 69, 361-366, 1985.
10. Cséplö A, Medgyesy P, Hideg E, Demeter S, Marton L, Maliga P: Mol. Gen. Genet. 200, 508-510, 1985.
11. Galun E and Aviv D: In: Methods in Enzymology, Academic Press, New York, 118, 595-611, 1986.
12. Gengenbach B, Green C, Donovan C: Proc. Natl. Acad. Sci (USA) 74, 5113-5117, 1977.
13. Glimelius K, Bonnett H: Theor. Appl. Genet. 72, 794-798, 1986.
14. Glimelius K, Chen K, Bonnet H: Planta, 153, 497-503, 1981.
15. Gressel J: SO Duke, ed, Weed Physiology. CRC Press, Boca Raton, FL, 159-189, 1985.
16. Heyn F: In: Proc. 5th Int. Rapessed Conf. Malmoe, 82-83, 1978.
17. Kemble R, Shepard J: Theor. Appl. Genet. 69, 211-216, 1984.
18. Kumashiro T, Asahi T, Komari T: Plant Sci. Lett. in press.
19. Li X: Thesis Orsay University, June 1987.
20. Menczel L, Polsby L, Steinback K, Maliga P: Mol. Gen. Genet. 205, 201-205, 1986.
21. Medgyesy P, Feyes E, Maliga P: Proc. Natl. Acad. Sci. (USA), 82, 6960-6964, 1985.
22. Medgyesy P, Pay A, Marton L: Mol. Gen. Genet. 204, 195-198, 1986.
23. Ogura H: Mem. Fac. Agric. Kagoshima Univ. 6, 39-78, 1968.
24. Ozias-Akins P, Pring D, Vasil I: Theor. Appl. Genet. 74, 15-20, 1987.
25. Pelletier G: Oxford Surveys of Plant Molecular and Cell Biology, 3, 97-121, 1986.
26. Pelletier G, Férault M, Goujaud J, Vedel F, Caboche M: Theor. Appl. Genet. 1987 (in press).
27. Rothenberg M, Boeshore M, Hanson M, Izhar S: Curr. Genet. 9, 615-618, 1985.
28. Schönfeld M, Yaacoby T, Michael O, Rubin B: Plant Physiol. 83, 329-333, 1987.
29. Sidorov V, Menczel L, Nagy F, Maliga P: Planta, 152, 328-332, 1981.
30. Thanh N, Pay A, Smith M, Medgyesy P, Marton L: Mol. Gen. Genet. (in press).
31. Vedel F, Chetrit P, Mathieu C, Pelletier G, Primard C: Curr. Genet. 11, 17-24, 1986.

PROTOPLAST FUSION BETWEEN ERUCA SATIVA AND BRASSICA NAPUS - ANALYSIS OF REGENERATED PLANTS

J. FAHLESON, L. RÅHLEN and K. GLIMELIUS

Department of Plant Breeding, Swedish University of Agricultural Sciences, Box 7003, S-750 07 Uppsala, Sweden.

1. INTRODUCTION

To investigate if somatic hybridization could be a method for introgression of desirable traits to one of the cultivated oilcrops, *Brassica napus*, protoplast fusions were carried out between this species and species from either of the related genera *Eruca*, *Diplotaxis*, *Raphanus* or *Sinapis*. From a first set of experiments it was observed that the combination *B. napus* + *E. sativa* gave the highest plating efficiency and the following experiments were therefore concentrated on this combination.

This presentation describes protoplast fusion between *B. napus* and *E. sativa* and the analysis of the regenerated plants obtained from this intergeneric species combination.

2. MATERIALS AND METHODS

2.1. Isolation, selection and cultivation

From *B. napus* protoplasts were isolated from 5-day-old hypocotyls while the protoplasts from the other parental species, *E. sativa*, were derived from leaves of *an vitro* grown plant. Isolation of protoplasts was performed as described by Glimelius (1). Fusion of isolated mixed protoplasts were induced with polyethylene glycol, according to Sundberg and Glimelius (2). Selection of heterokaryons were made using either a micromanipulator (2) or by flow cytometry and cellsorting (3). Hybrid cells were cultured in multivial dishes (Falcon^R or Costar^R) in a modified 8p medium (1) supplemented with 4.5 µM 2,4-dichlorophenoxyacetic acid (2,4-D), 2.2 µM 6-benzylaminopurine (BAP) and 0.5 µM 1-naphtylacetic acid (NAA). When frequent cell divisions were observed the cells were diluted with 8p medium without hormones. For further growth and differentiation, cell aggregates were transferred to K_3 solid medium containing 0.6 uM indole-3-acetic acid (IAA) and BAP and Zeatin at different concentrations (2.2-9.9 uM).

2.2. Isozyme analysis

Hybrid nature of regenerated plants was investigated by isozyme analysis as described by Sundberg and Glimelius (2).

2.3. Nuclear DNA content, chloroplast genotype and fertility

The nuclear DNA content of confirmed hybrids were quantitatively measured in a flow cytometer (Becton-Dickinson III with a Spectra-Physics model 2025 argon laser) (4). The chloroplast genotype was investigated in 13 hybrids following the procedure described by Sundberg et al (5). Fertility was investigated in hybrids transferred to the greenhouse and expressed as pollen viability and seeds/pollinated flower compared to *B. napus*.

2.4. Fatty acid composition

The fatty acid composition and mainly the erucic acid content ($C_{22:1}$) was studied. From each of 9 hybrids 5 selfpollinated seeds were individually analysed by a modified method of Johansson and Uppström (6). Seeds were crushed between two filter papers and the fatty acids were extracted by adding 200 ul of a solution containing 60% etanol, 40% hexane and 0.012 M NaOH. After 24 hours, 0.5 ml of distilled water and 1 ml of hexane were added. The hexane phase was withdrawn with a Pasteur pipette and concentrated. Of this concentrate, 2 ul was analysed by GLC with a glascolumn (2m x 2 mm internal diameter) containing 10% BDS on Chromosorb W (HP 80-100 mesh) on a Shimadzu gaschromatograph equipped with a FID-detector.

3. RESULTS

3.1. Isolation, selection and cultivation

On average the fusion ferquency was 9% estimated on the living population 2-3 hours after fusion. Selection of hybrid protoplasts by flow sorting gave on average a hybrid cell frequency of 65%. Totally 1066 calli were obtained from four experiments. Of these, 65 shoots from different calli were recovered.

3.2. Isozyme analysis

Three isozymes, LAP (leucine aminopeptidase), PGI (phosphoglucose isomerase) and 6-P (6-phosphogluconate dehydrogenase) were found to be suitable for hybrid identification. Of the 65 shoots obtained, 54 were investigated and in 34 of them hybrid nature was found for at least one isozyme. Uptil now, 24 hybrids (23 from different calli) have been transferred to the greenhouse. These were subjected to further isozyme analysis in order to obtain the isozyme pattern for all three isozymes. The results showed that 11 of these hybrids were partial hybrids. The nuclear DNA content of the hybrids transferred to the greenhouse varied from 2.4 to 5.3 pg DNA/cell and the chromosome number ranged from 40 to 90. One of the hybrids (J14b) contained cells with two different amounts of DNA (2.6 and 5.3 pg DNA/nucleus) and the chromosome numbers of these cells were found to be 46 and 85 respectively. Of 13 hybrids, 11 contained chloroplasts from *B. napus* while only two contained chloroplasts from *E. sativa*. The pollen viability was on average 43.4% of *B. napus*. When selfpollinated, the number of seeds/pollinated flower was on average 7% of selfpollinated *B. napus*. When backcrossed to *B. napus* the number of seeds/pollinated flower was increased to 26.4 % of crosspollinated *B. napus*.

3.3. Erucic acid content

The erucic acid content varied from 0-35% between different hybrids. Variation within the same hybrid was also observed.

REFERENCES

1. Glimelius K: High growth rate and regeneration capacity of hypocotyl protoplasts in some Brassicaceae. Physiol. Plant. 61:38-44, 1984.
2. Sundberg E and Glimelius K: A method for production of interspecific hybrids within *Brassicaceae* via somatic hybridization, using resynthesis of *Brassica napus* as a model. Pl. Sci. 43:155-162, 1986.
3. Glimelius K, Djupsjöbacka M and Fellner-Feldegg H: Selection and enrichment of plant protoplast heterokaryons of *Brassicaceae* by flow sorting. Pl. Sci. 45:133-141, 1986.
4. Fahleson J, Dixelius J, Sundberg E and Glimelius K: Correlation between flow cytometric determination of nuclear DNA content and chromosome number in somatic hybrids within *Brassicaceae*. Plant Cell Rep. (accepted).
5. Sundberg E, Landgren M and Glimelius K: Fertility, chromosome stability and chloroplast segregation in synthetic *Brassica napus* produced by protoplast fusion. Theor. Appl. Genet. (in press).
6. Johansson S-Å and Uppström B: Analys av oljeväxtfrö: Bedömning av fettsyrasammansättning. Sver. Utsädesför. Tidskr. 88:35-44, 1978.

FERTILITY AND CHROMOSOME STABILITY IN BRASSICA NAPUS RESYNTHESISED BY PROTOPLAST FUSION

E. SUNDBERG*, M. LANDGREN and K. GLIMELIUS

Department of Plant Breeding, Swedish University of Agricultural Sciences, Box 7003, S-750 07 Uppsala, Sweden.

INTRODUCTION

Somatic hybridization gives new possibilities for increasing the genetic variation in our crop plants by facilitating gene flow from one species to another. However, interspecific hybridizations are often connected with chromosome instability and lack of fertility. In the present study, we have resynthesised rapeseed, *Brassica napus*, through somatic hybridization between *B. oleracea* and *B. campestris*. Even though somatic hybrids between these species has been obtained earlier (Schenck and Röbbelen 1982, Sundberg and Glimelius 1986, Robertson et al 1987 and Terada et al 1987) only one investigation of fertility of the hybrids have been reported (Jourdan et al 1986).

MATERIAL AND METHODS

Protoplast isolation was performed according to Glimelius (1984). From one of the fusion partners 4-day-old hybpocotyls were used and from the other partner leaves of a 3-week-old plant grown *in vitro* were used to produce morphologically different protoplasts. The heterokaryons obtained after PEG induced fusion were isolated either with micropipettes using a micromanipulator (Sundberg and Glimelius 1986) or by flow sorting (Glimelius et al 1986). In the experiments where fusion products were enriched by flow sorting, the hypocotyl protoplasts were stained with carboxyflourescein diacetate (12ug/ml) during enzyme treatment. The heterokaryons were cultured according to Glimelius (1984) with some modifications. Hybrid character of the obtained plants was confirmed by isozyme analysis. Ploidy level of the hybrid plants was determined by chromosome counting and relative DNA-content analysis. Pollen viability of the hybrid plants was determined by applying a method for vital staining of cells using flourecein diacetate as a marker. Chloroplast-DNA was isolated according to a slightly modified method of Palmer (1986) using young greenhouse-grown plants. The chloroplast-DNA was digested with *BamH1*, *EcoR1*, *HindIII* and *Pst1* and the fragments were separated by electrophoresis in horizontal 0.5% agarose gels.

RESULTS

The amount of calli differentiating to shoots was 1.6% from the fished fusion products and 2.5% of the flow sorted. Twentythree hybrid plants were obtained. All of the shoots obtained from manually isolated fusion products showed hybrid character for the isozymes LAP and PGI (Sundberg and Glimelius 1986). Of the plants obtained from fusion products collected by flow sorting, 87% were hybrids while the rest were true B. oleracea.

The chromosome number varied considerably (Table 1). The sum of the chromosome number (38) from the two fusion partners were found in 30% of the hybrids, 9% had less and 61% had more chromosomes. According to measurements of relative DNA-content, two hybrids were chimeras. Pollen viability and seed set varied with ploidy level (Table 1). Seed set was very low for hybrids with chromosome number deviating from the expected 38. Of the 11 hybrids examined for chloroplast-DNA, 6 contained B. oleracea and 5 B. campestris chloroplasts, suggesting random segregation of chloroplasts.

Table 1

Chromosome number	number of hybrids	pollen viability % of rape	fertility % of rape
38	7	75	17
<38	2	33	-
48-58	11	19	0.2
76-86	1	-	-
chimeras	2	8	0

REFERENCES

1. H. R. Schenck and G. Röbbelen, Z. Pflanzenzüchtg. 89 (1982) 278-288.
2. E. Sundberg and K. Glimelius, Pl. Sci. 43 (1986) 155-162.
3. D. Robertson, E. D. Earle, P. S. Jourdan and M. A. Mutschler. In: Valentine, F. A. (ed) Progress and perspectives in forest and crop biotechnology. Springer Verlag (1987) 193-211.
4. R. Terada, Y. Yamashita, S. Nishibayashi and K. Shimamoto, Theor. Appl. Genet. 73 (1987) 379-384.
5. P. S. Jourdan, M. A. Mutschler and E. D. Earle, Cruciferae Newsletter 11 (1986) 84-85.
6. K. Glimelius, Physiol. Plant. 44 (1984) 273-277.
7. K. Glimelius, M. Djupsjöbacka and H. Fellner-Feldegg, Pl. Sci. 45 (1986) 133-141.
8. J. D. Palmer, Methods Enzymol. 118 (1986) 167-186.

INVOLVEMENT OF PROTOPLAST FUSION IN CAULIFLOWER BREEDING.

D. CARTER, D.L.ARNOLD, S.F.BERRY.

Twyford Plant Laboratories U.K.
Baltonsborough, Glastonbury.
Somerset BA6 8QG. U.K.

INTRODUCTION.

Efficient F1 hybrid production relies on good self-incompatibility or male sterility systems. The most suitable CMS system available for Brassica oleracea is the Ogura cytoplasm, originally derived from Raphanus sativus. This system is stable, and does not break down at high temperatures. It is available in a number of B.oleracea varieties. However, a serious disadvantage is the low temperature cholorosis associated with ogu chloroplasts. This problem cannot be overcome by conventional breeding methods. Protoplast fusion can be used to combine the CMS cytoplasm directly with the desired nuclear characters. In addition, it is possible to select individual regenerants which combine the CMS mitochondria with normal chloroplasts from the male fertile parent; thus correcting the low temperature chlorosis normally associated with the ogu cytoplasm. (Menczel et al 1987). By using protoplast source, it may be possible to select against transfer of the ogu chloroplasts. Examples of suitable material include dark-grown hypocotyls and cell suspension. The poster describes preliminary results of a programme to transfer CMS into cauliflower by protoplast fusion.

MATERIALS AND METHODS.

1. CMS protoplasts. 4-5 day old, dark-grown hypocotyls were used for protoplast isolation. The hypocotyls were cut into 1mm pieces, and incubated overnight on a slow orbital shaker in the enzyme mixture described by Robertson and Earle (1986). Protoplasts were purified following the method of Robertson and Earle (1986).
2. Protoplasts from male fertile cauliflowers.

2.1. Cotyledons. Surface sterilized seeds were sown in LS medium in 60ml screw-capped jars. Some problems were encountered with the non-expansion of the cotyledons if the jars were tightly sealed. This problem was partially solved by the addition of activated charcoal to the medium, or by covering the jars with cling film instead of using the screw-caps. Protoplasts were isolated from 10-12 day old cotyledons, as described by Robertson and Earle (1986).

2.2. In-vitro shoot cultures. Shoot cultures were established from seedling shoot-tips. These were innoculated onto LS medium + 1mg/l BAP. On this medium, shoot proliferation occurred. The shoots were then transferred to LS +1mg/l +5g/l activated charcoal; on which shoot proliferation was inhibited, but leaf expansion occurred. Protoplasts were isolated from leaves 10-14 days after transfer to charcoal containing medium. The protocol of Robertson and Earle (1986) was again followed.

3. RESULTS AND DISCUSSION.

3.1. Hypocotyls. Protoplasts could be successfully isolated and cultured from hypocotyls of cabbage CMS lines. Colonies were formed when the protoplasts were grown in KM8P (Kao and Michayluk 1874) medium. Typical yields were 2.4x105/gfw. Plating efficiency after 10 days in KM8P was 24%.

3.2. Cotyledons and shoot cultures. Protoplasts were successfully isolated from both cotyledons and leaves of in-vitro shoot cultures. Typical yields were:

cotyledons: 1.65×10^6/gfw.

shoot cultures: 3.68×10^6/gfw.

However, a suitable culture medium for sustained divisions has yet to be identified. Cotyledon and mesophyll protoplasts were viable, reformed cell walls and divided in KM8P medium of medium B of Pelletier et al (1983) but sustained division has not been observed. Different isolation/culture protocols are being tested to resolve this problem.

3.3. Cultured cell lines. Attempts have been made to establish cell suspension cultures. Callus was initiated from roots and hypocotyls of male sterile lines. The callus was hard and nodular, and when innoculated into liquid culture, did not breakdown to form a suspension but grew as nodules of callus in liquid. Further work using a range of different media to establish a more friable callus is being carried out. If successful, cell suspensions will represent an efficient alternative protoplast source to hypocotyls for the male sterile lines. Once the isolation and culture protocols for the different male fertile and male sterile lines have been established, protoplast fusion experiments can commence. Efficient selection of cytoplasmic hybrids is also important. The growth and development of the CMS parental protoplasts can be inhibited by treatment with irradiation. This will also ensure that the products of fusion are essentially diploid and enables fusion of different B.oleracea varieties (eg cabbage with ogu cytoplasm with cauliflower). Regenerants can then be screened for male sterility followed by more detailed analysis of the plastid complement of male sterile regenerants.

REFERENCES.

Menczel L, Morgan A, Brown S, Maliga P. Fusion-mediated combination of Ogura-type ctyoplasmic male sterility with Brassica napus plastids using X-irradiated CMS protoplasts. Plant Cell Reports (1987)6:98-101.

Robertson D, and Earle E.D. Plant regeneration from leaf protoplasts of Brassica oleracea var italica cv Green Comet broccoli. Plant Cell Reports (1986)5:61-64.

Kao K.N, and Michayhuk M.R. Nutritional requirements for growth of Vicia hajastana cells and protoplasts at a very low population density in liquid medium. Planta (Berl.) 126:105-110.

Pelletier G, Primard C, Vedel F, et al. Intergeneric cytoplasmic hybridization in Cruciferae by protoplast fusion. Mol.Gen. Genet. 191:244-250. (1983).

PROTOPLAST FUSION AS A METHOD OF INCREASING CYTOPLASMIC HETEROGENEITY IN BRASSICA NAPUS.

S.A. YARROW*, R.J. KEMBLE, AND T.L. BARSBY.

Agriculture Division, Allelix Inc., 6850 Goreway Drive, Mississauga, Ontario, L4V 1P1, Canada.

Protoplast fusion technologies are useful plant breeding aids as they allow the creation of products which are unattainable by conventional methods, and permit the completion of desired goals more rapidly or at reduced costs.

The best examples of this can be found in the production of cytoplasmic hybrids or cybrids in canola (Brassica napus). Commercial production of F_1 hybrid seed can be facilitated by both CMS and triazine resistance (Atr^R) traits being present in one of the parents (Beversdorf et al, 1985). CMS and Atr^R are controlled genetically by chloroplast DNA (cpDNA) and mitochondria DNA (mtDNA) respectively. Protoplast fusion can allow biparental transmission of these traits into cybrids.

Pelletier et al. (1983) regenerated B. napus CMS/Atr^R cybrids carrying 'ogura' CMS type mitochondria. Alternatively we have produced spring canola cybrids which carry either 'polima' or 'napus' CMS traits, in combination with Atr^R (Barsby et al., 1987a, Yarrow et al., 1986), as identified by RFLP analysis (Kemble, 1987).

Viable seed was obtained from almost all the cybrids, many of which were polyploid. These seed produced diploid plants from which further seed generations have been recovered and are being evaluated in field trials.

Greater yield advantages can be realised through heterosis in autumn-planted (winter) F_1 canola hybrids. Introducing the CMS character from spring into winter genotypes while retaining the winter habit requires several generations (approx. 3 years) of conventional crossing. We have eliminated this extensive backcrossing process by protoplast fusion to incorporate nuclei from winter inbred lines into 'Polima' CMS cytoplasms available in spring canola (Barsby et al., 1987b). The desired cybrids have been backcrossed up to three times. Progeny will be evaluated as female parents for hybrid seed production.

Fusion also provides the opportunity to produce unique cytoplasmic combinations incorporating organelles from related Brassica species. Alloplasmic substitutions can often lead to male sterility. New forms of CMS are always of interest to the breeder. Presently we are studying a cybrid of B. napus canola, containing wild type 'napus' chloroplasts and mitochondria from the species B. maurorum. Backcrossing in addition to identification of 'maintainers' and 'restorer' lines, will be necessary for establishing a new CMS system.

No organelle DNA recombination was observed in any of the cybrids. Other workers have reported extensive mtDNA recombination in their Brassica fusion products (Morgan and Maliga, 1987; Vedel et al., 1986). These reports are based on the 'Ogura' cytoplasm. The cytoplasms we studied (polima, napus and Atr^{R}) are perhaps not as susceptible to recombination. However, we have produced three cybrids that carried 'ogura' mitochondria, and these possessed a mtDNA RFLP profile identical to control 'ogura' lines which were not subjected to tissue culture. This suggests that mtDNA recombination is independent of cytoplasmic type and may be dependent on the tissue culture system employed.

Further analysis of our cybrids revealed that the Brassica mitochondrial DNA plasmid was often lost and sometimes transferred from the chondriome of one protoplast parent to the chondriome of the other. Since protoplasts can be fused with protoplasts from other species and with prokaryotic spheroplasts it is suggested that fusion may be a method for transforming plant mitochondria with naturally occurring mt-plasmids or maybe with recombinant vectors. Intergration into the mt-chromosome is likely if sequences homologous to regions in plant mtDNA are incorporated into the vectors.

REFERENCES

1. T.L. Barsby, P.V. Chuong, S.A. Yarrow, S.C. Wu, M. Coumans, R.J. Kemble, A.D. Powell, W.D. Beversdorf and K.P. Pauls. The combination of Polima cms and cytoplasmic triazine resistance in Brassica napus. Theor. Appl. Genet. 73 (1987a) 809-814.
2. T.L. Barsby, S.A. Yarrow, R.J. Kemble and I. Grant. The transfer of cytoplasmic male sterility to winter-type oilseed rape (Brassica napus L.) by protoplast fusion. Plant Science (1987b - in press)
3. W.D. Beversdorf, L.R. Erickson and I. Grant. Hybridization process utilizing a combination of cytoplasmic male sterility and herbicide tolerance. United States Patent and Trademark. Office serial No. 493,511 (1985).
4. R.J. Kemble. A rapid, single leaf, nucleic acid assay for determining the cytoplasmic organelle complement of rapeseed and related Brassica species. Theor. Appl. Genet. 73 (1987) 364-370.
5. A. Morgan and P. Maliga. Rapid chloroplast segregation and recombination of mitochondrial DNA in Brassica cybrids. Mol. Gen. Genet. 209 (1987) 240-246.
6. G. Pelletier, C. Primard, F. Vedel, P. Chetrit, R. Remy, P. Rouselle and M. Renard. Intergeneric cytoplasmic hybridisation in Cruciferae by protoplast fusion. Mol. Gen. Genet. 191 (1983) 244-250.
7. F. Vedel, P. Chetrit, C. Mathieu, G. Pelletier and C. Primard. Several different mitochondrial DNA regions are involved in intergenomic recombination in Brassica napus cybrid plants. Curr. Genet. 11 (1986) 17-24.
8. S.A. Yarrow, S.C. Wu, T.L. Barsby, R.J. Kemble and J.F. Shepard. The introduction of CMS mitochondria to triazine tolerant Brassica napus L., var. "Regent", by micromanipulation of individual heterokaryons. Plant Cell Reports 5 (1986) 415-418.

SELECTION OF A UNIVERSAL HYBRIDIZER IN SINAPIS TURGIDA DEL. AND REGENERATION OF PLANTLETS FROM SOMATIC HYBRIDS WITH BRASSICA SPECIES

K. TORIYAMA[1], T. KAMEYA[2] and K. HINATA[1]

[1]Faculty of Agriculture and [2]Institute for Agricultural Research, Tohoku University, Sendai 980, Japan.

If a double-mutant can be established with two markers that are dominant and counter-selectable, it can be serve as a universal hybridizer in selecting somatic hybrids exclusively after protoplast fusion with any other wild-type partner. Such a universal hybridizer was selected from the cell suspension of Sinapis turgida Del., members of the Brassicaceae. In a preliminary survey on tissue culture for about 50 species in Brassica and its wild allies, we found that S. turgida delivered cell suspension in a liquid culture (Toriyama et al 1987b). Following protoplast fusion between S. turgida and B. oleracea L., and between S. turgida and B. nigra (L.) Koch, somatic hybrids were selected and plantlets were obtained (Toriyama et al 1987a).

A double-mutant cell line, which was unable to grow in a medium with NO_3 as the sole nitrogen source and was resistant to 5-methyltryptophan (5MT), was selected from cell suspension of S. turgida by culturing the cells in AA medium (Toriyama and Hinata 1985) supplemented with 50 mM chlorate and 229 µM 5MT. The mutant had lost the ability to regenerate shoots. Protoplasts of this mutant were fused with mesophyll protoplasts of B. oleracea (cauliflower) by dextran method (Kameya 1981). Out of 4×10^5 fusion-treated protoplasts, six somatic hybrids were selected initially by culture in the NO_3 medium and then by transfer to the NO_3 medium supplemented with 229 µM 5MT (Fig. 1). The somatic hybrids produced embryoids, leaves and plantlets on a regeneration medium. These plantlets, however, showed abnormal development. They expanded several leaves about 1.5 cm long but never produced further leaves or stems and eventually died. Whole plants could not be obtained so far.

The hybrid characters were confirmed by investigations of acid phosphatase and peroxidase isoenzymes. Chromosome number of each hybrid was equal to the sum of that of S. turgida (2n=18) and B. oleracea (2n=18). The hybrids grew on the NO_3 medium, although the extent of growth was different among cell lines. The level of 5MT resistance in the hybrids was intermediate between that of the mutant of S. turgida and unselected cell line of S. turgida. Somatic hybrids between S. turgida and B. nigra were also obtained using the same method. These results indicate that the double mutant of S. turgida established here will be able to serve as a universal hybridizer.

In order to obtain normal whole plants, "Back-fusion" was carried out between the abnormal hybrids (S. turgida + B. oleracea, cell line 417-2) and B. oleracea (cauliflower). Protoplast fusion was carried out by dextran using protoplasts isolated from suspended cells of the original somatic hybrid (417-2, 2n=36) and from hypocotyls of cauliflower (2n=18). The fusion-treated protoplasts were cultured without selection of hybrid cells, and two whole plants, strain 505-2 and -3, were regenerated. Both of them showed intermediate morphology between cauliflower and *S. turgida* (Fig. 2). Starin 505-3 bore flowers in *vivo*, but the anthers were stunted. The chromosome number of these plants was 2n=38 for 505-2 and 2n=28 for 505-3. Although it is unclear whether these plants were originated back-fused protoplasts or the original hybrid (417-2) after elimination of some chromosomes, we considered that they were back-fused hybrids, because 417-2, by it self, never regenerated plants. Back-fusion techniques will be effectively used to obtain whole plants from abnormal somatic hybrids.

REFERENCES

1. K. Toriyama, T. Kameya and K. Hinata, Planta 170 (1987a) 308-313.
2. K. Toriyama, T. Kameya and K. Hinata, Plant Tissue Culture Lett. 4 (1987b) 75-78.
3. T. Kameya, M.E. Horn, J.M. Widholm, Z. Pflanzenphysiol. 104 (1981) 459-466.

Fig.1. Selection scheme for somatic hybrids between universal hybridizer (35-i) of *S. turgida* and other wild-type partners. Somatic hybrids are selected by NO_3 medium and then 5-methyltryptophan (5MT) medium.

Fig.2. Young plants of *S. turgida* (A), and cauliflower (B). Plantlet of somatic hybrid between *S. turgida* and cauliflower (C). Young plant (D) and flowering plant (E) of back-fused somatic hybrid between (*S. turgida* + cauliflower) and cauliflower. Scale: 1 cm.

TOWARDS THE INTRODUCTION OF CYTOPLASMIC MALE STERILITY (CMS) INTO BRASSICA NAPUS THROUGH PROTOPLAST FUSION

M.S. MCLELLAN, P. OLESEN[1] and J.B. POWER
Plant Genetic Manipulation Group, Department of Botany, University of Nottingham, Nottingham NG7 2RD, UK.
[1]De Danske Sukkerfabrikker, Biotechnology Section, 1 Langebrogade, P.O. Box 17, DK 1001, Copenhagen, Denmark.

1. INTRODUCTION

Genetically restorable cytoplasmic male sterility (CMS) introduced into *Brassica napus* (oil-seed rape) is desirable since it would aid breeders in the production of hybrid seed. *Diplotaxis muralis* has been used to introduce CMS into *B. campestris* by substituting, through sexual crossing and backcrossing, the *B. campestris* genome into the cytoplasmic background of *D. muralis* (1). Investigations are underway to assess the potential of *D. muralis* as a source for the introduction of CMS, via protoplast fusion, into *B. napus*.

2. MATERIALS AND METHODS

Stem epidermal strip protoplasts of *B. napus* cv Line and cotyledon protoplasts of *D. muralis* were isolated following a 4-5h incubation in 0.75% (w/v) Cellulase RS, 0.2% (w/v) Pectolyase Y23, CPW 13% (w/v) mannitol (pH 5.8). Protoplasts were isolated from the hypocotyls of 5 day-old dark-grown seedlings of *B. napus* and *D. muralis* following a 4-5h incubation in 0.5% (w/v) Rhozyme HP150, 1.0% (w/v) Cellulase RS, 0.2% (w/v) Pectolyase Y23, CPW 13% mannitol (pH 5.8).

Protoplasts were cultured, using the agarose bead technique (2), in KM8p medium (3) solidified with 1% (w/v) Seaplaque agarose. Calli (2mm diameter, 4 weeks) arising from cultures, were transferred to LS medium (4) with 0.1 mgl^{-1} 2,4-D, 0.1 mgl^{-1} NAA, 1.0 mgl^{-1} zeatin, 3% (w/v) sucrose and solidified with 0.4% (w/v) agarose (Sigma, Type I). After a further 2-3 weeks, colonies were transferred to LS medium with 1.0 mgl^{-1} zeatin, 1.0 mgl^{-1} BAP, 0.1 mgl^{-1} IAA, 3% (w/v) sucrose, 0.4% (w/v) agarose (Sigma, Type I). Shoots arising from calli were rooted on MS medium lacking phytohormones and with 0.7% (w/v) agar. Rooted plantlets were established in soil-less compost and maintained in the greenhouse.

The following protoplast systems were fused:-

D. muralis (cotyledon) with *B. napus* (hypocotyl)
B. napus (stem epidermal strip) with *D. muralis* (hypocotyl)

In both cases the hypocotyl protoplasts were stained with 12.5 $\mu g\ ml^{-1}$ fluorescein diacetate to allow for the identification of heterokaryons based on a dual FDA/chlorophyll fluorescence. Three methods of fusion were employed; small-scale and large-scale chemical fusion with 30% (w/v) PEG MW1540 and high pH-Ca^{++}, or electrofusion as based on the method of Watts and King (5). For electrofusion, protoplasts of the two species were suspended at a density of 1 x 10^5/ml in 9% mannitol with 0.5 mM $CaCl_2$ (pH 5.8), aligned in an AC field (40 V/cm and 500 KHz) and fused with 1 ms DC pulse of 400 V/cm.

A Coulter Electronics EPICS V (Cell Sorter) was used for heterokaryon isolation. The argon ion laser was operated at 457 nm, output 100 mW. The 90° light scatter photomultiplier (PMT) detected light of 430-485 nm, the green PMT 495-565 nm and the red PMT 630 up to 700 nm. The machine was fitted with a 200 µm nozzle and stable droplet formation was achieved at a sheath pressure of 7.5 psi with a piezoelectric drive frequency of 9.8 KHz.

3. RESULTS AND DISCUSSION

Protoplasts isolated from hypocotyls and stem epidermal strips of B. napus cv Line gave reproducibly high plating efficiencies and regeneration frequencies.

One approach for the transfer of cytoplasm of D. muralis to B. napus was the fusion of B. napus and D. muralis protoplasts followed either by mass regeneration, or by the isolation and subsequent culture of heterokaryons from a fusion mixture. The basis for this approach was the development of a cultural regime that allowed for a high plating efficiency and plant regeneration for protoplasts of B. napus but at the same time did not support the division of protoplasts of D. muralis. It was considered, that following the culture of heterokaryons, differences in the cultural characteristics of the parental protoplasts may allow for the segregation of the nuclear genomes and would result in the desired combination of the nuclear genome of B. napus plus the cytoplasm of D. muralis.

Techniques have been developed for the isolation of heterokaryons by micromanipulation and fluorescence activated cell sorting. In the latter case heterokaryons were sorted on log integrated green and log integrated red fluorescence signals. Heterokaryons were sorted into liquid KM8p medium (3) in 96 well microtiter plates and were cultured for 14 days prior to transfer to agarose solidified bead culture.

Within this study small-scale chemical fusion resulted in a higher frequency of heterokaryon formation (6.4%) as compared to large-scale chemical fusion (1.6%) or indeed electrofusion (4.7%). However, PEG induced fusion usually resulted in the aggregation/adhesion of parental protoplasts which hindered the sorting or micromanipulation of heterokaryons. For this reason electrofusion was found to be the more suitable porotoplast fusion method.

A greater % colony formation from isolated heterokaryons was seen following micromanipulation (20%) as compared to fluorescence activated cell sorting (6.7%). This was assumed to be due to the fact that micromanipulation allowed for an evaluation of the viability of heterokaryons to be selected. However cell sorting allowed for the isolation of high numbers of heterokaryons within a short period of time (60-100 heterokaryons/minute) whilst micromanipulation proved to be more time-consuming (20-40 heterokaryons/hour). For this reason fluorescence activated cell sorting was considered the most promising approach for the isolation of heterokaryons, with a large number of calli being recovered following the culture of sorted heterokaryons.

4. SUMMARY

The protocols outlined allow for the isolation of large numbers of viable protoplasts of B. napus and their subsequent manipulation to the whole plant level. This alongside techniques for the isolation of heterokaryons and their regeneration to the callus level will allow selected fusion products between B. napus and D. muralis to be regenerated.

5. REFERENCES

1. Hinata K, Konno N: Jap. J. Breed. 29: 305-311, 1979.
2. Shillito RD, Paszkowski J, Potrykus I: Plant Cell Rep. 2: 244-247,
3. Kao KN, Michayluk MR: Planta. 126: 105-110, 1975.
4. Linsmaier EM, Skoog F: Physiol. Plant. 18: 100-127, 1965.
5. Watts JW, King JM: Bioscience Rep. 4: 334-342, 1984.

PRODUCTION OF SOMATIC HYBRID PLANTS, 'BRASSICOMORICANDIA', THROUGH PROTOPLAST FUSION BETWEEN MORICANDIA ARVENSIS AND BRASSICA OLERACEA

K. HINATA[1], K. TORIYAMA[1], T. KAMEYA[2] AND R. ISHII[3]

[1]Faculty of Agriculture, [2]Institute for Agricultural Research, Tohoku University, Sendai 980, Japan.
[3]Faculty of Agriculture, Tokyo University, Tokyo 113, Japan.

Intergeneric hybrid plants were obtained through protoplast fusion between *Brassica oleracea* L. and *Moricandia arvensis* (L.)DC. (Toriyama et al 1987). *M. arvensis* is classified as a member of the subtribe Moricandiinae of the family Brassicaceae, and is known to be a C_3-C_4 intermediate species with respect to low CO_2 compensation point (Holaday et al 1981). We have been studying (i) the production of somatic hybrids between distantly related species, and (ii) the genetic regulation of photosynthetic systems in the hybrids.

Callus of *M. arvensis* was obtained by incubating the leaf segments on MS medium supplemented with 1 mg/l 2,4-D. Protoplast fusion was carried out by the dextran method (Kameya 1983) using protoplasts isolated from suspended callus cells of *M. arvensis* and from hypocotyls of red cabbage. The fusion-treated protoplasts were cultured in B5 medium with 2 mg/l 2,4-D and 1 mg/l BAP, without selection of hybrid cells. Out of 112 calli transferred to regeneration medium (MS medium with 1 mg/l NAA, 1 mg/l BAP and 100 mg/l casamino acids), 8 calli gave rise to plants. All the regenerated plants exhibited combined morphological characteristics of both parental species. Analysis of leaf isozymes for acid phosphatase and aspartate aminotransferase also indicated their hybridity. Even though no selection was performed for hybrid cells, all the regenerated plants were hybrids. This is attributed to the effects of hybrid vigor on shoot regeneration.

Most of the hybrid plants had double flowers (Fig.1). From the observation of the pollen mother cells of the single-flowering plants (466-1), the hybrid plants were revealed to be amphidiploid. Although the anthers did not contain any fertile pollen grains, offsprings were obtained by backcrossing with *B. oleracea*. Chloroplast (cp) DNA types of the hybrid plants were determined using restriction endonuclease *Xho* I. Out of 11 plants investigated, 5 plants had the cpDNA of *B. oleracea*, while 6 had that of *M. arvensis* (Table 1). Independently of cpDNA types, the CO_2 compensation point of these hybrids was as high as that of *B. oleracea* (Table 1). The C_4-like character of *M. arvensis* was not exhibited in the hybrids. Two somatic hybrid plants were also obtained between cauliflower and *M. arvensis*.

REFERENCES

1. K. Toriyama, K. Hinata and T. Kameya, Plant Sci. 48 (1987) 123-128.
2. A.S. Holaday, Y.J. Shieth, K.W. Lee and R. Chollet, Biochim. Biophys. Acta 637 (1981) 334-341.
3. T. Kameya, Cytologia 48 (1983) 873-878.

Table 1. Chloroplast (cp) DNA type, the CO_2 compensation point (Γ) and flower type in the somatic hybrid plants (466) between Brassica oleracea (BO) and Moricandia arvensis (MA). The hybrid plants derived from the same callus were grouped as a strain. For flower type, see Fig.1B.

Strain No.	Plant No.	cpDNA type	Γ (µl/l CO_2)	Flower type
BO		BO	50	a
MA		MA	20	b
466-1	1	BO	52	c
466-1	2	BO	-	c
466-3	1	MA	53	d
466-3	2	MA	-	d
466-4	-	-	-	d
466-6	1	BO	56	e
466-6	2	MA	55	e
466-6	3	BO	-	e
466-12	1	MA	-	d
466-12	2	MA	-	d
466-13	1	BO	-	d
466-14	1	MA	-	d
466-15	-	-	-	d

Fig.1. A: (left to right) Flowering plants of Red cabbage, somatic hybrid between Red cabbage and M. arvensis, and M. arvensis. B: Flowers of Red cabbage (a; yellow), M. arvensis (b, violet) and somatic hybrids between them (c; violet, d; violet, e; white). Scale: 10 cm (A) and 1 cm (B).

THE USE OF HAPLOID PROTOPLAST FUSION TO COMBINE CYTOPLASMIC ATRAZINE RESISTANCE AND CYTOPLASMIC MALE STERILITY IN BRASSICA NAPUS.

P.V. CHUONG, W.D. BEVERSDORF, A.D. POWELL and K. P. PAULS
Department of Crop Science, University of Guelph, Guelph, Ontario, CANADA

Introduction

This report describes the use of haploid protoplast fusion to create cytoplasmic atrazine resistant (CATR) and cytoplasmic male sterile (CMS) B. napus plants. The study was carried out in parallel with the diploid/diploid protoplast fusion work described by Barsby et al. (1987). Protoplasts from haploids were used because the fusion process reestablishes the diploid condition and precludes the need for nuclear inactivation or cytoplast production in one of the parental protoplast populations. The haploid protoplasts were also well suited for use with a flow cytometer/cell sorter to isolate heterokaryons from the fusion mixtures. The success of this approach was dependent on the development of procedures to isolate haploid lines from CMS and CATR B. napus (Chuong et al., 1987a) and the development of plant regeneration protocols from haploid B. napus protoplasts (Chuong et al., 1987b,c).

Material and Methods

Microspores extracted from CATR and Polima-CMS (Sernyk, 1983) plants were cultured as described by Chuong et al., (1987a)., Fusion experiments not utilizing the flow cytometer cell sorter were conducted with protoplasts isolated from stem peels of CATR and CMS plants (Chuong et al. 1987 b,c).

Cell sorter studies required that a non-pigmented protoplast be used as one of the fusion partners. Protoplasts from root callus (Powell et al., in prep) were used and were stained by adding 20 µl of a 0.2 mg/ml solution of fluorescein isothioyanate (FITC) in ethanol to 2 ml of protoplasts,suspended in CPW, and incubating the mixture for 30 min., at room temperature, prior to inducing fusion.

Both stem/stem and stem/root protoplast mixtures were induced to fuse by addition of PEG 8000. Stem/stem protoplast fusion mixtures were cultured in VN medium (Chuong et al., 1987 b,c). Stem/root protoplast fusion products isolated by flow cytometric sorting were cultured in 96 microwell plates containing one of the following: 1) microspores in 50 µl NLN medium (Chuong and Beversdorf, 1985); 2) 50 µl NLN medium; 3) microspores in 50 µl NLN containing 0.8% (w/v) agarose with 25 µl of VN medium on top or 4) 50 µl NLN containing 0.8% (w/v) agarose with 25 µl of VN medium on top. Microcalli which developed in both sets of experiments were cultured according to Chuong et al., (1987c).

A Coulter Epics V single laser flow cytometer with a 103 µm flow cell tip was used. (Sheath pressure -9 psi, crystal drive frequency-20kHz). Atrazine sensitivity was assessed by fluorimetry (Ali and Souza-Machado, 1981) and ploidy levels were determined by cytophotometry (Barsby et al., 1987).

Mitochondrial DNA (mt DNA) was extracted by the method of Kolodner and Tewari (1972). Chloroplast DNA was isolated after in organellar digestion (Kut and Flick, 1986). EcoRI restriction digests were resolved by electrophoresis using 0.8% agarose gels .

Results

Two hundred and sixty-one plants were regenerated from 10 stem/stem protoplast fusion experiments carried out with mixtures of haploid CATR lines and Polima-CMS lines. The plants were characterized with respect to their fertility, atrazine susceptibility and ploidy (Table 1). The regenerants included parental type plants as well as one cytoplasmic hybrid (cybrid) which was atrazine resistant/male sterile (CATR/CMS). Among the regenerants 87% were CATR/fertile, 53% were haploid, 43% were diploid and 3% were tetraploid.

TABLE 1:Ploidy of plants regenerated from haploid fusions

Phenotype	Type	Number of Plants n	2n	4n	Total	%
Atrazine resist./Male fertile	Parental	128	90	8	226	86.6
Atrazine resist./Male sterile	Cybrid	0	1	0	1	0. 4
Atrazine sensit./Male sterile	Parental	11	22	1	34	13.0
TOTAL		139	113	9	261	
Percentage		53.3	43.3	3.4		

The CATR/CMS cybrid was diploid and set seed normally when pollinated with a maintainer line (Fig. 1). It had typical Polima type flowers with markedly reduced stamens (Fig. 2) and was atrazine resistant. Further confirmation that this plant was a product of protoplast fusion was obtained from an analysis of organellar DNA (Fig. 3). The chloroplast DNA restriction fragments pattern of the cybrid was identical to the CATR chloroplast pattern and differed from the Polima-CMS pattern by possessing an additional fragment of 3.3 kb. The mitochondrial DNA (mtDNA) restriction fragment pattern was identical to the Polima-CMS mtDNA and differed from the CATR mtDNA pattern by possessing extra fragments of 5.4 and 5.1 kb but missing a fragment of 6.8 kb.

Stem/root protoplast fusions were used in the flow cytometry studies. Stem and root protoplasts fluoresced red (chlorophyll) and green/yellow (FITC) respectively. Fusion products between these two protoplasts had significant red and green fluorescence and could be visually distinguished from the parental protoplasts (Fig. 4). The two large peaks on the x- and y-axes represent cells which did not engage in the fusion process. The much smaller population present in the area between the axes represents the products

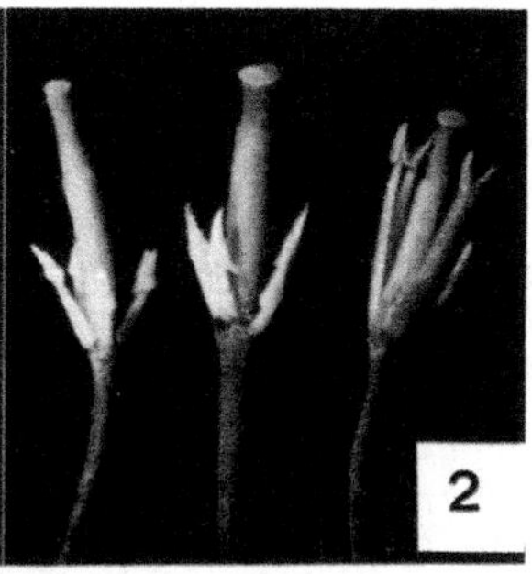

Fig. 1 Cybrid from stem/stem fusions. Fig. 2 Pistils and stamens of parental lines and cybrid. left, atrazine sens./male ster. parent; middle, haploid diploid atrazine resist./male ster. cybrid; right haploid, atrazine resist./male fert. parent.

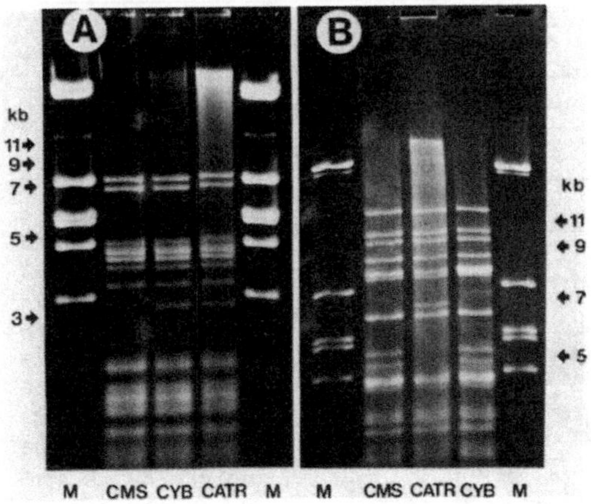

Fig. 3 Restriction fragment analysis of a) chloroplast DNA and b) mitochondrial DNA from parental lines and cybrid. CMS-DNA from ATR sensitive/male sterile parent; CYB-DNA from ATR resistant/male sterile cybrid and CATR-DNA from resistant/male fertile parent.

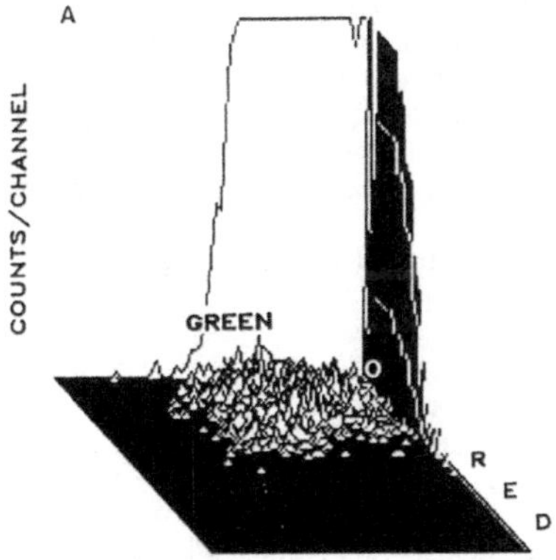

Fig. 4 3-D representation of fluorescence from the protoplast fusion mixture.

of protoplast fusion. Protoplasts which fluoresced red and green were sorted by the flow cytometer into microwell plates containing B. napus microspores. The microspores provided a nurse function to the sorted protoplasts since cell division and microcalli development only occurred in wells containing the microspore nurse culture (Fig. 5). For all the protoplast densities tested the liquid culture medium was superior to the liquid on agar medium in supporting the division of the sorted cells. Twenty-six plants from 7 calli were regenerated. All of the plants were male fertile and atrazine sensitive suggesting that they are cybrids. Most of the plants appeared to be diploid because of the high frequency of seed set when selfed. The diploid nature of 19 of the plants was confirmed cytophotometrically after staining root tip squashes with Feulgen stain.

Discussion

This report illustrates the utility of haploid protoplast fusion for altering the cytoplasmic makeup of Brassica napus. The most significant advantage of using this approach for intraspecific cybrid formation is the fact that there is a high probability of recovering diploid fusion products. The diploids from the fusions described in this report were completely female fertile. In contrast the tetraploid and pentaploid Brassica napus CATR/CMS cybrids obtained from fusions between untreated diploid gamma-irradiated protoplasts with diploid iodoacetate-treated protoplasts had very low female fertility (Barsby et al. 1987).

The use of haploid protoplasts also facilitated the utilization of flow cytometry to isolate heterokaryons. Because haploid protoplasts are smaller than diploid cells the fusion products derived from the

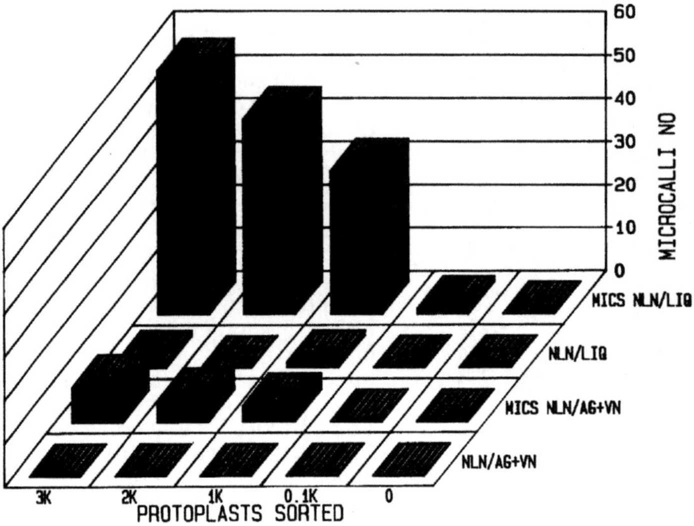

Fig.5 Microcalli development from sorted protoplasts in the presence or absence of a microspore nurse culture. Abbreviations- MICS= micro- spores; NLN= microspore culture medium; LIQ=liquid; AG= agarose; VN= protoplast culture medium.

former are rarely too large to pass through a 103 µm orifice in the flow cell tip of the instrument. The larger flow cell tips that are often required to sort heterokaryons from fusions between diploid protoplasts are more difficult to use and require crystal drive frequencies that are extremely irritating to the ear.

The CATR/CMS cytoplasm described previously (Barsby et al., 1987) and in this report is stable. All of the progeny of the cybrids pollinated with maintainer lines are atrazine resistant and male sterile. They are currently being evaluated for their usefulness as female parents for hybrid seed production (I. Grant, personal communication, Allelix Inc.).

REFERENCES

Ali, A. and Souza-Machado, V. (1981). Weed Res. **21**:142-147

Barsby, T.L., Chuong, P.V., Yarrow, S.A., Coumans, M., Kemble, R.J., Powell,A.D., Beversdorf, W.D. and Pauls, K.P. (1987). Theor. Appl. Genet. **73**:809-814

Chuong, P.V. and Beversdorf, W.D. (1985). Plant Sci. **39**:219-226

Chuong P.V., Pauls, K.P. and Beversdorf, W.D. (1987a). Can. J. Bot. In press.

Chuong, P.V., Beversdorf, W.D. and Pauls, K.P. (1987b). J. Plant Physiol. **130**:57-65

Chuong, P.V. Beversdorf, W.D. Powell, A.D. and Pauls, K.P. (1987c). Molec. Gen. Genet. In press.

Kolodner, R and Tewari, K.K. (1972). Proc. Nat. Acad. Sci. USA **69**:1830-1834

Kut, S.A. and Flick, C.E. (1985). Plant Molec. Biol Rep. **4**:48-55

Powell, A.D., Chuong, P.V., Beversdorf, W.D. and Pauls, K.P. (1987). In prep.

Sernyk, J.L. (1983). Proc. Brassica workshops on hybrids and analytical chemistry.Paris, p11

SOMATIC HYBRIDIZATION OF DIPLOID POTATO LINES, A TOOL IN POTATO BREEDING

L.SCHILDE-RENTSCHLER, G.BOOS and H.NINNEMANN
Medizinisch-Naturwissenschaftliches Forschungszentrum der Universität, D74 Tübingen.

INTRODUCTION

Although the combination of two different selected diploids to a tetraploid clone is an often existing breeding goal, and somatic hybridization of S.tuberosum with S.brevidens has been reported, protoplast fusion has not yet become a routinely used technique in potato breeding. Some problems are: isolation and culture of protoplats of ordinary breeding lines not selected for the response in cell culture; unreproducible fusion methods and selection methods of hybrids. We tried to overcome these problems and elaborated conditions which permit to utilize protoplast fusion as a tool for combining diploid breeding lines.

MATERIAL AND METHODS

Plant material: diploid lines of a commercial breeder.

Isolation of protoplasts: Leaves of in vitro plants after dissection digested by 2-step method (macerozyme, cellulase).

Culture of protoplasts: in liquid VKM (1) and modified MS-medium (2), often charcoal added. Dilution every 3 to 4 days. Transfer of colonies to VKM-agarose plates, for regeneration to MS13-medium (3).

Culture at low density: 1 vol. protoplasts mixed with 1 vol. 2.8-3.2% Na-alginate in 0.4M mannitol, then cultured in the presence of protoplast or cell suspension. After colony formation depolymerization with 20 mM Na-citrate, 0.3M mannitol.

Protoplast fusion: a. with PEG, at high pH and 50-80 mM Ca^{++}. b. By electric pulse in a homogenous electric field.

Identification of hybrids: Small amounts of tissues (100mg) of colonies, shoots or tubers were extracted (4) and applied to ampholine gels (LKB). IEF of esterases and in some cases peroxidases. PAGE of tuber proteins: in vitro formed or field grown tubers were extracted. PAGE at pH 7.9 and 8.9, gels stained for proteins, esterases or peroxidases.

RESULTS

Protoplast isolation andculture: Our procedure could be utilized with small modifications for most breeding lines tested so far. The preculture conditions of the in vitro plants are very important.

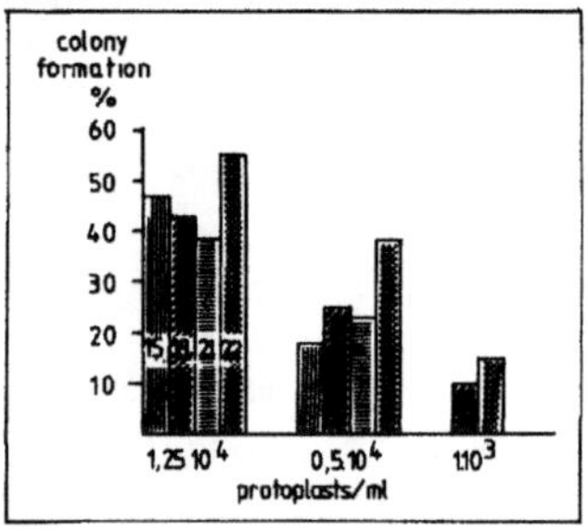

fig.1.Culture of protoplasts in alginat (4 resp. 2 clones)

Culture in alginate beads allowed to culture small volumes of protoplasts and also reduce the protoplast density in the in the beads (fig.1).

Protoplast fusion: Although we obtained hybrids, PEG-fusion was not satisfying: the conditions for exposure varied and fusion rate and size of fusion products could not be controlled sufficiently. By fusion with electric pulse it was possible to control in limits both by varying protoplast density and amplitude and period of the RF-field. Conditions can be found (fig.2) when the fusion products consist mainly of 2 or 3 protoplasts. The treatment under these optimal conditions has no influence on viability.

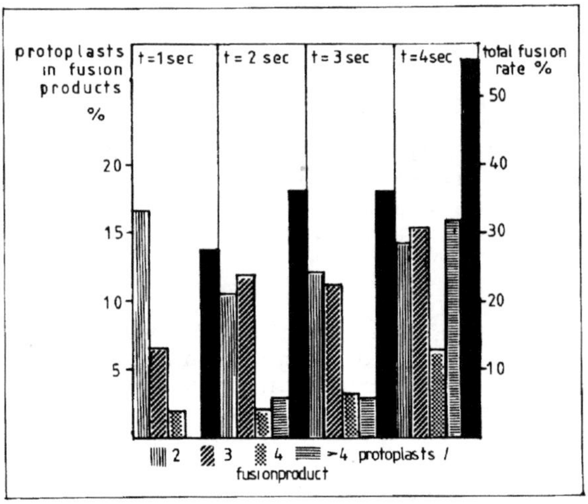

fig.2. Influence of the period of the RF-field on fusion rate and size of fusion products.■=total f.r.

Identification of hybrids: Small amounts of tissues (50 - 100 mg) were sufficient for several IEF and PAGE-runs. The data allowed to identify hybrids. The patterns were more reproducible with esterases than with peroxidases. Especially for one combination hybrids from different clones were studied morphologically. In some flower colour was intermediate betw ween the two partners, leaflets were broader, the tubers larger, and the shape intermediate. In others IEF and protein pattern looked as hybrids, but flowers were mostly white with only little violet, the tubers long and small, the leaflets curled. They will be analysed cytologically.

CONCLUSION

It is possible to obtain viable protoplasts and regenerated plants from ordinary breeding lines not selected for in vitro purposes. Electrofusion allows to obtain good percentages of fusion products containing mainly 2 or 3 protoplasts. Identification of probable hybrids at an early stage is possible, introduction of markers in breeding material therefore unnecessary.

REFERENCES

1. Binding H, Nehls R. Z.Pflanzenphysiol.85, 279-280 (1977).
2. Tavazza R, Ancora C. Plant Cell Reports 5, 243-246 (1986).
3. Behnke M. Theor. Appl.Genet. 56, 151-152 (1979).
4. Stegemann H, Löschcke V. Potato Res. 20,101-110 (1977).

This project was financed by the Ministry of Research and Technology together with the Association for the Promotion of Private Plant Breeding (GfP).

SOMATIC HYBRIDIZATION BETWEEN ANTHER-DERIVED DIHAPLOID LINES OF *SOLANUM TUBEROSUM* L. (POTATO)

SYLVIA B. KARLSSON AND TAGE ERIKSSON Institute of Physiological Botany of the University of Uppsala, Box 540, 751 21 Uppsala, Sweden

INTRODUCTION

A breeding program incorporating somatic hybridization has been suggested for potato (7). In order to make this technique available for potato breeders we have tried to develop a general method of fusion and identification of fusion products. We also tested the use of isoenzyme pattern as a method to identify the somatic hybrids produced.

PROCEDURE

Anther-derived dihaploid clones were produced by and according to Johansson (4). Dihaploid clone 67:9 (derived from Norweigan number variety Y-67-20-40) was chlorophyll bleached with the photosynthetic herbicide SAN 9789 (6) by placing nodal explants in shoot propagation medium (1) containing 3% sucrose and 2.8 mg/l SAN 9789. Green shoot cultures of dihaploid clone 198:2 (derived from cv. Maria) were subcultured as described previously by Carlberg et al. (1). Protoplasts were isolated and purified according to Carlberg et al. (1). Following PEG fusion, fusion frequencies up to 4% was observed (5). Fusion products could be identified up to 2 days after fusion using the dual fluorescence of carboxyfluorescein stained chlorophyll bleached mesophyll protoplasts and the autofluorescence of green mesophyll protoplasts.

Unfused and fused protoplasts were cultured together in a modified 8p medium containing 1.0 mg/l 2,4-D, 0.5 mg/l BA and 0.4 M glucose (for further details on the culture procedure see reference 2) and after 2 months of culture, fast growing calli were selected as showing hybrid vigour. Of these, 25 were transferred to shoot induction medium and plants were regenerated from 15 of these.

By screening 15 enzyme systems using both isoelectric focusing and starch gel electrophoresis, 6 enzyme systems were found in which isoenzyme pattern differed between dihaploid clones (Tab. 1), these were utilized for hybrid identification (5).
Out of the 15 calli regenerating shoots, 7 plants (2A, 2B, 2C, 8A, 8B, 8E and 22C) arising from 3 different calli (numbered 2, 8 and 22) were considered as somatic hybrids based upon the combination of isoenzymes of both parents. The remaining plants all showed isoenzyme pattern identical to dihaploid clone 67:9.
The DNA analysis of hybrid plants revealed that plants from calli 2 and 8 were stable tetraploids, while plant 22C was highly mixoploid containing nuclei ranging from 3x to 6x.

CONCLUSIONS

* Fusion products could be obtained from PEG fusions and they were identified by the dual fluorescence of stained chlorophyll bleached mesophyll protoplasts and green mesophyll protoplasts.

* Intraspecific hybrids could be identified using isoenzyme pattern as phenotypical marker.
* Hybrid plants and parental plants were obtained from chlorophyll bleached plants, which indicates that the herbicide as well as the fluorescent stain caused no damage to the protoplasts.

REFERENCES

(1) Carlberg I., K. Glimelius and T. Eriksson. 1983. Plant Cell Reports 2:223-225.
(2) Carlberg I. K. Glimelius and T. Eriksson. 1984. Plant Sci. Let. 35: 225-230.
(3) Glimelius K., T. Eriksson, R. Grafe and A. J. Muller. 1978. Physiol. Plant. 44:273-277.
(4) Johansson L. 1986. Potato Res. 29:179-190.
(5) Karlsson S. B., T. Eriksson and A. Wallin. 1987. In manuscript.
(6) Uhrig H. 1981. Mol. Gen. Genet. 18:403-405.
(7) Wenzel G., O. Schieder, T. Przewozny, S. Soporyand G. Melchers. 1979. Theor. Appl. Genet. 55:49-55.

Table 1. Isoenzyme systems used for hybrid identification.

Acid phosphatase
Esterases
Isocitrate dehydrogenase
Malate dehydrogenase
6-phosphogluconate dehydrogenase
phosphoglucose isomerase

FUSION OF DIHAPLOID SOLANUM TUBEROSUM PROTOPLASTS AND TRIALS OF HYBRID SELECTION WITH PHYTOTOXINS

C. Möllers, S. Deimling and G. Wenzel

Institute of Resistance Genetics, Federal Biological Research Centre for Agriculture and Forestry; 8059 Grünbach, F.R.G.

INTRODUCTION

In potato, breeding on the dihaploid level gaines increasing importance. One bottleneck in this procedure is the return from the dihaploid to the tetraploid level. Fusion of protoplasts is probably the most efficient technique to overcome this problem. The regeneration rate of protoplasts increased significantly even after fusion. Now there are realistic chances of regenerating somatic hybrids. If, however, the somatic hybridization shall be applicable by commercial plant breeders, the likelihood of regenerating hybrids must be increased as otherwise an enormous input of labour is necessary. Besides optimizing the fusion process a selection system based on the resistance to phytotoxins is under development. Potato clones, containing different phytotoxin resistances are fused and phytotoxins are added. This leads to heterosis and furthermore, existing differences in toxin sensitivity can be utilized to increase the regeneration rate of hybrids. Here the influence of Phytophthora toxins on the development of protoplasts from different clones with and without fusion is described.

MATERIALS AND METHODS

Protoplasts from tetraploid and dihaploid clones were screened for their toxin resistance (Tab. 1). Protoplasts of the clone H 75.1207/7, showing the highest resistance among the tested dihaploids, were fused with less resistant protoplasts from the clones H 80.1202/9 and HH 372. Protoplasts were isolated from axenic shoot cultures following the method of Haberlach et al. (1985). Fusion was carried out with PEG 6000. For plating in V/KM medium and regeneration the method of Binding et al. (1978) was used. The Phytophthora toxin was purified and 50fold concentrated by ultrafiltration (kindly provided by Dr. Schöber; BBA, Braunschweig), and added in different concentrations to the protoplasts (Tab. 1) immediately after plating. Every 10-14 days 50% of the medium was replaced by fresh medium containing the same toxin concentration.

RESULTS

Tab. 1 shows that the tested dihaploid potatoes were more susceptible to the applied toxin than the tetraploid ones. The Phytophthora toxin showed in no case an effect during early protoplast development, i.e. cell wall formation and beginning chloroplast degeneration. First obvious effects of susceptible clones treated with the toxin were a lower cell enlargement, a reduced first division frequency and with intensified toxin concentrations an increasing browning and protoplast conglutination. The number of formed microcalli decreased and they showed a more compact formation. In resistant clones these symptoms were not that strong and the regeneration rate of microcalli was reduced only at higher toxin levels. Fusion of protoplasts with subsequent toxin treatment caused a similar reduction in microcalli development.

Tab. 1: Influence of Phytophthora infestans toxin on the development of microcalli from protoplasts

Potato cv. /clone /fusion trial		Regenerated microcalli in % with toxin concentrations** of 0*	1:7	1:3.5	1:1.7	1:0.85
cv. Berolina	(4x)	100	112	120	121	70
SM 2	(4x)	100	114	110	68	38
cv. Achat	(4x)	100	110	60	50	nd
H 80.1202/9	(2x)	100	27	26	14	nd
H 75.1207/7	(2x)	100	44	40	30	nd
HH 372	(2x)	100	20	15	nd	nd
H 75.1207/7 + H 80.1202/9		100	30	25	18	nd
H 75.1207/7 + HH 372		100	27	19	nd	nd

* = number of microcalli in the control = 100%, ** = toxin concentration in Phytophthora infestans culture filtrate = 1

DISCUSSION

The different susceptibilities of in vitro tested potato clones against Phytophthora toxin confirm the results obtained in vivo (Stolle and Schöber, 1985). The initially missing effect of the toxin applied supports the results of Berggren (1987), who did not get any reaction of Phytophthora toxin or zoospores on the viability and early development of potato protoplasts; but in this experiment the toxin was not tested during later regeneration stages. The compact and brownish appearance of clones susceptible to the toxin is in accordance with Foroughi-Wehr and Stolle (1985), who observed a normal growth of calli during toxin treatment (which finally turned black) on a medium that results normally in a rather loose callus growth. The stronger resistance of the tetraploids to the toxin compared with the dihaploids can be explained by their higher vigour. This may justify the presumption that heterozygous 4x fusion products are also superior and thus can be enriched. The toxin treatment leads to a lower number of plants and only these have to be screened further for hybrid confirmation. Due to the higher vigour of the 4x plants and the different toxin susceptibilities of the dihaploid donor clones an increased rate of hybrids compared to previous experiments is expected. Screening trials are in progress for fusaric acid resistance enabling the use of two toxin resistances and combining their effects for an even more efficient preselection.

REFERENCES

Berggren, B., 1987: Sensitivity of potato protoplasts to Phytophthora infestans. In: Proceedings of European Association for Potato Research, 10th Triennial conference of the EAPR; Aalborg, Denmark, p. 220

Binding, H., R. Nehls, O. Schieder, K. Sopory & G. Wenzel, 1978: Regeneration of potato mesophyll protoplasts isolated from dihaploid clones of Solanum tuberosum. Physiol. Plant. 43: 52-54.

Foroughi-Wehr, B. and K. Stolle, 1985: Resistenzselektion in vitro am Beispiel des Systems Kartoffel/Phytophthora infestans (Mont.) De Bary. Nachrichtenbl. Deut. Pflanzenschutzd. 37: 170-173.

Haberlach, G.T., B.A. Cohen, N.A. Reichert, M.A. Baer, L.E. Towill and John P. Helgeson, 1985: Isolation, culture and regeneration of protoplasts from potato and several related Solanum species. Plant Science 39: 67-74.

Stolle, K. and B. Schöber, 1985: Wirkung eines Toxins von Phytophthora infestans (Mont.) de Bary auf Kartoffelblätter. Potato Research 28: 389-402.

PRODUCTION AND FIELD ASSESSMENT OF SOMATIC HYBRIDS BETWEEN DIHAPLOID SOLANUM TUBEROSUM and SOLANUM BREVIDENS

N. FISH[1] and M.G.K. JONES[2]

[1] Shell Research Ltd., Sittingbourne, Kent, ME9 8AG, U.K.
[2] AFRC Institute of Arable Crops Research, Biochemistry Department, Rothamsted Experimental Station, Harpenden, Herts. AL5 2JQ, U.K.

INTRODUCTION AND RESULTS

Following protoplast fusion, 32 somatic hybrids have been produced between a dihaploid S. tuberosum genotype (PDH 40) and the potato leaf roll virus (PLRV) resistant, sexually incompatible wild species, S. brevidens. Somatic hybrids have been regenerated after applying both chemical (1) and electrical (2) methods of protoplast fusion, and have been identified by examining their morphology, isoenzyme and cytological characteristics. Additional proof of hybridity has been obtained by analysis of DNA restriction fragment length polymorphisms (RFLP), using a cDNA probe from the patatin gene (2). Examination of PLRV resistance using ELISA has shown a wide spread of resistance in the somatic hybrids (3). Most of the flowering hybrids are extremely resistant to PLRV and are suitable sources for introduction into a potato breeding programme. Attempts to cross the somatic hybrids with commercial U.K. potato cultivars are currently in progress.

A selection of the somatic hybrids and the parental plants have been examined further in a small field trial on Rothamsted Experimental Farm, during the 1987 season. As the somatic hybrids are classified as genetically engineered plants in the U.K., approval for the field trial was sought from a government committee, the Advisory Committee on Genetic Manipulation (ACGM). Approval was given by ACGM but restrictions, such as the removal of flower buds were imposed. Examination of the somatic hybrids in the field showed that each line was morphologically distinct from every other line as well as both parents. Most hybrids had characters that were intermediate between those of either parent e.g. anthocyanin pigmentation, as well as completely new characters such as leaf shape.

REFERENCES

1. Fish, N., Karp, A. and Jones, M.G.K. (1987). Improved isolation of dihaploid *Solanum tuberosum* protoplasts and the production of somatic hybrids between *S. tuberosum* and *S. brevidens*. In Vitro Cellular and Developmental Biology, 23, 575-580.
2. Fish, N., Karp, A. and Jones, M.G.K. (1988). Production of somatic hybrids by electrofusion in *Solanum*. Submitted.
3. Gibson, R.W., Jones, M.G.K. and Fish, N. (1988). Resistance to potato leaf roll virus and potato virus Y in somatic hybrids between dihaploid *S. tuberosum* and *S. brevidens*. Submitted.

Fusion of *Solanum tuberosum* L. cv. Explorer(Potato) and *Solanum melongena* L. cv. Imperial Black Beauty(Eggplant) Protoplasts and Subsequent Callus Regeneration

G.G. Li* and F. Constabel
Plant Biotechnology Institute, National Research Council, Saskatoon, Saskatchewan, S7N 0W9 Canada
* Visiting Scientist, Academia Sinica, South China Institute of Botany, Guangzhou, China

Introduction

Successful induction of fusion of plant protoplasts by PEG has already been shown earlier (7,1). This communication describes an attempt to obtain a high frequency of protoplast fusion and viable fusion products of potato and eggplant protoplasts.

Materials and Methods

Seeds of *Solanum tuberosum* L.cv. Explorer(potato) were surface-sterilized and then germinated on medium(3). After 10 days of culture, the seedlings had grown to a length of 2-3cm. The seedlings were transferred to MS medium(5) supplemented with 0.04 mg/L NAA and solidified with agar. The plantlets (Figure 4) were incubated in a growth chamber(3).

Callus of *Solanum melongena* L.cv. Imperial Black Beauty(eggplant) were obtained according to Gleddie et al. (2). Suspension cultures were grown into 500 ml flasks containing 50 ml of liquid MS medium supplemented with 2,4-D at a concentration of 1mg/l. The cultures were maintained on a gyratory shaker at 100 rpm at 26°C under continuous fluorescent light (40 $\mu EM^{-2}S^{-1}$). Every week 10 ml of suspension was transferred into 50 ml of fresh medium.

Isolation of protoplasts: Protoplasts of potato were isolated by previously reported methods(3). The enzyme solution (3.5 ml) contained 1% cellulase Onozuka R10, 0.25% hemicellulase Rhozyme, 0.25% pectinase, 3 mM $CaH_4(PO_4)_2 \cdot H_2O$, 0.55M sorbitol and 1.5 ml of culture medium 1 (Table 1). Protoplasts of eggplant were prepared from cell suspensions, by incubating the cells with 3.5 ml of enzyme solution containing 1.5%cellulase Onozuka R10, 0.5% hemicellulase Rhozyme, 0.5% pectinase, 3mM $CaH_4(PO_4)_2 \cdot H_2O$, 0.55M sorbitol and 1.5 ml of culture medium 1. Digestion was carried out in the dark, at 26°C and for ca.5 h. Protoplasts were purified by the same way(3) in each case.

Fusion of protoplasts: 0.5 ml each of leaf protoplasts and suspension culture protoplasts were mixed and the cell density adjusted to about 2×10^3 to 10^4 protoplasts/ml. Approximately 100 µl of the protoplast suspension were pipetted onto a 22 X 22 mm cover slip. A tiny drop of silicone 200 fluid 100 cs was used to hold the cover slip in position. 3 drops of fusion solution were added. This solution consisted of 25 mg $CaCl_2 \cdot 2H_2O$ (3.4mM) and 5 mg KH_2PO_4(7.3 mM) per 50 ml; 19 g polyethylene glycol (PEG) 4000 was added and the solution adjusted to pH 5.7. After incubating this sample on the cover slip for 5 min, a solution containing 10% DMSO in addition to solution 3(8) was added twice for 15 min. After another 10 min the protoplasts were washed 4 times with culture medium 2 (Table 1). Finally, the protoplasts were culture in a thin layer of culture medium 2. The dishes were sealed with Parafilm. The culture containing the fusion products was maintained at 26°C, at a light intensity of 50-100 lx in a photoperiod of 16h/day.

Results and Discussion

Experiments were carried out to compare the effects of different DMSO concentration in PEG 4000 solution on protoplast fusion of eggplant and potato protoplasts. Five percent DMSO enhanced fusion frequency to a few percent, at 10% DMSO interspecific fusion products attained 40 to 50% of the surviving population. 20% DMSO concentration sustained similar percent, but the fusion products were damaged. Although the method of using DMSO and the frequency of protoplast fusion was different, the result was the same as reported previously for Brassica napus R 1466 hypocotyl and B. campestris mesophyll protoplasts (4).

The results in Table 2 indicated that frequencies of heterokaryocyte formation were considerably higher when PEG was eluted from the adhesion protoplasts with a high pH, high Ca^{2+} and DMSO solution than when it was eluted with a high pH, high Ca^{2+} or a protoplast culture medium.

Fusion products with well developed chloroplasts from the mesophyll parent of sterile plantlets were highly active. The fusion products underwent division(Fig. 1-2) after 2-3 days of culture. Within 8-10 days of culture, fusion products reached the 4-6 cell stage(Fig. 3). Small colonies consisting of up to 8-10 cells were obtained with 10-15 days of culturing. Overall division rate of the fusion products was approximately 70%. After a few days, as the culture medium was used up and cell colonies began to grow, an additional 2 drops of fresh culture medium containing 0.35 M carbohydrates and 0.25 M carbohydrates was added after another 10 days of culture. When visible colonies had formed, they were transferred to an agar solidified medium 3(Table 1) (Fig. 5). Investigations are underway to study the behavior of nuclei in the fused cells and also to culture callus arising from the heterokaryocytes.

References 1.Constabel F & Kao KN: Can.J.Bot.,1603-1606,1974. 2.Gleddie S et al.: Can. J.Bot.,61:656-666,1983. 3.Li GG & Constabel F: J. plant Physiol.,117:137-142,1984. 4.Menczel L & Wolfe K: Plant cell Rep., 3: 196-198,1984. 5.Murashige T & Skoog F: Physiol.Plant,15:473-497,1962. 6.Nelson RS et al., Plant Sci. Lett.,30:355-362,1983. 7.Kao KN & Michayluk MR: Planta,115:355-367,1974. 8.Kao KN:In:Wetter LR & Constabel F (eds.),Plant Tissue culture Methods.National Research Council of Canada 49,1982. 9.Shepard JF:Plant Sci.Lett.,18:327-333, 1980.

Table 2. Effect of various solutions for PEG elution on fusion of protoplasts induced with eggplant suspension cell and potato mesophyll protoplasts. Sample size Ca. 800-900 protoplasts taken from 3 dishes

Solution Medium	Frequency of Heterokaryocytes to Total Surviving Protoplasts					
	Expt 1	Expt 2	Expt 3	Expt 4	Expt 5	Expt 6
Medium 2 (Table 1)	1 2	3 1	2 4	5	3	2 6
PEG 4000 + High Ca^{2+} + high pH	20 4	9 8	11 7	14 2	16 5	19 2
PEG 4000 + high Ca^{2+} + high pH + DMSO	44 8	55 7	54	42 8	45 6	50 6

Sample size was about 800 900 protoplasts from 3 dishes

Table 1 Protoplast, heterokaryocyte and callus culture media

Compound	Protoplast Culture Medium Medium 1 (mg/l)	Heterokaryocyte Culture Medium Medium 2 (mg/l)	Callus Culture Medium Medium 3 (mg/l)
Macronutrients	After Nelson et al. (1983)	After Nelson et al (1983)	After MS (1962)
Iron and micronutrients	After Shepard (1980)	After Shepard (1980)	*
Vitamins and Glycine	*	*	*
Sucrose	34,200	34,200	30,000
Mannitol	18,200	18,200	
Glucose	36,400	36,400	
Xylose		40,000	
2,4-D		0 8	
ZEA	0 5	1 5	0 5
NAA	1 0	1 0	0 1
Casein hydrolygate	250	250	250
Coconut water	50 ml	50 ml	50
Agar			8,000
pH	5 7	5 7	5 7

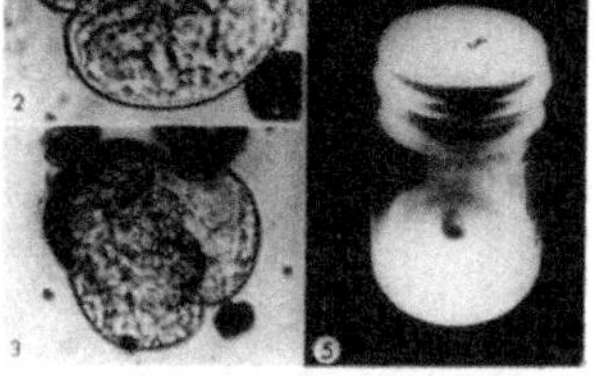

Fig.1-2.Fusion products underwent divisions. 3.Fusion products reached the 4-6 cell. 4.Stem used for protoplast isolation. 5.Visible colonies

"SOMATIC FUSIONS OF *SOLANUM* SPECIES"

John P. Helgeson, Geraldine T. Haberlach, John Pohlman and Sandra Austin

Agricultural Research Service, U.S. Department of Agriculture, Dept. of Plant Pathology, Univ. of Wisconsin, Madison, WI 53706, USA.

INTRODUCTION

Our research focuses on somatic hybridization of potato (*Solanum tuberosum*) and other *Solanum* species to incorporate utilizable new germplasm for disease resistances. Our emphasis is on combining potato breeding lines with wild *Solanum* species that are sexually incompatible with potato. In this paper we will summarize some of the results of our research. Many of the illustrations will be from the hybridization between *S. brevidens* (PI 218228), a wild, diploid South American species and *S. tuberosum* (PI 203900), a tetraploid potato line.

PROCEDURES

Plant lines used for experiments were obtained from Dr. R. E. Hanneman and associates at the Inter-regional Potato Introduction Station, Sturgeon Bay, WI or from Dr. H. De Jong of Agriculture Canada, Fredricton, N. B. Plants were micropropagated and maintained in vitro as described by Haberlach et al (5). For isolating and culturing protoplasts we used modification (5) of the floatation procedure devised by Shepard and Totten (8). Fusions were carried out by a modification (2) of the procedure of Melchers et al (7). Somatic hybrids were selected by differential sensitivities to media and by phenotype. Hybridity was verified by chloroplast DNA and isozyme analysis.

RESULTS AND DISCUSSION

Some of the hybrid combinations obtained in our studies are listed in Table 1. In most cases, several hybrid calli were obtained after protoplast fusion and each callus gave rise to multiple shoots. Thus we had a large group of somatic hybrids from each combination. Hybrids between diploid and tetraploid lines were generally hexaploid whereas hybrids between diploid and diploid lines were generally tetraploid. Field analyses of a large group of hybrids between *S. tuberosum* PI203900 and *S. brevidens* PI218228 demonstrated that morphological characteristics of the hybrids are intermediate between the two species (3). Aneuploid plants from fusions tended to have grossly abnormal flowers and leaves. Evaluations of plants from other fusion experiments support these conclusions. In addition, the hybrids often show enhanced vigor when compared with parent lines.

Table 1

Somatic hybrids obtained by Helgeson and associates*

1. S. brevidens [2x] + S. tuberosum [4x]
   [PI 218228]   [PI 203900]

2. S. brevidens [2x] + S. tuberosum [2x]
   [PI 218228]   [Superior haploid]

3. S. brevidens [2x] + S. tuberosum [2x]
   [PI 245763]   [Gp Phureja-Stonotomum De Jong 77-1]

4. S. brevidens [2x] + S. tuberosum [2x]
   [PI 218228]   [Gp Phureja-Stonotomum De Jong 77-19]

5. S. brevidens [2x] + S. tuberosum [4x]
   [PI 218228]   [Russet Burbank]

6. S. bulbocastanum [2x] + S. tuberosum [4x]
   [PI 243510]   [PI 203900]

7. S. brevidens [2x] + S. cardiophyllum [2x]
   [PI 245764]   [PI 283062]

8. S. tuberosum [2x] + S. tuberosum [2x]
   [US-W9545.99]   [US-W9310.3]

* PI numbers are from the inventory of species maintained by Dr. R. E. Hanneman and associates at the inter-regional potato introduction station, Sturgeon Bay, WI.

In spite of observed meiotic abnormalities in anthers, some of the hybrids are fertile as females (4). In particular, the hexaploid somatic hybrids (combination 1 in Table 1) cross well with pollen from tetraploid cultivars Katahdin and Norland. In one cross, for example, 548 seeds were obtained from 42 pollinations that resulted in 25 fruit. In contrast, only 8 seeds from 12 pollinations and 7 fruit were obtained from the best cross of a tetraploid somatic hybrid with a 4x S. tuberosum Gp. Andigena test line. Although preliminary experiments indicated that pollen sterility may be a problem, successes in selfing experiments and in more detailed crossing studies indicate that the hybrids may also serve as male parents in crosses. Seed set is likely to be low in these cases. Disease resistances of each of the fusion partners are expressed in the hybrids. For example, resistance to potato leaf roll virus (PLRV) (from S. brevidens) and to Phytophthora infestans (from S. tuberosum) are expressed in fusion hybrids of these two lines (6). Crosses of these hybrids with the susceptible cultivar Katahdin indicate that the disease resistances obtained from the wild species by somatic fusions can be passed on to progeny from normal sexual crosses.

CONCLUSIONS

Somatic hybridization of Solanum species appears to be a useful tool for incorporating germplasm from wild, sexually incompatible species into potato breeding lines. Genes from the wild species and the potato cultivars are expressed in the same plant and substantial vigor is imparted to the somatic hybrid by the union of the two species. The fate of the wild DNA through many sexual crosses is still to be determined, although passage through 2 crosses can be demonstrated. Also, much needs to be learned about the fate of the cytoplasmic organelles of the plants that result from somatic fusions. However, even with the problems yet to be solved, it would appear that some progress can be made in bringing economically useful genes from wild species into cultivars by this method. The utility of the method may not be limited to interspecific hybridization. Hybridization of S. tuberosum diploid lines yielded strong healthy plants that expressed genes from both parental lines. Thus, somatic hybridization with sterile potato haploid lines may provide a means of combining useful genotypes(1).

ACKNOWLEDGEMENTS

We thank Marlette Baer, Beth Howman-Combs, Mark Ehlenfeldt, and Greg Hunt for their contributions to the work cited above. Some of this work was supported by grants (84-CRCR-1-1512 and 86-CRCR-1-2142) to J. P. Helgeson from the USDA Competitive Research Grants Program.

REFERENCES

1. Austin, S., M. Baer, M. Ehlenfeldt, P. J. Kazmierczak and J. P. Helgeson. 1985. Intra-specific fusions in Solanum tuberosum. Theor. Appl. Genet. 71:172-175.
2. Austin, S., M. A. Baer and J. P. Helgeson. 1985. Transfer of resistance to potato leaf roll virus from Solanum brevidens into Solanum tuberosum by somatic fusion. Plant Sci. Lett. 39:75-82.
3. Austin, S., M. Ehlenfeldt, M. Baer and J. P. Helgeson. 1986. Somatic hybrids produced by protoplast fusion between S. tuberosum and S. brevidens: phenotypic variation under field conditions. Theor. Appl. Genet. 71:682-690.
4. Ehlenfeldt, M. K. and J. P. Helgeson. 1987. Fertility of somatic hybrids from protoplast fusions of Solanum brevidens and S. tuberosum. Theor. Appl. Genet. 73:395-402.
5. Haberlach, G. T., B. A. Cohen, N. A. Reichert, M. A. Baer, L. E. Towill and J. P. Helgeson. 1985. Isolation, culture and regeneration of protoplasts from potato and several related Solanum species. Plant Science 39:67-74.
6. Helgeson, J. P., G. J. Hunt, G. T. Haberlach and S. Austin. 1986. Somatic hybrids between Solanum brevidens and Solanum tuberosum: Expression of a late blight resistance gene and potato leaf roll resistance. Plant Cell Reports 3:212-214.
7. Melchers, G. M. D. Sacristan and A. A. Holder. 1978. Somatic hybrid plants of potato and tomato regenerated from fused protoplasts. Carlsberg Res. Commun. 43:203-218.
8. Shepard, J. F. and R. E. Totten. 1977. Mesophyll cell protoplasts of potato. Isolation, proliferation and plant regeneration. Plant Physiol. 60:313-316.

FLOW CYTOMETRIC SORTING OF POTATO HETEROKARYONS WITH INHIBITION OF CELL WALL SYNTHESIS BEFORE SORTING

K.J. PUITE[1]*, W.R.R. TEN BROEKE[1], J.G. SCHAART[1] and J.H.N. SCHEL[2]

[1]Research Institute Ital, Postbox 48, NL-6700 AA Wageningen and [2]Dept. of Plant Cytology and Morphology, Agricultural University, Arboretumlaan 4, NL-6703 BD Wageningen, The Netherlands.

INTRODUCTION

Tetraploidy is required for high yield potatoes. Breeding potato cultivars at the diploid level followed by somatic hybridization between two diploid lines will be a possible approach for obtaining tetraploids. Practical application of this technique will require high numbers of heterokaryons, because the resulting hybrid plants will show extensive variation and plants with a normal chromosome number have to be selected amongst the regenerants. Using a micromanipulator up to 100 heterokaryons/h can be collected 3 days after fusion. Identification of heterokaryons is based on red (chlorophyll) and yellow-green (fluorescein) fluorescence (Puite *et al*. 1986).

Because flow cytometric sorting may result in a high number of potato heterokaryons in a short time, a procedure was developed for efficient collection of these heterokaryons using a FACS IV flow cytometer. The fusion mixture was introduced into the flow system with a motor-driven syringe. Sorting was carried out using a 200 μm nozzle orifice. As sheath fluid the culture medium was used.

RESULTS

Initial sorting experiments of potato protoplasts were not successful. Therefore, various storage conditions until sorting for protoplasts and heterokaryons were studied. Optimal results (recovery of 40 %) for sorting potato protoplasts were obtained when 10 μg/ml 2,6 dichlorobenzonitrile (DB) was added during 3 to 5 h to the protoplasts in mannitol + culture medium (1:1). DB inhibits cell wall synthesis.

In Table 1 storage conditions are given after electrofusion of fluorescein diacetate stained diploid *Solanum tuberosum* (plantlets bleached with a herbicide) and green diploid *S. phureja* protoplasts. The use of DB doubles the percentage of recovery of the heterokaryons. Table 2 represents a typical sorting experiment. The percentage of heterokaryons before sorting was 6.1 ± 1.8%.

Table 1. Recovery percentages and standard deviation of potato heterokaryons

Storage condition	Recovery (%)
0.43 M mannitol	5
0.43 M mannitol + culture medium (1:1)	8 ± 3
idem, with 2 or 5 μg/ml DB added	11 ± 4
idem, with 10 μg/ml DB added	17 ± 3

Table 2. A typical example of a sorting experiment of potato heterokaryons

Number of parental protoplasts	Sorted objects	Sorting time (min)	Intact heterokaryons after sorting	Volume (ml)
0.9×10^5	4974	40	849	0.5
1.2×10^5	4600	40	662	0.4
1.2×10^5	5418	40	847	0.5
1.1×10^5	6457	40	1097	0.6

Sorting rates for intact heterokaryons ranged from 250 to 1500/h, depending on the quality of the material, the frequency of the heterofusions and the storage conditions chosen between fusion and sorting. The purity of the heterokaryons after sorting was estimated to be 84 %.

The heterokaryons were cultured in Falcon dishes with Nicotiana plumbaginifolia cell suspension in the outer ring. From 39 calli, shoots and plants were regenerated. The ploidy level was determined by measuring the nuclear DNA content with the flow cytometer. Fourteen tetraploid regenerants were obtained. Six have now been analysed for their hybrid character by Giemsa C-band pattern analysis of the chromosomes. These plants proved to be hybrids. (Fig. 1).

Figure 1. Tetraploid hybrid plants from flow sorted heterokaryons of diploid S. tuberosum and S. phureja

ACKNOWLEDGEMENT

The chromosome analysis was carried out by Dr. L.P. Pijnacker and Ms. M.A. Ferwerda, Dept. of Genetics, University of Groningen.

REFERENCE

K.J. Puite, S. Roest and L.P. Pijnacker, Plant Cell Reports 5 (1986) 262-265

AN IMPROVED PEG FUSION PROCEDURE FOR FLOW SORTING OF PLANT PROTOPLASTS

P.K. CHAND[1], E.C. COCKING, A. LISTER, M.R. DAVEY and J.B. POWER.

[1]Post-Graduate Department of Botany, Utkal University, Bhubaneshwar-751004, Orissa, India.
Plant Genetic Manipulation Group, Department of Botany, University of Nottingham, Nottingham NG7 2RD, U.K.

1. INTRODUCTION

Polyethylene glycol (PEG) has been used extensively to induce protoplast fusion (1). However, protoplasts of many plant species do not tolerate the inevitable osmotic transitions during PEG treatment. This toxicity may firstly be due to acidic impurities found in some commercial PEG preparations (2), or secondly may relate to the dehydrating capacity of PEG, thus making protoplasts irreversibly leaky (3). Protoplast aggregates often result from an inadequate rehydration during the elution of PEG which renders the protoplasts less amenable to flow cytometry.

A more efficient fusion method for use with flow cytometry has been developed for a range of protoplasts. This method involves purified PEG solutions, coupled with controlled deplasmolysis of the treated protoplasts.

2. MATERIALS AND METHODS

2.1. PEG Batch Assay. PEG samples from several sources were assessed by colorimetry (4) for their carbonyl content. For this assay, a standard 30% (w/v) solution of PEG, in distilled water, was prepared and the level of contamination with carbonyl compounds expressed as equivalent moles of acetaldehyde per litre.

2.2. Protoplast Fusion. Protoplasts (10^5 ml^{-1}) of each fusion partner were suspended (1:1) in a Ca^{2+}-free incubation medium. Aliquots (1.0 ml) of this mixture were placed (a) in the centre of a 5 cm Petri dish, (b) in each well of a 25 well plate (Sterilin, U.K.), or (c) in a centrifuge tube (16 x 120 mm). Protoplasts were allowed to settle for 10 min. One ml of a Ca -free solution of PEG (30% w/v in 75 mM Hepes buffer, pH 8.0) and containing a previously determined and neglible amount of carbonyl compounds (0.5-5.0 x 10^{-4} M l^{-1}) was added drop-wise to the settled protoplasts and left for 20 min. The PEG solution was eluted over a period of 2-4 min using a Ca^{2+}-free hypotonic solution (pH 5.8, 322 mOsm kg^{-1}) containing 5% w/v mannitol and 0.2% w/v bovine serum albumin. Protoplasts were washed and collected in a 9% w/v mannitol solution (pH 5.8) in which they were allowed to settle (2h, 4°C). Protoplasts were resuspended in CPW 13M (5) for flow cytometry.

2.3. Flow Cytometry and Heterokaryon Culture. Heterokaryons were isolated from mixtures of unfused parentals and homokaryons using a Coulter EPICS 541 flow cytometer equipped with an argon ion laser operated at 457 nm regulated at 100 mW output. Heterokaryons were identified by their dual fluorescence (630-700 nm red chlorophyll auto-fluorescence of the leaf mesophyll or cotyledon component; 495-565 nm yellow-green fluorescence of the FDA-labelled cell suspension partner). Sorting was achieved using CPW 13M medium as sheath fluid with a 200 µm nozzle operated at 6 psig (41 k Pa) and 8 kHz droplet formation frequency.

Heterokaryons were collected directly in 96-well (Nunclon) or 24-well (Falcon) multiwell plates, primed with a microculture of albino protoplasts (5 x 10^4 ml^{-1}) in K8P medium (5) solidified with 0.6% agarose (Sea Plaque).

3. RESULTS AND DISCUSSION

Of a range of PEG samples evaluated, those provided by Aldrich (M.W. 1500), Koch-Light (M.W. 1540, 6000) and Boehringer-Mannheim (M.W. 1500) had a low carbonyl content and produced viable heterokaryons at high frequency, e.g. 8-11% (determined optically immediately after fusion) for cotyledon protoplasts of S. viarum fused with cell suspension protoplasts of S. dulcamara. Heterokaryons exhibited a high post-fusion viability (60-80%) and mitotic division (25-45%). PEG solutions could be stored before fusion treatment at 4°C in amber bottles in order to reduce photo-oxidation. Autoclaved PEG preparations with significantly higher carbonyl content (2-50 fold increase over filter-sterilised PEG) were considerably detrimental to the post-fusion viability as compared to those that were filter-sterilised. PEG eluting solutions with a high $CaCl_2$ (10-50 mM) caused significant reduction in heterokaryon survival. By contrast, increased heterokaryon viability and division was obtained when a Ca^{2+}-free hypotonic elution was employed for a shorter duration, and then replaced with a Ca^{2+}-depleted 9% w/v mannitol solution. Heterokaryons of interspecific and intergeneric combinations involving Medicago, Nicotiana, Oryza, Petunia, Salpiglossis and Solanum produced using this fusion method were intact, free-floating and exhibited detectable fluorescence. The suitability of such chemically-produced heterokaryons for flow cytometry was comparable to that of electrofused protoplasts. Sorted heterokaryons between Solanum viarum and S. dulcamara and between Medicago sativa and M. falcata divided to the 10-cell colony stage when cultured in the presence of nurse protoplasts derived from albino cell suspensions of Salpiglossis sinuata or M. sativa respectively.

Filter-sterilised and optimally buffered PEG solutions containing negligible amounts of aldehydes and peroxides and maintained under optimal storage conditions were shown to be non-toxic to plant protoplasts. Divalent cations (e.g. Ca^{2+}) as normally used in elution solutions may act as bridges between the negatively charged multi-polarized groups of PEG molecules and the protoplast membrane. This might in turn hinder the process of PEG elution and account for the protoplast damage often observed in many protoplast systems. The use of suitable hypotonic solutions for PEG elution ensures an adequate rehydration and coalescence of the fusing protoplasts since this encourages an osmotic breakdown at adhering membrane interfaces. This, in turn, not only elevates the fusion frequency but also decreases protoplast clumping, thereby rendering most systems better suited to flow cytometry.

REFERENCES

1. Evans DA, Bravo JE, Gleba YY: in: Int. Rev. Cytology. Suppl. 16. (ed.) KL Giles, Academic Press, New York: 143-159, 1983.
2. Kao KN, Saleem M: J. Plant Physiol. 122: 217-225, 1986.
3. Hahn-Hagerdal B, Hosono K, Zachrisson A, Bornman CH: Physiol. Plant. 67: 359-364, 1986.
4. Lappin GR, Clark LC: Anal. Chem. 23: 541-542, 1951.
5. Power JB, Chapman JV, Wilson D: Laboratory Manual Plant Tissue Culture, Nottingham University, U.K.

REGENERATION AND EVALUATION OF SOMATIC HYBRID PLANTS BETWEEN *SOLANUM TUBEROSUM* AND *LYCOPERSICON PIMPINELLIFOLIUM*

M.Okamura
Plant Lab., Kirin Brewery Co.,LTD, 3377, Kitsuregawa, Tochigi, 329-14, JAPAN

INTRODUCTION

"Pomato" produced by Melchers et. al.(1) demonstrated a possibility of overcoming the crossing barrier between *Solanum* and *Lycopersicon*. *Lycopersicon pimpinellifolium*, a wild species of tomato has important traits for potato breeding such as resistance to bacterial wilt, late blight and high temperature (2), but is unable to be crossed sexually with *Solanum tuberosum*, the cultivated potato. This report details the production and evaluation of somatic hybrid plants between *S.tuberosum* and *L.pimpinellifolium*.

MATERIALS AND METHODS

Mesophyll protoplasts of both *S.tuberosum* (*St*) and *L.pimpinellifolium* (*Lp*) were isolated from *in vitro* shoot cultures and electrofused using an Electro Cell Manipulator 401A (BTX Inc., USA). The fusion-treated protoplasts containing heterokaryons were cultured in P medium (MS -NH4NO3, 1000mg/l inositol, 0.1mg/l 2,4-D, 0.5mg/l NAA, 0.5mg/l BA, 0.5% glucose, 1% sucrose, 4.6% mannitol). After ten days of culture, protoplast-derived colonies were transferred into C medium (MS 300mg/l NH4NO3, 0.1mg/l NAA, 0.5mg/l BA, 0.3% sucrose, 5% mannitol) and subcultured weekly in the same medium. Five or six weeks after the fusion treatment, the developing calli were transferred onto S medium (MS 300mg/l NH4NO3, 2% coconut milk, 0.3mg/l IAA, 2mg/l zeatin, 0.3% sucrose, 3.5% mannitol) for shoot induction. Regenerated shoots were transplanted to hormone-free MS medium for rooting. Hybridity was examined by morphological characteristics and malate dehydrogenase isozyme analysis.

RESULTS AND DISCUSSION

The maximun fusion frequency was obtained when the protoplasts were aligned in the field of 200 V/cm at 2 MHz and fused by a 40μs pulse of 1550 V/cm. Approximately 15% of the plated protoplasts containing heterokaryons underwent cell division in P medium and formed mini-calli in C medium. Three morphologically different types of calli were obtained: *St* type (green, compact), *Lp* type (yellow to light green, soft) and intermediate type (green, soft), respectively. After 1 or 2 months of culture in S medium, numerous shoots were formed from both *St* and intermediate types, but no shoot has been formed from *Lp* type.

Eighty-eight plants regenerated from 20 intermediate types exhibited combined characteristics of both parental species. Malate dehydrogenase isozymes of the putative hybrids were composed of both parents'. Protoclones regenerated from protoplasts of the hybrids also showed the hybridity.

A large number of hybrids grew vigorously in the field and flowered. *S.tuberosum* line used in this study is highly fertile and reciprocal crossing between parental *S.tuberosum* and the hybrids is under way.

Figs.1–8. Fig.1. A hybrid cell produced by electrofusion between *S.tuberosum* and *L.pimpinellifolium*. Fig.2. Division of a hybrid cell. Fig.3. Calli from the fusion-treated protoplasts. Fig.4. Regeneration of shoots from a hybrid callus. Fig.5. A somatic hybrid plant growing in the field. Fig.6. Inflorescence of a somatic hybrid with the flower having a long pistil of *S.tuberosum* type and leaf-like petals of *L.pimpinellifolium* type. Fig.7. Isozyme patterns of malate dehydrogenase in (from left to right) *S.tuberosum*, a somatic hybrid, and *L.pimpinellifolium*. Fig.8. Flower shape (upper row) and leaf shape (lower row) of (from left to right) *S.tuberosum*, a somatic hybrid, and *L.pimpinellifolium*.

REFERENCES

1. Melchers,G. Sacristan,M.D. and Holder,A.A. (1978) Carlsberg Res. Commun., 43: 203–218.
2. Rick,C.M. (1982) In: Vasil,I.K., Scowcroft,W.R., Frey,K.J. (eds) Plant improvement and somatic cell genetics. Academic Press, pp 1–28.

STUDIES ON GENE MAPPING IN THE POTATO USING SOMATIC HYBRIDS OF SOLANUM TUBEROSUM L. + NICOTIANA PLUMBAGINIFOLIA

S.E. De Vries[1], W.J. Van Everdink, M.A. Ferwerda, E. Jacobsen, L.P. Pijnacker and W.J. Feenstra
Department of Genetics, University of Groningen, NL-9751 NN Haren, The Netherlands

INTRODUCTION

Our purpose is to map potato (*Solanum tuberosum* L.) genes by interspecific somatic hybridization. Nitrate reductase (NR^-) mutants of *Nicotiana plumbaginifolia* are used as fusion partner for the potato. After random partial loss of potato chromosomes from the fusion product it might be possible to localize the compensating NR gene in the potato. *N.plumbaginifolia* and potato chromosomes can be distinguished (Pijnacker and Ferwerda 1984, Mouras et al. 1986, Pijnacker and Mouras 1986) and the potato chromosomes can be identified individually (Pijnacker and Ferwerda 1984).
In previous studies no preferential loss of potato chromosomes was observed in NR^+ somatic hybrids of the same species (de Vries et al. 1987). Research is at present focussed on the selection of NR^- somatic hybrid lines. Such lines may lack NR activity due to the loss of the potato chromosome bearing the compensating NR gene. Identification of the remaining potato chromosomes will reveal which chromosomes can be excluded as being involved in the compensation of the defect in the fusion marker used.

MATERIALS AND METHODS

NR^+ somatic hybrid calli were obtained after electrofusion of mesophyll protoplasts of the monohaploid potato clone H7322 (2n=x=12, kindly supplied by Prof. G. Wenzel, FRG) and callus protoplasts of *N.plumbaginifolia* NR^- mutant Nia 26 (2n=4x=40) obtained from Dr. I. Negrutiu, Belgium). Cell suspension cultures from 3 established NR^+ somatic hybrid lines were plated onto solid AA medium (Muller and Grafe 1978) supplemented with 1900 mg/l KNO_3 and 20, 50 or 80 mM $KClO_3$. NR^- lines are resistant to this compound which is reduced to the toxic chlorite by active NR. Putative NR^- sublines were examined in growth experiments. NR activity in vivo (Jaworski 1971) was determined as well. From 5 NR^- lines enzyme activities (cytochrome-c reductase activity (CcR) (Sanderson and Cocking 1964), xanthine dehydrogenase activity (XDH) (Scazzocchio et al. 1973) were determined. Chromosome analysis was carried out using aceto-orcein squashes or air dried slides (de Vries et al. 1987).

[1]present adress: Institute for Horticultural Plant Breeding, NL-6700 AA Wageningen, The Netherlands

RESULTS

Selection of NR^- sublines

In all, 32 NR^- sublines were obtained after plating cell suspensions onto chlorate containing culture medium. All the lines originated from the same somatic hybrid line, line SH753. The majority of the NR^- lines were selected on medium with 80 mM $KClO_3$. Only a prolonged culture (6 weeks or more) in the presence of this concentration $KClO_3$ made it possible to discriminate between NR^+ and NR^- calli. In subsequent growth experiments it was clearly visible that the selected NR^- sublines were not able to grow on medium containing nitrate as nitrogen source, but were proliferating well in the presence of chlorate (Figure 1).

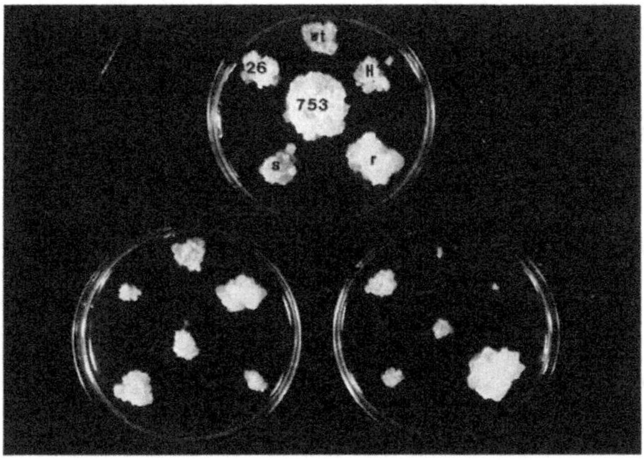

Figure 1. Callus growth on AA medium (upper row), KNO_3-containing (A) medium (left) and AA medium + $KClO_3$ (right) of _N.plumgaginifolia_ wild type (wt), Nia 26 (26), potato H7322 (H), NR^+ somatic hybrid line 753 (753), a NR^+ subline (s) and a NR^- subline (r).

Biochemical analysis

Biochemical analysis of the NR^- lines revealed a phenotype similar to Nia 26. All lines lacked NR in vivo and CcR activity. XDH activity was present in all examined lines (Table 1). In SH753 the mutant character of Nia 26 is compensated by the presence of the NR^+ trait from the potato. _S.tuberosum_ H7322 and SH753 both had NR in vivo, CcR as well as XDH activity.

Table 1. Induced[1] in vivo NR activity, induced NADH-CcR activity and XDH activity from callus of S.tuberosum H7322, N.plumbaginifolia Nia 26, NR^+ somatic hybrid line SH753 and 5 NR^- sublines.

Line	NR in vivo	CcR	XDH
H7322	+	+	+
Nia 26	-	-	+
SH753	+	+	+
NR^- 20C1	-	-	+
NR^- 20C4	-	-	+
NR^- 80C10	-	-	+
NR^- 80C11	-	-	n.d.[2]
NR^- 80C12	-	-	n.d.

[1] measured on AA medium + KNO_3
[2] not determined

Analysis of chromosomes

Chromosome analysis of the 5 NR^- lines revealed that potato as well as Nicotiana chromosomes were present. Lines 20C1, 20C4 and 80C10 were studied in detail. The total number of chromosomes varied between 40-150 and even higher numbers were observed. Polyploidization still took place. The lowest number of potato chromosomes present in a plate was 8. Chromosome aberrations had taken place in the potato chromosomes. This made identification of all the potato chromosomes impossible. Translocations between potato and Nicotiana chromosomes were observed as well. It is well known that chromosomes mutations do occur during a prolonged period in tissue culture. The short arms of chromosomes 2 and 4 were present in most of the metaphase plates.

CONCLUSIONS

- Loss of a specific trait (NR^+) could succesfully be induced in established somatic hybrids of S.tuberosum H7322 + N.plumbaginifolia Nia 26.
- The selected NR^- somatic hybrid lines have the biochemical phenotype of Nia 26. This is most probably due to loss of the compensating potato gene, c.q. chromosome.
- The short arms of potato chromosomes 2 and 4 might be excluded as bearing the nia^+ gene. Karyotypes of other independently selected NR^- lines have to be studied to confirm this conclusion.

ACKNOWLEDGEMENTS

We gratefully acknowledge Miss J. Oosterkamp and Th. Saat for technical assistance. This work is part of the research program of the Netherlands Foundation for Biological Research (BION) and is made possible by financial support from the Netherlands Technology Foundation (STW).

REFERENCES

1. Jaworsky EG (1971) Biochem Biophys Res Commun 43: 1274
2. Mouras A, Wildenstein C, Salesses GW (1986) Genetica 68: 197
3. Muller AJ, Grafe R (1978) Mol Gen Genet 161: 67
4. Pijnacker LP, Ferwerda MA (1984) Can J Genet Cytol 26: 415
5. Pijnacker LP, Mouras A (1986) Genetica 71: 137
6. Sanderson GW, Cocking EC (1964) Plant Physiol 39: 416
7. Schazzocchio C, Holl FB, Foguelman AE (1973) Eur J Biochem 26: 428
8. Vries SE de, Ferwerda MA, Loonen AEHM, Pijnacker LP, Feenstra WJ (1987) Theor Appl Genet (in press)

INTERSPECIFIC EXPRESSION OF AEC-RESISTANCE IN SOMATIC HYBRIDS OF AEC-RESISTANT SOLANUM TUBEROSUM AND NITRATE-REDUCTASE-DEFICIENT NICOTIANA PLUMBAGINIFOLIA.

MJ Tempelaar, LJ Drenth-Diephuis, A Hesseling-Meinders, R Hartman, A Duyst, E Jacobsen

Department of Genetics, Biological Centre, University of Groningen, Postbox 14, NL 9751 AA Haren, The Netherlands

INTRODUCTION

In fusion experiments involving potato (Solanum tuberosum),somatic interspecific hybrids of monohaploid potato clone AM 79.7322 (1) and chlorate-resistant and nitrate-reductase (NaR) deficient mutant Cnx-20 of Nicotiana plumbaginifolia (2) have NaR activity and are chlorate susceptible (3). In potato intraspecific hybrids, amino acid analogue-resistance markers of AEC-1 and 5MT-26 show dominant expression after fusion (4). The AEC marker is now being used in interspecific hybridization.

MATERIAL AND METHODS

Selective electric fusion (5) is applied to generate hybrids between N. plumbaginifolia Cnx-20 callus protoplasts and S. tuberosum AEC-1 suspension culture- or leafprotoplasts. Fusion conditions have been modified to increase the yield (6). Selectivity in the first week of culture is further enhanced by media selection against the potato protoplasts, which do not grow in N. plumbaginifolia culture media.

More selective steps have been applied in the next weeks in liquid culture (with mainly nitrate as N source to select for NaR complementation by potato) and in solid media (selection on colony and callus level with nitrate and AEC).

Colonies were screened for the activity of NaR and processed for chromosome analysis.

RESULTS AND DISCUSSION

Yield and analysis of fusionproducts shortly after fusion

In this particular fusion system, an overall fusion percentage of 25-40% is attainable. In an extensive series of experiments, microscopic observation suggested incompatibility between Cnx-20 and AEC potato suspension culture derived protoplasts, which could not be influenced by the use of different media.

With potato leaf protoplasts, eventually 5% of the fusion products started dividing. Accordingly, chromosome analysis revealed around 5% hybrids in 3 week old colonies. In the same preparations, cells typically had a low plastid content, which is a potato- rather than a N plumbaginifolia characteristic.

Selection and characterization

Attempts at chemical selection (AEC resistance and growth on nitrate) were singularly unsuccesful in the short run (weeks). Ultimately it transpired, that effects of selection were visible after 4-5 months. Hybrids were found to have a distinct morphology. This would again seem to point to potato rather than N plumbaginifolia characteristics. At this stage, 6 hybrids have been characterized in various ways. The poster describes analysis of growth patterns, chromosomes, NaR activity and AEC resistance. Both potato characteristics (NaR and AEC positive) are present in the hybrids tested. This is reflected by chromosome analysis, which indicates the presence of high numbers of chromosomes of both partners.

Examples from fusionproduct FL1:

growth on selective medium — chromosome analysis

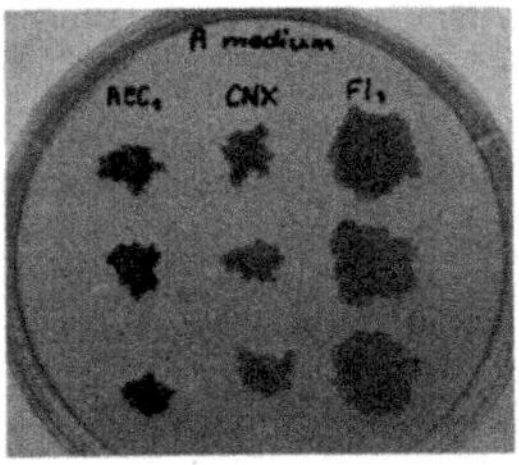

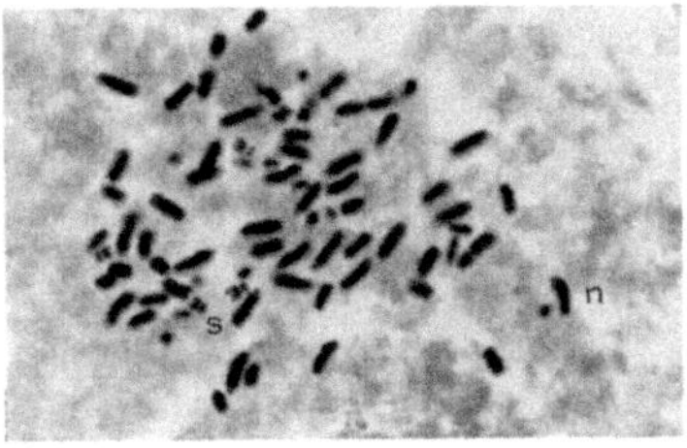

The preliminary conclusion of these experiments would be, that in fusions the AEC-1 strain dictates growth patterns (in some stages) of potato rather than N plumbaginifolia, does not respond readily to chemical selection and induces incompatibility reactions. Nevertheless, these problems can be circumvented so that somatic cell genetical work is feasible.

Further experiments are in progress

-to establish linkage of AEC and NaR properties by selecting/-screening for chlorate-resistance, indicating NaR deficiency by loss of potato chromosomes in sublines.

-to investigate possibilities and limitations of partial genome transfer after radiation-induced chromosome breakage.

REFERENCES

1 MJ Tempelaar, E Jacobsen, MA Ferwerda and M Hartog (1985), Z.Pflanzenzuecht 95: 193-200

2 I Negrutiu, R Dirks and M Jacobs (1983), Theor. Appl. Genet 66:341-347

3 SE de Vries, MA Ferwerda, AEHM Loonen, LP Pijnacker and WJ Feenstra, Theor. Appl. Genet, submitted

4 SE de Vries, E Jacobsen, MGK Jones, AEHM Loonen, MJ Tempelaar, J Wijbrandi and WJ Feenstra (1987), Theor. Appl. Genet 73: 451-458

5 MJ Tempelaar & MGK Jones (1985), Plant Cell Reports 4: 92-95

6 MJ Tempelaar, A Duyst, SJ de Vlas, G Krol, C Symmonds and MGK Jones (1987), Plant Science 48: 99-105

INTERSPECIFIC PROTOPLAST FUSION FOR THE TRANSFER OF ATRAZINE RESISTANCE FROM SOLANUM NIGRUM TO TOMATO (LYCOPERSICON ESCULENTUM.L) *

S. M. JAIN[2*], E. A. SHAHIN[1] and SAM SUN[1]
[1]ARCO, Plant Cell Research Institute, 6560 Trinity Court, Dublin, CA 94568, USA
[2]Department of Forest Science, Texas Agricultural Experiment Station, Texas A&M University System, College Station, TX 77843, USA

INTRODUCTION

Somatic cell fusion can be useful to transfer agriculturally important characters from one plant to another and to overcome sexual barriers between species, although fusions have so far worked mostly between closely related species. Atrazine resistance is due to modified functions of the Hill reaction in the chloroplasts, is maternally inherited and has been reported in many plant species (Gressel et al 1982). It has been transferred from the resistant *Brassica campestris* to *B. napus* through hybridization and backcrossing (Beversdorf et al 1980). Atrazine resistant plants have been regenerated from fused protoplasts of: haploid *Solanum tuberosum* L. and atrazine resistant *S. nigrum* (Binding et al 1982); cytoplasmic male sterile (cms) *Raphanus sativus* and atrazine resistant *B. campestris* (Pelletier et al 1983); and polima type cms *B. napus* and atrazine resistant *B. napus* (Barsby et al 1987).

We have transferred atrazine resistance from *S. nigrum* to tomato (*Lycopersicon esculentum* var. VF-36) by protoplast fusion using four approaches: 1) *S. nigrum* with tomato; 2) irradiated *S. nigrum* with tomato; 3) irradiated *S. nigrum* with iodoacetate-treated tomato; and 4) backfusions of tomato with a somatic hybrid plant.

PROCEDURE

Materials and Methods

Seeds of atrazine resistant and susceptible *S. nigrum* (4n = 48) were surface sterilized with 20% chlorox for 15 min and washed with sterile water 3 times. Seeds were transferred in B5 liquid medium (Gamborg et al 1968) and kept in a water bath for 5 h at 50°C and germinated at 25°C. After 2-3 weeks, hypocotyls were cultured on TM5 medium (Shahin, 1985) at 25°C ± 1°C, under 10,000 lux light intensity. Shoots were regularly subcultured on TM5 medium after every 2-3 weeks.

Tomato (*L. esculentum* var. VF-36) (2n=24) seeds were sterilized as described above and germinated in darkness on TM5 medium without phytohormones. All other procedures were similar to those described above for *S. nigrum*.

Iodoacetate and UV treatment

Tomato protoplasts were treated with 1mM iodoacetate for 20 min at 4°C in darkness. Subsequently, they were washed 2 times in a washing medium (Shahin, 1985) by centrifugation at 800 rpm for 5 min.

S. nigrum protoplasts were irradiated with UV light using a UV lamp (model UVS-54), 6 inches above the petri dish, for 1 min and then were kept in darkness for 4 h at 25°C.

Protoplast isolation and fusion

Atrazine resistant *S. nigrum* protoplasts were isolated by the procedure of Binding et al (1982). Isolation of tomato protoplasts was done according to Shahin (1985). Protoplasts of both the parents were mixed in a 1:1 ratio in a 15 ml centrifuge tube and the PEG and high pH fusion treatment was given as described by Binding et al (1982).

Protoplast culture, plant regeneration and cytology

After protoplast fusion treatments, a protoplast mixture of both parents was cultured according to Shahin (1985). Regenerated shoots were rooted on TM5 medium containing 0.1 mg/l IBA. Cytology of the regenerated plants was done by using protocol of Jain (1978).

Isozyme studies

Glutamate aspartic aminotransferase (GAT) isozyme pattern was studied according to Sun (1971) for the identification of somatic hybrids.

Atrazine test

Atrazine sensitivity was tested by culturing shoot tips on TM5 medium containing 1% sucrose and 10^{-4}M atrazine (2-chloro-4-ethylamino 6-isopropyl-amino-s-triazine). Sensitivity to atrazine was indicated by bleaching of green leaves after 10-15 days.

RESULTS AND DISCUSSION

The initial selection of hybrid shoots was done on the basis of leaf morphology and stem hair structures. This was followed by mass selection of shoots using the method of Binding et al (1982) after fusion treatment of haploid S. tuberosum and atrazine resistant S. nigrum protoplasts. S. nigrum had short knob-like hairs, atrazine resistance, rhomboid and denticulate leaves, white flowers, and 48 chromosomes, whereas tomato had long and pointed hair structures, pinnate and lanceolate leaves, yellow flowers, atrazine sensitivity, and 24 chromosomes (Table 2). All the regenerated plants were sterile and classified into four catagories (type I-IV) on the basis of leaf morphology, flower color, chromosome numbers, hair structures and atrazine response (Table 2). Types I and II appeared to be the products of parental protoplast fusants whereas types III and IV were due to nonfusants and single-parent fusants. Approach (1) - Our aim was to produce somatic hybrids between S. nigrum and tomato. A total of 2989 shoots were differentiated in 4239 randomly selected protoplast calli (p-calli) from 6 fusion experiments (Table 1), and from them, 2332 shoots rooted. Most of the regenerated shoots looked similar to S. nigrum and were discarded. They developed either from protoplast fusion (S. nigrum x S. nigrum) or from nonfusants. Twelve putative hybrids were retained and they were found to be resistant to 10^{-4}M atrazine. Ten of the 12 plants were similar to type III, 1 plant was similar to type I and another was similar to type II. Some of the plants were varigated and atrazine resistant.

Approach (2 and 3) - The objective was to produce cybrids between tomato and S. nigrum. From 12 fusion experiments, 2728 p-calli were cultured at random for shoot differentiation and 2050 formed shoots. A total of 1741 shoots developed roots and 20 plants were selected (Table 1). Two plants were similar to type I, 1 plant to type II and 17 plants to type III. All regenerated plants were nuclear chimeric and none of them were cybrids. It seems that UV radiation did not prevent transmission of the S. nigrum genome in the somatic hybrids.

Approach (4) - The aim of this study was to eliminate the S. nigrum nuclear genome from somatic hybrid (ST31) by backfusion with tomato protoplasts. A total of 728 p-calli were cultured for shoot induction from 3 experiments, of which 521 formed shoots. On TM5 medium, roots were induced on 472 shoots (Table 1). Regenerated plants (18) were selected; 8 of them were morphologically similar to tomato plants (type IV) and the others had either mixed characters of both parents or the S. nigrum type

Table 1. Protoplast fusion experiments using different approaches.

Protoplast fusion expt. (approach)	Total No. of expts.	No. of p-calli	No of p-calli formed shoots	No. of shoots rooted	No. of plants retained
S. nigrum x tomato (1)	6	4239	2989	2332	12
S. nigrum(UV)x tomato (2)	6	1828	1321	1120	10
S. nigrum(UV)x tomato (iodoacetate) (3)	6	900	729	621	10
Backfusions (4)	3	728	521	472	18

Table 2. Classification regenerants from protoplast fusion treatments.

Categories (fertility)	Atrazine response (10^{-4}M)	Leaf shape	Flower color	Hair Structure	Chromosome number
S.nigrum (fertile)	resistant	rhomboid, denticulate	white	short	48
Tomato (fertile)	sensitive	pinnate, lanceolate	yellow	long	24
Type I (sterile)	resistant	heart shaped, denticulate/ lanceolate	white/light yellow	very long	72
Type II (sterile)	resistant	lanceolate, denticulate/ lanceolate	white/light yellow	very long/ intermediate	72/73
Type III (sterile)	resistant	rhomboid, denticulate/ lanceolate	white	short	96/48
Type IV (sterile)	sensitive	pinnate, lanceolate	yellow	long	24

(type I - 1 plant; type II - 2 plants; and type III - 7 plants). All type IV plants were sensitive to 10^{-4}M atrazine (Table 2) and probably originated from nonfusant tomato protoplasts.

Identification of somatic hybrids

The hybrid nature of plants was confirmed by GAT isozymes using gel electrophoresis. GAT zymograms showed 3 bands (Rf 0.63, 0.67, and 0.71) in *S. nigrum* and 4 bands (Rf 0.41, 0.47, 0.58, and 0.67) in tomato

(Fig. 1). A band with the Rf of 0.67 was common in both the parents and was absent in all regenerated plants. In all, six hybrids (ST1, ST5, ST21, ST31, ST37, and ST38) were identified by GAT isozymes. ST1 and ST21 were regenerated by approach (1), ST5, ST38, and ST41 by approach (2), ST13 and ST31 by approach (3), and ST37 by approach (4). By backfusion of ST31 with tomato protoplasts, a new band with an Rf value 0.41 appeared and the intensity of band with the RF of 0.71 became lighter in the ST37 somatic hybrid plant (Fig 1). This plant had characteristics of type II (Table 2) (Fig 2). However, these results indicate that in ST37, S. nigrum genome is partially eliminated and tomato genome is partially transferred . By several backfusions, it may be possible to select atrazine resistant tomato plants by completely removing the S. nigrum genome.

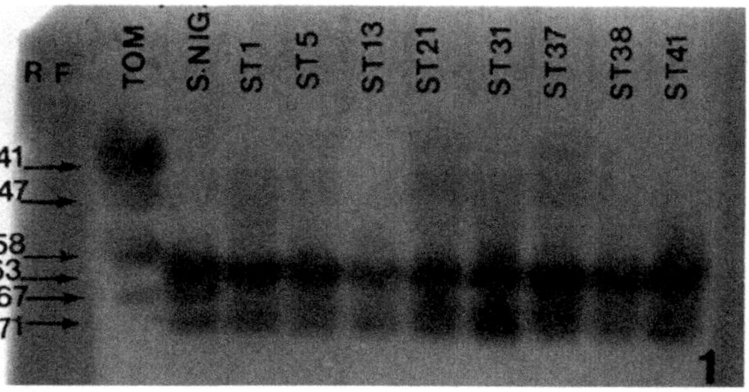

FIGURE 1. GAT isozyme patterns of somatic hybrids and the parents.

FIGURE 2. ST37, a backfusion regenerant.

REFERENCES

1. Barsby, T. V., P. V. Chuong, S. A. Yarrow, S. C. Wu, M. Coumans, R. J. Kemble, A. D. Powell, W. D. Beversdorf and K. P. Pauls. Theor. Appl. Genet. 73 (1987) 809-814.
2. Beversdorf, W. D., J. Weiss-Lerman, L. R. Eriksson and Z. Souza Machado. Can. J. Genet. Cytol. 22 (1980) 167-172.
3. Binding, H., S. M. Jain, J. Finger, G. Mordhorst, R. Nehls and J. Gressel. Theor. Appl. Genet. 63 (1982) 273-277.
4. Gamborg, O. L., R. A. Miller and K. Ojima. Exp. Cell Res. 50 (1968) 151-158.
5. Gressel, J., G. Ezra and S. M. Jain. In: McLaren, J. D. (ed) Chemical manipulation of crop growth and development. Butterworth, London (1982) pp 79-91.
6. Jain, S. M. Ph.D. thesis (1978) Jawaharlal Nehru University, New Delhi, India.
7. Pelletier, G., C. Primard, F. Vedal, P. Chetrit, R. Remy, P. Rouselle and M. Renard. Mol. Gen. Genet. 191 (1983) 244-240.
8. Shahin, E. A. Theor. Appl. Genet. 69 (1985) 235-240.
9. Sai-Ming Sun. M.Sc. thesis (1971) University of Hong Kong, Hong Kong.

*TAES Technical Article #23184

Somatic hybridization between tomato and other Lycopersicon or Solanum species1

A. Zelcer, S. Izhar, O. Soferman, B. Levy, D. Wolf

Division of Genetics and Plant Breeding, Institute of Field and Garden Crops
Volcani Center, Agricultural Research Organization
Bet-Dagan 50250, ISRAEL

Although many important traits originated in wild relatives have been introgressed into commercial tomato cultivars, the potential germplasm resources are severely limited in many instances by unilateral incompatibility and hybrid sterility among these crosses (Rick, In: Tomato Biotechnology , Eds: Nevins DJ, Jones RA) Alan R Liss, Inc, New York, pp17-26). Therefore, somatic hybridization is emerging as a complementary technique in the breeding of tomato (Melchers et al, Carlsberg Res Comm 43:203; Shepard et al, Science 219:683; Adams and Quiros, Pl Sci 40:209; O'Connell and Hanson, Theor Appl Genet 70:1, Theor Appl Genet 72: 59; Handley et al, Theor Appl Genet 71:691; Kinsara et al, J Pl Physiol 125:225). Moreover, somatic hybridization can generate novel combinations of nuclear and cytoplasmic determinants not previously available by conventional breeding . We report herein on several strategies for the generation of novel somatic hybrids and cybrids between tomato and additional Lycopersicon or Solanum species of interest.

THE TOOLS

Cytoplasmic-albino Large Red Cherry: this mutant was isolated and kindly supplied by **M. Hanson**.It is a fairly stable trait; only very infrequent reversions in the form of green islands in otherwise white leaves were observed. Mesophyll protoplasts are extremely fragile; therefore young cell-suspensions were utilized as source of protoplasts.

Protoplast isolation and culture: mesophyll and suspension protoplasts were isolated with Onozuka R-10 Cellulase (0.25-0.50 %) and Macerozyme (0.05-0.1 %). Protoplasts were cultured usually as described by Shahin (Theor Appl Genet 69: 235). Regeneration was achieved in TM-4 medium (op. cit.) or in MS medium, additioned with 2% sucrose, 1% mannitol, 2 ppm zeatin and 0.1-0.2 ppm IAA.

Protoplast fusion: parental protoplasts were resuspended in W5 solution, mixed at appropriate concentrations and fused with PEG-DMSO at high pH according to Menczel and Wolfe (Plant Cell Rept. 3:196).

Probbing of organellar DNA: the identity of organelles in somatic hybrids was analyzed by blot-hybridization with the following probes:

Chloroplast probes: Tobacco PstI-10 and PstI-2A (Fluhr et al, Gene 25:271).

Mitochondrial probes: Maize Pst I-3.2 kB fragment (Chao et al, Nucl. Ac. Res. 12:6629); Petunia P-88 (Sal I-2.9 kB fragment; Boeshore et al, Pl. Mol. Biol. 4:125)

Isozyme characterization: Activities of Pgi (phosphoglucose isomerase), Pgm (phosphoglucose mutase) and Got (glutamate oxaloacetate transaminase) were assayed as described by Rick (In: Isozymes in Plant Genetics and Breeding, eds: Tanksley and Orton, Elsevier Sci. Pub., part B, pp 147-165).

1Supported by the US-Israel Binational Agricultural Research and Development Fund (BARD), and by the Biotechnology Program, NCRD (Israel)-BMFT (Germany)

RESULTS AND DISCUSSION

An overview of the general scope in this project is depicted in accompanying figure. The different attempted fusion experiments are visually superimposed on a Lycopersicon-Solanum crossing polygon. In the following, advanced fusion experiments are individually reported and discussed.

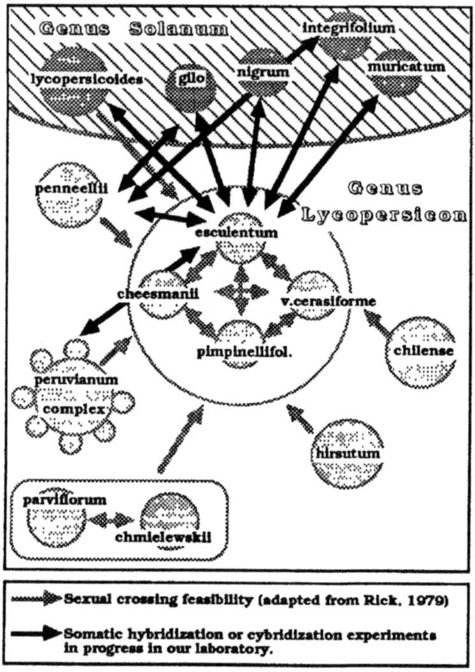

L. esculentum X L. peruvianum hybrids: Tomato (cv. Lukullus) mesophyll protoplasts were fused with L. peruvianum suspension protoplasts. The later were irradiated with a relatively low dose (5kR) of x-rays, in order to induce partial chromosome elimination. Fused cells were individually isolated and cultured above a feeder-layer of L. pennellii cells, as previously described by us (Shneyour et al, Pl Sci Letts 33:293). Hybridity was confirmed in regenerants by isozyme analyses (Pgm). Preliminary observations by blot hybridization with the Petunia P-88 probe, revealed in hybrids' mt-DNA the presence of L. peruvianum specific fragments, as well as novel fragments not to be found in parental restriction analyses. Several regenerated plantlets were transferred to greenhouse conditions. Flower development is abnormal; morphological aberrations in anthers and a low pollen count was observed. Parthenocarpic small fruits devoid of seeds develop occasionally. Attempts are being made to rescue fertile progeny by crosses with several cultivated or wild genotypes.

L. pennellii X Solanum integrifolium hybrids: Suspension-derived protoplasts from L. pennellii were fused with S. integrifolium mesophyll protoplasts. Fusion products were isolated and cultured as described above. Hybridity of regenerants was confirmed by their Pgm and Pgi activity pattern. Blot hybridization with mitochondrial probes revealed in the hybrids the presence of mt-DNA fragments from both parents, as well as novel fragments.

Hybrid plantlets seldom develop a normal root system. Special hormonal treatments and grafting on normal seedlings are being attempted.

L. esculentum X L. peruvianum cybrids: this experiment is conceptually different from the previously described. The **donor-recipient** scheme for hybridization (Zelcer et al, Z. Pflanzenphysiol. 90:397) was employed herein. Cytoplasmic donor's protoplasts (from L. peruvianum cell-suspensions) were irradiated before fusion with massive doses (100 kR) of x-rays in order to preclude cell division and transfer of nuclear determinants to fusion products. Cell suspension protoplasts from albino Large Red Cherry were utilized as cytoplasmic recipients. Green colonies (putative cybrids) resulting from these experiments were isolated and subsequently transferred to regeneration media. Regenerated plants seem morphologically similar to wild-type L. esculentum. Moreover, isozyme analyses of putative cybrids revealed a Pgi and Got pattern identical to the L. esculentum parent. Blot hybridization with maize Pst I-3.2 and petunia P-88 probes revealed in mt-DNA from putative cybrids fragments unique to the L. esculentum parent, as well as novel DNA fragments.

TRANSFER OF REGENERATION CAPACITY FROM LYCOPERSICON PERUVIANUM TO L. ESCULENTUM BY PROTOPLAST FUSION

J. WIJBRANDI, J.G.M. VOS AND M. KOORNNEEF

Department of Genetics, Agricultural University, 53 Generaal Foulkesweg, 6703 BM Wageningen, The Netherlands

1. INTRODUCTION

Asymmetric hybrids obtained by fusing protoplasts of a recipient genotype with irradiated protoplasts of a donor species can be useful to introduce properties from a wild species into a cultivated species such as tomato (Lycopersicon esculentum). Regeneration from established callus cultures and protoplasts of the latter species is relatively difficult, especially when compared with the excellent regeneration capacity of the related species L. peruvianum (1). This species which possesses several desirable agricultural traits (2), is very difficult to hybridize with the cultivated tomato. The present study describes the use of the regeneration capacity of L. peruvianum to select for somatic hybrids. Selection against the unfused and self-fused L. peruvianum protoplasts was achieved by an irradiation treatment or by their sensitivity to kanamycin when a resistent tomato genotype was used as recipient. The very good genetics of tomato will allow a detailed analysis of the (asymmetric) hybrids that were obtained.

2. MATERIAL AND METHODS

Kanamycin resistance was introduced into the tomato cultivar Bellina by leaf disc transformation. Protoplasts of this genotype were fused with non-irradiated protoplasts of L. peruvianum (PI128650). The symmetric, somatic hybrids were obtained by selection for plants which regenerated from fusion-derived, kanamycin resistant calli. In other experiments L. per. protoplasts were exposed to 50, 300 or 1000 Gy of gamma-rays prior to fusion with Bellina protoplasts. The asymmetric hybrids derived from shoot regenerating calli.

3. RESULTS

3.1. Plant regeneration

The well-growing, kanamycin resistant calli, selected after fusion of non-irradiated protoplasts, regenerated nearly as good as protoplast-derived calli of L. per. (Fig. 1). All plants, derived from these fusion calli, were shown to be hybrids (see later). After fusing with irradiated L. per. protoplasts, plant regeneration from the calli was less efficient and more slowly (see Fig. 1). The resulting plants were considered to be asymmetric hybrids.

3.2. Chromosome number

The symmetric hybrids were tetraploid (2n=48; mainly) or hexaploid (2n=72. See Fig. 2). The asymmetric hybrids showed much more variation. Chromosome numbers both below and above the tetraploid level were found (Fig. 2). Fragments couldn't be detected thus far. In some regenerants mixoploidy was observed in root tips.

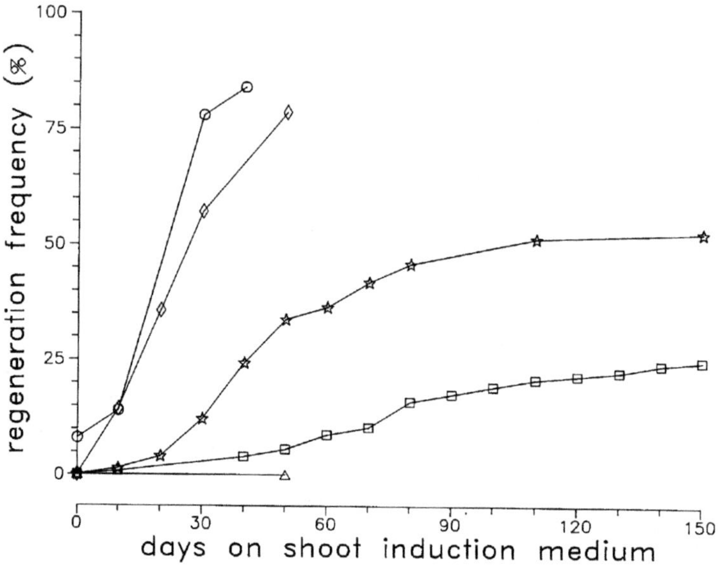

FIGURE 1. Percentage of calli regenerating on shoot induction medium. The calli are derived from individual protoplasts or heterokaryons of L. peruvianum (○ ; n=50), L. esculentum (△; n=121), putative symmetric hybrids (◇ ; n=14), putative low dose (50 Gy) hybrids (☆ ; n=74) and putative high dose (300 + 1000 Gy) hybrids (□ ; n=125).

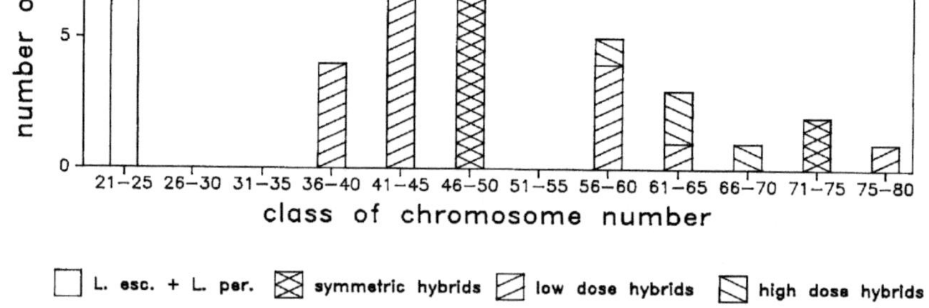

FIGURE 2. Chromosome numbers of parents (L. peruvianum and L. esculentum), symmetric hybrids, low dose (50 Gy) hybrids and high dose (300 + 1000 Gy) hybrids. Chromosomes were counted at metaphases in root tip preparations. The numbers are divided in classes of five.

3.3. Isozyme analysis

The hybrids were screened for the isozyme pattern of acid phosphatase, locus Aps-1 (located on chromosome 6). The L. per. parent is heterozygous for this gene, resulting in a three band pattern. L. esc. is homozygous for a different allele, which results in a single band (higher than the L. per. bands). All symmetric hybrids tested, had a pattern consisting of the bands from both parents and some additional bands due to formation of new heterodimers. The asymmetric hybrids had a 'symmetric hybrid pattern', a pattern with one L. per. band (plus one additional band), or only the L. esc. band (see Table 1).

TABLE 1. The presence of Aps-1 alleles of L. peruvianum in the somatic hybrids (indicated by the irradiation dose of the L. per. parent).

Irradiation dose	Number of Aps-1 alleles of L. peruvianum		
	2	1	0
0 Gy	30	0	0
50 Gy	16	6	2
300 + 1000 Gy	4	0	2

It can be concluded that several asymmetric hybrids lack one or both chromosome(s) 6 of the wild species.

3.4. Morphology

The morphology of the hybrids was described for plants growing in an unheated greenhouse under Dutch summer conditions.

3.4.1. Leaf morphology. The leaves of the symmetric hybrids were intermediate between the two parents (see Fig. 3a). The asymmetric hybrids showed much variation. Some had leaves that resembled the symmetric hybrid ones, others had tomato-like or aberrant leaves (see Fig. 3b).

3.4.2. Flower morphology. The two parents differ from each other for this character. The L. per. flowers have short sepals, dark yellow petals and a protruded style, whereas the tomato flowers have long sepals, pale yellow petals and a style as long as the stamens. Flowers of symmetric hybrids had sepals of intermediate length, dark yellow petals and a style which was not or a little protruded. The asymmetric hybrids were very variable, especially with respect to sepal length and petal colour. The style was, except for one hybrid, not protruded. In several hybrids it was shorter than the stamens.

3.4.3. Fruit morphology. The fruits of the symmetric hybrids were as big as or slightly bigger than L. per. fruits, but smaller than Bellina fruits and had a yellow colour (the parents have green respectively red fruits). The fruits of the asymmetric hybrids did vary in size (smaller than L. per. upto as big as Bellina fruits) and colour. The colour was green, yellow, orange (mostly) or red.

3.5. Fertility

Pollen fertility was investigated by staining pollen with fuchsin. The parents had 80 to 100 % stainable pollen, the symmetric hybrids 0 to 90 % and the low dose hybrids 0 to 20 %. Fruits of symmetric hybrids often had several seeds after selfing. Asymmetric hybrids thus far gave no seeds, neither after selfing nor after backcrossing with one of the parents.

FIGURE 3. a) The leaves of (from left to right) L. esculentum cv. Bellina, a symmetric hybrid and L. peruvianum. b) The leaves of twelve different asymmetric (low dose) hybrids together with on the left side a Bellina leaf and on the right side a L. peruvianum leaf.

4. DISCUSSION

Regeneration capacity can be a dominant selectable marker in somatic hybridization, as indicated by the fusion experiments with non-irradiated protoplasts of L. esculentum and L. peruvianum.
Gamma-irradiation induces loss of L. peruvianum chromosomes. This is demonstrated by the reduced regeneration capacity, the aneuploid chromosome numbers, the loss of L. peruvianum Aps-1 alleles and the morphological characteristics of the asymmetric hybrids. The phenotype of the low dose hybrids is ranging from that of symmetric hybrids to that of tomato. The few high dose hybrids are in general more tomato-like.
Chromosome specific, molecular markers (more isozymes, RFLP's) will be used for further characterization of the somatic hybrids.

REFERENCES

1. Koornneef M, Hanhart CJ and Martinelli L: A genetic analysis of cell culture traits in tomato. Theor Appl Genet 74:633-641, 1987.
2. Rick CM: The potential of exotic germplasm for tomato improvement. In: Vasil IK, Scowcroft WR and Frey KJ (eds): Plant improvement and somatic cell genetics, pp. 1-28. New York: Academic Press, Inc, 1982.

SOMATIC HYBRIDIZATION IN LYCOPERSICON; USING GAMMA-IRRADIATED NICOTIANA TABACUM SR-1 OR LYCOPERSICON PERUVIANUM.

F.H.M. Derks and C.M. Colijn-Hooymans.

Institute for Horticultural Plant Breeding,
P.O.Box 16, NL-6700 AA Wageningen, The Netherlands.

Introduction

The final aim of the research is to study the activity of chloroplasts originating from different species of Solanaceae into the cultivated tomato *Lycopersicon esculentum*. For this study we developed a reproducible method for fusion and regeneration of tomato protoplasts. Furthermore we examined the dose rate of irradiation for the inactivation of the nucleus of donor protoplasts.

Isolation and regeneration of protoplasts.

Protoplasts were isolated from in vitro grown plants of *L. esculentum* cv. Moneymaker. Leaves of four weeks old plants were incubated for 16 h in enzyme solution (0.5% (w/v) Cellulase R-10 and 0.25% (w/v) Macerozyme R-10). After filtration and floatation on 15 sucrose a yield of $1x10^6$ protoplast per gram leaf material was obtained. Protoplasts were plated in TM-2 medium (Shahin, 1985) in agarose droplets (Dons and Bouwer, 1986). A plating efficiency (PE*) of 8.6% was obtained. *L. esculentum* cv. Moneymaker protoplasts did not regenerate on TM-4 medium as described by Shahin (1985), however when the sucrose concentration in the TM-4 medium was lowered to 5 g/l 8% of the calli developed at least one shoot.
For the isolation and regeneration of *L. peruvianum* protoplasts the same procedure was followed except for the enzyme concentration: 0.25% (w/v) Cellulase R-10 and 0.05% Macerozyme R-10. The PE of *L. peruvianum* protoplasts was 12.1% and the regeneration frequency on modified TM-4 medium was 88%.
Protoplasts of *Nicotiana tabacum* SR-1 were isolated according to Nagy and Maliga (1976) with a PE of 8.8% and a regeneration frequency of 100%.

Inactivation of the donor protoplasts.

To establish the optimal conditions for transfer of the cytoplasm protoplasts of *L. peruvianum* and protoplasts of *N. tabacum* SR-1 were treated with different doses of irradiation (50, 100, 300, 1000 and 2500 Gy, ^{60}Co source). Five, 26, 72 and 120 h. after irradiation treatment of the protoplasts, their viability was studied by staining with Fluoresceine Diacetate (FDA) (Widholm, 1972). Just after 72 h the effect of irradiation on the viability of the protoplasts was obvious. 25% of the protoplasts of *N. tabacum* SR-1 (compared to the control) were dead after irradiation with 300 Gy and only 1% of those protoplasts divided once or twice but no microcalli were formed. *L. peruvianum* protoplasts were more sensitive to irradiation than the *N. tabacum* SR-1 protoplasts because even after treatment with only 50 Gy 25% of the protoplasts were dead. At that dose no microcalli (more than 4 cells) were formed after plating of the irradiated protoplasts. Based on these results we choose 300 Gy as the irradiation dose for the inactivation of the *N. tabacum* SR-1 donor protoplasts and 50 Gy as the dose for *L. peruvianum* protoplasts.

Asymmetric fusion of irradiated N.tabacum SR-1 protoplasts and L.peruvianum protoplasts.

For this fusion we used streptomycin resistance present in the chloroplasts of N. tabacum SR-1 as selection system. The selective streptomycin concentration for L. peruvianum calli was 500 mg/l. PEG mediated fusions (according to Negrutiu, 1986) were performed with irradiated protoplasts of N. tabacum SR-1 and protoplasts of L. peruvianum. The protoplast fusion suspension was selected for fusion products on streptomycin (500 mg/l). Up to now only five green calli were obtained and one of them regenerated one shoot. This very low number of green shoots could be due to the incongruity of the far-related species used in this fusion.

Asymmetric fusion of irradiated L.peruvianum protoplasts and tomato protoplasts.

Also PEG mediated fusions were done with irradiated L. peruvianum protoplasts and L. esculentum cv. Moneymaker protoplasts. The fusion products were plated on modified TM-4 medium (Shahin, 1985) with the lowered sucrose concentration. Shoots were obtained after 11 weeks, which is approximately 2 weeks earlier than the regeneration of L. esculentum protoplasts and 4 weeks later than the regeneration of untreated L. peruvianum protoplasts. The regeneration frequency (29%) of this fusion experiment was higher than the regeneration frequency of L. esculentum cv. Moneymaker. These results suggest that the shoots obtained from this experiment are possible fusion products. In this fusion experiment no selection conditions were available. Therefore the regenerated shoots we have obtained so far have to be analysed at the plant level for their cybrid characters.

PE*= number of divided cells on day 7/ number of protoplasts initially plated.

REFERENCES

1. J.J.M. Dons and R. Bouwer. In: Nuclear techniques and in vitro culture for plant improvement. Proceedings of an international symposium on nuclear techniques and in vitro culture for plant improvement. IAEA Vienna (1986) 497-504.
2. J.I. Nagy and P. Maliga. Z. Pflanzenphysiol. 78 (1976) 453-455.
3. I Negrutiu, D. de Brouwer, J.W. Watts, V.I. Sidorov, R. Dirks and M. Jacobs. Theor. Appl. Genet. 72 (1986) 279-286.
4. E.A. Shahin. Theor. Appl. Genet. 69 (1985) 235-240.
5. J.M. Widholm. Stain Technology 47 (1972) 189-194.

REGENERATION OF SOMATIC HYBRID PLANTS BETWEEN TOBACCO (NICOTIANA TABACUM L.) AND BLACK NIGHTSHADE (SOLANUM NIGRUM L.)

X.S.WAN[1], F.T.WANG[1], G.Y.WANG[1], X.F.YENG[1], C.A.HSIA(ZHEN-AO XIA)[1], X.M.CHANG[2], F.N.LIU[2]
1 Shanghai Institute of Plant Physiology, Academia Sinica, 300 Fonglin Road, Shanghai, China,200032.
2 Institute of Tobacco, Xuchang,Henan, China.

1. INTRODUCTION

The aim of this work is to test whether any useful character could transfer from *Solanum nigrum*(S) to *Nicotiana tabacum* cv. Ge-Xin No.1(T) by protoplast fusion.

2. MATERIALS AND METHODS

2.1.Mesophyll protoplasts of T or S were isolated enzymatically. One of them was labelled with FITC or RITC. PEG(m.w. 6000) and high pH high Ca^{++} method was used in fusion.

2.2.Culture media

2.2.1.Protoplast. Modified Dudits medium(1976), (NAA 0.186, ZT 0.11, 2,4-D 0.5, KT 0.125 mg/l).

2.2.2.Callus.MS1D, (2,4-D 1 mg/l).

2.2.3.Shoot differentiation. MSIK (IAA 1, KT 2 mg/l).

2.2.4.Root formation. MS_0 (without hormone).

2.3.The regenerated plants(TS) were compared with parents(T or S) by morphological, cytological and biochemical characters.

3. RESULTS AND DISCUSSION

49 regenerated plants among five hundreds were selected to identification. The results of two of them (TS-28 and TS-33) were as follows:

(1)Plant height, leaf size and shape, flower morphology of TS-28 and TS-33 were different from T or S. (Fig. 1-3).

(2)Chromosome numbers(PMC II) of TS-28(n=46-52) or TS-33(n=48) were around the sum of T(n=24) and S(n=24).(Fig.5-7).

(3)Peroxidase and malic dehydrogenase isozyme patterns of TS-28 or TS-33 were different from T or S. (Fig. 8,9).

(4)The contents of total sugar,total nitrogen,protein,nicotine and nicotine-N in mature leaves of TS-28 or TS-33 were intermediate between that of T or S. (Tab. 1).

(5)TS-28 and TS-33 were susceptible to *Phytophora* disease but resistant to TMV.

(6)The flue-cured quality of mature leaves of TS-28 or TS-33 was worse than that of tobacco. (Fig.4).

From these results, TS-28 and TS-33 had been identified as somatic hybrids between *Nicotiana tabacum* and *Solanum nigrum.* The offerspring of them were propagated and some were back-crossed sexually with tobacco for further selection.

REFERENCES

1. Wang GY, Hsia CA: Acta Batanica Sinica 25, 111-114,1983.

T

TS-33
Regenerated plant

leaves

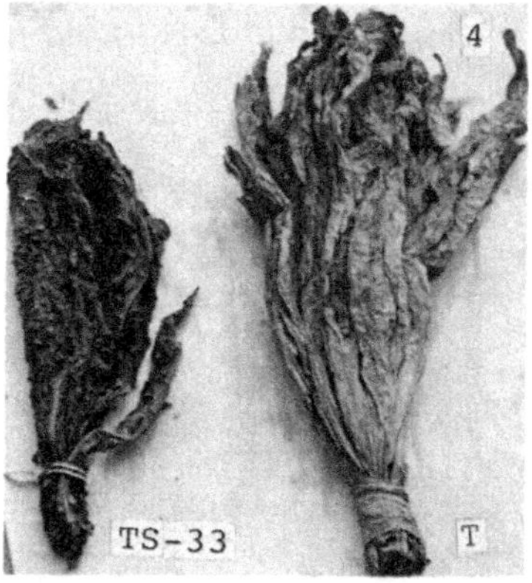

flue-cured leaves quality

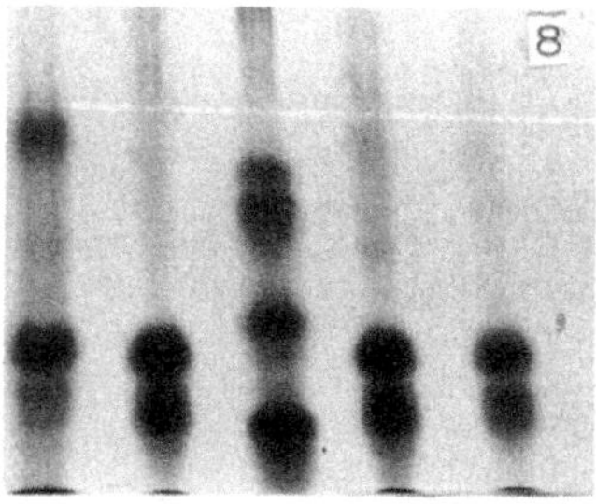

TS-33 S T
TS-33 TS-28
peroxidase isozyme

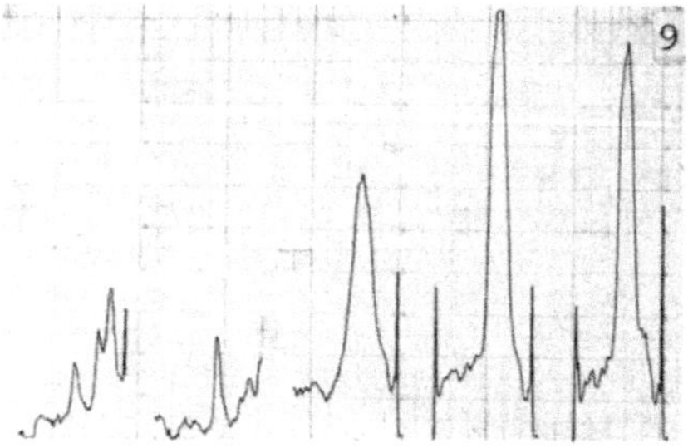

TS-33 TS-33 S TS-28 T
malic dehydrogenase isozyme

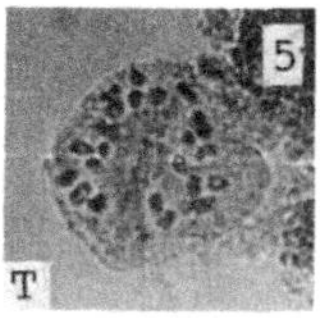

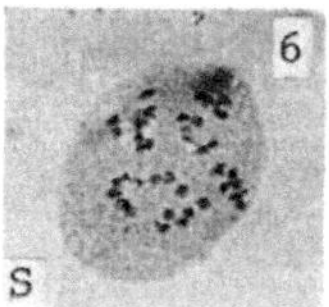

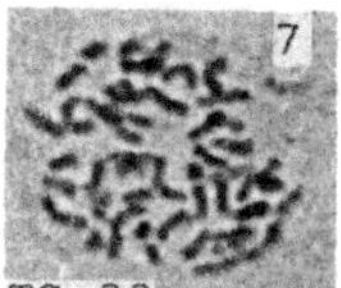

PMC II
chromosomes

Tab.1 Analysis of leaf contents(%)

material	total sugar	total nitrogen	protein	nicotine	nicotine nitrogen
TS-28	6.11	2.03	12.36	1.76	0.30
TS-33	4.34	2.30	14.20	0.92	0.16
T	18.25	1.34	8.00	2.01	0.35
S	3.94	2.49	15.53	0.16	0.03

THE USE OF STREPTOMYCIN RESISTANCE AND CHLOROPHYLL DEFICIENCY FOR SELECTION OF SOMATIC HYBRIDS BETWEEN NICOTIANA TABACUM AND N.RUSTICA.

Seiichi Toki and Toshiaki Kameya
Institute for Agricultural Research, Tohoku University, Sendai. Japan

Recently, many somatic hybrids have been obtained by using various selection markers. In particular, resistant to amino acid analogs, antibiotics resistance or chlorophyll deficiency have been used as useful selection markers. However,if both positive and negative selection markers were incorporated into a plant, that is called a universal hybridizer, since it can be hybridized with any wild type plants.

In order to produce a universal hybridizer, we have sexually crossed between a streptomycin resistant plants of Nicotiana tabacum SR1 (Maliga et al. 1973) and a chlorophyll deficient plants of N.tabacum A1 (obtained in our laboratory as a spontaneous mutant and grafted on wild type N.tabacum). From their offspring of this cross, a double mutant (desinated as SA) was selected (Fig.1) and characterised with respect to streptomycin resistance and the ability to synthesize chlorophyll. Calli initiated from SA showed resistance to streptomycin and the chlorophyll content in SA leaves was lower than that of normal green plants(Toki and Kameya 1987).

To prove the usefulness for somatic hybridization of a N.tabacum double mutant SA, protoplasts obtained from SA suspension cultures were fused via the dextran method (Kameya 1983) with protoplasts isolated from N.rustica L. mesophyll cells. Prospective somatic hybrids were selected for streptomycin resistance and the ability to produce chlorophyll in regenerated plants(Fig.2).

By using this selection system, green plants were recovered from 5 colonies. Hybrid nature of these plants was confirmed by morphological studies(Fig.3) and isoelectric focusing of the Fraction 1 protein (Rubisco)(Fig.4). The present results show evidence for the potential use of the double mutant in studies of somatic hybridization (Toki and Kameya 1987). We have got plants from hybridization studies with Salpiglossis sinuata using this selection system. Furthermore, we have established a triple mutant by sexual crosses between kanamycin resistant N.tabacum(by transformation) and SA.

REFERENCES

1. T. Kameya, Cytorogia. 48 (1983) 873-878.
2. P.Maliga, A. Sz.-Breznovits, L. Marton and F.Joo, Nature New Biol. 244 (1973) 29-30.

3. S. Toki and T. Kameya, Heredity, 59(2) (1987) 223-226.
4. S. Toki and T. Kameya, Heredity,(1987) in press.

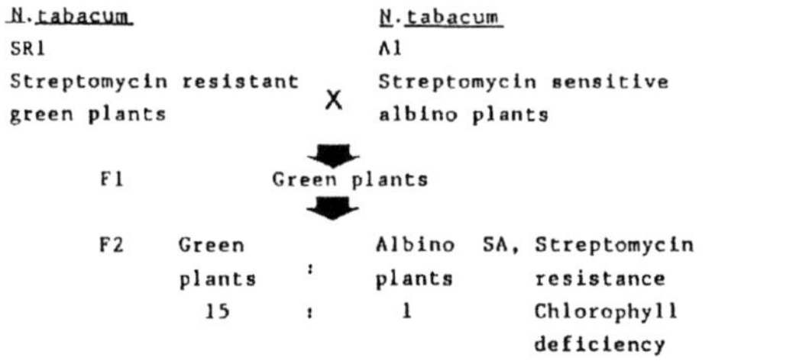

Figure 1 Production of a chlorophyll deficient streptomycin resistant mutant of Nicotiana tabacum SA.

Figure 3. Leaf and flower morphology of somatic hybrids and parents.
1:N.tabacum SA. 2:N.rustica.
3,4:somatic hybrids.

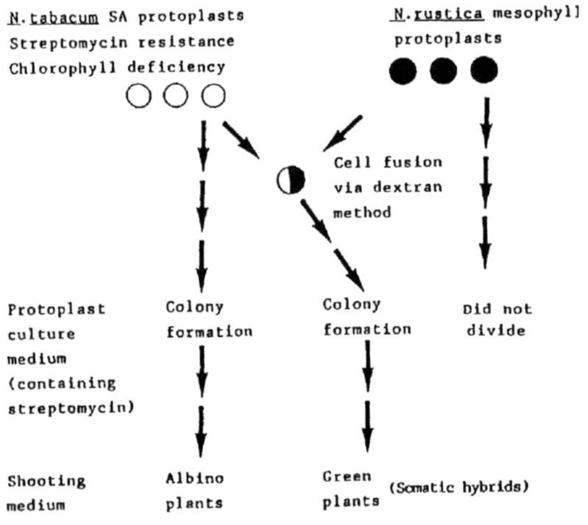

Figure 2 Selection scheme for hybrids between N.tabacum SA and N.rustica.

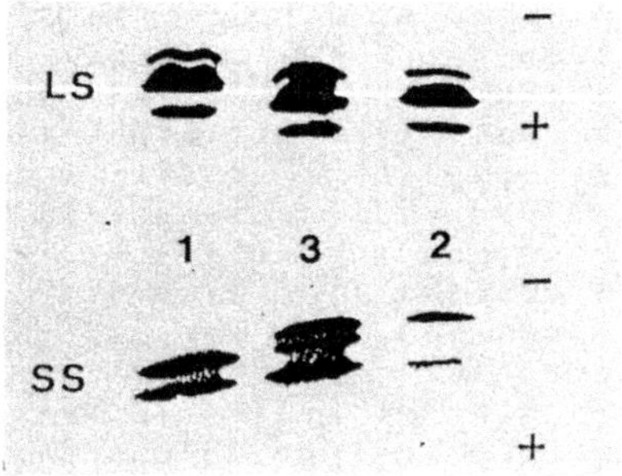

Figure 4. Large and small sub-unit polypeptides of Fraction I protein resolved by IEF.
1:N.tabacum SA. 2:N.rustica.
3:a somatic hybrid.

INTRA- AND INTERSPECIFIC GAMETOSOMATIC HYBRIDISATION WITHIN THE GENUS PETUNIA

C.H. LEE and J.B. POWER
Plant Genetic Manipulation Group, Department of Botany, University of Nottingham, Nottingham NG7 2RD, U.K.

ABSTRACT

Following PEG and high pH induced fusion, intraspecific gametosomatic hybrid plants (pollen tetrad protoplasts of a normal purple flowered variety of P. hybrida fused with cell suspension protoplasts of a nuclear albino mutant of the variety Blue Lace) and interspecific gametosomatic hybrid plants (tetrad protoplasts (as above) fused with cell suspension protoplasts of a nuclear albino mutant of P. parviflora) were recovered. Hybrid plants of both combinations possessed an intermediate vegetative and floral morphology with chromosome numbers of 2n=3x=21 and 2n=3x=25 respectively. Hybrid cells were in both systems identified as green colonies against an albino background as a result of complementation to chlorophyll proficiency. Pollen tetrad protoplasts did not divide. The production of such plants at the intra- and interspecific level in Petunia has shown that the concept of gametosomatic hybridisation can be extended to genera other than Nicotiana. An alternative selection strategy is available to that as used earlier for Nicotiana.

INTRODUCTION

There is one report [1] which described the production of gametosomatic hybrid plants following the fusion of tetrad protoplasts of Nicotiana glutinosa with mesophyll protoplasts of a nitrate reductase deficient (NR^-) mutant of N. tabacum. In this system, the gametosomatic hybrid had a pentaploid chromosome number and was fertile.

It was expected that the gametosomatic hybrid produced between two very closely related species or two varieties of the same species, would be sterile as a result of complete trivalent formation at meiosis, reflecting a high degree of chromosomal homology. Conversely, it was anticipated that a gametosomatic hybrid of two species of different genera would have a predictable level of fertility. In this case unpaired chromosomes might also be lost during meiosis giving offspring with a reduced haploid parental chromosome complement yet retaining the basic diploid genome.

It was important therefore to firstly demonstrate that this would in fact be the situation and secondly to show that alternative selection strategies can be utilised for the production of these novel types of hybrids. In this latter respect the readily available albino mutants of Petunia were chosen.

MATERIALS AND METHODS

Isolation of protoplasts

Pollen tetrad protoplasts were isolated from greenhouse-grown plants of a purple-flowered intervarietal hybrid of P. hybrida (accession R51 x V23, 'Monsanto') using the method of Pirrie and Power [1]. Cell suspension protoplasts were isolated from nuclear albino mutants of P. hybrida cv. Blue Lace and P. parviflora [2].

Fusion of protoplasts

Cell suspension protoplasts of albino P. hybrida cv. Blue Lace or P. parviflora at a density of 2.5 x 10^5/ml in CPW 13M medium [3] were mixed with an equal volume of the appropriate wild type tetrad protoplasts in their respective enzyme solution and at the same density. Aliquots (1.0 ml) of the mixtures were placed in 5 cm Petri dishes and the protoplasts allowed to settle (20 min). Two drops of PEG fusion solution [22.5% (w/v) PEG, MW 6000, 1.8% (w/v) sucrose, 1.54 g/l $CaCl_2.2H_2O$, 95.2 mg/l KH_2PO_4, pH 5.8] were added diametrically opposite so that they coalesced with the protoplasts. After 20 min. (room temp.) the PEG was removed and replaced with 2.0 ml of a filter-sterilised high pH fusion solution [3.75 g/l glycine, 7.4 g/l $CaCl_2.2H_2O$, 9% (w/v) mannitol, pH adjusted to 10.4 with NaOH] for a further 20 min. Protoplasts were washed twice, with 10 ml of an 11% (w/v) sucrose solution containing 7.4 g/l $CaCl_2.2H_2O$ and 3.75 g/l glycine (pH 5.8), by flooding the dish and allowing the protoplasts to settle for 20 min. This was followed by two washes with KP8 medium [4]. Protoplasts were suspended in 3.0 ml/dish of this medium. For P. hybrida var. Monsanto fused with P. parviflora, two out of four fusion-treated dishes for each experiment were nursed with albino cell suspension protoplasts (2.5 x 10^4/ml) of P. parviflora in order to enhance division of heterokaryons. The other two dishes were cultured on their own. Dishes were sealed with Nescofilm and kept at 27°C with a continuous illumination (1,000 lux; daylight fluorescent tubes). Unfused control dishes were handled in a similar way.

Culture, selection and regeneration of gametosomatic hybrids

The osmotic pressure of the protoplast culture medium was reduced by adding after 7 days, 1.0 ml/dish of K8 medium [4], and after 14 days a further 1.0 ml of K8 medium to each dish having first removed 1.0 ml of the old medium. The latter step was repeated weekly until after 4 weeks, green colonies (3-4 mm diam.) which had appeared in the fusion dishes were transferred to MSP1 medium [MS medium [5] with 2.0 mg/l NAA, 0.5 mg/l BAP, 3% (w/v) sucrose, 0.8% (w/v) agar, pH 5.8] with 5,000 lux continuous illumination. After 3-4 weeks selected calluses were transferred to and subcultured monthly on MSZ regeneration medium [MS medium with 1.0 mg/l Z, 3% (w/v) sucrose, 0.8% (w/v) agar, pH 5.8]. Regenerated shoots were rooted (2-4 weeks) in MSP2 medium [MS medium with 0.1 mg/l NAA, 3% (w/v) sucrose, 0.8% (w/v) agar, pH 5.8]. Putative gametosomatic hybrid plants were grown on in soil-less compost in greenhouse conditions at high humidity (2 weeks). Plants were flowered in the open greenhouse. Lateral shoots (5 cm) of the regenerated plants were detached and rooted (2 weeks) in hydroponic tanks which contained quarter-strength Hoaglands' [6] salts solution plus 1.0 mg/l IAA. Roots for somatic chromosome number determinations were pretreated with a saturated aqueous solution of α-bromonaphthalene (3h; 4°C), fixed in acetic alcohol (18h; room temp.), hydrolysed with 1N HCl (10 min.; 60°C) and stained in Schiff's reagent. Soluble protein extracts were prepared from leaf tissues of the interspecific gametosomatic hybrid and subjected to peroxidase isoenzyme analysis using standard techniques.

RESULTS

Gametosomatic hybrids: *P. hybrida* var. Monsanto with *P. hybrida* cv. Blue Lace.

Experiments were repeated ten times of which seven were negative. Three experiments gave several green calluses, two of which regenerated on MSZ medium to give a total of 20 plants. All control dishes were negative. The regenerated plants all had flowers that were intermediate between the two parental varieties with respect to colour distribution, colour intensity and petal fringing (Fig. 1). A branched, prostrate growth habit, pale foliage and fringing leaf morphology in the hybrid was inherited from *P. hybrida* cv. Blue Lace, coupled with narrow, long leaves typical of *P. hybrida* var. Monsanto. A random sample of plants of separate experiments all had a chromosome number of 2n=3x=21. This was the expected situation since both diploid parents were 2n=2x=14. Pollen of all gametosomatic hybrids was inviable as judged by staining in acetocarmine and the plants could not be reciprocally back-crossed to either parental variety or other varieties of *P. hybrida* [7].

Fig 1. Flowers of (L. to R.) *P. hybrida* var. Monsanto (2n=2x=14), the gametosomatic hybrid (2n=3x=21) and Blue Lace (2n=2x=14). Flower colour of Monsanto corresponds to cyclamen-purple 74A and the gametosomatic hybrid and Blue Lace to violet group 87D (light regions) and 83A (dark regions) (Royal Horticultural Society Colour Chart).

Gametosomatic hybrids: *P. hybrida* var. Monsanto with *P. parviflora*.

Twenty experiments were performed of which ten were positive giving fifteen colonies. Eleven of these were from the nursed fusion dishes. Control dishes were negative. Two green calluses gave 40 plants on MSZ medium and all the regenerated plants had flowers that were intermediate with respect to size (Fig. 2a) plus the branched, prostrate growth habit and flower colour of *P. parviflora*. Leaf size was also intermediate. Cytological analysis of the gametosomatic hybrid of these two *Petunia* species gave a root chromosome count of 2n=3x=25 which was the expected number. These hybrids when reciprocally backcrossed to *P. hybrida* var. Monsanto or other commercial varieties set seed. Leaf peroxidase isoenzyme analysis showed that these plants had both parental profiles plus extra hybrid bands.

Fig.2. a. Flowers of (L. to R.) P. hybrida var. Monsanto (2n=2x=14), the gametosomatic hybrid (2n=3x=25) and P. parviflora (2n=2x=18). Flower colours, as determined using the Royal Horticultural Society Colour Charts, were for Monsanto, cyclamen-purple 74A, for the gametosomatic hybrid, bishops violet 81A and for P. parviflora, purple group 77B. b. Plants of P. hybrida var. Monsanto, the gametosomatic hybrid and P. parviflora.

DISCUSSION

The production of gametosomatic hybrids in Petunia has shown that albino mutants can be substituted for the previously used nitrate reductase deficient mutant as in Nicotiana [1]. Complementation to chlorophyll proficiency in the gametosomatic hybrid can therefore be effected by the fusion of a non-dividing pollen tetrad protoplast system (wild type) with an actively dividing nuclear albino somatic protoplast system. Nurse culture, using albino cell suspension protoplasts, after fusion was effective in terms of increasing the survival of heterokaryons. Hybrids can be readily visualised against the background of colourless, dividing cells from the unfused or homokaryon protoplasts of the albino parent. This approach is therefore likely to be of general applicability not only for the production of gametosomatic hybrids but, provided the parental species are not closely related, for the generation of fertile, interspecific gametosomatic hybrids such as those in the genus Nicotiana.

REFERENCES

1. Pirrie A, Power JB: Theor. Appl. Genet. 72: 48-52, 1986.
2. Power JB, Berry SF, Chapman JV, Cocking EC, Sink KC: Theor. Appl. Genet. 55: 97-99, 1979.
3. Power JB, Chapman JV, Wilson D: Laboratory Manual: Plant Tissue Culture, University of Nottingham, UK.
4. Kao KN: Mol. Gen. Genet. 150: 225-230, 1977.
5. Murashige T, Skoog F: Physiol. Plant. 15: 473-497, 1962.
6. Hoagland DR, Arnon DT: Calif. Agr. Expt. Sta. Cir. 347, 1950.
7. Lee CH, Power JB: Plant Cell Rep. (in press).

THE TRANSFER OF CYTOPLASMIC MALE STERILITY (CMS) BY INTRASPECIFIC SOMATIC HYBRIDISATION IN *PETUNIA HYBRIDA* (HORT) VILM.

J.M. CHARTIER-HOLLIS, M.R. DAVEY and J.B. POWER.
Plant Genetic Manipulation Group, Department of Botany, University of Nottingham, Nottingham NG7 2RD. U.K.

1. INTRODUCTION

The expression of cytoplasmic male sterility (CMS) in *Petunia* is governed by the nuclear genome and a single plasmagene(S), believed to be located in mitochondria. In the absence of nuclear restorer genes, the S plasmon causes breakdown of microsporogenesis at meiosis. Fertility is restored if a sufficient number of multiple alleles (mfr) are present; when some alleles are present, the abortion of pollen will occur at different stages (Izhar, 1984).

In this study, a normal purple flowered line of *P. hybrida* (cv. R51 x cv. V23) was used as the male parent in sexual crosses with *P. hybrida* cv. Blue Bedder (CMS). The F1 hybrid was male sterile, but the point of breakdown of pollen formation indicated that the normal *P. hybrida* parent (cv. R51 x cv. V23) had a low level of fertility restorer alleles. Protoplasts, derived from both plants, were fused to produce somatic hybrids in order to study the result of mixing of cytoplasms from male sterile and male fertile lines. The mitochondrial inhibitor, Rhodamine 6G, was also incorporated in some experiments in order to promote a unidirectional inheritance of CMS mitochondria (Aviv *et al*, 1986).

2. MATERIALS AND METHODS

Leaf mesphyll protoplasts were isolated from greenhouse grown plants of *P. hybrida* (cv. R51 x cv. V23) (2n = 2x = 14) using an enzyme mixture which contained 2% Rhozyme HP 150, 2% Meicelase and 0.03% Macerozyme R10 in CPW13M, pH 5.8 (Power *et al*, 1976). In some cases, these protoplasts were treated with 20 µM Rhodamine 6G for 30 mins at room temperature and then washed several times in CPW13M medium (Power *et al*, 1976) prior to fusion. Suspension culture protoplasts (Power *et al*, 1979) of *P. hybrida* cv. Blue Bedder (CMS; 2n = 2x = 14) were stained with fluorescein diacetate and fused using the small-scale PEG high pH Ca^{2+} method (Power *et al*, 1984). Fusion frequency was assessed by a dual fluorescence in the heterokaryons and was typically 8-10%. Fusion dishes were incubated in the dark at 4°C for 24h and then at 25°C for a further 5 days, at which point the regenerating protoplasts were plated at a density of 2.5×10^4 cm^{-3} in MSP1 9M agarose medium (Power *et al*, 1984) and maintained under constant illumination of 5,000 lux provided by daylight fluorescent tubes.

Colonies were transferred initially to MSP1 OM medium and subsequently to MSZ medium (Power *et al*, 1984). The shoot regeneration medium, MSZ, also prevented further growth of the Blue Bedder parental colonies and callus. Regenerated plants were screened for pollen viability at flowering and those plants which showed a consistently low value, coupled with a chromosome number of 2n = 4x = 28 were selfed, where possible and the F2 population was analysed for male sterile and male fertile individuals.

3. RESULTS

The following classes of plants were distinguished following regeneration from the fusion treated protoplasts:-

Type A Fertile, pollen viability (p.v)>50%. Frequency of occurance: 9 in 10 of the regenerated plants.

Type B Homoplasmic male sterile, p.v <30%, 2n = 4x = 28. Female fertile but no seed set upon selfing. Frequency of occurance: 1 in 18 of the regenerated plants.

Type C Heteroplasmic male sterile, p.v<40%, 2n = 4x = 28. Seed set afer selfing. F2 segregated male fertile and male sterile individuals. Frequency of occurance: 1 in 18 of the regenerated plants.

Type D Unstable heteroplasmic male sterile individuals with some plants producing occasional fertile branches. Frequency of occurance: 1 in 53 of the regenerated plants.

When the leaf mesophyll protoplasts of P. hybrida (cv. R51 x cv. V23) were pretreated with Rhodamine 6G prior to fusion 1 in 4 of the regenerated plants were found to be Type B.

4. DISCUSSION

Analysis of the regenerated plants, following intraspecific somatic hybridisation, showed there to be a high frequency of occurance of plants which were either homoplasmic or heteroplasmic male sterile. The recovery rate of similar plant types following interspecific somatic hybridisation in the genus Petunia was found to be lower (Izhar et al, 1983). At the intraspecific level, the somatic hybrid nucleus, containing insufficient restorer alleles to affect fertility, is genetically very similar to both of the parental nuclei. The resultant effect therefore, on the mixed cytoplasm is to provide equivalent nuclear organelle co-operativity for both normal and CMS mitochondria. Regenerated plants will be homoplasmic or heteroplasmic male sterile depending upon the relative number of CMS-type or normal mitochondria within the callus cells from which the plants are ultimately derived. The use of Rhodamine 6G further increased the frequency of recovery of male sterile plants by promoting a unidirectional inheritance of CMS-type mitochondria.

The ability to produce homoplasmic male sterile somatic hybrid plants from a mixture of cytoplasms present in the heterokaryon has important consequences, not only with respect to the production of new CMS lines, but also for the unidirectional inheritance of mitochondria carrying other useful agronomic traits. The use of antimetabolites, such as herbicides, directed specifically against chloroplasts, could further extend the usefulness of this approach which permits a controlled inheritance of desired cytoplasmic factors.

REFERENCES

1. Izhar S: In: Sink KC (Ed). Monographs in Theoretical and Applied Genetics. No 9. Petunia pp. 77-82, 1984.
2. Aviv D, Chen R, Galun E: Plant Cell Rep. 3: 227-230, 1986.
3. Power JB, Frearson EM, George D, Evans PK, Berry SF, Haywood C, Cocking EC. Plant Sci. Lett. 7: 51-55, 1976.
4. Power JB, Berry SF, Chapman JV, Sink KC, Cocking EC: Theor. Appl. Genet. 5: 97-99, 1979.
5. Power JB, Chapman JV, Wilson D: Laboratory Manual: Plant Tissue Culture, Nottingham University, U.K.
6. Izhar S, Schlicter M, Swartzberg D: Mol. Gen. Genet. 190: 468-474, 1983.

APPLICATION OF PROTOPLAST FUSION TO AN ORNAMENTAL BREEDING PROGRAMME.

D.L.ARNOLD, S.F.BERRY, D.CARTER.

Twyford Plant Laboratories Ltd,
Baltonsborough, Glastonbury,
Somerset. BA6 8QG. U.K.

1. INTRODUCTION.

The exploitation of wider gene pools in breeding programmes is frequently prevented by the existence of breeding barriers between closely related species and genera. Improvement of *Petunia* is envisaged by the combination of floral characteristics of *Salpiglossis* with the growth habit of *Petunia*. *Petunia* and *Salpiglossis* demonstrate a prezygotic mode of incompatibility. Bud and *in-vitro* pollination, together with embryo culture techniques have so far failed to overcome incompatibility barriers. The alternative approaches to introgression of the desired traits are by somatic hybridization and limited genome transfer following protoplast fusion. Before a programme of protoplast fusion is initiated it is necessary to establish protoplast isolation and culture protocols for both species and investigate irradiation doses and fusion conditions.

2. MATERIALS AND METHODS.

2.1. Protoplast isolation and culture. Mesophyll protoplasts were isolated from shoot cultures of a number of *Petunia* lines. *Salpiglossis* protoplasts were isolated from cell suspensions established from mesophyll callus. Protoplasts were cultured in KM8P liquid over agar. The plating efficiency (%PE) of the protoplasts was calculated as the percentage of the original protoplast population which had undergone division after 12 days. Cultures were diluted regularly with KM8 and when visible colonies developed they were transferred to regeneration media.

2.2. Regeneration trials. Regeneration trials on a range of media were conducted with leaf callus and protoplast callus. The regeneration efficiency (%RE) was calculated as the percentage of calli producing one or more shoots.

2.3 Irradiation trials. Protoplasts were isolated from three *Salpiglossis* lines and exposed to a range of γ -irradiation doses (^{60}Co source). The effect of irradiation dose on %PE was assessed after 12 days in culture.

2.4. Fusion techniques. *Petunia* mesophyll and *Salpiglossis* cell suspension protoplasts were isolated and fused using a PEG/high pH method and cultured in a KM8P liquid medium. Two days after fusion the liquid protoplast culture was layered over KM8P agar. Heterokaryon frequency was calculated 24 hours after fusion as the percentage of heterokaryons in the total protoplast population. Colonies developing following fusion were transferred to regeneration medium.

2.5. Selection. All plants regenerated from protoplast fusion are to be established in the glasshouse and screened at flowering for desired traits and fertility.

3. RESULTS.

3.1. Protoplast isolation and culture. *Salpiglossis* protoplasts from all three lines gave rise to colonies with high regeneration capacity (Table 1). Colonies have been obtained from Petunia protoplasts of all three lines. Shoot regeneration trials have not been fully assessed, but regeneration capacity appears to be genotype dependent.

3.2. Irradiation trials. The effect of irradiation dose on %PE of Salpiglossis protoplasts is shown in Fig:1. A dose of 20 Krad inhibits protoplast division.

Table 1.

		%PE	%RE
Petunia line	1	12	
	2	13.5	NA
	3	3.3	
Salpiglossis line	1	7.4	74
	2	13.5	70
	3	14.4	88

NA = not assessed.

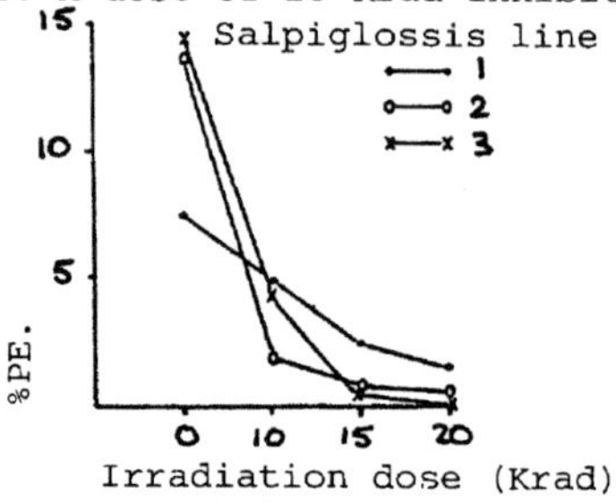

3.3. Protoplast fusion. A heterokaryon frequency of 2% was observed following PEG fusion between unirradiated protoplasts of Salpiglossis and Petunia. Hybrid cell selection was not applied and therefore parental protoplasts divided alongside putative somatic hybrids. Colonies transferred to regeneration medium have given rise to shoots. These will be rooted and established in the glasshouse for flowering and screening.

4. DISCUSSION. The establishment of protoplast regeneration protocols for Petunia and Salpiglossis will allow for a full programme of somatic hybridization and limited genome transfer between the two genera. Selection of Petunia and Salpiglossis hybrid cells based on complementation of chlorophyll deficiency and/or growth requirements, did not result in the recovery of somatic hybrid plants (Power and Chapman, 1983). Application of complementation selection at the cell level may eliminate a proportion of nuclear genome combinants. Selection at the whole plant level should recover a range of complete and incomplete genome combinations. Partial genome transfer has been demonstrated following intergeneric fusion between Aegopodium podagraria and Daucus carota (Dudits et al, 1979). It is also possible to effect limited genome transfer by irradiation of the donor protoplast prior to fusion (Gupta et al, 1982). Modification of Petunia requires the stable transfer of a limited amount of genetic material from Salpiglossis. Lines selected for possession of desirable traits will be employed in a Petunia breeding programme and hence fertility is of prime importance.

REFERENCES.

1. Power,J.B. and Chapman,J.V. Intra- and intersubfamilial somatic hybridization within the Solanaceae. Poster Proceedings, 6th International Protoplast Symposium, Basel. August 12-16, 1983. pp 276-277.
2. Dudits,D., Hadlaczky,G.Y., Bajszar,C.Y., Koncz,C.S., Lazar,G. and Horvath,G. 1979. Plant regeneration from intergeneric cell hybrids. Plant Science Letters. 15: 101-112.
3. Gupta,P., Gupta,M., and Schieder,O. 1982. Correction of nitrate reductase defect in auxotrophic plant cells through protoplast mediated intergeneric gene transfer. Molec. Gen. Genet. 188: 378-383.

FORMATION OF CARROT CYBRIDS BASED ON METABOLIC COMPLEMENTATION BY FUSION BETWEEN X-IRRADIATED CELLS AND IODOACETAMIDE-TREATED CELLS

H. ICHIKAWA, L. TANNO-SUENAGA AND J. IMAMURA
PLANTECH RESEARCH INSTITUTE, 1000, Kamoshida-cho, Midori-ku, Yokohama 227, JAPAN

INTRODUCTION

The genomes of two cytoplasmic organelles, plastids and mitochondria, encode such important traits as herbicide resistance or cytoplasmic male sterility (CMS), respectively. In higher plants, cytoplasmically-encoded traits are usually inherited maternally by crossing. However, in somatic cell fusion, the cytoplasms from both parents are also mixed as well as nuclei in a fused cell. In somatic hybrid or cybrid plants regenerated from fused cells, sorting out of one or the other parental chloroplasts and rearrangement of mitochondrial DNAs (mtDNAs) have been frequently observed (Maliga 1986). It is possible to create plants possessing novel organelle compositions. It has been also reported that the maternally inherited traits such as CMS were introduced through protoplast fusion in <u>Nicotiana</u> (Zelcer et al. 1978, Aviv et al. 1980, Medgesy et al. 1985 Kumashiro and Kubo 1986), <u>Petunia</u> (Boeshore et al. 1983) and <u>Brassica</u> (Pelletier et al. 1983, Yarrow et al. 1986, Menczel et al. 1987).

X-or gamma-irradiation of protoplasts is useful for the transfer of alien cytoplasmic traits through cell fusion. Since the irradiation treatment prevents colony formation from protoplasts and the irradiated nucleus is eliminated from a fused cell (Zelcer et al. 1987, Menczel et al. 1982), cybrids are formed through protoplast fusion between irradiated cells (cytoplasm donor) and iodoacetate-treated cells (recipient) by metobolic complementation (Sidorov et al. 1981). We report here the selection of <u>Daucus</u> cybrid plants with the double selection method, and show the successful transfer of a CMS trait. Mitochondrial genomes of those cybrid plants are also analysed.

MATERIALS AND METHODS

Protoplast isolation, inactivation, fusion and culture

Carrot protoplasts were prepared from 3- to 5-day old subcultures of a donor line (<u>D. carota</u> 28K a CMS line) or recipient line (<u>D. carota</u> cv. Kikuyo gosun (K5)). Protoplasts of a donor line ($2x10^6$) were X-irradiated with a total dosage of 60 Krad (1 Krad/min). Recipient protoplasts (10^6/ml) were inactivated by incubation in 15 mM iodoacetamide for 10 min. The iodoacetamide- treated protoplasts and X-irradiated protoplasts were mixed in a 1:1 ratio and fused. Fusion mixtures were cultured in a liquid MS medium containing 0.3 M sorbitol and 2,4-D for about a month. The colonies were gradually substituted with MS + 0.1 M sorbitol without 2,4-D. Putative cybrid plants were regenerated through somatic embryogenesis on MS medium and transferred into soil.

Isolation and restriction endonuclease analysis of mtDNAs from carrot suspension cultures

Cell suspensions of the fusion-derived plants were prepared from their shoot segments. Mitochondria were prepared from 5-10 g of the cell suspensions by differential centrifugation, and treated with DNAse I and then Proteinase K as described elsewhere(Ichikawa et al. 1987). MtDNAs were extracted from the lysates twice with Phenol-chloroform.

MtDNAs were digested with restriction enzymes for at least 4 h at 37°C. The restriction fragments were separated on 0.5-0.7% agarose gels, stained with ethidium bromide and photographed.

RESULTS AND DISCUSSION

Selection of *Daucus* cybrids between X-irradiated *D.carota* 28K and iodoacetamide-treated *D.carota* K5

No colony formation of irradiated 28K protoplasts were observed after 1 month of culture for more than 40 Krad X-ray exposure. More than 10 mM iodoacetamide treatment inhibited *D. carota* K5 protoplast division. From these results, pre-fusion treatments of protoplasts were performed as 60 Krad exposure to donor and 15 mM iodoacetamide treatment to recipient (Ichikawa et al. 1987). A mixture of both kinds of protoplasts showed no growth without PEG treatment, whereas many colonies were obtained after fusion treatment of mixed parental protoplasts.

Since the leaf morphological traits could not be distingusihed between both parents, it was impossible to characterize those of the regenerated plants. Both parental cell lines possessed chromosome numbers of 18 (2n). Chromosome numbers of six fusion-derived plants were the range of 17-46 (Table 1). Three of the six regenerated plants had 18 chromosomes.

MtDNAs from both parental cell lines and the regenerated plants were analysed (Fig.1, Table 1). Both parental mtDNA restriction patterns were distinguished from each other. MtDNAs from all of 11 fusion-derived plants investigated showed novel restriction patterns compared to those of both parental lines. Two (A-2-18, A-7-33) and four (D-1-27, D-2-14, E-2-8, E-2-14)in those plants possessed similar, but not identical, recombinat mt DNA patterns to the donor (recombinant-D) and the recipient (recombinant-R), respectively (Table1). The other five plants showed unique recombinant patterns (recombinant-U) different from the six plants above mentioned. From these results, the selection system based on metabolic complementation could be useful for the production of carrot cybrid plants.

Some of the regenerated plants passed the winter and flowered. 28K, a donor parent, has CMS flowers with brown anthers. Flower morphology of the regenerated plants was either sterile type homologous to 28k brown anther type or half-sterile type with red anthers, The ratios of normal pollen in morphology were 0% in the former and about 50% in the latter. A tendency was observed. Namely, the mtDNAs of half male-sterile plants were analogous to the mtDNA of a recipient, while those of the male-sterile plants were unique recombinat types or similar to that of a donor. Aviv and Galun (1987) also pointed out a similar relationship between mtDNAs and male sterility in Nicotina cybrids. These results indicated that a CMS trait in carrot was transferred successfully from donor cells to recipient cells by protoplast fusion.

In future, the relationship between the mtDNA type and flower characteristics, and an inheritance of the male-sterility to the second generation will further be investigated.

Table 1. Characterization of regenerated plants from D. carota (28K) + D. carota (K5) fusion.

Plant no.	Chromo-some no.	Type of mtDNA pattern[a]	Flower
A-2-18	-	recombinant-D	sterile
A-7-33	-	recombinant-D	sterile
D-1-27	17	recombinant-R	-
D-2-14	-	recombinant-R	half-sterile
E-2- 5	-	recombinant-U	sterile
E-2- 8	18	recombinant-R	-
E-2- 9	18	recombinant-U	-
E-2-14	18	recombinant-R	half-sterile
E-2-26	46	recombinant-U	-
A-8- 1[b]	24	recombinant-U	-
A-1-20[b]	-	recombinant-U	sterile

a. Types of recombinant DNAs in the regenerated plants, D or R, and U correspond to that analogous to donor (28K) or recipient (K5), and that with a unique pattern, respectively.

b. These two plants were derived from the other experiment.

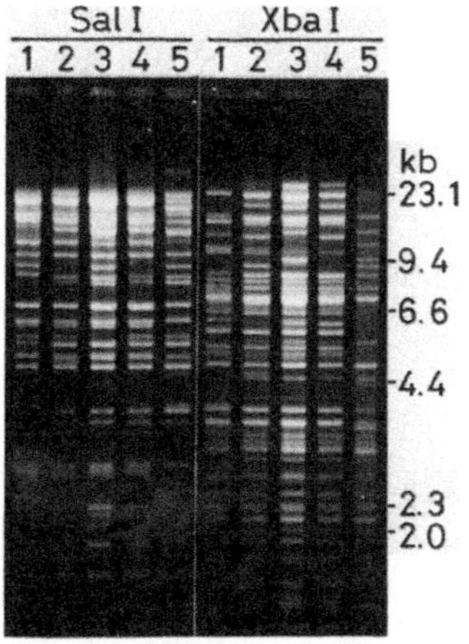

Fig.1. Sal I and Xba I restriction fragment patterns of mtDNAs from D. carota 28K, D. carota K5 and their cybrid plants. 28K (lane 1), cybrid plants: E-1-6 (2), E-2-9 (3) and A-1-20 (4), and K5 (5).

REFERENCES

1. Aviv D, Galun E (1987) Theor Appl Genet 73:821-826
2. Aviv D, Fluhr M, Galun E (1980) Theor Appl Genet 56:145-150
3. Boeshore ML, Lifshitz I, Hanson MR, Izhar S (1983) Mol Gen Genet 190:459-467
4. Ichikawa H, Tanno-Suenaga L, Imamura J (1987) Theor Appl Genet 74:746-752
5. Kumashiro T, Kubo T (1986) Japan J Breed 36:39-48
6. Maliga P (1986) Cell fusion to introduce genetic information coded by chloroplasts and mitochondria in flowering plants. In: Bogorad L (ed) Molecular developmental biology. Alan R Liss, New York, pp45-53
7. Medgyesy P, Golling R, Nagy F (1985) Theor Appl Genet 70: 590-594
8. Menczel L, Galiba G, Nagy F, Maliga P (1982) Genetics 100: 487-495

9. Menczel L, Morgan A, Brown S, Maliga P (1987) Plant Cell Reports 6:98-101
10. Pelletier G, Primard C, Vedel F, Chetrit P, Remy R, Rousselle P, Renard M (1983) Mol Gen Genet 191:244-250
11. Sidorov VA, Menczel L, Nagy F, Maliga P (1981) Planta 152: 341-345
12. Yarrow SA, Wu SC, Barsby TL, Kemble RJ, Shepard JF (1986) Plant Cell Reports 5:415-418
13. Zelcer A, Aviv D, Galun E (1978) Z Pflanzenphysiol 90:397-407

OPTIMIZATION OF CONDITIONS FOR SOMATIC CYBRIDIZATION OF DAUCUS CAROTA L.

S. ROEST, K.J. PUITE and M.J. VAN STAVEREN
Research Institute ITAL, P.O. Box 48, 6700 AA Wageningen, The Netherlands.

INTRODUCTION

Hybrid seed production of carrot (*Daucus carota* L.) via cross breeding is hampered by the practical impossibility of hand-emasculation to prevent self-pollination. In addition the introduction of cytoplasmic male sterility (cms) into a cytoplasmic male fertile (cmf) cultivar via cross breeding may be complicated by crossing barriers and recurrent back-crossing. On the other hand, the transfer of cms might be achieved through somatic cybridization in only one generation.

In this regard, we have established conditions for isolation, culture and regeneration of cell suspension protoplasts, for inactivation of the nuclear genome of the cms line (donor) and the cytoplasm of the cmf line (recipient) and for electrofusion of inactivated (stained) protoplasts of both parental lines. The first plantlets were regenerated after fusion.

RESULTS

Isolation, culture and regeneration of cell suspension protoplasts

Callus was induced on various types of explants on B_5-medium (1) supplemented with 0.9% agar, 2% sucrose and 0.1 mg/l 2,4-D. The callus was suspended and the cell suspensions were weekly subcultured in liquid B_5-medium in the presence of 2% sucrose and 0.3 mg/l 2,4-D. Protoplasts were isolated from cell suspensions 3 days after subculture. The enzyme treatment (based on 1% cellulase and 0.125% pectolyase) was carried out overnight at 27 ^{o}C in darkness on a rotary shaker at 30 rpm. After washing with mannitol and purification in sucrose the protoplasts were cultured at about 2.10^4/ml in ½V-KM-medium (2). Immediately after plating, 0.10-0.15% agarose was added to the culture medium to prevent agglutination. The first cell divisions were observed after 3 days of culture. The plating efficiency (PE, the number of cell aggregates as percentage of the initial number of plated protoplasts) was determined after 3 weeks of culture. The PE, which ranged from 0 to 8%, was strongly affected by both the genotype and the age of the cell suspension culture.

Developing cell aggregates were plated on top of a solidified MS-medium (3) supplemented with 2% sucrose after 3 weeks of culture. Subsequent callus growth resulted in the differentiation of embryoids and shoots. Plantlets were produced after transfer to soil and they were grown to maturity.

Electrofusion of inactivated (stained) cell suspension protoplasts

Cell suspension protoplasts of the donor cms line were irradiated with 0.5-0.7 kGy of γ-rays and stained with 1.5 μg/ml fluorescein diacetate (FDA). Protoplasts of the recipeint cmf line were treated with 0.1-0.2 mM iodoacetate and stained with 90 μg/ml rhodamine isothiocyanate (RITC).

Protoplasts were mixed in a ratio of 1:1 at a density of about 4.10^4/ml in 0.42 M mannitol and subjected to electrofusion (4) in a 1 ml multi-electrode fusion chamber. Heterokaryons were identified by red (RITC) and green-yellow (FDA) fluorescence. Immediately after fusion an adapted culture medium was added to the fusion mixture resulting in the normal density of the protoplasts and the normal composition of the culture medium.

Irradiation of protoplasts of the cms line never resulted in sustained cell divisions and iodoacetate-treated protoplasts of the cmf line seldom gave rise to the formation of aggregates. Some aggregates developed after electrofusion, and they continued growth giving rise to calli. Eventually, these calli differentiated into embryoids and shoots. The putative cybrids are being analysed on the basis of their mt-DNA restriction fragment patterns. The selection for cms will be carried out after flowering.

REFERENCES

1. Gamborg OL, Miller RA, Ohyama K: Exp. Cell Res. 50 (1986) 151-158.
2. Bokelmann GS Roest S: Z. Pflanzenphysiol. 109 (1983) 259-265.
3. Murashige T, Skoog F: Physiol. Plant. 15 (1962) 473-497.
4. Puite KJ, Wikselaar P van, Verhoeven H: Plant Cell Reports 4 (1985) 274-276.

ISOLATION AND FUSION OF APIUM GRAVEOLENS (CELERY) AND LEVISTICUM OFFICINALE (LOVAGE) MESOPHYLL PROTOPLASTS

A.DONOVAN, H.A.COLLIN, S.ISAAC
Department of Botany, University of Liverpool, P.O.Box 147, Liverpool, L69 3BX, U.K.

1. INTRODUCTION

Late blight is a major disease of celery caused by Septoria apiicola. As no celery varieties are immune to infection, breeders are now looking at wider crosses using protoplast fusion to introduce resistance. A closely related species of the Umbelliferae, Levisticum officinale (lovage), was considered to be a suitable partner for somatic fusion since it was morphologically similar to celery and resistant to late blight. By fusing protoplasts of such closely related species it was hoped to obtain a fertile hybrid.

2. PROCEDURE

2.1. Protoplast isolation

Celery protoplasts were isolated from leaf mesophyll tissue by the method of Tseng et al (1975). Lovage protoplasts were obtained by incubating chopped leaf material with a lytic enzyme mixture containing 0.1% Pectolyase Y23, 3% Meicelase 3% made up in 0.6M mannitol, pH 5.6, for 5h. Protoplasts were washed x3 with 0.6M mannitol and finally resuspended in 0.6M mannitol containing 2mM calcium.

2.2 Protoplast culture

The source of mesophyll tissue was critical for the successful culture of celery and lovage mesophyll protoplasts. Only protoplasts derived from either sterile seedlings, or from plantlets regenerated from callus cultures showed wall regeneration and division. The source material was grown under conditions of high light and humidity.

A maximum of 8% celery mesophyll protoplasts showed division when cultured on the NTK medium of Scowcroft and Larkin (1980) supplemented with 0.5 $mg.l^{-1}$ 2,4-D and 0.6 $mg.l^{-1}$ kinetin. Lovage mesophyll protoplasts divided (7.0%) on the same basic medium used for celery except the 2,4-D level was increased to 2 $mg.l^{-1}$ and kinetin to 2.4 $mg.l^{-1}$. Both the numbers of celery and lovage protoplasts were increased on their respective media by 1-2% by addition of 100mM calcium and the omission of ammonium salts.

2.3. Protoplast fusion

Celery and lovage mesophyll protoplasts were fused using the Zimmermann Cell Fusion apparatus in a 200 ul open fusion chamber with an alignment frequency of 2 MHz and an alignment voltage of 5V. Protoplasts aligned to form pearl chains. A fusion voltage of 30V given in two pulses of 90 usec duration then gave high rates (40.8%) of celery:lovage fusion (Fig.1).

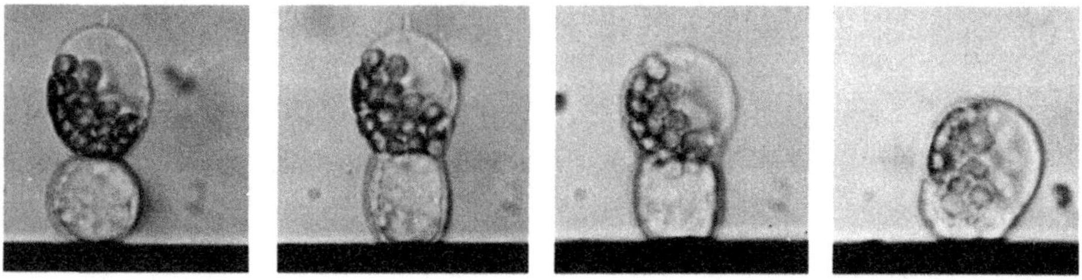

FIGURE 1. Electrofusion of celery and lovage protoplasts. a)Protoplast alignment, b)30 sec after high intensity DC pulse, c,d)Fusion sequence at 90 and 150 sec.

3. CONCLUSIONS

a)A protocol was developed for isolating protoplasts from lovage mesophyll tissue and for regenerating celery and lovage protoplasts to form callus.

b)A high percentage fusion was obtained between celery and lovage mesophyll protoplasts using the Zimmermann Cell Fusion system.

REFERENCES

1. Scowcroft WR and Larkin PJ: Isolation, culture and plant regeneration from protoplasts of Nicotiana debneyi. Australian Journal of Plant Physiology, 7, 653-644, 1980.
2. Tseng TC, Liu DF and Shiuo SY: Isolation of protoplasts from crop plants. Botanical Bulletin of Academic Science 16, 55-60, 1975.

ELECTROFUSION OF PLANT AND FUNGAL PROTOPLASTS

P.T. LYNCH*, H.A. COLLIN and S. ISAAC
Department of Botany, University of Liverpool, P.O. Box 147, Liverpool L69 3BX, U.K.
* Present address, Plant Genetic Manipulation Group, Department of Botany, University of Nottingham, Nottingham NG7 2RD, UK.

1. INTRODUCTION

Electrofusion provides a technique for initiating fusion between plant and fungal protoplasts. It was necessary to develop methods for the separate electrofusion of celery (Apium graveoleus) and Aspergillus nidulans BDUN 33 protoplasts, before determining a technique for fusion of celery protoplasts with A. nidulans protoplasts. After fusion the effect of an electrofusion treatment and plant fungal fusion was examined.

2. PROCEDURE

2.1. Materials and methods

2.1.1. Protoplast isolation.

Celery - Cells from 3 day old suspension cultures were incubated overnight in 0.6M mannitol plus 2% Cellulysin and 0.1% Pectolyase Y23. Protoplasts were resuspended in 0.6M mannitol and 0.5 mM $CaCl_2 2H_2O$. Prior to fusion with fungal protoplasts, celery protoplasts were incubated at 30°C for up to 1 hr in 0.6M mannitol and 0.5M $CaCl_2 2H_2O$ plus 1 mg ml^{-1} Pronase E.

A. nidulans - Mycelium from 18 hr cultures was incubated for 3 hrs in 0.4M mannitol, 0.2M phosphate buffer pH 5.8, plus 1.0 mg ml^{-1} NovoZym 234. Protoplasts were resuspended in 0.6M mannitol plus 0.5 mM $CaCl_2 2H_2O$.

2.1.2. Protoplast fusion

Electric fields were generated by a Zimmermann Cell Fusion™ System in both flat and helical fusion chambers. Parameters were established for the separate fusion of celery protoplasts and A. nidulans protoplasts and fusion between celery and A. nidulans protoplasts (Table 1).

2.1.3. Protoplast viability and respiration

Protoplast viability was assessed by exclusion of Evans Blue, and oxygen consumption was measured using a Rank Oxygen Electrode, both during the 24 hr period after fusion.

3. RESULTS

3.1. Fusion conditions

Alignment and fusion of A. nidulans protoplasts required greater field strengths than fusion of celery protoplasts (Table 1). Intermediate values were necessary for alignment and fusion of plant and fungal protoplasts.

In an AC field celery and A. nidulans protoplasts formed pearl chains containing both protoplast types (Fig. 1a). DC pulse treatment resulted in fusion between the two protoplast types (Fig. 1) protoplasts subsequently rounded up. Pronase E treatment appeared to stabilise celery protoplasts when subjected to higher DC voltages, and lowered the DC voltage required for fusion between plant and fungal protoplasts.

3.2. Physiological changes

A decrease in the number of viable protoplasts occurred during fusion treatment and in the subsequent 24 hrs. During the first 3 hrs after fusion of celery protoplasts with A. nidulans protoplasts, the protoplasts exhibited greatly elevated respiration rates, after which the values were comparable with control treatments.

TABLE 1. Parameters used to stimulate maximum fusion between protoplasts

Electrical Parameter	celery-celery protoplast fusion	A. nidulans-A. nidulans protoplast fusion	Celery-A. nidulans protoplast fusion
Alignment frequency (MHz)	1.5	1.0	3.0
Alignment voltage (V/cm)	200	1000	400
Fusion voltage (V)	15	50	25
Number of DC pulses	2	2	2
Pulse duration (μsec)	99	50	75
Time between pulses (sec)	1.0	0.5	1.0

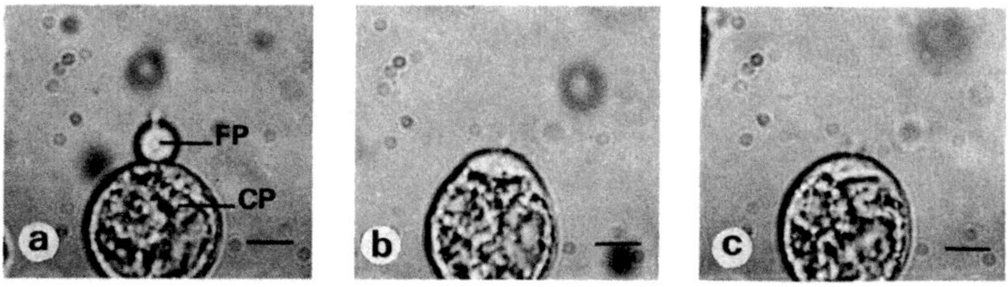

Fig.1. Celery, A.nidulans heterokaryons rounding up after fusion treatment
a. Chain of celery and A. nidulans protoplasts in an AC field
b. Immediately after fusion pulses, A. nidulans and celery protoplasts fused; alignment field present
c. 5 minutes after fusion pulses heterokaryon starting to round up.
CP - celery protoplast, FP - fungal protoplast.
Bar markers represent 10 μm.

4. DISCUSSION

The field strengths required to fuse the parental protoplasts alone and together was related to the difference in protoplast sizes. This inequality in field strengths resulted in lysis of the celery protoplasts during plant, fungal fusion.

Increased protoplast respiration rates after fusion between celery and A. nidulans protoplasts may have been due to an additive effect of the fungal mitochondria and/or a stimulatory effect of mixing fungal and plant cytoplasm.

Fusion of A. nidulans and celery protoplasts results in heterokaryon formation and hence the possibility of nuclear fusion and hybrid cell formation.

REGENERATION AND ANALYSIS OF CALLUS FROM FLOW-SORTED HETEROKARYONS OF SOYBEAN (GLYCINE MAX) AND G. CANESCENS

N. Hammatt[1], R.S. Nelson[1], A. Lister[2], E.C. Cocking[2] and M.R. Davey[2]

[1]Shell Research Ltd., Sittingbourne Research Centre, Sittingbourne, Kent ME9 8AG, UK.
[2]Department of Botany, University Park, Nottingham NG7 2RD, UK.

INTRODUCTION

Perennial, wild Glycine spp. harbour a wealth of genetic traits useful for the agronomic improvement of the soybean (G. max). Postzygotic incompatibility has necessitated the development of techniques for hybrid embryo rescue and somatic hybridisation, via protoplast fusion, to facilitate gene flow between these species.

This paper describes the production of callus from flow-sorted heterokaryons of soybean and G. canescens, and evidence for somatic hybridity.

MATERIALS AND METHODS

Protoplast isolation, culture and plant regeneration - The protocols used have been described previously (1,2). Soybean hypocotyl protoplasts were labelled with fluorescein by the addition of 50µl of a 1mg ml^{-1} fluorescein diacetate solution in acetone, to each 10ml aliquot of enzyme mixture prior to incubation with plant tissue.

Electrofusion - Protoplasts of G. max cv. HP-20-20 and G. canescens G1171 were suspended in a solution consisting of $110gl^{-1}$ mannitol and $147mgl^{-1}$ $CaCl_2.2H_2O$, pH5.8, at a density of $1.0 \times 10^5 ml^{-1}$, and mixed in a 1:1 ratio. Fusion was carried out using an electrode similar to that described (3). Protoplasts were aligned in an AC field (500KHz, $30Vcm^{-1}$, 30-60s), and fused with a 0.2ms DC pulse ($400Vcm^{-1}$).

Flow cytometry - Heterokaryons were isolated from parental protoplasts and homokaryons, 3-4h after fusion, using an EPICS V flow cytometer (Coulter). Sorting was carried out using a 200µm nozzle, a crystal drive frequency of 8.0 KHz, and CPW13M (1) as sheath fluid, at a pressure of 6psi. The argon laser was operated at 457nm wavelength using a power output of 100mW. Heterokaryons were sorted on the basis of the fluorescent emissions of fluorescein (495-565nm, peak 520nm) and chlorophyll (630-700nm, peak 685nm).

Heterokaryon culture - Heterokaryons were sorted at a rate of 1000 per flat-bottomed well, of a Sterilin 96 well microtitre plate, each well of which contained 30µl of liquid K8P culture medium (1). After sorting, heterokaryons were allowed to settle, and the medium was carefully removed, in 20 µl volumes, to avoid any disturbance. 40-60µl of agarose solidified culture medium (1), were added to each well to give a final heterokaryon density of 200 per 30µl medium. The heterokaryon suspension was dispensed as 20µl drops onto a premarked area on the bottom of a 9cm diameter Petri dish. The remainder of the dish was filled with 30µl droplets (3ml total per dish) of G. canescens G1171 cotyledon protoplasts, also in agarose-solidified medium at a density of 6.0 x 10^3 ml^{-1}, as nurse. After agarose gelling, the dish was filled with 10ml liquid culture medium (1), and the protoplasts cultured as described (2) for plant regeneration.

Isoenzyme analysis - Proteins were extracted, electrophoresed on a 0.6% acrylamide gel at 15mA (2-4h), and stained as described previously (4).

RESULTS AND DISCUSSION

Plant regeneration - Plants have been successfully recovered from cotyledon protoplasts of *G. argyrea* G1626, *G. canescens* G1171 and *G. clandestina* G1231 using the protocols described.

Protoplasts isolated from dark-grown soybean hypocotyls have been cultured to callus using liquid and agarose-solidified media. Following the regeneration protocol developed for perennial spp., it has not, to date, been possible to recover plants from protoplast-derived soybean tissues.

Electrofusion - Hypocotyl protoplasts from dark-grown soybean seedlings, labelled with fluorescein (green fluorescence) and cotyledon protoplasts of *G. canescens* G1171 (red autofluorescence) fused electrically. Heterokaryons with dual green and red fluorescence, formed between 2-4% of the protoplast population following fusion.

Flow cytometry - Heterokaryons isolated using the flow cytometer, were cultured to cell colonies that could be transferred to solid medium. Colonies transferred to agar K8 medium (1), gave rise to fast-growing callus lines. Five percent of colonies transferred to, and regularly subcultured for six months on SC2 medium (2) produced shoot buds.

Isoenzyme analysis - When proteins from callus maintained on K8 medium were electrophoresed and stained for aspartate amino transferase activity, two of the 12 cell lines exhibited bands additional to those seen in parental extracts. It is proposed that these hybrid-specific bands represent heteromeric enzyme formation (5). This suggests that flow-sorted *Glycine* heterokaryons have given rise to somatic hybrid tissue.

Further biochemical (isoenzymes; RFLPS) and cytological analyses will be carried out on heterokaryon-derived shoots to determine their hybridity.

REFERENCES

1. Hammatt N, Davey MR: Isolation and culture of soybean hypocotyl protoplasts. In vitro (submitted).
2. Hammatt N, Kim H-I, Davey MR, Nelson RS, Cocking EC: Plant regeneration from cotyledon protoplasts of *Glycine canescens* and *G. clandestina*. Plant Sci. 48, 129-135, 1987.
3. Watts JW, King JM: A simple method for large-scale electrofusion and culture of plant protoplasts. Biosci. Rep. 4, 335-342, 1984.
4. Hammatt N: Plant regeneration and protoplast fusion in the Genus *Glycine*. Ph.D. Thesis, Univ. Nottingham, 1986.
5. Schiavo FL, Guiliano G, Terzi M: Identifying natural and parasexual hybrids. In: Isozymes in Plant Genetics and Breeding, (eds.) SD Tanksley, TJ Orton, Elsevier, Amsterdam: 305-312, 1983.

INTERSPECIFIC SOMATIC HYBRIDIZATION STUDIES IN THE GENUS *LOTUS*

M.A. AZIZ[1], M.R. DAVEY AND J.B. POWER.

[1]Department of Agronomy and Horticulture, Malaysian University of Agriculture, 43400 UPM, Serdang, Selangor, Malaysia.

Plant Genetic Manipulation Group, Department of Botany, University of Nottingham, Nottingham NG7 2RD, U.K.

1. INTRODUCTION

Lotus species, are a group of perennial forage legumes utilised for production of pasture and high quality hay and silage. Interspecific sexual hybridization within the genus, has only been partially successful, due to a high incidence of post-fertilisation failure of endosperm development and abortion of F_1 hybrid embryos (1). The application of embryo rescue techniques and endosperm transplantation (2,3), to overcome post-zygotic incompatibility, has only been of limited success. Somatic hybridization is an approach to overcome such incompatibility aimed at combining desirable characters of different species, such as, increased tannin content (*L. corniculatus*, *L. pedunculatus*), drought tolerance (*L. corniculatus*, *L. tenuis*) or resistance to insect attack (*L. pedunculatus*). Prolific shoot regeneration from protoplast-derived tissues of *L. corniculatus* (4), makes this species a useful partner in somatic hybridization. Successful recovery of somatic hybrids, following protoplast fusion, of *L. corniculatus* and *L. coimbrensis* has been reported (5).

This report describes interspecific protoplast fusion of *L. corniculatus*, with *L. tenuis* and *L. pedunculatus* by electrical methods, successful sorting of heterokaryons by flow cytometry and the growth of heterokaryons to putative hybrid calli.

2. MATERIALS AND METHODS

2.1 *Isolation and Electrofusion of Protoplasts*. Green cotyledon protoplasts of *L. corniculatus* were isolated from 7 days old *in vitro* grown seedlings. Protoplasts of *L. tenuis* and *L. pedunculatus* were isolated from 4-5 days old cell suspensions.

The protoplast-enzyme mixture was passed through a 64 μm nylon sieve, and the protoplasts pelleted by centrifugation (80 x g, 5 min). Protoplasts were resuspended in CPW21S medium (6), washed in CPW13M solution (6) (2 changes) and resuspended in electrofusion medium (EF) (0.5 mM $CaCl_2 2H_2O$, 11% mannitol, pH 5.8) (2 changes).

A 1:1 mixture (1.0 ml) of protoplasts of both species, suspended at a density of 1×10^5 ml^{-1} in EF medium, was pipetted into each of the inner nine wells of a 5 x 5 grid dish (Sterilin U.K). A multi-electrode block was placed into each well, and protoplasts aligned in an AC field (40-60Vcm^{-1}, at 500-600 kHz) and subsequently fused by a DC pulse (0.83-1.33 kV cm^{-1}, for 1.0-2.0 msec.). Fusion-treated protoplasts were resuspended in CPW13M solution for flow cytometry.

2.2 *Flow Cytometry and Heterokaryon Culture*. Heterokaryons were isolated using a Coulter EPICS 541 flow cytometer, equipped with an argon ion laser operated at 457 nm and regulated at 100 mW power output. Heterokaryons were identified by their dual fluorescence (630-700 nm red chlorophyll autofluorescence of green cotyledon protoplasts; 495-565 nm yellow green fluorescence of FDA labelled cell suspension protoplasts). Sorting was achieved using CPW13M as sheath fluid, with a 200 µm nozzle operated at 6psig (41 kPa) and 8 kHz droplet formation frequency. Heterokaryons were collected directly in 24 well (FALCON) or 96 well (NUNC) multiwell plates.

Several culture methods for the growth of heterokaryons were assessed: in liquid culture medium in the presence of albino protoplasts of *Medicago sativa* as nurse, in agarose (1.2% w/v) solidified medium surrounded by nurse protoplasts in liquid K8P medium (6), and in microchambers (7) with the nurse protoplasts surrounding each microchamber but physically separated from the heterokaryons.

3. RESULTS AND DISCUSSION

A heterokaryon frequency of 6.0-7.0% was obtained in the fusion of *L. pedunculatus* cell suspension protoplasts with *L. corniculatus* green cotyledon protoplasts (Lp_{cs} + Lc_{gc}). Fusion of *L. tenuis* cell suspension protoplasts with *L. corniculatus* green cotyledon protoplasts (Lt_{cs} + Lc_{gc}) produced a lower heterokaryon frequency (4.4-5.0%). Alignment of protoplasts in both combination, occurred within 20 sec of application of the AC field. Subsequent application of a single DC pulse resulted in the formation of heterokaryons. In both combinations, the maximum frequency of heterokaryon formation was observed using a DC pulse of 1.0-1.17 kV cm , for 2.0 msec. Heterokaryons exhibited a high post-fusion viability (65-80%). The percentage of intact heterokaryons sorted was higher for Lp_{cs} + Lc_{gc} (11.2-19.0%) than for Lt_{cs} + Lc_{gc} (8.3-12.0%). However, division of heterokaryons, observed after 14 days in culture, was lower for Lp_{cs} + Lc_{gc} (9.8-11.1%) than for Lt_{cs} + Lc_{gc} (15.5-16.7%). A 4.0-6.0% contamination by parentals was observed in all experiments for both combinations.

Micro-chambers were the most suitable for culture of low numbers of heterokaryons, the latter dividing to form putative hybrid colonies (10-15 cells) after 5-6 weeks. Division and growth of heterokaryons was, however, slower than that of their parental protoplasts.

The above results highlight the efficiency of electrofusion for somatic hybridization studies in *Lotus*. Flow sorting of heterokaryons, though not completely free from contaminating parental protoplasts, is rapid and less tedious than micromanipulation. Nurse protoplasts were useful in stimulating heterokaryon division at low plating densities. The microchamber technique facilitated monitoring of heterokaryon division and growth, while dilution of nurse protoplasts outside the chambers, is easily accomplished without disturbing the dividing heterokaryons.

REFERENCES

1. Williams E, White DWR: New Zeal. J. Bot. 14: 307-314, 1976.
2. De Latour G, Jones WT, Ross MD: New Zeal. J. Bot. 16: 61-68, 1978.
3. Williams EG, De Latour G: Bot. Gaz. 141 (3): 252-257, 1980.
4. Ahuja PS, Hadiuzzamaan S, Davey MR, Cocking EC: Plant Cell Report 2: 101-104, 1983.
5. Wright RL, Somers DA, Mc Graw RL: International Congress of Plant Tissue and Cell Culture, Abs. VI: 86, 1986.
6. Power JB, Chapman JV, Wilson D: Laboratory Manual, Plant Tissue Culture, Nottingham University, U.K.
7. Gilmour DM, Davey MR, Cocking EC, Pental D: J. Plant Physiol. 126: 457-465, 1987.

INTERSPECIFIC HYBRID LINES OF M. SATIVA AND M. ARBOREA BY PROTOPLAST ELECTROFUSION.

S. ARCIONI, F. PUPILLI, M. PEZZOTTI, E. FALISTOCCO, F. DAMIANI
Centro Miglioramento Genetico Piante Foraggere, CNR - Borgo XX Giugno, 74 - Perugia (Italy)

INTRODUCTION

Interspecific hybridization in the genus Medicago has been considered as good means of improving cultivated alfalfa germplasm. Among the wild Medicago species, M.arborea, characterized by shrub habit, drought resistance and winter growth, is an interesting species to be crossed with lucerne. Because the two species are cross incompatible, somatic hybridization can be used to bypass the sexual barriers.
This report concerns electrofusion between mesophyll protoplasts of M. arborea and cell suspension protoplasts of M.sativa., with the characterization of hybrid cell lines.

Materials and Methods

Mesophyll protoplasts of M.arborea were isolated as described previously (1) and those of M.sativa were derived from a cell suspension (2) maintained in the presence of fluorescein isothiocyanate (5 gml^{-1}) during enzymatic incubation. Protoplast electrofusion was carried out in a chamber with parallel electrodes mounted 3 mm apart, 0.5 ml capacity, following the procedure previously reported (3), with the exception of pulse voltage (920 $V\ cm^{-1}$). After electrofusion, the content of the fusion chamber was transferred into a petri dish (3 cm) and 2 ml of KM8P liquid medium was added. Heterokaryons were identified with fluorescent light on the basis of their colour: half red (from mesophyll) and half green (from cell suspension). When heterokaryons had regenerated the cell wall, the cultures were diluted with 1 ml of double strength mixture of KM8P and KM8 (3:1) and solidified with 0.8 ml of 2.5% (w/v) agarose. The position (coordinates) of dividing heterokaryons was recorded and when they reached the 15-30 cell stage, they were transferred to a cylinder made of KM8 medium solidified with 0.5% agarose which was fixed to the bottom of a 5 cm petri dish. Around this cylindrical chamber, M.sativa cells were cultured as nurse cells in KM8 agarose solidified medium. Hybrid minicolonies were cultured in 1 ml of a mixture (2:1) of KM8P and KM8 liquid media and weekly the osmotic pressure was lowered adding 0.1 ml of KM8. Hybrid calli were finally grown on B_5 medium and examined for peroxidase isozymes(3)and cytological characteristics.

Results

Electrofusion yielded 4% heterokaryons. In liquid medium (6.10^4 prot.

ml^{-1}), heterokaryons regenerated cell walls one day after fusion and started mitotic divisions a day later; however, within two days of culture, about half of them were dead. At this stage, cells aggregated into clumps and gentle squeezing was necessary to separate parental and hybrid cells before embedding them in agarose solidified medium. In this culture medium (3.5 cells and prot.ml^{-1}), the position of heterokaryons, far enough from the surrounding cells, was recorded. For each experiment (8 electrofusions), about 200 hybrid cells were taken into account and after 8 days of culture 34% of the heterokaryons were viable but at the single cell stage; six percent had undergone the first and 25% the second and third mitotic divisions. The hybrid minicolonies were manually picked-up and cultured individually in an agarose nurse chamber where about 40% grew into small calli. Seven to eight months after fusion the hybrid nature of some calli was confirmed by electrophoretic separation of peroxidase. The hybrid calli showed some bands typical of M.sativa and others of M.arborea calli and their chimerical origin was excluded by the absence of some parental bands. Caryological analysis carried out on 3-months-old calli showed cells with 64 chromosomes and the hybrid origin of calli was confirmed by the presence of the four nucleolar chromosomes of M.sativa and the two nucleolar chromosomes of M.arborea (Fig.1). Chromosome structural (Fig.2) changes were demonstrated by the occurrence of dicentric, telocentric and fragment chromosomes. Iso-chromosomes were also observed. The same calli examined four months later showed only the presence of polyploid cells with a chromosome number ranging from 120 to 250. The high chromosome number observed in the hybrid calli cannot be due only to in vitro treatment because the alfalfa cell suspension and the leaf derived calli of M. arborea displayed most the cells with the normal chromosome number. Up to now all the efforts in inducing plant regeneration were unsuccessful and probably this was due to the high chromosome number of the hybrid cells. As a matter of fact, the cell line of M.sativa used in this study is characterized by high regeneration ability through somatic embryogenesis and almost all regenerants had the normal chromosome number.

References

1) Mariotti D.;Arcioni S.;Pezzotti M.(1984) Plant Sci. lett. 37: 149-156
2) Pezzotti M.;Arcioni S.;Mariotti D.(1984) Genet. Agr. 38: 195-208
3) Damiani F.;Pezzotti M.;Arcioni S. (1987) J. Plant Physiol. (in press)

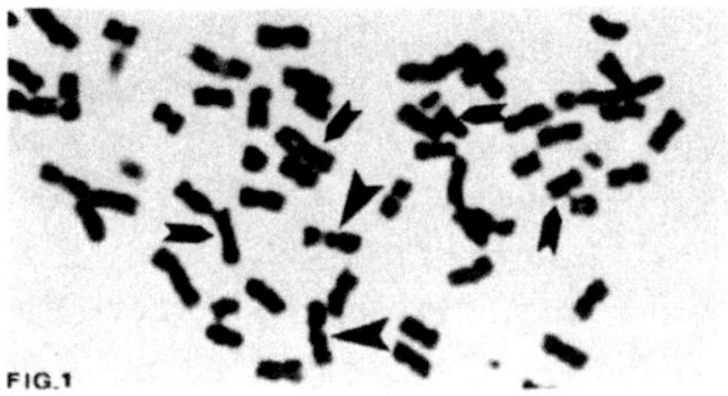

Fig.1. Nucleolar chromosomes of ➤ M.sativa and M.arborea ➤

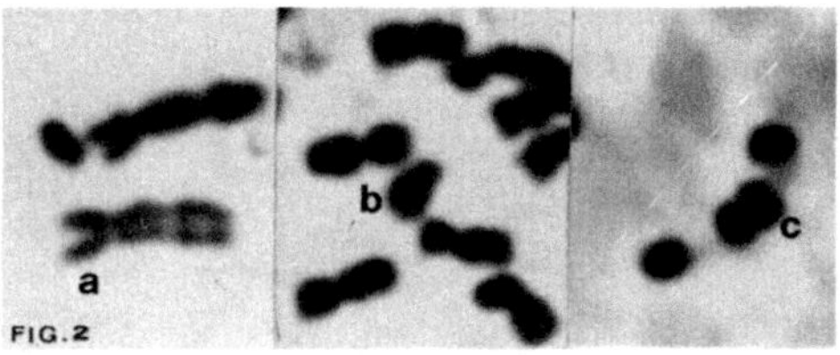

Fig.2. a) Dicentric, b) Telocentric, c) Iso-chromosome.

ELECTROFUSION AND ELECTROPORATION OF SUGAR BEET (*BETA VULGARIS* L.) PROTOPLASTS

C. Eady[1], G. Warren[2], K. Lindsey[1] and M.G.K. Jones[1].

[1] AFRC Institute of Arable Crops Research, Biochemistry Dept., Rothamsted Experimental Station, Harpenden, Herts. AL5 2JQ. U.K.
[2] Wolfson Institute of Biotechnology, University of Sheffield, U.K.

INTRODUCTION

Electrofusion and electroporation techniques have been successfully applied to crop plants to produce somatic hybrids, e.g. potato (1), or stable, specific transformants, e.g. maize (2). Using the Zimmermann rectangular pulse electrofusion apparatus, analytical electrodes and a mass fusion chamber, sugar beet protoplasts have been electrofused and electroporated, and a range of parameters studied.

RESULTS

A. Electrofusion

Initial studies were carried out using analytical fusion electrodes, and results from these fusions were applied to the production of large numbers of fusion products in a mass fusion chamber (3). To test the effect of alignment voltage, pulse duration-fusion response curves were produced for leaf protoplasts aligned at different voltages at 1 MHz, until the average chain length was 5-10 protoplasts. An alignment voltage of 200-300 Vcm^{-1} gave the highest fusion frequencies, whereas higher voltages (e.g. 400Vcm^{-1}) induced protoplast damage.

Leaf protoplasts fused at D.C. pulse voltages of 2000 Vcm^{-1} produced optimum fusion frequencies, while considerable damage was observed at 2500 Vcm^{-1}. For maximum binary (1:1) fusions, short pulse durations (40 µs) at 2500 Vcm^{-1} were used and produced up to 15 binary fusion products per 100 aligned protoplasts. The presence of 1.5 mM Ca^{++} ions in the fusion medium resulted in an increase in fusion frequency of up to 50%, and the insertion of negatively-charged synthetic glycolipid (based on fetuin) into the plasmamembrane of suspension culture protoplasts resulted in a small but significant increase in fusability.

Sugar beet leaf protoplasts of various genotypes showed similar fusion frequencies. However, leaf protoplasts were more fusogenic than suspension protoplasts, despite their smaller size.

In order to identify heterokaryons, morphologically distinct cell types were fused. Maximum heterofusions were produced at ratios of 5 suspensions:1 leaf protoplast, 19 binary heterokaryons being produced from 100 aligned leaf protoplasts. High frequency heterokaryon formation was obtained in a mass fusion chamber, and fusion products were purified on a sucrose cushion, mechanically isolated and grown up in nurse culture systems.

B. Electroporation

Changes in plasmamembrane permeability after electroporation were measured as the differential uptake and retention of the hydrophilic dye phenosafranine, which was determined spectrophotometrically (4). Time course studies indicate that the half life of the 'pores' was temperature dependent, and at 25°C, maximum resealing took ca. 40-60 min. Both cells and protoplasts retained the highest levels of phenosafranine when supplied with a series of 5 pulses of 50 µs, 2500 Vcm^{-1}. If these parameters were exceeded, viability of the population was reduced. The pattern of accumulation and retention, by protoplasts, of radioactivity from [^{3}H] pABDI, a modified bacterial plasmid, was similar to that of phenosafranine.

The transient expression of the introduced marker gene chloramphenicol acetyltransferase (CaMV35S promoter, nos terminator, assay at 45h) was demonstrated (5), and activity was optimum when 3 pulses of relatively long duration (99.9µs), low field strength (250-750 Vcm^{-1}) and 9.9 second intervals between pulses, were applied. These conditions were correlated with the improved long-term survival of the protoplasts. Activity is also influenced by other factors, including the exogenous plasmid concentration and the physiological state of the cells (stage of the growth cycle or tissue origin) from which the protoplasts were isolated.

REFERENCES

1. Fish, N., Karp, A. and Jones, M.G.K. (1988). Production of somatic hybrids by electrofusion. Submitted.
2. Fromm, M.E., Taylor, L.P. and Walbot, V. (1986). Stable transformation of maize after gene transfer by electroporation. Nature 319, 791.
3. Eady, C. and Jones, M.G.K. (1988). Electrofusion parameters of sugar beet (*Beta vulgaris* L.) protoplasts. Submitted.
4. Lindsey, K. and Jones, M.G.K. (1987). The permeability of electroporated cells and protoplasts of sugar beet. Planta 172, 346.
5. Lindsey, K. and Jones, M.G.K. (1987). Transient gene expression in electroporated protoplasts and intact cells of sugar beet. Plant Molecular Biology (in press).

THE PRODUCTION AND ELECTROFUSION OF BETA CYTOPLASTS

R.D. HALL, F.A. KRENS.
Foundation for Agricultural Plant Breeding, P.O. Box 117 Wageningen, The Netherlands

1. Introduction

It would be of great benefit to sugarbeet breeders if a technique could be devised whereby cytoplasmically-encoded traits could be rapidly transferred, at will, to any selected breeding line without altering its nuclear genetic complement. One potentially rapid and efficient means to achieve this is through the use of protoplast fusion techniques. However, in order to apply such a method it is first necessary to 'inactivate' the nucleus of the cytoplasm donor to avoid the production of a somatic hybrid. This can be done by irradiation, chemical treatment or nuclear elimination through the production of cytoplasts. As only the last technique guarantees that no internuclear genetic exchange will take place, an investigation was initiated into the production of sugarbeet cytoplasts for use as the cytoplasm donor in fusion experiments.

2. Materials and Methods

Cell suspension cultures of *Beta vulgaris* (genotype N144) were grown in PG_o medium (1) supplemented with 1 mgl^{-1} 2,4-D and 0.5 mgl^{-1} kinetin. Cells collected 2-3 d after subculture were resuspended in CPW-13M(2). After 1h the medium was replaced with CPW13M containing 1% cellulase R-10, 0.5% Macerozyme, 0.05% Driselase (pH 5.8) and the mixture incubated overnight at 25 ^oC on a slow shaker (50 rpm). After filtration through 88 um and 37 um nylon mesh the protoplasts were collected by centrifugation at 35 x g for 5 min, washed once in CPW-13M, once in CPW-21S and finally in 6.4% mannitol containing 0.1 M $CaCl_2$. After resuspension in 6.4% M/0.1MCa^{++}, the population density was determined and adjusted to $c1x10^6 ml^{-1}$ and 3 ml of the mixture layered onto an appropriate gradient for centrifugation at high speed (see below). The resulting bands produced within the gradient were carefully removed and analysed for cell viability, (FDA staining), cell population density and the presence of nuclei (DAPI staining).

3. Results and Discussion

A number of published methods, with modifications were tested for their suitability for producing sugarbeet cytoplasts. The results have been combined in Table 1. The use of Percoll gradients proved unsuitable for these protoplasts. In each of the three methods tested, viability was found to be reduced after centrifugation and the yield of cytoplasts was either poor or contained an unacceptable proportion of nucleated contaminants. The loading capacity of this method was also severely limited. The published method (3) based on non-isoosmotic layers of mannitol and sucrose proved most successful and several modifications were tested in order to improve cytoplast yields in association with improved enucleation percentages. The best results were obtained by centrifugation for 60 min at 50,000 x g at 23 ^oC in gradient 1. Under

Table 1. The effieiency of cytoplast production by various methods

Gradient* method	time (min)	force (g)	temp. (°C)	cytoplast yield (%)**	viability (%)	enucleation (%)
1	60	23500	23	7	88	93
1	60	30000	23	14	82	92
1	60	50000	23	31	90	92
1	120	50000	23	31	83	91
1	60	50000	4	20	83	92
2	60	50000	23	19	85	93
3	60	90000	23	13	69	86
4	60	90000	23	13	65	91
4	60	50000	23	8	67	92
4	60	23000	4	<1	71	86
5	60	30000	12	1	75	47

* (1) 8ml 80%S,8ml 33%S,8ml 18.2%M (2) 6ml 80%S,6ml 33%S,6ml 21%S, 6ml 18.2%M (3) 25ml Percoll cont. 0.1M $CaCl_2$,5mM MES pH 7,9%M (4) As 3 but with 50% Percoll in H_2O (5) Isoosmotic Percoll/Mannitol layers as Lorz et al. (1981) S = sucrose M = mannitol

** (of original cytoplasts) in main cytoplast fraction.

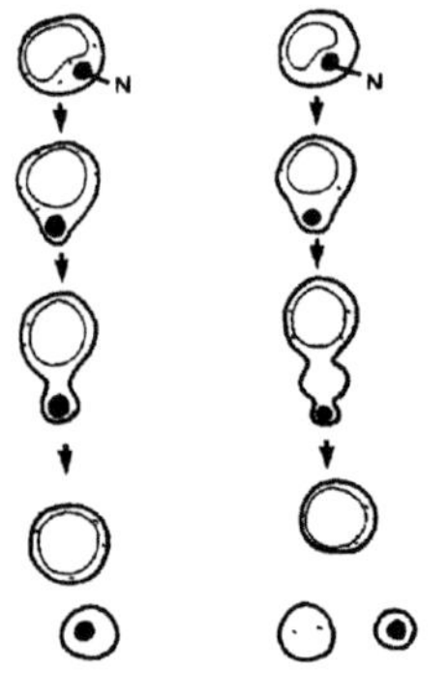

Figure 1. Two mechanisms for cytoplast formation from *Beta* protoplasts
N = nucleus

these conditions >30% of the protoplasts were enucleated and the resultant cytoplasts could be obtained in a fraction containing only 5-8% protoplasts. It was therefore possible to produce in a single centrifuge run several million cytoplasts, well in excess of requirements.

Interestingly, two cytoplast types were observed in these preparations. The main cytoplast fraction consisted almost entirely of vacuolate cytoplasts which were positioned at the 6.4%/18.2% mannitol interface. This is in contrast to previous findings (3) where it was observed that most *Solanum nigrum* cytoplasts banded at the 18.2% M/33% S interface. In addition, non-vacuolate cytoplasts were also produced from sugarbeet protoplasts at a frequency of c 10% and occurred in association with protoplasts and some mini-protoplasts at the 18.2% M/33% S interface. The proposed mechanisms responsible for the production of both cytoplast types are presented in Fig. 1. Cells at each of the stages represented were often observed immediately after centrifugation. As evacuolate cytoplasts are likely contain more organelles and prove more stable than vacuolate cytoplasts attempts were made, to isolate these components free from nucleated contaminants but these proved unsuccessful.

Many cytoplasts were found to be lost on washing in Ca^{++} free media and experiments are presently underway to modify the above method by removing Ca^{++} from the protoplast medium so that freshly-isolated cytoplasts may be used directly in electrofusion experiments. In the first such experiment it has been found that these cytoplasts appear quite stable to electrofusion and under conditions suitable for beet mesophyll cell fusion (30 V AC, 50 usec pulse 1500 V DC) few cytoplasts were observed to lyse and c 30% of those aligned were involved in fusion events. It is considered therefore that these cytoplasts should prove most useful for the production of *Beta* cybrids by electrofusion and this work is now underway.

REFERENCES

(1) de Greef, W., Jacobs, M. (1979) Pl. Sci. Lett. 17 55-61
(2) Frearson, E.M., Power, J.B., Cocking, E.C. (1973) Dev. Biol. 33 130-137
(3) Lesney, M.S., Callow, P.W., Sink, R.C. (1986) Pl. Cell Rep. 5 115-118
(4) Lorz, H., Paszkowski, J., Dierks-Ventling, C., Potrykus, I (1981) Phys. Plant 53: 385-391

This project was funded by INPLA-G86-02

INACTIVATION OF SUGARBEET PROTOPLASTS USING ACRIDIN ORANGE, AN AGENT FOR LATE SELECTION OF FUSION PRODUCTS.

H.C. Pedersen *, A.B.-Larsen, K. Vamling, B. Keimer.
MARIBO SEED, P.O.B. 29, DK-4960 Holeby, Denmark.

INTRODUCTION

Inactivation of protoplasts has been used for several purposes of which selection of somatic hybrids and cybrids by inactivation of one or both parent protoplasts is one of the more elegant applications.

Inactivation of protoplasts can be done either chemical (1) or by irradiation (2). Attempts to select fusion products by using two differently acting chemical agents only had limited success (3).

Acridin Orange (AO) interchalates with DNA in nuclei and plastomes (4). Light induces the breakdown of the AO-DNA complex, resulting in single and double strand breaks (5).

The aim of this work was to investigate the action of AO on sugarbeet (Beta vulgaris L.) protoplasts.

MATERIALS AND METHODS

Protoplasts were isolated from cell suspensions of sugarbeet. The freshly isolated protoplasts were suspended in W5 or culture medium containing AO (50, 100 or 200 mg/l) and incubated in dark for 1, 2 or 24 hours. The protoplasts were washed several times in W5-medium and cultured at $5\cdot10^4$ protoplasts/ml. After 1, 2, 3, 7 and 14 days cultures were exposed to white light (growth room, approx. 1000 lux) for 4 hours and placed in dark again for further growth.

As a measure for AO-effect we used the ability to form protoplast colonies after 4 weeks in liquid culture or calli after plating after 6 weeks. Microscopical observations (UV, Zeiss filter: Blau 450-490) were made on probes taken during culture.

PEG-fusions (6) were made between iodoacetic acid treated protoplasts (7) and protoplasts treated with AO. After 24 hrs. the cultures were exposed to light as described.

RESULTS

As can be seen from table 1, the Acridin Orange treatment of protoplasts affects the viability of protoplasts. Viability is calculated as percentage of colony forming protoplasts in relation to control without AO-treatment. The viability depends both on the duration of AO-treatment, the AO concentration and on the time for light treatment.

Cultures exposed to light after more than 7 days shows almost the same division pattern as the dark grown AO-treated controls.

In UV-microscope the protoplasts appears light green with darker green or red nuclei. Often also red spots in the cytoplasm can be seen. The pale green colour of the cytoplasm holds for several weeks, whereas the red dots and darker green nuclei disappears during colony formation.

Geno-type	AO-conc. (mg/l)	light treatment after (days) 1	2	3	7	14	dark
87	50	0	0	-	12	-	40
	100	0	0	-	15	-	28
204	50	0	1	13	60	71	69
	100	0	0	9	48	75	57
	200	0	0	1	21	21	14
249	50	1	6	45	68	102	100
	100	0	1	41	63	109	93
	200	0	0	6	26	50	38

Table 1. Viability of protoplasts judged from ability to form colonies in liquid (204, 249) or after plating on solid medium (87).

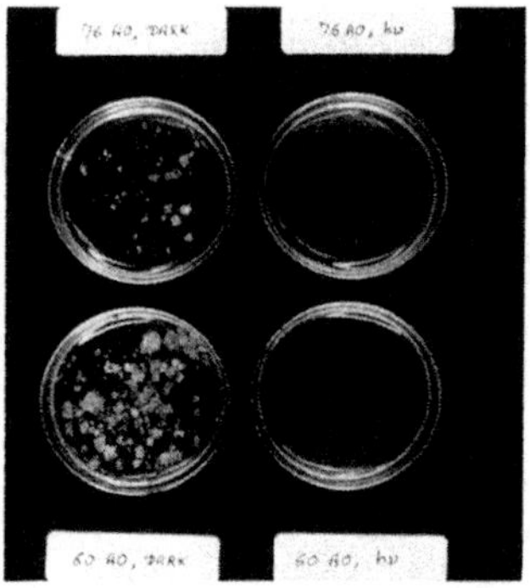

Fig.1. Platings after treatment with 100 mg/l AO and cultured in dark (left) or exposed to light two days after treatment (right).

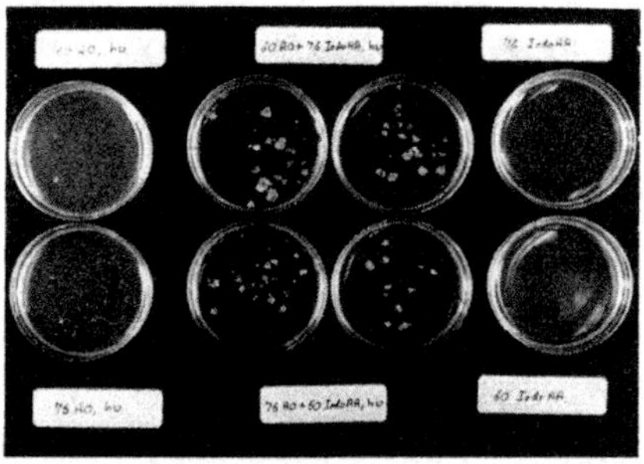

Fig.2. Plating of fusions between AO-treated and iodoacetic acid treated protoplasts incl. control fusions, exposed to light 24 hrs. after fusion.

Preliminary results from fusions between iodoacetic acid inactivated protoplasts and AO-treated protoplasts shows colony formation in the fusion dish, whereas all iodoacetic acid treated protoplasts and most of the AO-treated protoplasts are unable to grow up to colonies (fig. 2).

DISCUSSION

The action of light on the AO-stained dividing cells is presumably a breakdown or mutation of chromosomes to a stage, where no further life is possible. When the protoplasts have divided a few times, exposure to light has no longer a dramatic toxic effect, but may still have mutational effect. The cells and colonies hold the pale green colour for several weeks. This phenomenon is of value in verification and mechanical selection of fusion products. Complementation by fusion of AO-treated and iodoacetic acid treated protoplasts is possible. This avoid the need for special mutants (albino, auxotrophs etc.) in selection of fusion products. Further studies on the action of AO will include cytological investigations of the regenerated calli and fusion products.

REFERENCES

1. Medgyesy P, Menczel L, Maliga P (1980) Mol Gen Genet. 179:693-698.
2. Zelcer A, Aviv D, Galun E (1978) Z.Pflanzenphysiol. 90:397-408.
3. Nehls R (1978) Mol Gen Genet. 166:117-118.
4. Gupta HS, Deepesh NDe (1983) Proc Ind Nat Sci Acad. B49(6):653-660.
5. Dhar A, Chaudhuri U (1974) Arch Biochem Biophys. 162:310-312.
6. Glimelius K, Wallin A, Eriksson T (1978) Physiol Plant. 44:92-96.
7. Cella R, Carbonera D, Iadarola P (1983) Z.Pflanzenphysiol. 112:449-457.

MITOCHONDRIAL DNA AMPLIFICATION IN SOMATIC HYBRID CELLS OF GRASSES

P. Ozias-Akins[1], Z. Tabaeizadeh[4], D.R. Pring[3], and I.K. Vasil[2]

[1]Department of Horticulture, University of Georgia Coastal Plain Experiment Station, Tifton, GA 31793, USA
[2]Department of Botany and [3]USDA/ARS, Department of Plant Pathology, University of Florida, Gainesville, FL 32611, USA
[4]Department of Biology, McGill University, Montreal, Quebec, Canada H3A1B1

1. INTRODUCTION

Mitochondrial DNA (mtDNA) frequently undergoes rearrangement in somatic hybrid cells so that novel restriction fragment patterns are observed. Intergenomic recombination after protoplast fusion has been demonstrated in Petunia (3) and Brassica (7). Infrequent intragenomic recombination events may generate sub-stoichiometric mtDNA fragments (4). We have identified and cloned sequences with altered stoichiometry from somatic hybrid cells of grasses.

2. MATERIALS AND METHODS

2.1. Cell lines. Suspension cultures of Pennisetum americanum (Pa), Panicum maximum (Pm), Saccharum officinarum, and somatic hybrid cell lines were maintained in a modified Murashige and Skoog medium containing 2,4-D (1, 5).

2.2. Mitochondrial DNA analysis. Isolation of mtDNA, restriction endonuclease digestion, electrophoresis, Southern blotting, and hybridization have been described (2, 6). Restriction fragments were cloned into pUC8 after isolation from preparative agarose gels with NA-45 DEAE cellulose membrane (Schleicher and Schuell).

3. RESULTS AND DISCUSSION

A "novel" 5.1 kb-BamHI mtDNA fragment from two Pennisetum americanum + Panicum maximum somatic hybrid cell lines was detected with a probe for atpA (2). This fragment was cloned and hybridized to parental and hybrid mtDNAs (Fig. 1a). Weak hybridization with a 5.1 kb fragment from both parental DNAs was observed. Scanning densitometry established that the 5.1 kb fragment had been amplified 15-30 fold in the somatic hybrid cells. An 11 kb fragment from P. maximum showed strong homology with the 5.1 kb probe. The 11 kb fragment was not detectable in any of the somatic hybrids. The cloned 5.1 kb and 11 kb DNAs were mapped after digestion with PstI, EcoRI, XbaI, HindIII, and HincII. The two cloned fragments share a repeated sequence of approximately 4.5 kb (Fig. 1b).

In one somatic hybrid cell line of P. americanum + Saccharum officinarum, an apparently "novel" 2.4 kb-PstI fragment was detected with cosmid 11H7 (6). The cloned 2.4 kb fragment hybridized weakly with S. officinarum indicating that this fragment was amplified in the somatic hybrid cells. However, no other sequences homologous to the 2.4 kb cloned DNA were observed in parental or somatic hybrid DNAs.

We conclude that amplification can account for a part of the variation between mtDNA restriction fragment patterns of parental and somatic hybrid cell lines. Stability of amplified sequences over time is being investigated.

One somatic hybrid cell line of P. americanum + P. maximum is being used to investigate somaclonal variation in mtDNA. Twenty-five protoclones of this cell line have been isolated and PstI restriction fragments which differ from the original cell line have been identified (Fig. 1c). Hybridization with cosmid clones of sorghum mtDNA will be carried out to possibly reveal additional variation. Variant fragments will be cloned and analyzed further.

REFERENCES

1. Ozias-Akins P, Ferl RJ, Vasil IK: Mol. Gen. Genet. 203:365-370, 1986.
2. Ozias-Akins P, Pring DR, Vasil IK: Theor. Appl. Genet. 74:15-20, 1987.
3. Rothenberg M, Boeshore ML, Hanson MR, Izhar S: Curr. Genet. 9:615-618, 1985.
4. Small ID, Isaac PG, Leaver CJ: EMBO J. 6:865-870, 1987.
5. Tabaeizadeh Z, Ferl RJ, Vasil IK: Proc. Natl. Acad. Sci. USA 83:5616-5619, 1986.
6. Tabaeizadeh Z, Pring DR, Vasil IK: Plant Mol. Biol. 8:509-513, 1987.
7. Vedel F, Chetrit P, Mathieu C, Pelletier G, Primard C: Curr. Genet. 11:17-24, 1986.

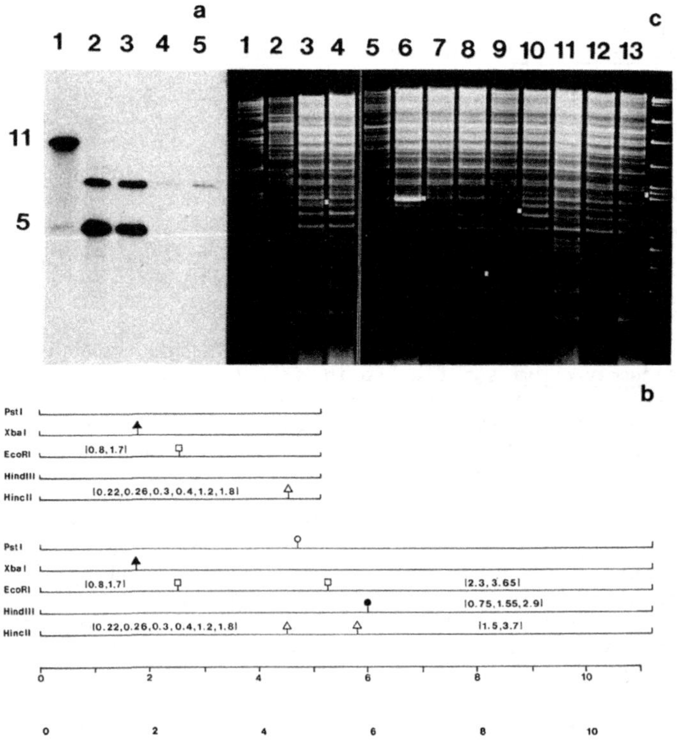

FIGURE 1. a) Cloned 5.1 kb-BamHI fragment hybridized to mtDNA of Pm (1), Pa (5), and 3 somatic hybrids (2-4). b) Partial restriction map of 5.1 and 11 kb cloned DNAs. c) PstI-restricted mtDNA from Pa (1), Pm (2), and somatic hybrid protoclones (3-13). White dots indicate variant bands. Marker lane - PstI and HindIII digests of lambda DNA.

PLANT REGENERATION FROM ELECTROFUSED PROTOPLASTS OF CHRYSANTHEMUM MORIFOLIUM AND SALPIGLOSSIS SINUATA

N. KHALID[1], C.H. LEE, J.B. POWER and M.R. DAVEY
[1]Department of Botany, Faculty of Science, University of Malaya, 59100 Kuala Lumpur, West Malaysia.
Plant Genetic Manipulation Group, Department of Botany, University of Nottingham, Nottingham NG7 2RD, UK.

1. INTRODUCTION

Chrysanthemum morifolium is an important species in the floriculture industry. In order to apply new methods of cultivation, allied to genetic manipulation technologies, it is necessary to develop this species as a protoplast system capable of whole plant regeneration. In this latter context, interfamilial protoplast fusion of *C. morifolium* with *Salpiglossis sinuata* was attempted primarily to trigger plant regeneration in *Chrysanthemum* protoplasts, a hitherto recalcitrant system. This in turn could lead to the generation of protoclonal variation in this ornamental. In the anticipated absence of stable intergeneric somatic hybridisation, the electrofusion treatment was seen to induce plant regeneration from mesophyll protoplasts of *C. morifolium*.

2. MATERIALS AND METHODS

Seeds of *C. morifolium* var. Early Charm (Suttons Seeds Ltd., Torquay, UK) were germinated in Levington soilless compost (Fisons, Ipswich, UK) and seedlings grown in the glasshouse under natural daylight supplemented with fluorescent illumination of 10,000 lux (daylight tubes), a 16h day-length and a day/night temperature of 25/22°C.

In vitro shoot cultures were established by surface sterilising shoot tips of glasshouse-grown plants (8% v/v "Domestos" solution for 20 min followed by 6 washes in sterile tap water) and placing these on BGS medium (1) under continuous illumination of 500 lux at 23 ± 2°C. Leaf protoplasts were isolated from 4 weeks old axenic shoot cultures using an enzyme mixture consisting of 1% w/v Cellulase R10 and 0.25% w/v Macerozyme R10 dissolved in CPW salts solution with 13% w/v mannitol (pH 5.8) following standard procedures (2). Cell suspensions of albino *Salpiglossis* were used as a source of protoplasts (3).

Chrysanthemum mesophyll protoplasts and cell suspension protoplasts of albino *Salpiglossis* were mixed 1:1 and electrofused (4) and fusion-treated protoplasts plated at a density of 5 x 10^4 ml^{-1} in liquid K8P medium (5) followed by dilution every 7 days of the protoplast culture medium with K8 medium (5). The osmotic pressure was reduced every 7 days by adding 1 ml of liquid K8 medium (5) and the total protoplast culture media volume was maintained at 4 mls throughout. Control dishes containing unfused, self fused and mixture of self fused protoplasts were handled in a similar way. All cultures were maintained at 500 lux (25 ±2°C). Green calli which developed in the fusion dishes were grown on for 2 weeks in agar-solidified K8 medium before transfer (for 4 weeks) to MSP1 medium (MS medium (6) with 2.0 mg/l NAA, 0.5 mg/l 6-BAP) (2). Shoots were induced on the protoplast-derived calli following transfer to RM medium (MS medium with 0.05 mg/l NAA and 1.0 mg/l 6-BAP). In order to induce stem elongation, detached shoots were kept on to MS medium with 0.5 mg/l 6-BAP alone for one month prior to rooting on BGS medium (2). Plants were ultimately transferred to the glasshouse. The somatic chromosome number of regenerated plants were determined using root tips coupled with an isoenzyme analysis of leaf material of these plants and both C. morifolium and S. sinuata. Leaf tissue (2.0g) were ground in 1.0 ml of an ice cold buffer (50 mM Tris Hcl (pH 6.8), 1% (w/v) leupeptin, 18% (v/v) glycerol, 0.5% (v/v) mercaptoethanol). Gels (9% acrylamide separating gel, 3% acrylamide stacking gel) were run on an LKB 2001 vertical slab-gel electroporesis unit for 6h at 11°C (current, 10 mA). Gels were stained (30 min, in the dark) for peroxidase in a solution of (12 g/l Tris (pH 7.2), 23 g/l NaCl, 2.5% (w/v) 4-chloro-1-naphthol in methanol with 2% (v/v) hydrogen peroxide) and fixed in 8% (v/v) acetic acid.

3. RESULTS AND DISCUSSION

The plating efficiency of protoplasts in the fusion dishes, based on green calli/colonies only was 2% after 4 weeks. These green calli/colonies were either of heterokaryon or Chrysanthemum (unfused, homokaryon) protoplast origin. Chrysanthemum mesophyll protoplasts did not divide when plated on their own in liquid K8P medium but were capable of division in non-fused mixtures with albino Salpiglossis.

Shoots were produced as the green calli (only from fusion dishes) was placed on RM medium. Calli from non-fused protoplast preparations and mixture contents did not regenerate. Regenerated shoots produced abnormal leaves but following their transfer to MS medium with 6-BAP alone, growth reverted to normal. Morphologically, the regenerants grown on into the glasshouse resembled the parental C. morifolium type.

The somatic chromosome number of the regenerated plants was that of C. morifolium (2n = 6x = 54). Leaf peroxidase isoenzyme analysis showed that the regenerated plants contained a summation of the banding patterns of both the Chrysanthemum and Salpiglossis parent species less one band associated with the latter species. One additional band could not be attributed to either of the fusion partners.

To date, it has been difficult to induce morphogenesis in protoplast derived callus of Chrysanthemum, with only one previous report (7) of plant regeneration from mesophyll protoplasts. Even in the absence of nuclear hybridisation of Chrysanthemum with Salpiglossis, the electrofusion treatment had clearly stimulated regeneration in protoplasts of C. morifolium. The fact that the isoenzyme profile of the regenerants was similar to both parental species would suggest that the regenerants had the heterokaryon as their origin.

4. REFERENCES

1. Binding H: Physiol. Plant. 35: 225-227, 1975.
2. Power JB, Chapman JV and Wilson D: Laboratory Manual, Plant Tissue Culture, University of Nottingham, 1984.
3. Chapman JV: Ph.D. thesis, Univ. of Nottingham, 1987.
4. Watts JM, King JM: Bioscience Rep. 4: 335-342, 1984.
5. Kao KN: Molec. Gen. Genet. 150: 225-230, 1977.
6. Murashige T, Skoog F: Physiol. Plant. 15: 473-497, 1962.
7. Otsuka H, Suetmatsu N, Tode M: Bull. Shiznoka Agr. Exp. Stn. 30: 25-33, 1985.

ELECTROFUSION OF CHEMICALLY-AGGREGATED PROTOPLASTS DERIVED FROM MESOPHYLL TISSUE AND HABITUATED CELLS OF CATHARANTHUS ROSEUS

C. Perronnet*, C. Denécheau, M. Rideau, J.C. Chénieux.

Plant Biotechnology Laboratory, Faculty of Pharmaceutical Sciences,
2bis, Bd Tonnellé - 37042 Tours - France.

1 - INTRODUCTION

Different methods have been developed to induce fusion of plant protoplasts, the most promising of which is the process of electrically mediated fusion.

To study the influence of habituation in the secondary metabolite production, we should get "hybridoma" produced by electrofusion of leaf mesophyll cells and habituated-cell suspensions of Catharanthus roseus.

2 - MATERIALS AND METHODS

Protoplast isolation

Mesophyll protoplasts from Catharanthus roseus G. Don (cv "little pinckie") were isolated from 8 to 10 week old plants. After removing the lower epidermis, isolation was according to Constabel and al., (1979).

Habituated strain C20A protoplasts were isolated as reported by Perronnet and al., (1987).

Electrofusion

Protoplasts were suspended in a 1 : 1 ratio in 0.5 M mannitol to give a total concentration of 5 x 10^5/ml. 75 µl drop was placed in plastic Petri dish (3 x 55 mm) and let 10 min. to settle. Then, 75 µl of 20 % (w/v) polyethylene glycol (P.E.G. 6000), 1 mM $MgSO_4$, 7 H_2O, 50 mM NaCl, 0.5 M mannitol (pH 5.9) were added. After 30 min. at room temperature, electrical fusion was carried out : two flat stainless-steel electrodes were dipped into the aggregated protoplast suspension ; thereafter, electric field necessary to induce protoplast fusion were applied (Electrofuser ATEIM-CNRS - France). After the pulse, the electrodes were removed and the protoplast suspension was mixed with, (0.13 M $CaCl_2$ - 0.27 M mannitol, pH 5.5) and incubated at room temperature for 1hr. After incubation, protoplasts were suspended in protoplast culture medium (Perronnet and al., 1987). After 7 days in the dark at 25° C, the protoplasts were transferred to light. 24 hours after electroporation procedure, a random sample of 500 protoplasts was examined among the electroporated protoplasts (the experiments were repeated 3 to 5 times).

The viability of electrofused-protoplasts and the efficiency of the method were determined by considering the percentage of divisions after 5 days of culture determined by microscopic observation under inverted microscope.

3 - RESULTS AND DISCUSSION

Our method requires firstly the aggregation of the plant protoplasts by a chemical agent and then fusion by means of short d.c. electric pulses (electrofuser, ATEIM-CNRS, France : Chapel and al., 1986). The percentage of fusions induced by electropulsation is dependent on the cell developmental stage, the voltage applied and the fusion medium. Best results were obtained with one 100 microsecond, 1500 volts per centimeter pulse.

At a concentration of 15 % (w/v) Dextran 500.000 solution used as aggregation agent, 3 % heterofusion was achieved but dextran was shown to induce protoplast lysis one day after the electrofusion pulse.

The bests results were obtained with a PEG 6000 aggregating solution (see materials and methods). PEG showed a beneficial effect on the survival and development of protoplasts. Purification of commercial PEG (mixed-bed resin AG501.X.8-D) ([Kao and Saleem, 1986] did not significantly change the frequency of protoplast electrofusion and viability compared to PEG commercial control. Two facilitators : pronase E and protease XXI (Sigma) [Ruzin and Mac Carty, 1986] have been tested. We did not found those to be effective in increasing the frequency in heterofusion production. We observed an increase in the heterokaryon production when 0.025 mM Ca^{++} were added in the fusion medium (x1,6) and at this concentration calcium was found to promote cell division.

In optimal fusion condition, 15 % heterofusion were observed and most of them began to divide after 5 days in the culture medium.

REFERENCES

1 - M. Chapel, M.H. Montane, B. Ranty, J. Tessie, G. Alibert, FEBS Letters, 196 (1986)) 79-86.
2 - K.N. Kao, M. Saleem, J. Plant Physiol. ; 122 (1986) 217-225.
3 - C. Perronnet, C. Denecheau, M. Rideau, J.C. Chénieux, (1987) (this symposium).
4 - S.G. Ruzin, S.M. Carty, Plant Cell Reports, 5 (1986) 342-345.
5 - F. Constabel, H. Koblitz, J.W. Kirkaptrick, S. Rambold, Can. J. Bot., 58 (1979) 1032-1034.

STUDIES ON THE TRANSFER OF THE MITOCHONDRIAL GENOME THROUGH PROTOPLAST FUSION IN RICE

H.S. GUPTA, E.C. COCKING[1] and B.J. MULLIGAN[1]
Division of Plant Breeding, ICAR Research Complex for North-Eastern Hills Region, Shillong-793003, India.
[1]Plant Genetic Manipulation Group, Department of Botany, University of Nottingham, Nottingham NG7 2RD, UK.

1. INTRODUCTION

Cytoplasmic male sterility (CMS) is a failure to produce or release pollen due to an incompatibility between nuclear and mitochondrial genes. CMS transfer is normally only attainable in five to six years through traditional methods of back-crossing. By contrast, protoplast fusion-mediated transfer of CMS is rapid and is effected through the transfer of the mitochondrial genome (1). CMS can also arise from a recombination between the mitochondrial genomes of male fertile plants (2). The present investigation reports on transfer of the mitochondrial genome from rice strain A58 to Taipei 309 using the donor-recipient method.

2. MATERIALS AND METHODS

2.1. Protoplast isolation and prefusion treatments

Embryogenic cell suspensions of A58 and Taipei 309 were maintained in an amino-acid medium supplemented with 2.0 mg l^{-1} 2,4-D (AA2). Protoplasts were isolated using the method of Thompson et al. (3) except that the enzyme mixture was diluted (1:3) and the incubation period was extended to 16hr. In order to label the protoplasts of A58 and Taipei 309 differently, the enzyme mixtures used with A58 and Taipei 309 were labelled with fluorescein diacetate(FDA) and rhodamine 6G (R6G) respectively (both at 12 μg ml^{-1}). After isolation, protoplasts of A58 were gamma-irradiated (11Krad).

2.2. Electrofusion

Irradiated donor (A58) and R6G-treated Taipei 309 protoplasts were resuspended in 11% mannitol at 2 x 10^5 ml^{-1}. The protoplasts were mixed in an equal ratio and 2 ml aliquots of the mixture dispensed into each well of a 5 x 5 repliplate. Protoplast alignment and short chain formation were obtained by inserting a multiblade electrode into the well and applying an AC current (25V 600 KHz). The amplitude of the AC was zeroed, and a single pulse DC current (400V) applied to induce fusion. After fusion, protoplasts were cultured in KPR liquid medium (3) at 4 x 10^5 ml^{-1}. Some of the selected colonies were transferred to AA2 liquid medium to initiate cell suspensions and some to N60 medium for plant regeneration.

2.3. Extraction and analysis of mitochondrial DNA

Mitochondrial DNA (Mt-DNA) was isolated from 20g of actively growing cell suspension (4). Two to three μg of DNA was digested with PstI according to the supplier's instructions. Two μg of restricted and 1.5 μg of uncut Mt-DNA were electrophoresed on 0.8% agarose. Lambda phage DNA digested with Hind III was used for size markers.

3. RESULTS

3.1. Isolation and inactivation of protoplasts

Protoplasts were isolated from embryogenic cell suspensions of A58 and Taipei 309. Twelve μg FDA ml^{-1} of enzyme mixture was optimal for staining the protoplasts of A58. Subsequent irradiation of these protoplasts with gamma rays prevented division, which did not occur even after 20 days in culture. The treatment of Taipei 309 cell suspensions with R6G caused mitochondrial inactivation as evident from the absence of division in the protoplasts even after 30 days in culture. Occasionally, a few protoplasts underwent one division; rarely, one or two protoplasts formed 4-8 celled colonies, but these did not develop further.

3.2. Protoplast fusion

Irradiated donor (A58) and R6G-treated recipient (Taipei 309) protoplasts aligned and formed short chains when an AC field (25V; 600 KHz) was applied for 20-25 sec. Fusion was induced by a single 400V DC pulse of 3 msec. Subsequently an AC field (20V; 600 KHz) was applied for 20-30 sec during which rounding up of the fused protoplasts occured. The fusion frequency was 5-7%.

3.3. Culture of electrofused protoplasts

Electrofused protoplasts entered division after 6-8 days in KPR medium, and formed microcolonies which developed to callus. The putative cybrid calli were transferred to hormone-free N6 medium for plant regeneration. Some calli were also transferred to AA2 liquid medium to produce cell suspensions. The latter will be used for extraction and analysis of Mt-DNA to assess the cybridity of selected tissues.

3.4. Analysis of Mt-DNA of donor and recipient

The electrophoretic banding patterns of uncut Mt-DNA from A58 and Taipei 309 showed one band of high molecular weight in DNA from both strains. The restriction patterns of total Mt-DNA digested with Pst I showed similar fragments common to both the mitochondria. Unique fragments specific to A58 and Taipei 309 Mt-DNA have been identified which will serve as markers in the analysis of putative cybrids formed as a result of donor-recipient fusion.

4. DISCUSSION

Previous reports on transfer of CMS (5,6) through the use of the donor-recipient method, have demonstrated the transfer the mitochondrial genome from the donor. In addition to transfer of the mitochondrial genome, novel recombinant mitochondria have also been reported (6). In the present investigation, gamma irradiation was expected to inactivate the donor nuclei, whereas R6G treatment may inactivate and eliminate the mitochondria of the recipient. Thus, a putative cybrid may possess the recipients' nuclus plus either the mitochondrial genome of the donor (A58) or novel recombinant mitochondrial genomes. However, this can be confirmed only after analysis of the Mt-DNA of putative cybrid. Such analyses are in progress.

ACKNOWLEDGEMENTS

The authors are grateful to the Rockefeller Foundation and the Overseas Development Administration for financial support in part. HSG is a Commonwealth Scholar.

REFERENCES

1. Zelcer A, Aviv D, Galun E: Z. Pflanzenphysiol. 90: 397-407, 1978.
2. Aviv D, Arzee-Gonen P, Bleichman S, Galun E: Mol. Gen. Genet. 196: 244-253, 1984.
3. Thompson JA, Abdullah R, Cocking EC: Plant Sci. 47: 123-133, 1986.
4. Kemble RJ: Theor. Appl. Genet. 73: 364-370, 1987.
5. Pelletier G, Primard C, Vedel F, Chetrit R, Remy R, Rouselle P, Renard M: Mol. Gen. Genet. 191: 244-250, 1983.
6. Robertson D, Palmer JD, Earle ED, Mutschler MA: Theor. Appl. Genet. 74: 303-309, 1987.

INCREASE OF GENETIC VARIABILITY BY ASYMMETRIC CELL HYBRIDIZATION AND ISOLATED CHROMOSOME TRANSFER

D. DUDITS
Institute of Genetics, Biological Research Center, Hungarian Academy of Sciences, H-6701 Szeged, POB 521, Hungary

1. INTRODUCTION

Barriers to crossability between species considerably limit the use of advantageous genes with diverse origin to improve the nutritional quality and increase the total yield potential of crop plants. As the result of progress in development of plant tissue culture techniques and of molecular and cell genetics parasexual manipulation of plant genome has gained an increasing significance to release new possibilities for genetic recombination and extension of genetic variability. Nowadays there are several major approaches to the introduction of foreign genes into somatic plant cells. The transfer of single isolated genes can be efficiently accomplished by using one of the advanced DNA transformation systems based on Agrobacterium (Schell, 1987) or direct plasmid uptake into protoplasts through chemical and physical treatments (Paszkowski et al, 1984) or with help of microinjection (Reich et al, 1986). In addition to production of transgenic plants (Abel et al, 1986; Hilder et al, 1987) with possible use in breeding programs there is a need to manipulate genetically uncharacterized or multigenic traits. For these purposes the methods of cell fusion or organelle transfer offer an alternative solution. The aim of the present paper is to analyse the available experimental results about genomic constructs of hybrid genotypes after cell fusion and isolated chromosome transfer. As we discussed previously (Dudits et al, 1980; Dudits and Praznovszky, 1985) reduction of somatic incompatibility through induced chromosome elimination can be a basic requirement for production of intergeneric hybrid plants. The successful use of asymmetric cell hybrids for limited alien gene transfer depends on tools to control chromosome elimination and genomic recombination in fusion products. Characterization of the introduced new traits in somatic hybrids and in their sexual progenies can help to outline the potentials of these approaches in plant breeding.

2. SPONTANEOUS AND IRRADIATION INDUCED CHROMOSOME ELIMINATION FROM FUSION PRODUCTS

Chromosomal behaviour of two parental genomes in cell hybrids is strongly influenced by phylogenetic distance between the fusion partners. Maintenance of double diploid or amphidiploid chromosome sets was characteristic for several intra- and interspecific somatic cell hybrids with related parents as Nicotiana glauca + N.tabacum (Evans et al, 1980),

Datura innoxia + D.stramonium (Schieder, 1978); Daucus carota + D.capillifolius (Dudits et al, 1977); Solanum tuberosum + S.brevidens (Ehlenfeldt and Helgeson, 1987). In few exceptional fusion combinations e.g. Nicotiana tabacum+N.knightiana (Maliga et al, 1977) high degree of spontaneous genomic instability was detected after interspecific fusion. Integration and maintenance of complete parental genomes was also achieved in some of the intergeneric somatic hybrids as Solanum tuberosum + Licopersicon esculentum (Melchers et al, 1978) Arabidopsis thaliana+Brassica campestris (Gleba and Hoffmann, 1978); Datura innoxia+Atropa belladonna (Krumbiegel and Schieder, 1979);Atropa belladonna+ Nicotiana chinensis (Gleba et al, 1982); Lycopersicon esculentum+Solanum rickii (O'Connell and Hanson, 1986); Sinapis turgida Del + Brassica oleracea (Toriyama et al, 1987); Lycopersicon esculentum+Solanum lycopresicoides (Handley et al, 1986). The stable coexistence and coordinated functioning of the two genomes, especially after plant regeneration can be considered as an indication for a closer phylogenetic relation between the fusion partners. In some of these symmetric hybrids the long exposition to in vitro culture conditions could induce a delayed and gradual loss of parental chromosomes (Hoffmann and Adachi, 1981; Krumbiegel and Schieder, 1981). Despite the fact that the mentioned examples clearly demonstrate the possibility of intergeneric hybrid production the practical significance of this plant material is significantly reduced by high degree of morphological and developmental abnormalities as well as by the failure in production of sexual progenies. The amphidiploid nature of genomes can mainly be responsible for negative characteristics of regenerated hybrid plants. To meet the requirements of practical breeding programs there is a need to produce asymmetric hybrids with complete genome of the recipient species plus a few chromosomes or integrated chromosome segments from a donor species. Asymmetric nature of nuclear genomes in cell hybrids originates from spontaneous or induced genomic instability with preferential loss of chromosomes belonging to one of the parental species. Considerable spontaneous chromosome elimination was detected in several remote cell hybrids with unrelated parents as soybean+Nicotiana glauca (Kao, 1977); soybean+N.tabacum (Chien et al, 1982); Vicia faba+Petunia hybrida (Binding and Nelhs, 1978); Daucus carota+Nicotiana tabacum (Hauptmann et al, 1983); Daucus carota+Oryza sativa (Sala et al, 1985). The degree and direction of chromosome loss is higly dependent on selection of fusion partners. In most of the cell hybrids the spontaneous genomic instability was insufficient to reduce somatic incompatibility and to allow regeneration of normal hybrid plants. Therefore enforced and directed chromosome elimination can be essential for hybrid plant production in several wide fusion combinations especially for breeding purposes.

Since introgression of desirable genes into the target crop plants is generally based on the extensive and time consuming backcrossing programs the possibility to control chromosome elimination experimentally in somatic cell hybrids offers a great potential for cell fusion in intra- and interspecific gene transfer as well. As an example the somatic hybridization between Solanum tuberosum (2x; 4x) and S.brevidens (2x) can be mentioned. The tetraploid and hexaploid somatic hybrids between these species have retained the required resistance against potato leaf roll virus with several negative traits from the wild species. Therefore further sexual crosses are needed to transfer of desirable characteristics into breeding material (Austin et al, 1986); Ehlenfeldt and Helgeson, 1987). In this case asymmetric hybridization might improve fertility of somatic hybrids and shorten the time requirement for production of useful breeding material. Up to now the following two approaches were found to be successful to induce preferential chromosome loss:

2.1. Fusion between dividing and mitotically inactive protoplasts

Protoplasts from fast growing suspension of an albino nuclear mutant from Daucus carota were fused with leaf protoplasts of Aegopodium podagraria (Dudits et al, 1979). This fusion resulted in hybrids exhibiting carrot phenotype supplemented with more than one Aegopodium specific characters. Plants regenerated from these green fusion products possessed 2n=18 chromosomes similar to carrot parent. Lack of cytological detectable additional chromosome suggested integration of subchromosomal fragment from Aegopodium to carrot. To interpret cytological mechanisms resulted in this type of hybrid genome we have to analyse the complex interaction between cells with various cell cycle stages. Cytological studies proved that the mitotic partner can activate the non-dividing cells (Kao and Michayluk, 1974). Futhermore in interphase-mitotic fusion premature chromosome condensation (PCC) can generate pulverized or fragmented nuclei (Szabados and Dudits, 1980). Formation of small chromosome fragments can be seen on Fig. 1. At present, available experimental data are insufficient to explore the potentials in using synchronized cell populations for asymmetric hybrid production

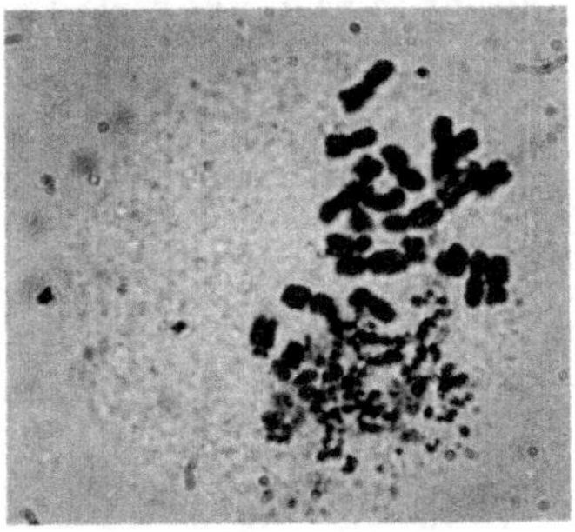

Figure 1. Fragmentation of Triticum monococcum nucleus in heterophasic homokaryon

2.2. High doses of irradiation as a tool to induce genome instability in fusion products

On the basis of experiments with mammalian cells irradiation of one of the parental protoplasts population was successfully used to induce selective organelle transfer in production of cybrids (reviewed by Galun and Aviv, 1986) or to achieve nuclear gene transfer via asymmetric hybridization (Dudits et al, 1980; Gupta et al, 1982; Itoh and Futsuhara, 1983; Gupta et al, 1984; Somers et al, 1986; Bates et al, 1987; Imamura et al, 1987; Dudits et al, 1987). Both cytological and molecular analysis of selected hybrids recorfirmed that irradiation could significantly reduce the amount of chromosomal material from irradiated fusion partner. The regenerated intergeneric hybrids as Daucus carota+Petroselium hortense (Dudits et al, 1980); Datura innoxia+Physalis minima (Gupta et al, 1984) or interspecific hybrid as N.tabacum+N.plumbaginifolia (Bates et al, 1987) have retained only one or two chromosomes and few phenotypic traits from the donor parent. Imamura et al (1987) demonstrated by use of species specific repetitive DNA probes that in Hyoscyamus+Nicotiana hybrids despite of clear reduction in amount of N.tabacum DNA, relatively large amounts of the donor DNA were still present in hybrid calli. The authors could not find differences in hybrids developed after 20 or 40 Krad treatment of Nicotiana protoplasts as the amount of introduced foreign DNA was concerned. Also in Nicotiana+Daucus hybridization the degree of chromosome loss and the applied irradiation dosage were not correlated (Dudits et al, 1987). These findings suggest the significance of other factors than irradiation in induction of somatic segregation. In the case of Nicotiana tabacum+N.plumbaginifolia fusion combination the chromosome loss can reflect a natural genomic instability as it was found in sexual hybrids of these two species (Moav, 1961). In intergeneric cells hybrids the irradiation effect can be combined with inbalanced chromosome behaviour caused by somatic incompatible reactions. Despite of gradual loss of one parental chromosome set the transient coexistence of two diverse genomes in hybrid cells provides the opportunity for recombination of chromosomes or a few genes from the eliminated partner. Recombination is also a key factor in gene transfer mediated by asymmetric hybridization. The irradiation induced repair system or the artificially maintained high cell division rate in the presence of plant hormones or specific effect of some selective agent e.g. methotrexate can influence the recombination events. At present we have no informations about the molecular and cytological mechanisms which finally can result in hybrid plant which resembles the recipient partner as the overall phenotype is concerned but new traits from the donor partner can be detected on the basis of morphological, biochemical or molecular markers. Important feature of asymmetric hybrids between tobacco and carrot was that their genomic constitution has made the transmission of resistance character into the next sexual generation possible (Dudits et al, 1987).

3. EXPRESSION OF INTRODUCED NEW CHARACTERISTICS IN ASYMMETRIC SOMATIC HYBRIDS

Since in two of the fusion experiments with irradiated protoplasts the correction of albino defects served as marker for gene transfer, quantitative data are availeble about expression of introduced genes. Interestingly in both asymmetric hybrids between carrot+parsley (Dudits et al, 1980) as well as between *Datura*+*Physalis* (Gupta et al, 1984) the restoration of chlorophyll synthesis was incomplete. Significantly lower amounts of chlorophylls were detected in the hybrids than in wild type parents. These findings call the attention to possible irregularities in expression of transferred genes into a new genetic environment.

Considerable instability was characteristic for wide hybrids between tobacco and carrot. The carrot specific resistance markers as methotrexate and 5-methyltryptophan resistance continuously segregated towards sensitive phenotypes during callus propagation and plant regeneration (Dudits et al, 1987). The frequency of somatic segregants was significantly lower in carrot+parsley hybrid (Dudits et al, 1980). While after fusion between *N.tabacum*+*N.plumbaginifolia* (gamma-irradiated, kanamycin resistance) the segregation of kanamycin resistance was not detected (Bates et al, 1987).

4. ISOLATED PLANT CHROMOSOMES AS POSSIBLE VECTORS FOR GENE TRANSFER

Successes in use of mammalian chromosomes to transfer nuclear genes by integration of subchromosomal fragments into recipient genomes (Klobutcher and Ruddle, 1981) suggests a similar approach to develop a system for gene transfer in higher plants. After the fist attempts (Malberg and Griesbach, 1980; Szabados et al, 1981) significant progress has been made in isolating intact plant chromosomes from mitotic protoplasts (Hadlaczky et al, 1983). Treatment of a mixture between chromosome suspension and plant protoplasts with polyethylene glycol (PEG) can cause an uptake of single chromosome into the cells (Szabados et al, 1981). To follow the fate of foreign chromosomes T. Praznovszky et al (unpublished) have introduced chromosomes from a methotrexate resistant carrot into protoplasts of *Triticum monococcum* cell suspension. The cytological characterization of the recovered methotrexate resistant *Triticum* clones showed that the majority of isolates carried no additional carrot chromosomes. However, in one of the lines, the presence of a single small carrot chromosome was detected after three months of culturing in methotrexate medium. Recently chromosomes from *Petunia alpicola* were introduced into *Petunia hybrida* protoplasts via micorinjection (Griesbach, 1987). These results encourage further studies to work out an efficient system which utilizes plant chromosomes as vectors in gene transfer.

4. CONCLUSIONS

The increasing number of examples clearly show that asymmetric cell hybridization can result in limited gene transfer. This approach based on experimental control of chromosome elimination has significance in increasing the efficiency of intra- and interspecific cell hybridization with breeding purposes or in combining characters from widely divergent species. The hybrid genotypes resulting from fusion with irradiated protoplasts represent a genomic constitutitons which are generally required for practical breeding. After further methodical improvements the chromosome mediated gene transfer can be an additional tool for genetic manipulation of agronomically import characters in the lack of cloned gene.

REFERENCES

1. Abel PP, Nelson RS, De B Hoffmann N, Rogers SG, Fraley RT & Beachy RN: Science 232, 738-743, 1986.
2. Austin S, Ehlenfeldt MK, Baer MA & Helgeson JP: Theor.Appl.Genet. 71, 682-690, 1986.
3. Bates GW, Hasenkampf CA, Contolini CL & Piastuch WC: Theor.Appl.Genet. 74, 718-726, 1987.
4. Binding H & Nehls R: Mol.Gen.Genet. 164, 137-143, 1978.
5. Chien JC, Kao KN & Wetter LR: Theor.Appl.Genet. 62, 301-314, 1982.
6. Dudits D & Praznovszky T. In:Biotechnology in Plant Science:Relevance to Agriculture in the Eighties" eds. Zaitlin M, Day P, Hollander A, pp 115-127, Academic Press New York, 1985.
7. Dudits D, Hadlaczky Gy, Levi E, Fejer O, Haydu Z & Lazar G: Theor.Appl.Genet. 51, 127-132, 1977.
8. Dudits D, Hadlaczky Gy, Bajszar Gy, Koncz C, Lazar G & Horvath G: Plant Sci. Lett. 15, 101-112, 1979.
9. Dudits D, Fejer O, Hadlaczky G, Koncz Cs, Lazar GB & Horvath G: Mol.Gen.Genet. 179, 283-288, 1980.
10. Dudits D, Hadlaczky Gy, Lazar G & Haydu Zs. In:"Plant Cell Cultures:Results and Perspectives"eds. Sala F, Parisi B, Cella R & Ciferri O. pp. 207-214, Elsevier/North Holland, 1980.
11. Dudits D, Maroy E, Praznovszky T, Olah Z & Gyorgyey J: Proc.Natl.Acad.Sci. USA 84, 1987. (in press)
12. Ehlenfeldt MK & Helgeson JP: Theor.Appl.Genet. 73, 395-402, 1987.
13. Evans DA, Wetter LR & Gamborg OL: Physiol. Plant. 48, 225-230, 1980.
14. Galun E & Aviv D: In:"Plant Molecular Biology. Methods in enzymology. eds. Weissbach A, Wessbach H. Vol. 118. pp. 595-611, Academic Press, Orlando, USA, 1986.
15. Gleba JJ & Hoffmann F: Mol.Gen.Genet. 165, 257-264, 1978.
16. Gleba JJ, Momot VP, Cherep NN & Skarzynskaya MV: Theor.Appl.Genet. 62, 75-79, 1982.
17. Griesbach RJ: Plant Science 50, 69-77, 1987.
18. Gupta PP, Gupta M & Schieder O: Mol.Gen.Genet. 188, 378-383, 1982.
19. Gupta PP, Schieder O & Gupta M: Mol.Gen.Genet. 197, 30-35, 1984.
20. Hauptmann P, Kumar P & Widholm I: In:"Protoplasts 1983", eds. Potrykus I, Harms CT, Hinnen A, Hutter R, King PJ & Schillito RD. pp. 92-93. Birkhausen, Basel, 1983.

21. Hilder VA, Gatehouse AMR, Sheerman SE, Baker RF & Boulter P: Nature 300, 160-163, 1987.
22. Hoffmann F and Adachi T: Planta 153, 586-593, 1981.
23. Imamura J, Saul MW & Potrykus I: Theor.Appl.Genet. 74, 445-450, 1987.
24. Itoh K & Futsuhara J: Jpn. J.Genet. 58, 545-553, 1983.
25. Kao KN: Mol.Gen.Genet. 150, 225-230, 1977.
26. Kao KN & Michayluk MR: Planta 115, 355-367, 1977.
27. Klobutcher LA & Ruddle FH: Ann.Rev.Biochem.50, 533-554,1981.
28. Krumbiegel G & Schieder O: Planta 145, 371-375, 1979.
29. Krumbiegel G & Schieder O: Planta 153, 466-470, 1981.
30. Malberg R & Griesbach RJ: Plant Sci.Lett. 17, 141-147, 1980.
31. Maliga P, Kiss ZR, Nagy AH & Lazar G: Mol.Gen.Genet. 163, 145-151, 1978.
32. Melchers G, Sacristan MD & Holder AA: Calsberg Res. Commun. 43, 203-218, 1978.
33. Moav R: Am.J.Bot. 47, 87-93, 1961.
34. O'Connell MA & Hanson MR: Theor.Appl.Genet. 72, 59-65, 1986.
35. Pakowski J, Schillito RD, Saul M, Mandak V, Hohn T, Hohn B & Potrykus J: EMBO J. 3, 2717-2722, 1984.
36. Reich TJ, Iyer VN & Miki BL: Biol. Technology 4, 1001-1004, 1986.
37. Sala C, Biasini MG, Morandi C, Nielsen E, Parisi B & Sala F: J. Plant Physiol. 118, 409-419, 1985.
38. Schell J: Science 237, 1176-1183, 1987.
39. Schieder O: Mol.Gen.Genet. 162, 113-119, 1978.
40. Somers DA, Narayanan KR, Kleinhofs A, Cooper-Bland S & Cocking EC: Mol.Gen.Genet. 204, 296-301, 1986.
41. Szabados L & Dudits D: Exp.Cell Res. 127, 442-446, 1980.
42. Szabados L, Hadlaczky Gy & Dudits D: Planta 151, 141-145, 1981.
43. Toriyama K, Kameya T & Hinata K: Planta, 1987. (in press)

CELL-MEDIATED GENE TRANSFER.

I. FAMELAER [1], YU. GLEBA [2], A. MOURAS [3], S. HINNISDAELS [1], D. CAMMAERTS [1], I. NEGRUTIU [1], M. JACOBS [1].

1. Laboratorium voor Plantengenetica, Institute of Molecular Biology, 1640 St.-Genesius-Rode (Belgium).
2. Institute of Botany, Academy of Sciences, Kiev (USSR).
3. Universitè de Bordeaux II, Laboratoire de Biologie Cellulaire, Avenue des Facultès, 33405 Talence (France).

A potential application of parasexual hybridization is to induce asymmetric somatic fusion products in which only a part of the donor plant genome has been transferred.
Somatic hybridization was shown to make possible the crossing of species belonging to distinct genera, tribes, or families (reviewed by Gleba and Sytnik, 1). In such hybrids the evolution of the caryological state is relatively slow, chromosome elimination is almost exclusively random, and the process of plant regeneration, if any, produces only sterile individuals, of questionable practical value.
Additional methods which may facilitate gene transfer and integration in the recipient genome and also avoid incompatibility barriers are therefore highly desirable (2).
Fusion of protoplasts with a gamma-irradiated donorpartner provides new opportunities of somatic gene exchange between the fused nuclei simultaneously with unidirectional elimination of the irradiated genome.
Selectable and molecular markers (3, 4, 5) that moreover can be traced by their biochemical and genetic properties, provide efficient tools in cell mediated gene tranfer (CMGT) methodologies. This technique, with as objective the transfer of small genetic attributes (chromosomes, chromosome fragments or even a few genes) should be most useful whenever dealing with multigenic traits, multiple copy genes, etc.
The objective of this study was to attempt the transfer of genetic information from a gamma-irradiated donor protoplast belonging to a species with various phylogenetic relatedness (N. sylvestris, A. belladonna, D. innoxia, P. hybrida, A. thaliana, D. carota, M. sativa, H. vulgare, B. napus and sugarbeet) into nitrate reductase deficient (NR^-) mutant protoplasts from N. plumbaginifolia (cnx20 or Nx26).
Corrected phenotypes for nitrate reductase were recovered at several doses in different fusion combinations.
In the first fusion combination with N. sylvestris (V-42, 2n = 24) as donorpartner, corrected phenotypes were recovered within a dose range of 10^2 to 10^3 Gy at frequencies between $2.3x10^{-4}$ to $2.2x10^{-5}$; which is much above the reversion frequencies of the established marker.

The second CMGT experiment with a N. sylvestris mutant, characterized by aminoethylcysteine resistance, constitutive expression of octopine (oct^+) and NR activity (cnx wild type), has been set up to attempt the transfer of several independant markers at once. A high number of fusion products were recovered (345) with a corrected phenotype for NR activity at 5.10^2 and 10^3 Gy, with correction frequencies between 1.3 to 2.9×10^{-3}.
Caryological analyses revealed the following :
- as a rule fusion products are recovered with a tetraploid chromosome number of N. plumbaginifolia and several donorchromosomes some of which are fragmented;
- donorchromosomes tend to be lost after passage through meiosis and mitosis.
- chromosome translocations could not yet be observed.

Cotransfer of other genetic traits could also be observed by the presence of donorisoenzymes (e.g. shikimate dehydrogenase, phosphoglucomutase,...) or by the transfer of the octopine marker.
Screening for the expresssion of this gene revealed it to be present in 27 plants out of 220 independently regenerated clones. The presence or loss of biochemical markers can also be followed after passage through meiosis and mitosis.
Progeny analyses for the NR-marker of self-fertile plants showed aberrant segregation ratios. This is obviously the result of the aneuploid nature of the asymmetric hybrids, which we can consider therefore as potential hybrids.
Fusion combinations with the other donor partners were only successful with A. belladonna and sugarbeet.
The fusion combination of N. plumbaginifolia cnx^- + B. vulgaris resulted in several fusion products which gave rise to highly abnormal plants, some of which are being analysed cytologically and biochemically.
In the fusion combination with A. belladonna (2n=2x=72) corrected phenotypes for the nitrate metabolism were produced by each irradiation dose (10^2, 3.10^2, 5.10^2 and 10^3 Gy). The introduction of genes from the donorpartner into the recipient genome has been shown by isoenzymes (esterase, hypoxanthine dehydrogenase) and by cytological analysis . Regenerated plants however were not fertile and T^1 progeny has been obtained by backcrossing with the parental recipient plant and embryo rescue.
The CMGT experiments reported here demonstrate the transfer of genetic information in several interspecific fusion combinations. Physical agents such as X- or gamma-rays causing chromosome breakage prove not always to be efficient enough to avoid incompatibility barriers, to eliminate excessive donor genetic material and to produce stably transformed regenerants.
CMGT experiments with more widely separated species met with much less success, but also demonstrate that gene transfer remains possible between some very unrelated species (e.g. B. vulgaris, 5).

Refinement of CMGT methodologies is obviously necessary to enhance efficient transfer and introgression of alien genes into recipient plant genomes.

REFERENCES

1. Gleba Yu, Sytnik KM: Protoplast fusion. Genetic Engineering in higher plants. In: Monographs on Theoretical and Applied Genetics, 8. Springer Verlag, 1984.
2. Gupta et al. Mol. Gen. 197, (1984).30-35.
3. Negrutiu et al. Plant Mol. Biol. 3 (1984) 289-302.
4. Negrutiu et al. Mol. Gen. Genet. 199 (1985) 330-337.
5. Somers et al. Mol. Gen. Genet. 204 (1986) 296-301.

PROTOPLAST FUSION OF KANAMYCIN RESISTANT, CNX⁻ NICOTIANA PLUMBAGINIFOLIA WITH STREPTOMYCIN RESISTANT N. TABACUM (SR1) AND THE EFFECT OF IRRADIATION OF THE TABACUM PARENT.

MAARTEN KOORNNEEF, BART DEN BOER, PATTY VAN LOENEN-MARTINET, PETRA VAN ROGGEN and JELLE WIJBRANDI
Department of Genetics, Agricultural University,
53 Generaal Foulkesweg, 6703 BM Wageningen, The Netherlands

1. INTRODUCTION

Nicotiana species provide a useful model system to study the genetic consequences of somatic hybridization because of their efficient protoplast culture system and the availability of cell-selectable markers. A recessive nitrate reductase mutant (CNX 20) and the dominant resistances to the antibiotics streptomycin (cp-DNA encoded) and kanamycin (encoded by the nuclear genome) were used to investigate, the effect of irradiation on the loss of genetic markers that were not used for the selection of the fusion products.

2. MATERIAL AND METHODS

In the CNX 20 mutant, received from Dirks et al. (2), kanamycin resistance was introduced by leaf disc transformation using Agrobacterium tumefaciens. For this ammonium succinate (10 mM) and $Na_2MoO_4-2H_2O$ (0.2 mM) was added to the plant media to allow growth of the cnx mutant. The protoplast isolation and fusion procedures and culture media (K3 modified with respect to the nitrogen source) were as described by Maliga (3).
SR1 protoplasts were irradiated with 0, 50, 300 and 1000 Gy rays prior to fusion. Fusion derived microcalli were subjected to selection on medium A or B.

A: Medium containing streptomycin (1 gl^{-1}) and kanamycin (50 mgl^{-1}) together with ammonium succinate.

B: Medium containing kanamycin (50 mgl^{-1}) and only nitrate as nitrogen source.

Calli and regenerating shoots surviving on medium A were tested for growth on media containing only nitrate. Shoots regenerated from fusion products on medium B were subjected to a test for streptomycin resistance.

3. RESULTS AND DISCUSSION

3.1. The Cnx phenotype of fusion products selected on medium A.

Without irradiation all calli resistant to both streptomycin and kanamycin grow on nitrate only, indicating that the percentage of cybrids is relatively low (Table 1).

TABLE 1. Characterization of fusion products selected on medium A with respect to their Cnx phenotype.

Irradiation dose (Gy)	Number of fusion clones that are: Cnx⁺	Cnx⁻	Total
0	45	0	45
50	3	7	10
300	0	87	87
1000	0	49	49

Due to irradiation the number of cybrids and asymmetric hybrids that lack cnx^+ alleles increases. However, elimination of the SR1 chromosomes is far from complete at an irradiation dose of 50 Gy.

3.2. The effect of irradiation on chloroplast composition of fusion products selected on medium B

By selection on medium B, fusion products could be isolated at all doses although at strongly reduced frequences at 300 and 1000 Gy. Streptomycin resistant and sensitive shoots are not represented in equal numbers among the hybrids (Table 2). The percentage of resistant shoots is somewhat reduced when irradiation was applied, although this reduction is statistically not significant. This shows the relative insensitivity of the chloroplast genome to irradiation damage.

TABLE 2. Characterization of shoots from fusion products selected on medium B for streptomycin resistance.

Irradiation dose (Gy)	Number of fusion clones that are: Resist.	Resist. + Sens.	Sens.	Total
0	9	6	29	44
50	0	2	19	21
300	2	5	8	15
1000	0	2	5	7
	11	15	61	87

3.3. The phenotype of hybrids selected on medium B

Somatic hybrids obtained without irradiation were intermediate in phenotype between the parental species. Some variation was present among the different hybrids especially with respect to flower morphology. Deviations were similar to the variation attributed to mt-DNA recombination in other *Nicotiana* somatic hybrids (1). From one callus a plant was regenerated that resembled the SR1 parent more than the other hybrids. This plant had a hybrid pattern for peroxidase isozyms but the SR1 isozym pattern for GOT suggesting a loss of genetic information of *N. plumbaginifolia*. Hybrids obtained from treatments where irradiation was applied, were more difficult to regenerate into plants than symmetric hybrids. These plants were also much less vigorous. Variation for morphological traits was very large. Table 3 shows the loss of the SR1 alleles of one of the GOT isozymes with increasing irradiation dose. However the hybrid peroxidase pattern was present in all symmetric and most asymmetric hybrids suggesting that a peroxidase gene may be loosely linked to the *cnxA* gene, which was the selected trait from SR1.

TABLE 3. The classification of fusion products selected on medium B for a GOT and peroxidase isozym pattern polymorphic between the two species.

Irradiation dose (Gy)	*N. tabacum* type GOT	Perox.	*N. plumbaginifolia* type GOT	Perox.	Hybrid type GOT	Perox.
0	1	0	0	0	25	27
50	0	0	1	1	1	4
300	0	0	3	1	0	3
1000	0	0	0	1	0	1

REFERENCES

1. Aviv D, Anzee-Gonen P, Bleichman S, Galun E: MGG 196: 244-253 (1984).
2. Dirks R, Negrutiu I, Sidorov V, Jacobs M: MGG 201: 339-343 (1985).
3. Maliga P: Plant Mol. Biol. Newsletter 3: 88-94 (1982).

SELECTON FOR RESISTANCE TO PHOMA LINGAM USING PYCNOSPORE INFECTIONS OF PLANTS IN-VIVO AND SIRODESMIN PL AS SELECTIVE AGENT IN-VITRO.

Sjödin, C. and K. Glimelius.

Department of Plant Breeding, Swedish University of Agricultural Sciences, Box 7003, S-750 07 Uppsala, Sweden.

The aim of the study was to find resistant accessions within *Brassicaceae* to *Phoma lingam* and to evaluate whether resistance could be expressed on protoplast or cellular level in presence of the toxin, sirodesmin PL. A gene transfer system was designed, for transfer of genes for resistance to *B. napus* via protoplast fusions, using the toxin as a selective agent in the screening for resistant hybrid cells.

MATERIAL AND METHODS

Screening for resistance.
In total, 96 accessions within *Brassicaceae* and 17 isolates of *P. lingam* with differences in origin and pathogenicity, were used in the study.
Initally, seedlings were sprayed with 2 ml of a spore suspension (10^6/ml) prepared from each of 3 isolates. After disease rating of 300 seedlings/accession, the material with the highest level of resistance was further tested. Pycnospores (10^5/ml) from 17 isolates were inoculated in cotyledons and adult leaves according to Williams (8), and in stem bases of 6-8 leaved plants. The disease symptoms obtained in the different leaf tests were rated according to Delwiche (1) and the results obtained were based on 25-36 observations/accession and isolate. Susceptibility was controlled by presence of the pathogen after surface sterilization and *in-vitro* culture of inoculated leaves. Pycnospores found were reinoculated on susceptible *Brassica napus* for control of relevant disease symptoms.

Selective agent.
Toxin producing isolates of *P. lingam* and culture conditions promoting toxin production have been. Toxic compounds have been isolated and purified from the fungal culture medium (5). Viability of protoplasts isolated as described by Glimelius (2) and necrosis on plantlets were used as bioassays to identify the toxic fractions. The effects of the toxic compounds were further studied on protoplasts, cells and plants of *B. napus*, *Nicotiana tabacum* and *Solanum tuberosum*.

Selection method.
The effects of sirodesmin PL on different accessions of *Brassica* resistant and susceptible to *P. lingam* on the cell aggregate stage were assessed. The *in-vitro* selection was carried out in bead-type cultures (4) where the cells derived

from protoplasts were embedded in 0.4% agarose and surrounded with 8p culture medium (3). After a culture period of 14 to 20 days after protoplast isolation, when the formation of starshaped aggregates had started, the toxin was added to the cultures in a range of 1-15 uM. Corresponding investigation was made on cotyledons and adult leaves on greenhouse grown plants.

Gene transfer system.
Transfer of genes for resistance were made by the protoplast fusion method developed by Sundberg and Glimelius (6). As the recipient, *B. napus* cv. Hanna was used, and as donors, the different resistant *Brassica* accessions. The donor material was treated with X-ray (7-15 kRad) before fusion. Hypocotyl protoplasts were stained with carboxyflourescein diacetate (CFDA) and fused with mesophyll protoplasts. Due to the red autofluorescence from the chlorophyll and CFDA staining of hypocotyls fluorescent markers were obtained, suitable for enrichment of heterokaryons by a fluorescent activated flow sorter (3). Identification of the hybrids have been made with isozyme analysis as described by Sundberg and Glimelius (6) and furthermore, the ploidy level, fertility and chromosome analysis have been performed (7).

RESULTS

Resistant sources.
All material classified as seedling-resistant and 7 accessions with slight susceptibility in the initial seedling test and 6 rapid-cycling *Brassica* species were subjected to extended studies. The number of fungal isolates used was increased to 17 and both cotyledons and adult leaves were inoculated. All representatives with resistance at the cotyledon stage were found to be resistant at the adult, too. The accessions showing a high to complete resistance in the leaf inoculation tests were subjected to additional checking by applying stem base injections and re-isolation controls. No lesions could be detected in the injected stems, and pycnospores could not be isolated from the inoculated leaves. A large variability in virulence was found among the different isolates of the pathogen demonstrating the importance of using a broad spectrum when screening for resistance. All together, 5 accessions of *B. juncea*, 2 of *B. nigra* and 2 of *B. carinata* could be classified as resistant.

Selective agent.
From a fungal culture medium 2 toxic compounds were purified, sirodesmin PL and deacetylsirodesmin PL, for which the molecular structures were determined. A concentration of 1 uM sirodesmin PL was enough to prevent protoplasts derived from *B. napus* to regenerate cell aggregates whereas 20 uM deacetylsirodesmin PL was needed to cause the same effect. The same toxic effects were also found for protoplasts isolated from potato, tobacco and 6 different *Brassica* species. When sirodesmin PL was added to cells and cell aggregates, however, cells of *B. napus* which is susceptible to the fungus

remained highly sensitive, but cells of non-hosts to the fungus e.g. tobacco and potato showed less sensitivity to the toxin. This difference was even more pronounced with increased time in culture. Also when sirodesmin PL was added to leaves of greenhouse grown plants clear differences in response between susceptible hosts and non-hosts to the pathogen were detected.

Selection system.
In concentrations of 1-5 uM sirodesmin PL, cell aggregates derived from resistant _B. carinata_, _B. juncea_ and _B. nigra_ showed a survival of 90-80% compared to untreated control cultures, while cell aggregates derived from susceptible _B. napus_ varieties died when treated with corresponding toxin concentration. To investigate whether the obtained differences in response to the toxin were species specific or related to resistance, cell aggregates derived from accessions of _B. carinata_, _B. juncea_ and _B. nigra_ susceptible to the pathogen were treated with the toxin in a similar way. All these different cultures showed the same sensitivity to the toxin as the cultures from the susceptible _B. napus_ varieties. The effects of the toxin were furthermore investigated on intact plantlets of susceptible and resistant _Brassica_ accessions under greenhouse conditions. When low concentrations of the toxin were used (1-3 uM), small chlorotic spots which quite rapidly turned necrotic, were obtained on the susceptible plant material. However, when higher concentrations were used, the same response to the toxin was obtained both in the resistant and the susceptible material, and the only difference was the delay in time found in the resistant material.

Gene transfer.
Experiments for transfer of genes were performed in two different ways. 1) Production of partial hybrids, where the donor material was treated with X-ray and 2) complete hybrids, with untreated donor material. In both cases selection using the toxin was performed. Shoots and plants of 3 combinations have been obtained. Complete hybrids were obtained after fusion of _B. napus_ cv. Hanna and _B. nigra_ cv. Junius and partial hybrids of _B. napus_ cv. Hanna and _B. juncea_ "38" and _B. napus_ cv. Hanna and _B. carinata_ "Etiopia". None of these combinatons were selected with the toxin. Of 30 plants obtained from fusions between rapeseed and _B. nigra_ cv. Junius, 24 were true hybrids according to isoenzyme and chromosome analysis, 3 had more than 54 chromosomes and 3 had between 46-48 chromosomes. All hybrids having 54 or more chromosomes were resistant to _P. lingam_ when tested on adult leaf stage, whereas 1 of the 3 having less than 54 chromosomes was susceptible. Nine plants being partial hybrids between rapeseed and B. juncea "38" and 2 shoots of rapeseed and _B. carinata_ "Etiopia" are for the moment analysed in the same respect as the true hybrids. Resistance tests have so far not revealed any resistance among these hybrids. This result indicates that a selection step enhances the probability for obtaining resistant hybrids. Hybrid calli selected with toxin have recertly been transferred to differentiation media, for initiation of shoots.

CONCLUSION

Toxic compounds produced by *P. lingam* in culture has been identified and characterized. The toxin, sirodesmin PL, had selective properties making it possible to distinguish susceptible *B. napus* from the non-hosts *N. tabacum* and *S. tuberosum*. These differences were found on cell aggregate level as well as on intact plants. The differences in sensitivity *in-vitro* are only expressed on differentiated cells, not on protoplasts. Plant material resistant to *P. lingam* has been identified in *B. juncea*, *B. nigra* and *B. carinata*. When treated with toxin at the cell aggregate stage, the resistant material survived where the susceptible material died. Similar results were obtained on intact plant material. These results show that the toxin has host-specific properties and that a correlation between toxin insensitivity and resistance to the pathogen exist and thus may be involved in the infection process of the pathogen. Based on these studies, protoplast fusion experiments have been performed to transfer resistance to *B. napus*. Sirodesmin PL has been used as selective agent when screening for resistant hybrid cells.

REFERENCES

1. Delwiche, P. A. 1980. Genetic aspects of blackleg (*Leptosphaeria maculans*) resistance in rapeseed (*Brasssica napus*) Ph.D. Thesis, University of Wisconsin, Madison, 144pp.
2. Glimelius, K. 1984. High growth rate and regeneration capacity of hypocotyl protoplasts in some *Brassicaceae*. Physiol. Plant. 61:38-44.
3. Glimelius, K., Djupsjöbacka, M. and H. Fellner-Feldegg. 1986. Selection and enrichment of plant protoplast heterokaryons of *Brassicaceae* by flow sorting. Pl. Sci. 45: 133-141.
4. Shillito, R.D., Paszkowski,J. and Potrykus, I. 1983. Agarose plating and a bead type culture technique enable and stimulate development of protoplast-derived colonies in a number of plant species. Plant Cell Rep. 2: 244-247.
5. Sjödin, C., Nyhammar, T. and K. Glimelius 1987. Effects of toxic metabolites produced by *Phoma lingam* (Tode ex Fr) Desm. on protoplasts, cells and plants of hosts and non--hosts to the pathogen. J. Phys. and Mol. Pl. Path. (In press).
6. Sundberg, E. and K. Glimelius. 1986. A method for production of interspecific hybrids within *Brassicaceae* via somatic hybridization, using resynthesis of *Brassica napus* as a model. Pl. Sci. 43:155-162.
7. Sundberg, E., Landgren, M. and K. Glimelius 1987. Fertility and chromosome stability in *Brassica napus* synthesized by protoplast fusion. Theor. Appl. Genet. (In press).
8. Williams, P.A. 1985. Crucifer Genetics Cooperative (CrGC) Resource Book, Dept. of Plant Pathology, Univ. of Wisconsin, Madison, USA.

INDUCTION OF MICRONUCLEI BY AMIPROPHOS-METHYL AND THEIR APPLICATION FOR TRANSFER OF SINGLE, SPECIFIC CHROMOSOMES IN PLANT CELLS

K. SREE RAMULU, H.A. VERHOEVEN, L.J.W. GILISSEN, P. DIJKHUIS AND H.C.P.M. VAN DER VALK

Research Institute Ital, Postbox 48, NL-6700 AA Wageningen, The Netherlands.

1. INTRODUCTION

In mammals, a number of somatic cell genetic techniques have been developed for the transfer of genes: whole cell hybridization, microcell-mediated gene transfer (MMGT), chromosome-mediated gene transfer (CMGT) and DNA-mediated gene transfer (DMGT) (1, 2). In plants, DMGT has become realistic by the recent progress in molecular and cell biological techniques. In the case where a monogenic trait has been characterized at the molecular level, the identified gene can be cloned and transferred into plant cells via DNA vectors or by direct DNA uptake. With respect to polygenic traits or non-identified genes, fusion of whole cells (protoplasts) offers the possibility for gene transfer. However, the hybrid cell lines or plants contain several donor chromosomes and undesired characters. Moreover, such somatic hybrids constitute a complex genetic system which is difficult to manipulate experimentally. In particular, intraclonal karyotypic heterogeneity and instability of the hybrid chromosome compliments are factors limiting their practical utility. A conceptually ideal collection of somatic hybrids would be one in which each clone of a hybrid panel retained a single, specific donor chromosome which was maintained in the population by direct selective pressure. In the case of CMGT the transferred chromosomes integrate only as fragments (2). On the other hand, MMGT provides the necessary means to transfer single, intact chromosomes. As microcells contain interphase chromosomes and an intact plasmamembrane, their integration into a recipient genome can be more stable than that of the isolated metaphase chromosomes. In mammals, MMGT has been shown to be a powerful tool for mapping selectable markers to specific chromosomes and to study the regulation of gene expression (3).

2. MICRONUCLEATION

Interference with the separation of chromosomes into an anaphase configuration or with cytokinesis can cause the formation of cells containing many micronuclei. The agents inducing micronucleation can be broadly divided into i) poisons and physical factors that interfere with the assembly of microtubules and the mitotic spindle and ii) chromosome breaking agents. Cold shock can induce aberrant chromosome segregation and cytokinesis resulting in cytoplasmic fragments with one or several micronuclei. Micronuclei can also be induced by cytochalasin-B applied at low concentrations (2). X-rays or chemical mutagens cause chromosome breakage and the resulting chromosome fragments can form micronuclei (4). Of all these, the best known agents which induce micronuclei on a mass-scale suitable for MMGT in a variety of mammalian cell lines are colchicine and its diacetyl derivative, colcemid (1, 5). in plant cells, however, the spindle toxins or chromosome breaking agents so far induced only low frequency of micronuclei (4, 6). Recently, we found that the herbicide, amiprophos-methyl (APM), a potent anti-microtubule agent, induces high frequency of micronuclei in plant cell suspension cultures (7-10). The results are briefly discussed below.

3. AMIPROPHOS-METHYL

Amiprophos-methyl was introduced in 1975. Since then several authors investigated the effect of APM, especially on microtubule (MT) polymerization (11, 12). Microtubule depolymerizing agents are useful tools for studies on cytoskeleton. Rapid acting and quickly reversible anti-MT drugs are highly desirable in plants. Colchicine has only a low and often incomplete MT-depolymerizing activity even at high concentrations, unlike in animals. By contrast, APM at very low concentrations (1-3 μM) can depolymerize cortical and mitotic microtubules in 1 h in cell cultures of various plant species (11). In comparison, a 2 h treatment with 3 mM colchicine had no effect on mitotic arrays.

4. MITOTIC ARREST, CHROMOSOME GROUPING AND MICRONUCLEATION

To determine if various cell culture lines are differentially sensitive to APM it was applied at various concentrations (4-64 μM) to Nicotiana plumbaginifolia, Solanum tuberosum, Haplopappus gracilis and Daucus carota cultures. After 2-5 h, metaphase chromosomes lost their regular arrangement on the spindle and several cells showed well scattered chromosomes all over the cytoplasm. Metaphase chromosomes were also arranged in groups of two or

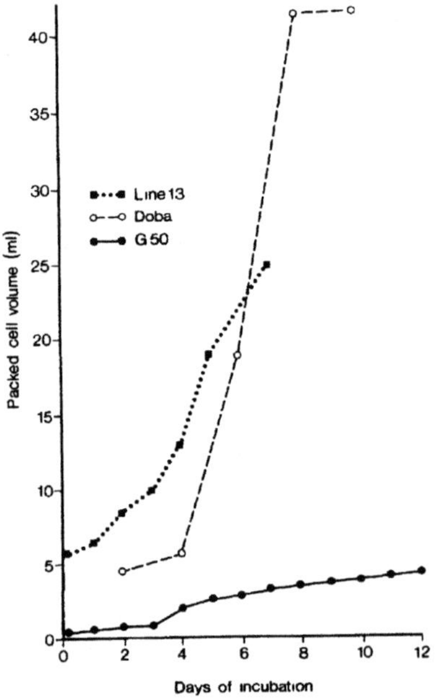

FIGURE 1. Growth of cell cultures of N. plumbaginifolia (Doba) and S. tuberosum (line 13, G50).

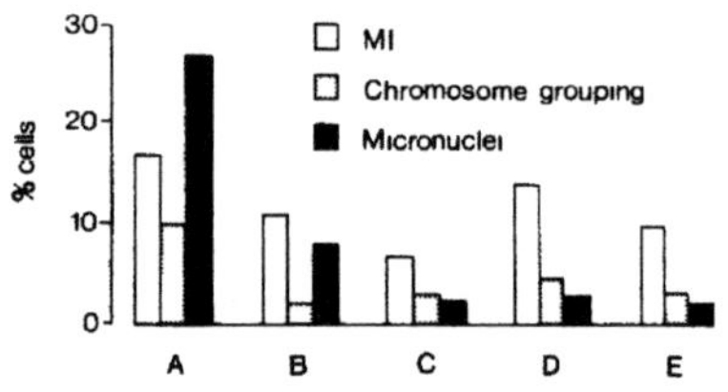

FIGURE 2. Maximum mitotic indices (MI) and % cells with chromosome grouping and micronuclei after APM (32 μM) treatment in N. plumbaginifolia (A), line 13 (B), G50 (C), H. gracilis (D) and D. carota (E) (9, 10).

more chromosomes. The chromosomes showed to centromere division or separation of chromatids (devoid of anaphase). They decondensed, developed nuclear membranes and formed micronuclei. Multinucleate cells were observed from 6 h onwards in N. plumbaginifolia and from 10 h in other species. The degree of mitotic arrest, chromosome grouping and micronucleation varied according to the concentration and duration of APM treatment, the species and the cell cycle characteristics (9, 10). The untreated cell suspension culture of N. plumbaginifolia (kanamycine resistant Doba line) grew faster than those of S. tuberosum (line 13 transformed by Agrobacterium rhizogenes, galactose variant line G50) (Fig. 1) and of H. gracilis and D. carota. After treatment with APM, it responded with higher mitotic index, and frequencies of cells with chromosome grouping and micronuclei (Fig. 2).

5. RECOVERY AFTER WASHING AND SUBCULTURING

The cells with chromosome grouping and micronuclei gradually disappeared after washing out APM and subculturing. As there was no chromatid separation (segregation) during APM treatment, the chromosome number doubled in the next cell division (Fig. 3). The cells continued to grow and divide normally as in the control cultures (8-10).

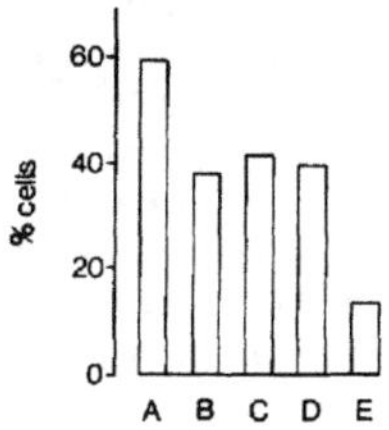

FIGURE 3. Relative % of cells with doubled chromosome numbers in N. plumbaginifolia (A), line 13 (B), G50 (C), H. gracilis (D) and D. carota (E) (9, 10).

6. DNA CONTENT OF MICRONUCLEI

Flow cytometry (FCM) was used to monitor mass induction of micronuclei after APM treatment, and for sorting and characterization of micronuclei. The sorted fractions were measured by microdensitometry for their DNA content. The results suggest that many micronuclei containing one or two chromosomes can be obtained APM treatment (9).

7. MICROCELL-MEDIATED GENE TRANSFER

The various steps in MMGT are briefly illustrated in Fig. 4: a) the induction of micronuclei by APM in donor cells carrying selectable markers, b) the isolation of microcells and c) the fusion of microcells with recipient cells generating microcell heterokaryons. Under appropriate selective conditions, those microcells heterokaryons containing the chromosome with the marker gene, proliferate to yield microcell hybrid clones. Microcells (micronuclei with a rim of cytoplasm and intact plasmamembrane) were isolated from multinucleated N. plumbaginifolia protoplasts by continuous gradient centrifugation in the presence of cytochalasin-B, and subdiploid microcells were characterized by FCM and microdensitometry (Verhoeven et al. unpubl.). The

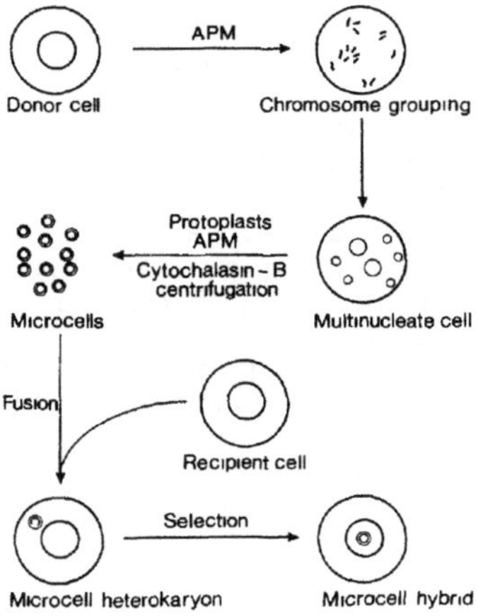

FIGURE 4. Schematic illustration on micronucleation after APM treatment and on microcell hybridization.

efficiency of microcell hybridization which depends on a number of factors (micronucleation of donor cells, viability of microcells, fusion ability), is crucial for the realization of microcell hybrid clones at high frequency.

8. SUMMARY

Amiprophos-methyl (APM) is an anti-microtubule agent, which reversibly inhibits the polymerization of plant microtubules, and is a potent inducer of micronuclei in plant cell cultures. The treatments with APM resulted in high mitotic indices and well scattered metaphases, offering good prospects for cytology and chromosome isolation. Metaphase chromosomes entered directly into telophase forming several micronuclei. Micronucleation of cells varied according to the species and the rate of cell growth and mitosis. The fast growing cell suspensions of N. plumbaginifolia responded with higher frequency of micronuclei than the relatively slow growing cultures of S. tuberosum, H. gracilis and D. carota. After removal of APM, micronucleated cells gradually disappeared and cell cultures continued normal growth and mitosis. Many sub-diploid micronuclei containing one or two chromosomes were obtained after APM treatment, and they were sorted by FCM. Microcells have been isolated from multinucleated protoplasts for transfer to the recipient partners. Microcells would be valuable tools for the transfer of single, specific chromosomes, gene mapping and to study the regulation of gene expression in plant cells.

ACKNOWLEDGEMENTS

Part of this work was supported by funds of the "Biotechnology Action Programme" of the CEC, contract BAP-0083-NL (GDF). We wish to thank Dr. L. van Vloten-Doting, Dr. W.J. Stiekema and Prof. B. de Groot for their interest and encouragement during the course of this investigation.

REFERENCES

1. Fournier REK: Techniques of Somatic Cell Genetics (J.W. Shay, ed.) Plenum Press, New York, 1982, p 309.
2. Mc Bride OW, Peterson JL: Ann. Rev. Genet., 14 (1980) 321.
3. Lugo TG, Fournier REK: Gene transfer (R. Kucherlapati, ed.), Plenum Publishing Corp., New York, 1986, p79.
4. Nuti Ronchi V, Bonatti S, Turchi G: Environ. Expt. Bot., 26 (1986) 115.
5. Levan A: Hereditas 40 (1954) 1.
6. Lo Schiavo F, Nuti Ronchi V, Terzi M: Theor. Appl. Genet., 58 (1980) 43.
7. Verhoeven HA, de Laat AMM, Sree Ramulu K, Dijkhuis P, van Vloten-Doting L: Report of the CEC meeting, Lovain-la-Neuve, 1986, p 119.
8. Laat de AMM, Verhoeven HA, Sree Ramulu K, Dijkhuis P: Planta (in press).
9. Sree Ramulu K, Verhoeven, Dijkhuis P: Theor. Appl. Genet., (in press).
10. Sree Ramulu K, Verhoeven, Dijkhuis P, Gilissen LJW: Plant Sci. (submitted).
11. Falconer MM, Seagull RW: Protoplasma 136 (1987) 118.
12. Morejohn LC, Fosket DE: Science 224 (1984) 874.

ISOLATION OF MICROPROTOPLASTS.

H.A. Verhoeven, K. Sree Ramulu, H.C.P.M. van der Valk, P. Dijkhuis and J. Blaas. Research Institute Ital, P.O.Box 48, 6700 AA Wageningen, the Netherlands.

Introduction

Formation of micronuclei, induced by colchicine or colchemid, followed by the separation of micronuclei into microcells, is the only way to transfer intact chromosomes into mammalian cells. However, these chemicals show poor induction of micronuclei in plant cells, and therefore, microcell mediated chromosome transfer (MMCT) has received little attention in plants. Recently, we found that the herbicide Amiprophos-methyl (APM) induces micronuclei at a high frequency in cell suspension cultures of N.plumbaginifolia (1-3). In this paper we describe preliminary results of the development of microprotoplast mediated chromosome transfer, aiming at the stable transfer of a single donor chromosome into a recipient genome. This would open up ways to construct monochromosomal hybrid panels for gene mapping and expression studies.

Materials and methods

A fast growing cell suspension culture of N.plumbaginifolia, was subcultured every fourth day. One day after subculture, cells were synchronized by the addition of 10 mM of hydroxyurea (HU), which was removed after 24 h of incubation. Together with fresh medium, 20 ppm APM was added to induce micronucleation. After 24 h, protoplasts were isolated as described previously (4), and incubated with 10 ppm cytochalasin B for 15 min, before the protoplasts were loaded onto a preformed Percoll gradient (0.4 M of mannitol in a Percoll solution, 100,000 g, 1 h), and centrifuged at 100,000 g during 1 h. Fractions of this gradient were collected, and the cellular DNA content was determined by flow cytometry, after staining with 10 mg/l ethidiumbromide in the presence of 0.02% Triton X-100.

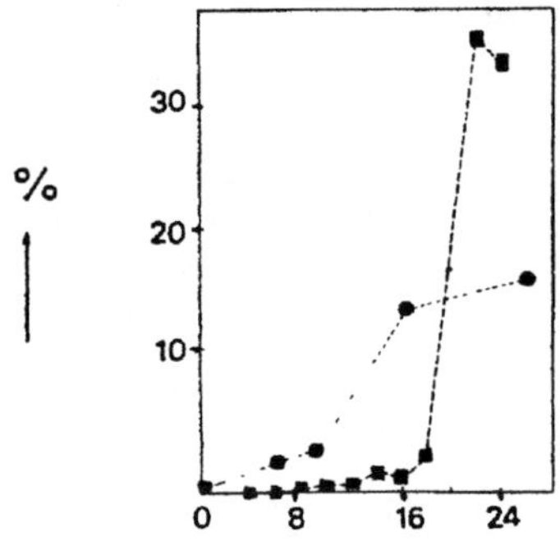

Fig. 1: Influence of synchronization on the occurrence of micronuclei: ● without HU; ■ with 10 mM HU, 24 h.

Results and discussion

1. Micronucleation.

The synchronization of the cell suspension culture results in a better control over the development of micronuclei. When compared with the addition of APM without HU pretreatment, two effects were obvious: the strong synchronous rise of the number of micronuceated cells, and the higher frequency of micronucleated cells (35 versus 25%) (Fig. 1). Another advantage is the fact that up to 80% of the cells have the same chromosome number at the peak of micronucleation.

2. Microcell isolation.

After protoplast isolation, the protoplasts still contained micronuclei. This was demonstrated by electroporation (750 V/cm, 100 microsec.) in the presence of 1 mg/l ethidium-homodimer (Mol. Probes).

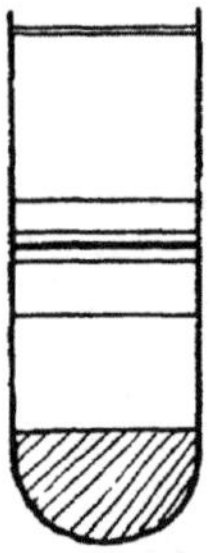

1 Fig. 2: Percoll gra
2 dient after proto-
3 plast fractionation
4 100,000g, 1 h): in-
5 dicated are the vi-
6 sible bands.
7
8
Solid Percoll

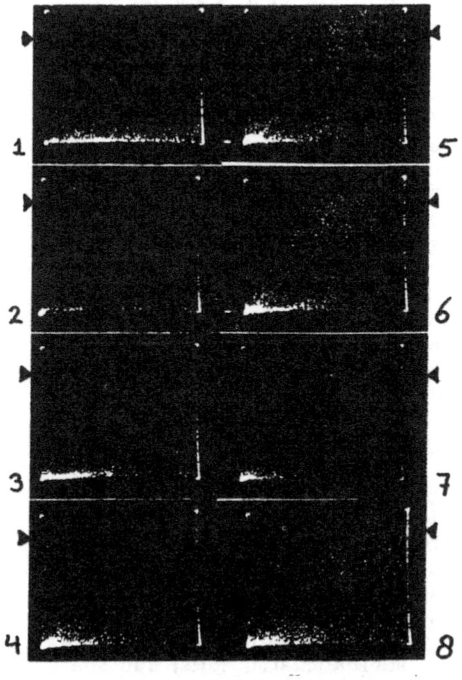

Fig. 3: Two parameter flowcytometer analysis of 0.5 ml gradient fractions: 1 top; 8: bottom. Vertical: DNA content, arrow indicates 4C (G1) level; horizontal: size in arbitrary units.

After this treatment, the nuclei could be visualized by fluorescence microscopy (data not shown). The Percoll gradient (Fig. 2) was analyzed in fractions of 0.5 ml by flow cytometry. Red fluorescence (stoichiometric with DNA content) is on the vertical axis, whereas the size of the protoplast is on the horizontal axis. It is obvious from these diagrams, that fraction 1 and 2 contain only vacuoplasts, and that fractions 5 to 8 contain more protoplasts with an intact genome. Fractions 3 and 4 contain up to 20% of microprotoplasts. These observations are consistent with the microdensitometric analysis of the the DNA content of the fractionated protoplasts in the different fractions of the gradient (data not shown). The transfer of chromosomes by fusion of microprotoplasts with recipient protoplasts is currently under investigation.

Conclusions

The development of a technique for the mass isolation of microprotoplasts, opens up the way for the application of MMCT in plants.

Acknowledgements

Part of this research was supported by funds of the "Biotechnology Action Programme" of the CEC, contract BAP-0083-NL.

References

1. H.A. Verhoeven, A.M.M de Laat, K. Sree Ramulu, P. Dijkhuis, and L. van Vloten-Doting, in: CEC meeting, Louvain-la-Neuve (1987), pp 119-120.
2. A.M.M. de Laat, H.A. Verhoeven, K. Sree Ramulu and P. Dijkhuis, Planta, 1987. (in press).
3. K. Sree Ramulu, H.A. Verhoeven and P. Dijkhuis, Theor. Appl. Genet., 1988 (in press).
4. A.M.M. de Laat and J. Blaas, Pl. Sci. 50: 161-169, 1987.

DIRECT CELL TO CELL TRANSFER OF MITOCHONDRIA BY MICROINJECTION.

Harrie A. Verhoeven and Jan Blaas.
Research Institute Ital, P.O.Box 48, 6700 AA Wageningen, the Netherlands.

Introduction.

For the injection of plant cells and protoplasts, a number of systems have been published, which employ small diameter glass capillaries for DNA injections. No systems for the handling of large organelles (chloroplasts, mitochondria, chromosomes), have been available. In this paper, we present a microinjection technique which offers the opportunity to take the desired organelles out of a donor protoplast and inject them directly into a recipient.

The injection system

The injection system is based on the large diameter bevelled glass capillaries, described before (1). To achieve suction with those needles, the tip of the needle is loaded with a small drop of mercury. The mercury can be forced into the tip of the needle by increasing the air pressure in the injection system, and decreasing the air pressure will result in the creation of an underpressure in the opening of the needle. In this way, suction and injection of very small volumes (from less than 10 fl up to 1 pl) can be achieved by varying the air pressure over several bar (Fig. 1).

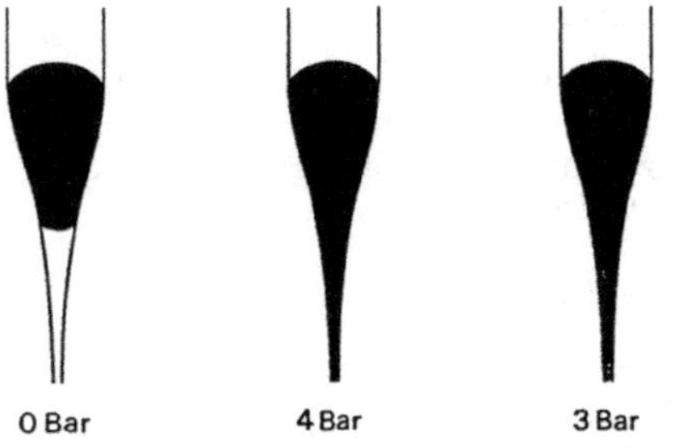

Fig. 1: Illustration of the injection system, showing the mercury (black) at different pressures.

Materials and methods

Protoplasts were isolated from a suspension culture of N.plumbaginifolia. Donor protoplasts were evacuolated on a Percoll gradient, and stained with 1 mg/l dihexyloxocarbocyanine (Serva), to stain the mitochondria (Fig. 2). The recipient protoplasts were immobilized as described previously (1), cultured for one or two days, and then the donor material was immobilized in the same Petri dish. Fluorescently labelled cytoplasm was taken out of the donor protoplasts, and injected into the recipient protoplasts.

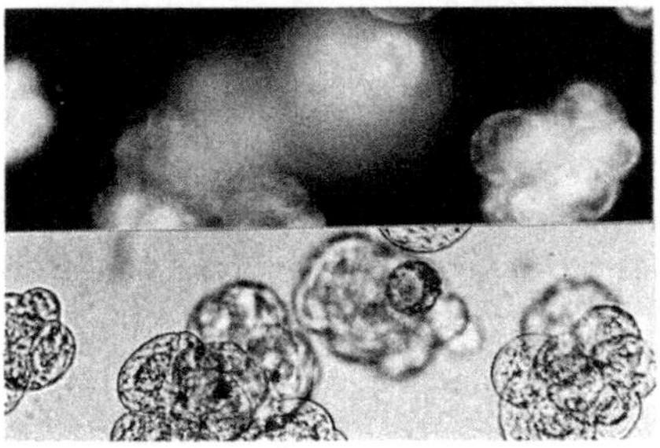

Fig. 2: Microcallus, derived from a stained protoplast, demonstrating the viability of stained protoplasts.

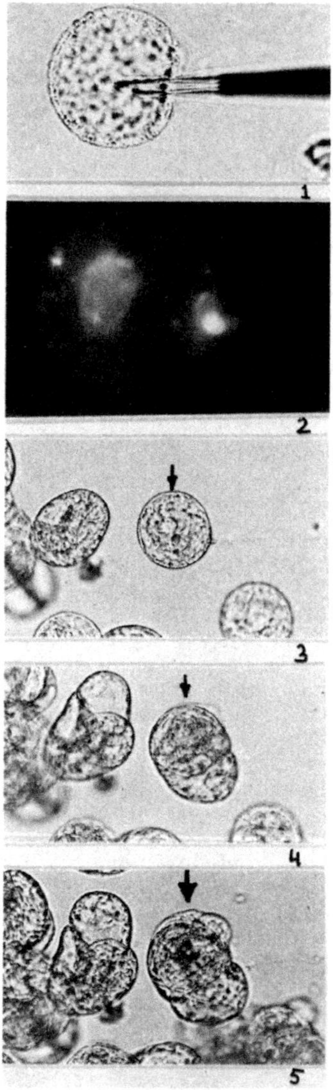

Fig 3: Sequence of micrographs, made during the injection procedure.
1: collection of donor material; 2: fluorescent donor cytoplasm injected into recipient cell; 3: 2 days; 4: 5 days: first division; 5: 10 days: microcallus. The arrow indicates the injected protoplast derived cell.

Results and Discussion

The series of micrographs, presented in this paper (Fig. 3), demonstrates the possibility to transfer cytoplasm, containing organelles, directly from donor to recipient. The survival rate of 20 pooled experiments, is about 30%. In individual experiments, percentages of up to 60% have been achieved. The visual preselection of a suitable recipient, after two days of culture, is crucial for this high recovery. It proved to be impossible to inject with those large needles (diameter > 5 micron) into freshly isolated protoplasts: a cell wall is necessary to protect the protoplast when the needle is removed. Calculations have shown that with those needles, about 50% of the cytoplasm of a 40 micron diameter protoplast can be transferred (0.8 pl).

The large opening of the needles and the high air pressure make the system less sensitive to blocking. Recent experiments (data not shown) have shown that this injection technique is also suitable for the injection of isolated,suspended organelles.

The improved control over the injection of small volumes, and the decreased sensitivity for blocking, make this system ideal for the introduction of isolated organelles, e.g after *in vitro* transformation.

Acknowledgements

Part of this project was supported by funds of the 'Biotechnology Action Programme' of the CEC, contract BAP-0083-NL.

References

1. A.M.M. de Laat and J.Blaas: Pl. Sci. 50: 161-169, 1987.

CLASSIFICATION AND SORTING OF PLANT CHROMOSOMES USING FLOW CYTOMETRY

J. Conia[1]*, C. Bergounioux[1], C. Perennes[1], P. Muller[2], S. Brown[2] and P. Gadal[1]

[1] Physiologie Végétale Moléculaire, CNRS UA 1128, Université de Paris-Sud, Bât. 430, 91405 Orsay Cedex, France.

[2] Service de Cytométrie, CNRS, 91190 Gif-sur-Yvette, France.

Flow karyotyping and chromosome sorting has been applied to the solanaceous Petunia hybrida (2n=14). Substantial synchronization of the earlier cell divisions and readily-disrupted membranes made protoplast-derived cells suitable material for chromosome isolation (Conia et al). The medium in which the cells were disrupted and the chromosomes dispersed was derived from the protoplast culture medium or was the $MgSO_4$ medium (van den Engh et al). The chromosomes were stained with Hoechst 33342 DNA dye and analysed under UV light.

On a histogram of fluorescence intensities, three peaks of metaphase chromosomes were observable in the Petunia hybrida flow karyotype (peak A on the logarithmic scale fig.I; peaks A1-3 on the linear scale fig.II), with only chromosome number I forming a distinct peak (A1). A duplication of the flow karyotype was provoked by monochromatid chromosomes (peak B fig.I). Reproducible peaks corresponding to re-association of various chromosomes were obtained (peaks C1-3 fig.I). Heterogeneous cellular debris detracted from chromosome resolution (peak D fig.I). Most of this debris was eliminated from analysis by selecting chromosomes through a window on Forward-angle light scatter versus Fluorescence intensity. An accumulation of dye-binding debris and appearance of micronuclei resulted from chromosome breakage and decondensation. On the Integral-fluorescence versus Peak-fluorescence representation, well condensed chromosomes were identified on a linear relationship while other fluorochrome binding debris was dispersed.

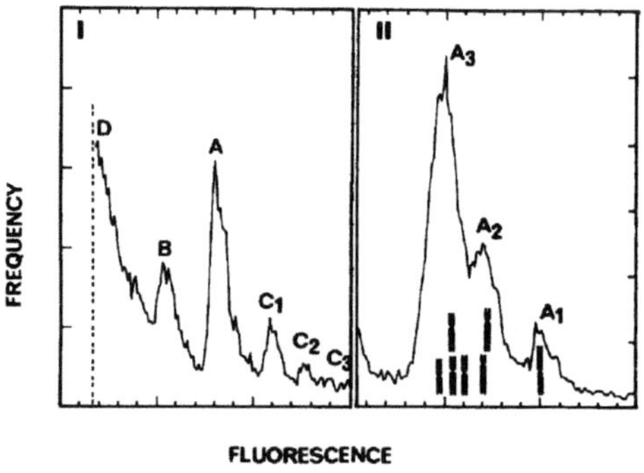

In order to determine whether flow cytometry and chromosome sorting can be applied theoretically to other plant systems, a flow karyotype computer model based on the relative chromosome lengths has been introduced (Conia and Muller). Flow karyotype models have been generated for different plant species. The theoretical flow karyotypes show that it would be possible to separate chromosomes from Arabidopsis thaliana, Zea mays, whereas the flow karyotype models of Secale cereale, Hordeum vulgare and Nicotiana plumbaginifolia result in broad peaks obviously not conducive to single chromosome sorting. Experimental histograms and sorted fractions agreed with the theoretical flow karyotype of Petunia hybrida when CV= 2.5%. When the CV is higher than 2.5%, no single chromosome is distinct as a discontinuous peak. The theoretical chromosome distribution shows that a reduction of CV to about 1.5% would resolve 6 of the 9 chromosomes of Daucus carota.

1. J. Conia, C. Bergounioux, C. Perennes, P. Muller, S. Brown and P. Gadal, Cytometry (1987) in press.
2. J. Conia and P. Muller (submitted).
3. G. J. van den Engh, B. J. Trask, J. W. Gray, R. G. Langlois, L.-C. Yu, Cytometry 5 (1984) 108-117.

APPLICATION OF FLOW CYTOMETRY FOR THE ISOLATION OF METAPHASE CHROMOSOMES FROM PROTOPLASTS OF *NICOTIANA PLUMBAGINIFOLIA*.

Henry C.P.M. van der Valk and Harrie A. Verhoeven
Research Institute Ital, P.O. Box 48, 6700 AA Wageningen, The Netherlands.

Introduction

Genetic manipulation of plant cells employs an array of methods and techniques for the transfer of genetic elements and complexes, varying from well defined DNA fragments incorporated into plant DNA vectors to complete nuclei by means of cell fusion. In between the two extremes, optimal transfer of e.g. naked DNA, chromosomes and cell organelles requires the further development of cellular engineering followed by investigations into integration and expression of the transferred material in the recipient cell. Moreover, chromosome transplantation may be a technique to map and identify plant genes.

Chromosome-mediated gene transfer implies the isolation and purification of intact chromosomes. Highly synchronized cell suspension cultures are excellent sources of large quantities of metaphase chromosomes. To establish optimal conditions for cell cycle synchronization of a cell suspension of *N. plumbaginifolia*, flow cytometric analysis of cellular DNA content was applied. Large quantities of chromosomes were sorted on the basis of their DNA content.

Procedure

A cell suspension culture of *N. plumbaginifolia* (Doba) was cultured at 26 °C in the dark and subcultured at weekly intervals. To synchronize the population, cells were incubated with increasing concentrations (0-10 mM) of hydroxy urea (HU) or aphidicolin (APH) for 24 h, to block the cells in S-phase. Subsequently, cells were washed free from HU or APH and arrested in metaphase by the addition of 10 ppm amiprophosmethyl (APM), immediately or 6-8 h after washing.

DNA histograms were determined and mitotic indices were calculated from lactopropionic orcein-stained squash slides of the cells. Also the occurrence of micronuclei was determined from the slides.

Chromosomes were isolated from synchronized cultures by flow cytometric sorting (De Laat et al. 1987).

Results

Cells incubated with 10 mM HU for 24 h were arrested in S-phase (Fig. 1). Lower concentrations (1 and 2.5 mM) were not sufficient to inhbiit DNA synthesis. Higher concentrations resulted in slower recovery of the HU block after washing the cells free from HU.

After washing, and in the presence of 10 ppm APM, cells continued the cell cycle until they reached metaphase. At this point, passage to anaphase was completely prevented by the action of APM on the spindle, causing the scattering of metaphase chromosomes throughout the cell.

After prolonged exposure to APM, the chromosomes formed small groups, decondensed and formed micronuclei. Because micronuclei interfere with the isolation of metaphase chromosomes, the effect of delayed addition of APM after removal of HU

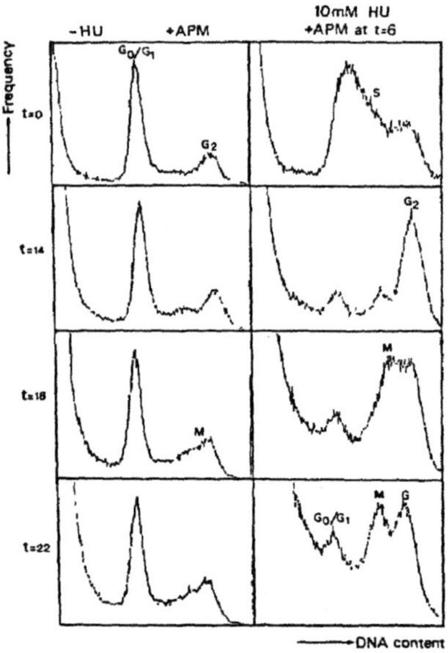

FIGURE 1. Frequency distribution of DNA content of protoplasts from HU and APM treated suspension cells, at different times after washing out HU (M, cells in mitosis; S, cells in S-phase).

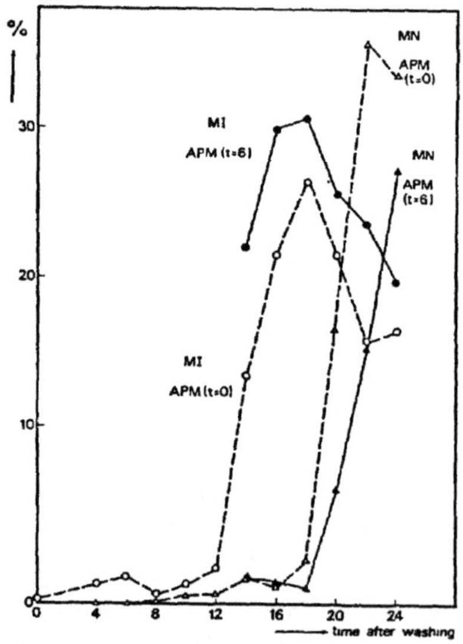

FIGURE 2. Time course of mitotic index (MT) and occurrence of micronuclei (MN) in suspension cells during synchronization.

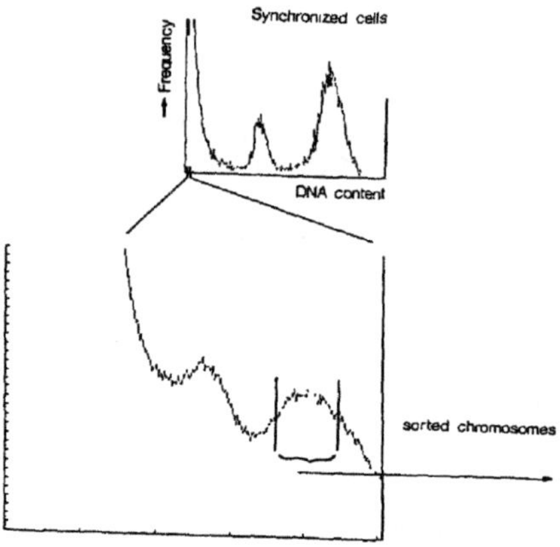

FIGURE 3. Frequency distribution of DNA content of protoplasts (top) and chromosomes (bottom) from APH and APM treated suspension cells.

was investigated. It appeared that APM addition could be delayed by 6-8 h, without affecting the mitotic index, whereas the amount of micronucleated cells was substantially reduced (Fig. 2).

For chromosome isolation, protoplasts were isolated 17 h after washing out HU or APH, and lysed in chromosome buffer for flow cytometric sorting. Large quantities of chromosomes were sorted on the basis of their DNA content (Fig. 3).

With this technique, it is possible to isolate large numbers of metaphase chromosomes for transfer into recipient cells.

Acknowledgements

This research was supported by funds of the "Biotechnology Action Programme" of the CEC, contract BAP-0083-NL.

References

De Laat AMM, Verhoeven HA, Sree Ramulu K and Dijkhuis P: Planta 1987, in press.

ELECTROPORATION FOR DNA DELIVERY TO PLANT PROTOPLASTS

George W. Bates, William Piastuch, Charles D. Riggs,* and Dietmar Rabussay[1]
Department of Biological Science, Florida State University, Tallahassee, FL, USA, and [1]Bethesda Research Laboratories/Life Technologies, Inc., Gaithersburg, MD, USA. *current address: Department of Biology, University of California at San Diego, La Jolla, CA, USA.

1. INTRODUCTION

Electroporation uses high-voltage electric-field pulses to make cells permeable. This technique is effective with a wide variety of species and can be used to induce the uptake of almost any substance to which cells are otherwise impermeable. Among plant biologists, the recent demonstrations of electroporation-mediated DNA uptake (1-4) and transformation (5, 6) of monocots and other species resistant to Agrobacterium (7) have generated great excitement. Electroporation is also being applied to the uptake of plant viruses (8-11) and reporter dyes (12, 13), and the successful electroporation of intact cells (11, 14, 15) suggests a possible approach for the transformation of species that are difficult to regenerate from protoplasts.

Although the subject of this paper is the use of electroporation for introducing DNA into plant protoplasts, we will focus on experimental approaches. Three main protocols use electroporation for DNA delivery. One involves the application of short, very-high-voltage pulses (5-200 µs, 2-10 kV/cm), which can be either square waves or exponentially decaying capacitive discharges, to protoplasts suspended in a low-salt medium (8, 16-18). The second approach, which we presently use, employs longer pulses (1-50 ms) of lower voltage (200-800 V/cm) delivered by capacitive discharges to protoplasts in a high-salt medium such as phosphate-buffered saline (1, 2, 19). Some groups also use intermediate conditions (12, 14). The third experimental approach involves a combination of chemical treatments (PEG plus calcium or magnesium) and electroporation (20, 21).

There have been some detailed recent studies of direct gene transfer both by electroporation and by chemical treatments (19-22). However, a thorough comparison of the various experimental approaches has yet to be done, and differences in plasmid construction, protoplast source, and data presentation make it difficult to compare the results of different laboratories. Thus, it is not clear whether one method is more effective than the others or whether different approaches are better suited to different species or for different applications. In our own work, we have examined both transient and stable expression of DNA delivered to carrot and tobacco protoplasts by application of low-voltage, long pulses. To a limited extent, we have compared this approach with the use of high-voltage, short pulses. Here, we will give an overview of our results and where possible compare them with those of other groups. We hope that this approach will highlight those areas most needing further work so that the uncertainties about the effectiveness of different protocols can be resolved.

2. TRANSIENT EXPRESSION IN CARROT PROTOPLASTS

The transient expression of genes introduced into protoplasts can provide a rapid method for assessing the effects of various experimental parameters on DNA delivery (1, 2, 14, 19). Transient expression can also be useful for assaying the effectiveness of promoters and gene constructs without interference from the position effects that complicate stable expression (3, 4, 18, 23-25). Four different reporter genes are available for measuring transient expression in plant protoplasts: chloramphenicol acetyltransferase or CAT, neomycin phosphotransferase or NPT, luciferase, and glucuronidase. We have compared CAT and NPT transient expression and have found the CAT assay simpler and easier to quantify.

In the experiments described below, protoplasts isolated from a carrot suspension culture were washed into HBS (Hepes Buffered Saline; 150 mM KCL, 4 mM $CaCl_2$, 0.2 M mannitol, 10 mM HEPES, pH 7.2) and resuspended at a density of 2×10^6 protoplasts/ml in HBS containing 20 μg/ml supercoiled plasmid DNA. After poration, the protoplasts were held for 3 min to allow for DNA uptake and then were diluted 10 fold into culture medium (19).

The plasmid used, pCaMVCAT (the gift of Drs. Fromm and Walbot, Stanford University), contains the CAT coding region under the control of the CaMV 35S promoter. Twenty-four to 48 hr after electroporation, the protoplasts were assayed for CAT activity as described by Fromm et al. (1). Quantitative estimates of CAT expression were obtained from the percentage of the total radioactivity in ^{14}C-chloramphenicol that was converted to acetylated chloramphenicol by cell extracts.

2.1. Pulse length and voltage

The most important experimental parameters in electroporation are pulse duration and voltage. When individual pulses of 250 V/cm, ranging from 8 to 50 ms in duration, were delivered to the carrot protoplasts, low levels of CAT activity were detected for pulses of 30 ms or longer (Fig. 1, closed squares). However, increasing the voltage to 750 V/cm resulted in much greater CAT expression, with a peak in activity in response to an 8-ms pulse (Fig. 1, closed circles). Cell viability was determined 24 hr after electroporation by the Evans Blue exclusion test. As seen in Fig. 1, high levels of CAT expression were associated with a reduction in cell viability of 50% or greater.

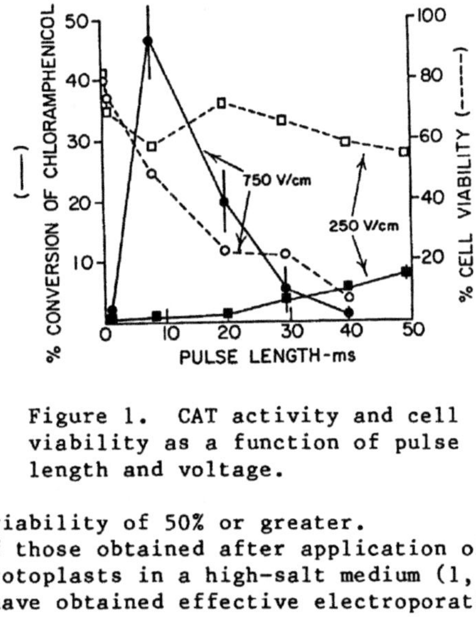

Figure 1. CAT activity and cell viability as a function of pulse length and voltage.

Results of this sort are typical of those obtained after application of a single, long, low-voltage pulse to protoplasts in a high-salt medium (1, 2). On the other hand, some groups have obtained effective electroporation with better retention of protoplast viability by applying three to ten short, high-voltage pulses to protoplasts in a low-salt medium (8, 14). This will be discussed further below (section 2.4).

Time courses indicated that CAT activity can be readily detected in the protoplasts as little as 6 hr after electroporation. Maximal activity was observed at 24 to 36 hours, after which CAT activity rapidly declined (although some activity was still obtained two weeks after electroporation).

2.2. Importance of temperature

The work of Potter et al. (26) indicates that the electroporation of animal cells is more effective when carried out on ice, as predicted by arguments that pulse-induced pores in the membrane should remain open longer in chilled cells (27). For these reasons, we and others (1, 16, 17) performed our initial protoplast electroporations with prechilled protoplasts and cold electroporation media. Subsequently we found maximal CAT expression when electroporation was carried out at room temperature. For example an 8-ms pulse (750 V/cm) resulted in 57% CAT conversion when applied at room temperature but only 16% when applied in ice-cold media. In all cases tested, pulses delivered at room temperature resulted in greater CAT expression except when the energy of the pulse was so great as to cause massive heat-induced cell death. Holding the protoplasts on ice in the presence of DNA for 10 min after electroporation did not improve transient expression.

Several groups include a heat shock step in their electroporation protocol (7, 20, 25), but heat shock did not improve transient expression in our system. Heat shock may be beneficial for some species and not others, as is suggested by the work of Negutriu et al. (21).

2.3. DNA form and concentration

When comparing supercoiled and linearized forms of plasmid DNA, we found that the former gave better transient expression. Preliminary experiments indicate that the difference may be related to the restriction site chosen for linearizing the plasmid. Thus plasmids cut at sites near the CAT gene gave lower expression than those cut at more distant sites. Two studies of the stable transformation of plant protoplasts after DNA uptake by combined chemical and electroporation treatments indicate that linear DNA is more effective than the supercoiled form (20, 21). Whether this difference between their results and ours is due to an effect of the chemicals used to assist in DNA uptake or to a difference between transient expression and stable integration is unclear.

We also found that doubling the DNA concentration approximately doubled subsequent CAT expression. This result verifies that of Fromm et al. (1). Carrier DNA is also used by many groups, and we found that adding 50 μg/ml heat-denatured salmon-sperm DNA doubled CAT expression. In systems giving low expression these could become important variables.

2.4. Medium composition

When protoplasts were electroporated in HBS lacking Ca^{2+}, we observed no transient expression and a complete loss of cell viability. However, expression was obtained when protoplasts were electroporated in a medium containing only mannitol (0.4 M) with no added salts. With this medium, the pulse length had to be shortened (< 1 ms) and the voltage increased to at least 2000 V/cm. These conditions are closer to those used for electrofusion. However, our preliminary results suggest that transient expression is greater when high-salt media and low-voltage, long pulses are used (HBS; 750 V/cm, 8 ms). Interestingly, no expression was obtained when protoplasts in salt-free media were porated with the lower-voltage, long pulses that are effective for protoplasts suspended in HBS. This interaction of the composition of the medium with pulse length and voltage suggests that the biophysical basis of electroporation is more complex than has previously been supposed (see 27 for review).

Several groups have been successful in obtaining transient and stable expression of DNA after electroporation in low-salt media such as CPW salts or plain mannitol (7, 14, 17, 18). Often a series of short, high-voltage pulses are applied. Whether the degree of DNA uptake is as great as when a single, long, low-voltage pulse is applied to protoplasts in a high salt medium is unclear. However, work on electroporation-induced virus uptake suggests that it may not be as great. Nishiguchi et al. (8) were able to induce the uptake of TMV RNA but not TMV particles using a series of short, high-voltage pulses and a low-salt medium. However, they (9) and Okada et al. (10) have demonstrated the uptake of both TMV RNA and TMV particles after the application of long, low-voltage pulses. Apparently long, low-voltage pulses and a high-salt medium result in larger transmembrane pores. Of course, larger pores also result in lower protoplast viability.

3. STABLE TRANSFORMATION BY ELECTROPORATION

Because they are easy to regenerate, we have used tobacco mesophyll protoplasts for the recovery of stable transformants (17). The plasmid used in these experiments was pMON200 (a gift of Drs. Fraley and Horsch, Monsanto Co.), which contains the selectable marker gene NPT under the control of the nopaline synthase promoter. Expression of this gene confers kanamycin resistance on plant cells.

3.1. Electroporations and recovery of transformants

For ease of discussion, we have reversed the order in which the stable transformation and transient expression experiments are presented. The former were actually performed before the latter. The electroporation protocol for these stable transformations used high-voltage, short pulses (2000 V/cm, 250 µs) delivered to protoplasts in ice-cold, salt-free media (0.5 M mannitol). The transforming plasmid, pMON200, was linearized at a unique EcoRI site close to the 3' end of the NPT gene. Based on our transient expression experiments, this protocol was probably not optimal. After electroporation, stable transformants were recovered by growth in agarose-bead cultures in the presence of 100 µg/ml kanamycin. The transformation efficiency, as judged from the number of kanamycin-resistant calli recovered 4-5 weeks after electroporation, exceeded 2×10^{-4} in many experiments. This value, which was calculated from the number of protoplasts present in each sample before electroporation, compares favorably with the absolute transformation frequencies reported elsewhere for electroporation (16), chemical treatments (28), and combination techniques (21). We are now using the improved protocol that has emerged from our work on transient expression to repeat these stable transformations. It will be of interest to see whether the new protocol improves the efficiency with which stable transformants are recovered and whether it alters the pattern of plasmid integration described below.

3.2. Analysis of plasmid integration

Kanamycin-resistant calli were regenerated into plants, and genomic DNA isolated from leaves of these plants was analyzed by Southern blotting. A 1.3-kb EcoRI/BamHI fragment of pMON200 that includes the NPT gene was used as a probe. Untransformed tobacco plants gave no signal, whereas all of the kanamycin-resistant plants recovered following electroporation that were tested showed hybridization to the probe. Analysis of the restriction patterns obtained indicated that the free ends of the plasmid were altered during integration in all cases. Some of the transformants showed more substantial internal rearrangments of the plasmid as well. In addition, one plant showed multiple bands of hybridization, suggesting the integration of more than one plasmid copy. All of the independent transformants displayed different integration patterns, suggesting random integration.

Plasmid rearrangements have been observed for all the transformed plants regenerated after direct gene transfer (e.g. see 5, 16, 20, 29), contrasting with the results obtained with *Agrobacterium*, where T-DNA is usually integrated intact. However, in both direct-DNA and *Agrobacterium*-

mediated transformations, the site of DNA integration appears to be random, and tandem insertions can occur.

3.3. Inheritance of kanamycin resistance

Of the 53 transformed plants we regenerated, all but two were morphologically normal; most proved to be fertile. After self fertilization, seed from 19 of the plants was tested for the inheritance of kanamycin resistance. Most of the transformants displayed inheritance ratios not statistically different from 3:1 (resistant:sensitive). These results are consistent with the inheritance of a single dominant gene. One plant had a ratio of nearly 2:1, perhaps suggesting a plasmid integration event in a vital house-keeping gene. Several plants displayed ratios consistent with the presence of two or more independent, functional copies of the plasmid. Again, this pattern of integration is similar to what has been observed before for direct gene transfer (30) as well as for *Agrobacterium*-mediated transformation.

4. CONCLUSIONS AND PROSPECTS

Although much work remains to be done before the biophysical basis of electroporation can be understood and optimized for DNA delivery, the effectiveness of the technique has already been clearly demonstrated. Electroporation is well suited for transient-expression assays, and the recent demonstrations of the electroporation of intact cells may allow the transformation of species that are difficult to regenerate from protoplasts. However, in terms of stable transformation, the greater plasmid rearrangements found in plants recovered after direct gene transfers than in those undergoing *Agrobacterium*-mediated transformation is a difficulty that deserves further investigation.

REFERENCES

1. Fromm M, Taylor LP, Walbot V: Expression of genes transferred into monocot and dicot plant cells by electroporation. Proc. Natl. Acad. Sci. USA 82:5824-5828, 1985.
2. Hauptmann RM, Ozias-Akins P, Vasil V, Tabaeizadeh Z, Rogers SG, Horsch RB, Vasil IK, Fraley RT: Plant Cell Rep. 6:265-270, 1987.
3. Junker B, Zimny J, Luhrs R, Lorz H: Transient expression of chimaeric genes in dividing and non-dividing cereal protoplasts after PEG-induced DNA uptake. Plant Cell Rep. 6:329-332, 1987.
4. Ou-Lee TM, Turgeon R, Wu R: Expression of a foreign gene linked to either a plant-virus or a *Drosophila* promoter, after electroporation of protoplasts of rice, wheat, and sorghum. Proc. Natl. Acad. Sci. USA 83:6815-6819, 1986.
5. Fromm ME, Taylor LP, Walbot V: Stable transformation of maize after gene transfer by electroporation. Nature 319:791-793, 1986.
6. Rhodes C: Personal communication (Sandoz Crop Protection Corp., Palo Alto, CA, USA).
7. Guerche P, Charbonnier M, Houanin L, Tourneur C, Paszkowski J, Pelletier G: Direct gene transfer by electroporation in *Brassica napus*. Plant Sci. 52:111-116, 1987.
8. Nishiguchi M, Langridge WHR, Szalay AA, Zaitlin M: Electroporation-mediated infection of tobacco leaf protoplasts with tobacco mosaic virus RNA and cucumber mosaic virus RNA. Plant Cell Rep. 5:57-60, 1986.
9. Nishiguchi M, Sato T, Motoyoshi F: An improved method for electroporation in plant protoplasts: infection of tobacco protoplasts by tobacco mosaic virus particles. Plant Cell Rep. 6:90-93, 1987.

10. Okada K, Nagata T, Takebe I: Introduction of functional RNA into plant protoplasts by electroporation. Plant Cell Physiol. 27:619-626, 1986.
11. Morikawa H, Iida A, Matsui C, Ikegami M, Yamada Y: Gene transfer into intact plant cells by electroinjection through cells walls and membranes. Gene 41:121-124, 1986.
12. Gilroy S, Hughes WA, Trewavas AJ: The measurement of intracellular calcium levels in protoplasts from higher plant cells. FEBS Letts. 199:217-221, 1986.
13. Cutler AJ, Saleem M: Permeabilizing soybean protoplasts to macromolecules using electroporation and hypotonic shock. Plant Physiol. 83:24-28, 1987.
14. Lindsey K, Jones MGK: Transient expression in electroporated protoplasts and intact cells of sugar beet. Plant Mol. Biol. (in press).
15. Mishra KP, Joshua DC, Bhatia CR: In vitro electroporation of tobacco pollen. Plant Sci. 52:135-139, 1987.
16. Langridge WHR, Li BJ, Szalay AA: Electric field mediated stable transformation of carrot protoplasts. Plant Cell Rep. 4:355-359, 1985.
17. Riggs CD, Bates GW: Stable transformation of tobacco by electroporation: evidence for plasmid concatenation. Proc. Natl. Acad. Sci. USA 83:5602-5606, 1986.
18. Koncz C, Olsson O, Langridge WHR, Schell J, Szalay AA: Expression and assembly of functional bacterial luciferase in plants. Proc. Natl. Acad. Sci. USA 84:131-135, 1987.
19. Rabussay D, Uher L, Bates G, Piastuch W: Electroporation of mammalian and plant cells. Focus 9(3):1-5, 1987.
20. Shillito RD, Saul MW, Paszkowski J, Muller M, Potrykus I: High efficiency direct gene transfer to plants. Biotechnology 3:1099-1103, 1985.
21. Negrutiu I, Shillito R, Potrykus I, Biasini G, Sala F: Hybrid genes in the analysis of transformation conditions. Plant Mol. Biol. 8:363-373, 1987.
22. Chu G, Hayakawa H, Berg P: Electroporation for the efficient transfection of mammalian cells with DNA. Nuc. Acids. Res. 15:1311-1326, 1987.
23. Ow DW, Wood KV, Deluca M, de Wet JR, Helinski DR, Howell SH: Transient and stable expression of the firefly luciferase gene in plant cells and transgenic plants. Science 234:856-859, 1986.
24. Werr W, Lorz H: Transient gene expression in a Gramineae cell line. Mol. Gen. Genet. 202:471-475, 1986.
25. Boston SR, Becwar MR, Ryan RD, Goldsbrough PB, Larkins BA, Hodges TK: Expression of heterologous promoters in electroporated carrot protoplasts. Plant Physiol. 83:742-746, 1987.
26. Potter H, Weir L, Leder P: Enhancer-dependent expression of human k-immunoglobulin genes introduced into mouse pre-B lymphocytes by electroporation. Proc. Natl. Acad. Sci. USA 81:7161-7165, 1984.
27. Bates GW, Saunders JA, Sowers AE: Electrofusion: principles and applications. In: Cell Fusion. A. E. Sowers (ed). Plenum Press, New York, pp. 367-395, 1987.
28. Lorz H, Baker B, Schell J: Gene transfer to cereal cells mediated by protoplast transformation. Mol. Gen. Genet. 199:178-182, 1985.
29. Czernilofsky AP, Hain R, Herrera-Estrella L, Lorz H, Goyvaerts E, Baker BJ, Schell J: Fate of selectable marker DNA integrated into the genome of Nicotiana tabacum. DNA 5:101-113, 1986.
30. Potrykus I, Paszkowski J, Saul M, Petruska J, Shillito RD: Molecular and general genetics of a hybrid foreign gene introduced into tobacco by direct gene transfer. Mol. Gen. Genet. 199:169-177, 1985.

ELECTROMANIPULATION OF PROTOPLASTS

MJ Tempelaar[1] & MGK JONES[2]

[1]Dept of Genetics, Biological Center, Univ. of Groningen, Postbox 14, 9751 AA, Haren, the Netherlands.

[2]AFRC Institute of Arable Crops Research, Biochemistry Department, Rothamsted Experimental Station, Harpenden, Herts. AL5 2JQ, U.K.

Report of the workshop.

In this workshop, various aspects of electric field effects on protoplasts have been reviewed and salient points from several poster contributions were highlighted and contrasted.

On the issue of electrofusion, it can now be stated, that this technique has matured since the last Protoplast Symposium into an accepted practice.
For preparative purposes, it is being used with a variety of electrode designs, most of them permitting work with large numbers of protoplasts on a scale comparable to chemical methods. In a reverse trend, attention has been roused by electrical procedures on a similar scale as pioneering biophysicists studies in the early eighties [4].

Basic conditions of preparative electrofusion have clearly been established: protoplasts are aggregated by high frequency (1 MHz or above) alternating fields in media of low conductivity. Fusion is done by supercritical (exceeding membrane breakdown voltage) fields, most often consisting of rectangular pulses of short duration (microseconds). It has been pointed out, that addition of Ca ions can enhance fusion frequency and production of 1:1 fusionproducts. This effect is more pronounced with protoplasts that have poor fusogenic properties. The procedures have succesfully been applied on a variety of species, and as an example the generation of somatic hybrids of potato, which could not easily have been done with conventional chemical fusion, has been presented [2] to demonstrate the ease of adjustment of electrical parameters and the degree of control of the electric field manipulations.

Although fusion yields can be high enough for obtaining hybrids without any selection, as in the case of potato, attention has been drawn to the possibility of further modulating and directing the fusion process. Combining aspecific (or specific) chemical aggregation of protoplasts, followed by electric fusion [1,3] have been illustrated and discussed.
Such approaches have practical implications for performing fusion with simpler equipment, in more conductive media or with

higher specificity.

Other findings with electrofusion may have a bearing on electric field mediated uptake of molecules. Fusion frequencies can be analyzed directly after initiation of the fusion process in sufficiently extensive series to determine fusogenic properties. When related to pulse duration and voltage, it now appears that a basic biophysical parameter (membrane breakdown voltage) can be deduced from the data [10]. Examples have been shown suggesting that patterns in transient expression do confirm these biophysical extrapolations from fusion results [8]. In conjunction with dye-uptake studies [9,10], the expectation is that the basics of uptake of molecules into protoplasts under the influence of electric fields can be clarified. This would have the benificial practical effect of arriving at more unified procedures for this process, which, in contrast to electrofusion, is still being done under a variety of conditions such as pulsshapes/durations and media [for examples see abstracts on applications]. Experiments have been suggested and variables mentioned for electric field mediated DNA uptake: direct monitoring might be done by in-situ hybridisation, fluidity of membranes and proplast properties other than size may be important, variability in transient expression, cellcycle effects, tissue specificity are variables to be reckoned with, cf [5-10].

Even without the basic conditions sorted out however, it is clear that electric field mediated DNA uptake can be in an equal order of magnitude as chemical procedures [12,13]. Comparisons of systems as presented must take in account the best conditions for both chemical and electrical procedures for fusion or transformation. This may present a problem as conditions are often not fully optimized to such an extent to allow a fair comparison.

Finally it was noted, that transformation of intact cells, in contrast to protoplasts, still seems to be a controversial issue.

Despite the debate on optimum conditions, comparison of systems and alternative steps in procedures, electric field mediated methods have succesfully been applied with protoplasts to generate hybrids and transformants. As examples, studies on total and partial hybrids, on cotransformation, on testing of molecular constructs can be singled out from the contributions to this symposium (see references).

The last topic to be mentioned is that of a putative stimulatory effect of electric fields on division and growth. Field manipulations on protoplasts were reported to have increased division frequency and to have improved regeneration [11], but the design of the experiments did not permit conclusions on the origin these effects.

To sum up, it appears that the use of biophysical techniques have brought a greater degree of control and flexibility in fusion and transformation . This is true for preparative experiments as well as for research into fusion and uptake properties of protoplasts. Electric field mediated techniques should very likely prove to be important tools for fundamental and applied protoplast research.

List of abstracts of posters and lectures to be consulted on topics mentioned:

I Advances in control, understanding and application of electric field mediated processes:

-fusion

1. Eady C, Warren G, Lindsey K & Jones MGK, Electrofusion and electroporation of sugar beet protoplasts.
2. Fish N & Jones MGK, Production and field assessment of somatic hybrids between Solanum tuberosum and Solanum brevidens
3. Perronnet C, Denecheau C, Rideau M & Chenieux JC, Electrofusion of chemically aggregated protoplasts derived from mesophyll tissue and habituated cells of Catharantus roseus.
4. Spangenberg G & Neuhaus G, Gene transfer by individual manupulation of protoplasts.

-uptake

5. Bates GW, Electroporation for DNA delivery.
6. Both MTJ de & Jones MGK, Electroporation of tabacco protoplasts as a model system to study plant gene expression.
7. Denecke J, Reynaerts A, D'Halluin K & Botterman J, Quantification of transient expression levels of genes transferred in plant protoplasts by electroporation.
8. Jones H, Symonds CJ & Jones MGK, Transient expression of genes electroporated into protoplasts of cultivated and wild potato.
9. Montane MH, Teissie J & Alibert G, Plant protoplasts electroperm eabilization: contribution of electrical and biological parameters.
10. Steege G van der, Verhoeven H, Hartman R & Tempelaar M, The fusogenic state of plant protoplasts as related to dye/DNA uptake.

-stimulation

11. Rech EL, Ochatt SJ, Chand PK, Power JB & Davey MR, Electro-enhancement of plant protoplast division and plant regeneration

II Comparisons between chemical and electrical systems

12. Aziz MA, Interspecific somatic hybridization within Lotus species.
13. Dix PJ, Marton L, Medgyesy P & Feher A, Comparison of direct gene transfer into Nicotiana protoplasts by electroporation and Mg2+/PEG methods.
14. Yang H, Zhang HM, Davey MR & Cocking EC, Assessments of rice transformation.

III Examples of routine applications of electric field methods

15. Bates GW, Hasenkampf CA, Contolini CS & Piastuch WC, Asymmetric hybridization of Nicotiana species by electrofusion of gamma-inactivated protoplasts.
16. Diaz I, Serra M, Wicke B, Avila-Rincon MJ, Alonso E, Garcia-Luque I & Diaz-Ruiz JR, Infection of tabacco protoplasts with RNA of a pepper strain of tabacco mosaic virus by electroporation.
17. Donovan AM, Collin HA & Isaac S, Isolation and fusion of Apium graveolens L. (Celery) and Levisticum officinale (Lovage) mesophyll protoplasts.
18. Gupta HS, Studies on the transfer of cytoplasmic male sterility through protoplasts fusion in rice.
19. Hall RD & Krens FA, The interspecific transfer of cytoplasmic male sterility in the genus Beta through the production of cybrids.
20. Lee BT, Murdoch K, Topping J, Both MTJ de, Wu QS, Karp A, Steele S, Symonds C, Kreis M & Jones MGK, Isolation, culture and morphogenesis from wheat protoplasts, and study of expression of DNA constructs by direct gene transfer.
21. Lynch PT, Collin HA & Isaac S, Electrofusion of plant and fungal protoplasts.
22. Masson J, Bellini C & Pelletier G, Introduction of selectable markers into different diploid potato (Solanum tuberosum) genotypes by direct gene transfer.
23. Pezzotti M, Arcioni S, Damiani F, Piccirilli M & Falistocco E, Somatic hybridization between Medicago sativa L and Medicago arborea L.
24. Roest S, Puite KJ & Staveren MJ van, Optimization of conditions for somatic cybridization of Daucu carota L.
25. Schilde-Renschler L & Ninnemann H, Somatic hybridization of diploid potato lines, a tool in potato breeding.
26. Seguin A & Lalonde M, Gene transfer by electroporation in Alnus and Betula protoplasts.
27. Tagu D, Bergounioux C, Perennes C, Cretin C & Gadal P, Direct gene transfer in Petunia hybrida protoplasts: evidence for co-transformation.
28. Tempelaar MJ, Drenth-Diephuis LJ, Hesseling-Meinders A, Hartman R, Duyst A & Jacobsen E, Interspecific expression of AEC-resistance in somatic hybrids of AEC-resistant Solanum tuberosum and nitrate-reductase-deficient Nicotiana plumbaginifolia.

PLANT PROTOPLAST ELECTROPERMEABILIZATION: ROLE OF ELECTRICAL AND BIOLOGICAL PARAMETERS

MONTANE M.H.(1-2), ALIBERT G.(1), TEISSIE J.(2)
1- INP-ENSAT, Lab. de Biotechnologie et Amélioration des Plantes
145, Avenue de Muret, 31076 Toulouse France
2- CRBGC du CNRS, 118 rte de Narbonne, 31062 Toulouse France

1. INTRODUCTION

Although direct gene transfer in protoplasts and efficient viable plants somatic hybrids formation methods using electric fields pulses have been reported, a clear understanding of the basic processes and of the role of the physical and physiological parameters on plant protoplast electropermeabilisation is still missing in order to achieve reproducible experiments. According to the theory (1,2), the permeation state is due to an increase of the membrane potential. It depends on both the field intensity and the duration of the pulse. It is also a function of the size and the physiology of the pulsed cell. In this way, we attempted to evaluate electrical and biological contribution on electropermeabilization of tobacco mesophyll protoplasts in order to detect the main factors that modulate electropermeabilization levels. In this way, permeation state was determined by means of exogenous material uptake and plating efficiencies of pulsed protoplasts.

2. MATERIAL AND METHODS

2.1 Protoplast isolation.

Mesophyll protoplasts of *Nicotiana tabacum* var. Xanthi were obtained according to Bourgin et al.(3). A purification step was added by floating freshly isolated protoplasts on a Ficoll 400 discontinuous gradient (20/10/0% w/v) during 15 min. at 1000 g. After washing twice, they were resuspended in the pulsation medium containing 0.56 M mannitol, 1mM $MgCl_2$, 5mM HEPES buffer pH 7.2.

2.2 Pulsation method.

Protoplasts were pulsed with a square wave pulse apparatus according to Montané et al.(4) with 2.5 mM calcein dye or cultivated according to (3).

2.3. Calcein uptake.

After washing, fluorescent protoplasts were counted under a fluorescent microscope .

2.4. Plating efficiency.

During the first week on culture, survival rate was determined by counting the percentage of dividing protoplasts.

3. RESULTS

3.1. Influence of the field intensity.

The behaviour of a protoplast population treated with one pulse of 150 microseconds duration is illustrated in the figure 1. Survival rate is not disturbed until the field reaches a " lethality" threshold of 0.8 kV/cm. Above this value, protoplasts begin to burst. In the same time, dye uptake is detectable when the field gets a threshold permeation value of 0.5 kV/cm.

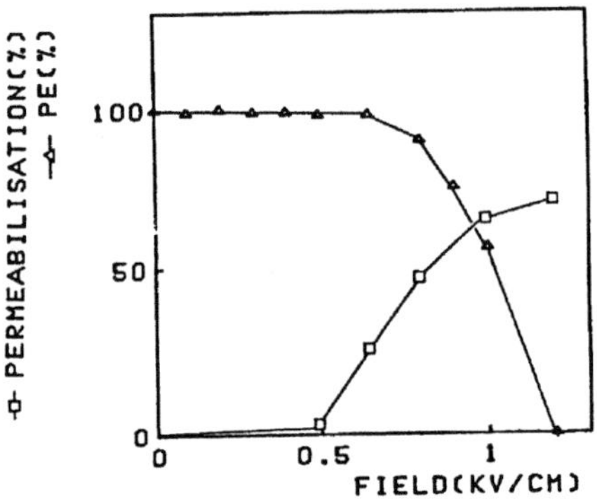

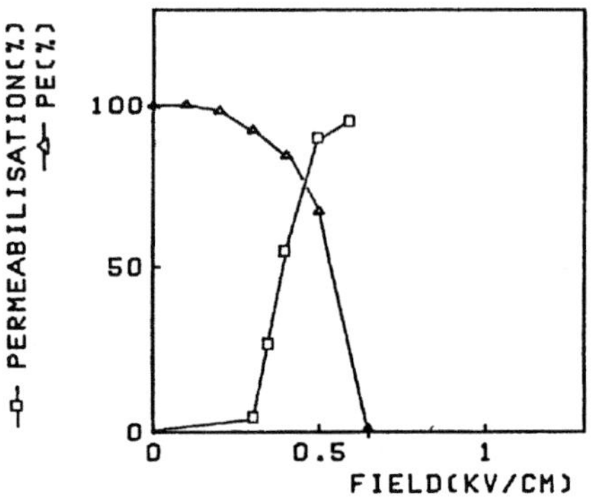

FIGURE 1. Permeation vs field intensity plot: 0.15 ms duration

FIGURE 2. Permeation vs field intensity plot: 3ms duration

3.2. Influence of the pulse duration.

The response of the protoplasts treated with a long pulse of 3 milliseconds duration is shown in the figure 2. This permeation versus field intensity plot has the same pattern as that obtained with a shorter pulse but differs greatly as regard to the two thresholds values that are lowered. The permeation percentage and the slopes of the plating efficiency and of the permation graphs are higher indicating that permeation under viable conditions can be modulated by field/duration combinations.

3.3 Influence of the biological parameters

As regard to the size factor effect involved in the creation of the permeation state, mean radius of the protoplast population was shown to be critical together with the size distribution. Then, both lethality and permeation thresholds values increased when the mean protoplast size decreased (data not shown). Physiological state of the protoplasts (by means of the culture conditions of the donor plants, for example) greatly influences the permeation and survival percentages even when the population has a narrow distribution. The survival rate is strongly affected by the stresses which affected the plant before protoplast isolation.

4. CONCLUSION

Optimization of protoplast electropermeabilization conditions for somatic fusion and exogenous material uptake experiments greatly depends on both electrical and biological parameters. As a consequence, in order to optimize permeation conditions for somatic fusion and exogenous material uptake methods, one should work on subpopulations with a narrow size distribution and select precisely intensity and duration of the pulses.

REFERENCES

1. Kinosita K, Tsong T.Y.(1977) BBA 471 227-242
2. Chernomordik L.V.(1987) BBA 902 360-373
3. Bourgin J.P., Chupeau Y., Missonier C.(1979) Physiol. Plant. 45 288-292
4. Montané M.H., Alibert G., Teissié J.(1987) Studia Biophysica 119 89-92

THE FUSOGENIC STATE OF PLANT PROTOPLASTS AS RELATED TO DYE/DNA UPTAKE

G van der Steege*, H Verhoeven'. R Hartman and MJ Tempelaar

Department of Genetics, Biological Center, Univ. of Groningen, Postbox 14, 9751 AA Haren, The Netherlands
'Institute ITAL, P0box 48, 6700 AA Wageningen, The Netherlands

INTRODUCTION

The mechanism for electric-field-mediated DNA uptake of plant protoplasts (electroporation) is none too clear. Both electro- fusion and electroporation are based on induced reversible membrane breakdown events. If the electroporation process is related to fusogenic events, then fusion properties of proto- plasts may be used as indicators for uptake capabilities.

In studying uptake properties of protoplasts as related to fusion properties dye and DNA uptake experiments have been done under fusogenic conditions.

PROCEDURE

Electrofusion was carried out as described by Tempelaar et al. (1985, 1987). Quantitative determinations of the fusion frequency were done under the microscope.

Dye uptake (ethidium bromide) was monitored by flow cytometry, uptake of DNA by measurement of transient expression (GUS system, Jefferson et al. (1987)) and stable transformation with various plasmid constructs by selection.

Electronic equipment with an independant choice of waveforms and pulses (rectangular and exponentially decaying pulses, and combinations of these) was used to investigate the effect of electric field conditions.

RESULTS AND DISCUSSION

The duration of the fusogenic state after application of an electric field of short duration is one parameter to be determined. In both suspension culture derived Solanum tuberosum and callus culture derived Nicotiana plumbaginifolia protoplasts, which have very different fusogenic properties, there is no evidence for a prolonged fusogenic condition after short pulses.

Fusion response / voltage or pulse duration curves have been generated. These curves are highly reproducable and each type of protoplasts has its own characteristics (shape and position of the curves). From the fusion/voltage curves the critical membrane breakdown voltage for fusion (Vcf) can be estimated (Jones et al., 1987), below which no fusion occurs. (Figure).

Extrapolations from fusion results closely parrallel results from biophysical membrane measurements and electrolyte leakage experiments in literature.

Critical voltage for fusion (Vcf) vs pulse duration.

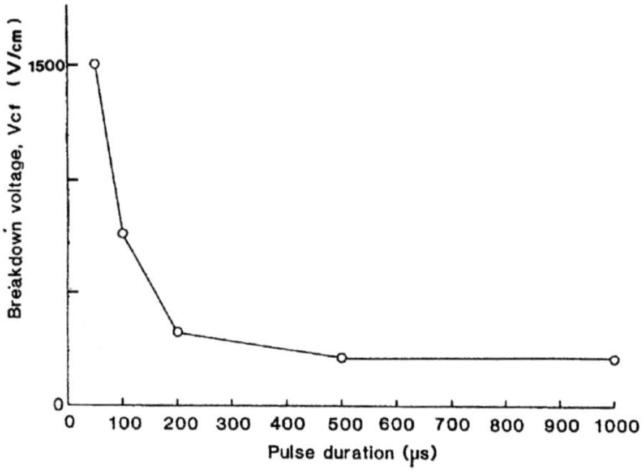

With the duration- and critical fusion voltages-results as reference point, uptake of molecules in the same kinds of protoplasts have been determined. Dye uptake experiments have been carried out on suspension culture derived Solanum tuberosum protoplasts using different pulses and different voltages above and below Vcf. Results showed that, using a short, 50μs, square pulse, maximum dye uptake was already reached below Vcf.

DNA uptake experiments, measuring the transient GUS expression in potato-suspension protoplasts, revealed that hardly any detectable GUS-activity could be obtained using short, 50-100 μs rectangular pulses and field strenghts above Vcf. Under these conditions more than 40% of the protoplasts fuse. Stable transformations however have been obtained from DNA uptake experiments under short-pulse, fusogenic conditions in Nicotiana plumbaginifolia.

Long pulses were more effective in uptake of molecules. Dye uptake and transient expression could be increased using long, RC-pulses or complex fields constructed from super- and subcritical Vcf voltages.

CONCLUSIONS

- Electrofusion can be used to characterize membrane-breakdown events.
- For effective uptake of molecules, pores may have to be open for longer durations than necessary for fusion.
- Optimal conditions may be a combination of super- and subcritical (Vcf) voltages of various duration.

REFERENCES

-Jefferson RA, Kavanagh TA and Bevan MW. (1987) EMBO Journal (submitted).
-Jones H, Tempelaar MJ and Jones MGK. (1987) Oxford Surveys of Plant Molecular and Cell Biology (in press).
-Tempelaar MJ and Jones MGK. 1985, Planta 165:205-216
-Tempelaar MJ, Duyst A, de Vlas S, Krol G, Symmonds C and Jones MGK. 1987, Plant Science 48:99-105

GENETIC MANIPULATION OF PLANT CELLS AND ORGANELLES WITH A LASER MICROBEAM

G. WEBER[1], S. MONAJEMBASHI[2], K.O. GREULICH[2], J. WOLFRUM[2]

[1] Max-Planck-Institut für Zellbiologie, Rosenhof, D-6802 Ladenburg
[2] Physikalisch-Chemisches Institut, Im Neuenheimer Feld 253, Ruprecht-Karls-Universität, D-6900 Heidelberg

1. INTRODUCTION

Direct introduction of cloned genes into eukaryotic cells has proven to be a powerful tool to study the molecular mechanisms of gene regulation. Direct transformation was mainly accomplished by delivering cloned genes into protoplasts or tissue (1, 2). Protoplasts, however, are comparatively fragile. Furthermore, in many species they cannot be regenerated into plants (3). Introduction of DNA into organelles like chloroplasts is even more difficult. DNA destined for the chloroplast has to be delivered through plasma membrane and chloroplast envelope into the stroma of the organelle.

Using laser microbeams (4) new ways have been developed to approach genetic manipulation of cells. Mammalian cells were transformed with DNA (5, 6). With a UV laser microbeam cell fusion has been induced selectively (7) and DNA was introduced into isolated chloroplasts (8).

In the following we describe the introduction of DNA through the cell wall into plant cells and chloroplasts with a UV laser.

2. MATERIALS AND METHODS

Seeds were germinated aseptically from Brassica napus (L.) cv. Lindora. Protoplasts were isolated from hypocotyl sections (9) and cultured in medium according to Glimelius (10). Microspores were obtained from immature anthers and cultured (11). DNA of the plasmid pBR322 was fluorescence labeled with bisbenzimide (12). Unbound dye was removed by precipitation with ethanol. DNA was suspended in buffer containing 0.4 M sorbitol for incorporation into cells and organelles. Etched microscope slides were used for localizing treated cells (13). The laser apparatus used throughout the experiments was described previously (7). For detection of fluorescence a video-imaging system was attached to the microscope (Hamamatsu Photonic Microscope System).

3. RESULTS

The rigid plant cell wall hinders the access to the plasma membrane, for instance, as necessary for introducing DNA into the cytoplasm. With a laser microbeam, however, holes can be generated. Hypocotyl tissue of B. napus was placed in the path of a laser beam to test whether plant cell walls can be selectively punctured. Under visual control of the microscope the focus of the laser was positioned at the wall. With a single pulse a hole was punctured through the cell wall (Fig. 1). Due to the characteristics of light a laser may even be used in areas obstructed to other means of manipulation.

FIGURE 1. Perforation of plant cell walls with a UV-laser microbeam. A single laser pulse generated a hole in the wall of a cell of B. napus (arrow).

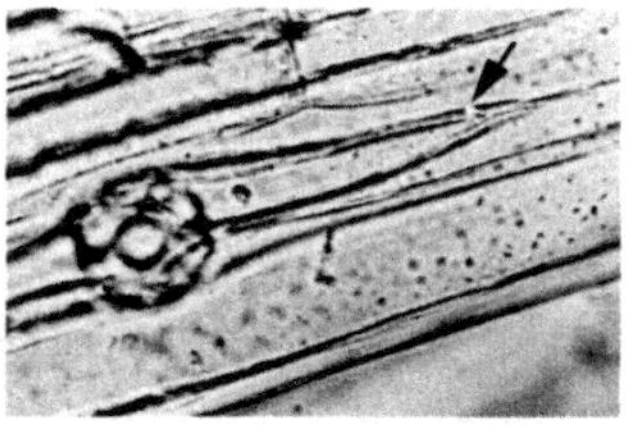

Uptake of material may be facilitated by incubating cells in hypotonic buffer. The plasmolized cells can then be punctured with single laser pulses. Developing pollen grains of B. napus were placed in hypotonic buffer to test this approach for DNA uptake. DNA labeled with bisbenzimide was suspended in the buffer. After ca. 10 min viable cells had plasmolyzed (Fig. 2a). With a laser pulse the cell wall was perforated. A second pulse directed at the plasma membrane punctured a hole. In less than 5s this hole had closed again. Through the temporary opening in the membrane buffer together with DNA entered the cell. The uptake of liquid became evident by the enlarged diameter of the cell (Fig. 2b). The increase in volume of treated cells was estimated to exceed 20%. Incorporated DNA became visible in UV-illumination. Fluorescence was associated with the cytoplasm (Fig. 2c).

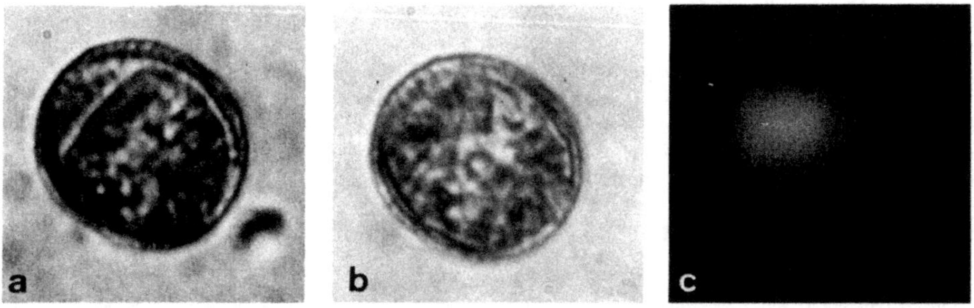

FIGURE 2. Uptake of DNA into microspores. Microspores of B. napus were plasmolyzed (a). Through a hole punctured by a laser pulse buffer entered the cell (b). DNA uptake became visible as fluorescence in UV illumination (c).

The capability to focus a laser beam to very small dimension, i.e. less than 100 nm diameter, makes it feasable to manipulate organelles inside of living cells. DNA could be introduced into chloroplasts. Protoplasts were isolated from B. napus hypocotyl and plated into agarose medium. With a conventional microcapillary fluorescence labeled DNA was injected into the cytoplasm (Fig. 3a). In UV-illumination no fluorescence was associated with the organelles. Then the laser was focussed on the membrane of individual chloroplasts. With one pulse a temporary hole was generated in the outer membrane. DNA entered the irradiated organelles as indicated by fluorescence (Fig. 3b).

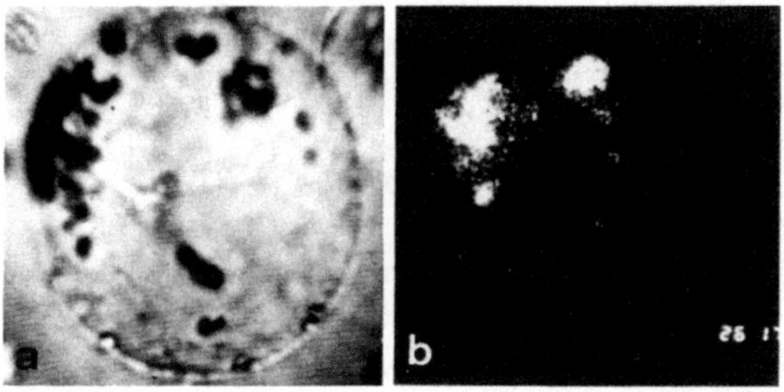

FIGURE 3. Incorporation of DNA into chloroplasts inside of B. napus cells. First DNA was injected into protoplasts (a). After irradiating individual chloroplasts with a laser pulse fluorescence was associated with these organelles indicating incorporation of DNA (b).

DISCUSSION

Using a UV laser microbeam DNA can be incorporated into whole plant cells. Furthermore, DNA was incorporated into chloroplasts inside of a cell. Unlike mechanical or electrical devices for transformation, a laser beam can be used inside of tissues, cells, and organelles. The diameter of the laser focus may be preselected to less than 100 nm. Thus manipulations in the range of nanometers are possible. Furthermore, the short duration of the laser pulse (15ns) minimizes disturbances to cells or organelles. In fact, cells divide readily and develope further after having been treated to incorporate DNA. Furthermore, a large number of cells (ca. 1000 x h^{-1}) can be irradiated. Semiautomatic laser irradiation has been demonstrated in transforming mammalian cells (5).
Under visual control a focussed UV-laser beam can now be employed for rapid genetic manipulation of plant tissue, individual cells and organelles.

The skillful technical help of M. Stanke is gratefully acknowledged.

REFERENCES

1. Potrykus I, Paszkowski J, Saul MW, Negrutiu I, Shillito RD: Direct gene transfer to plants: Facts and future. In Plant Tissue culture, p. 289-302 (C.E. Green et al. eds.) New York, Liss 1987.
2. Klein TM, Wolf ED, Wu R, Sanford JC: High-velocity microprojectiles for delivering nucleic acids into living cells. Nature 327, 70-73 (1987).
3. Halperin W: Attainment and retention of morphogenetic capacity in vitro. In Cell Culture and Somatic Cell Genetics of Plants (F.K. Vasil ed.) p. 3-37, Academic Press 1986.
4. Berns MW, Aist J, Edward J, Strahs K, Girton J, McNeill P, Rattner JB, Kitzes M, Hammer-Wilson M, Liaw LH, Siemens A, Koonce M, Peterson S, Brenner S, Burt J, Walter R, Bryant PJ, Van Dyk D, Coulombe J, Cahill T, Berns GS: Laser microsurgery in cell and developmental biology. Science 213, 505-513 (1983).
5. Tsukakoshi M, Kurata S, Nomiya Y, Ikawa Y, Kasuya T: A novel method of DNA transfection by laser microbeam cell surgery. Appl. Phys. B. 35, 135-140 (1984).
6. Tao W, Wilkinson J, Stanbridge EJ, Berns MW: Direct gene transfer into human cultured cells facilitated by laser micropuncture of the cell membrane. Proc. Natl. Acad. Sci. USA 84, 4180-4184 (1987).
7. Wiegand R, Weber G, Zimmermann K, Monajembashi S, Wolfrum J, Greulich K-O: Laser-induced fusion of mammalian cells and plant protoplasts. J. Cell Sci. 88, 145-149 (1987).
8. Weber G, Monjembashi S, Greulich K-O, Wolfrum J: Uptake of DNA in chloroplasts of Brassica napus (L.) by means of a microfocussed laser beam. Eur. J. Cell Biol. 43, 63 (1987).
9. Weber G, de Groot E, Schweiger H-G: Synchronization of protoplasts from Glycine max (L.) Merr. and Brassica napus (L.). Planta 168, 273-280 (1986).
10. Glimelius K: High growth rate and regeneration capacity of hypocotyl protoplasts in some Brassicaceae. Physiol. Pl. 61, 38-44 (1984).
11. Lichter R: From microspores to rape plants: A tentative way to low glucosinolate strains. In Advances in the production and utilization of cruciferous crops, p. 268-277 (H. Sørensen, ed.), Martinus Nijhoff/Dr. W. Junk Publishers, Dordrecht, Boston, Lancaster, 1985.
12. Cesarone CF, Bolognesi C, Santi L: Improved microfluorometric DNA determination in biological material using 33258 Hoechst. Anal. Biochem. 100, 188-197 (1979).
13. Lin P-F, Ruddle F: Photoengraving of coverslips and slides to facilitate monitoring of micromanipulated cells or chromosome spreads. Exp. Cell Res. 134, 485-488 (1981).

ELECTRIC FIELD MEDIATED DNA UPTAKE AND GENE EXPRESSION IN MESOPHYLL, SUSPENSION CULTURE, AND TUBER PROTOPLASTS OF WILD AND CULTIVATED POTATO

H. JONES and M.G.K. JONES

AFRC Institute of Arable Crops Research, Biochemistry Department, Rothamsted Experimental Station, Harpenden, Herts. AL5 2JQ

INTRODUCTION

The *Agrobacterium* technique has been used successfully to transform a wide variety of dicotyledonous crop plants including potato. However, there are some inherent limitations in the use of the technique. The expression of the transferred gene may depend on neighbouring DNA sequences (positional effects) and analysis of gene expression is made more difficult because the technique involves the regeneration of plants from transformed cells, and may therefore be subject to somaclonal variation. This is particularly true in the potato since previous work from this laboratory has shown the species to be extremely susceptible to somaclonal variation (1). Direct gene transfer methods involving the transient expression of the introduced gene may therefore circumvent these problems and yield results faster since there is no need to regenerate plants from transformed cells. We report here on the development of electroporation as a method for studying transient gene expression in different protoplast types from potato.

MATERIALS AND METHODS

Mesophyll protoplasts were isolated from *in vitro* grown *Solanum brevidens* and *Solanum tuberosum*. Suspension culture protoplasts of *Solanum brevidens* were isolated as described by Tempelaar and Jones (2). Tuber protoplasts were isolated from *in vitro* grown mini-tubers using a new method (manuscript in preparation). The protoplasts were electroporated using a workshop built capacitor discharge system capable of delivering 0 to 450 V/cm pulses set at a range of capacitances from 50 to 2000 µF. The pulse medium consisted of 20mM KCl with mannitol as osmoticum, pH 7.2. Most of the experiments were performed using the reporter gene coding for chloramphenicol acetyl transferase (CAT) under the control of CAMV 35S promotor and the nopaline 3' termination sequence. Co-transformation experiments with the CAT gene and with a new reporter gene coding for β-glucuronidase (GUS) under the control of the same promotor and termination sequence as the CAT gene were also carried out. Transient expression was detected in CAT and GUS assays following published procedures.

RESULTS AND DISCUSSION

Optimum CAT expression was obtained in mesophyll protoplasts of *Solanum brevidens* and *Solanum tuberosum* within a narrow range of field strengths between 200 and 250 V/cm. Background activity of CAT was low. Expression at 225 V/cm corresponded to approximately 50% of cell death. The capacitance was set at 50 µF with an RC time constant of 85 ms. There was considerable reduction in survival of protoplasts on increasing the capacitance, depending on voltage and the number of pulses applied. At 225 V/cm, increasing the capacitance also reduced CAT expression. There was no stimulatory effects of Ca^{2+} ions on DNA uptake and expression. Gene expression with the GUS system closely resembled that of CAT, with optimal expression between 200 and 275 V/cm. Background activity of GUS was also low.

Suspension culture and tuber protoplasts had a much lower field strength optimum compared to mesophyll protoplasts. For suspension culture protoplasts the field strength optimum was between 150 and 225 V/cm and was between 100 and 200 V/cm for tuber protoplasts. The difference in expression can be explained on the basis of the much larger size of suspension (mean diameter=35µm) and tuber protoplasts (mean diameter=65µm) compared to mesophyll protoplasts (mean diameter=25µm). The data obtained are consistent with biophysical theory in terms of the properties of cells and membranes in electric fields (3).

It is shown that electroporation is an extremely efficient and rapid technique for studying gene expression in potato. Both CAT and GUS genes had low endogenous activity compared to electrloporated samples and can be used with high efficiency for promotor, tissue specific, and developmental studies of gene expression. The successful isolation of tuber protoplasts and the demonstration for the first time of gene expression allows studies of important tuber genes such as that coding for the major tuber protein patatin.

Fig. 1. Comparison of CAT expression in mesophyll (A) and tuber protoplasts (B)

A.

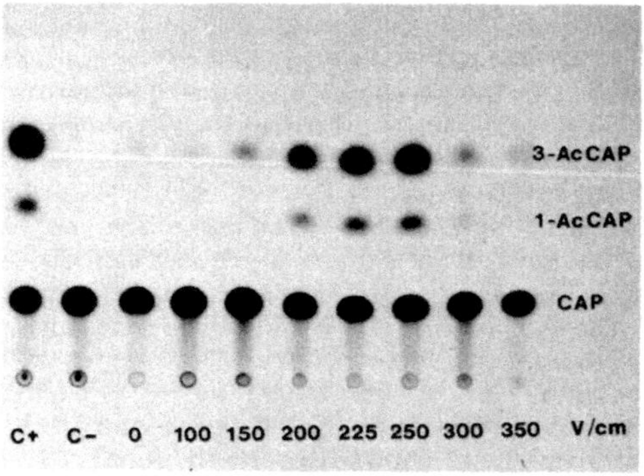

B.

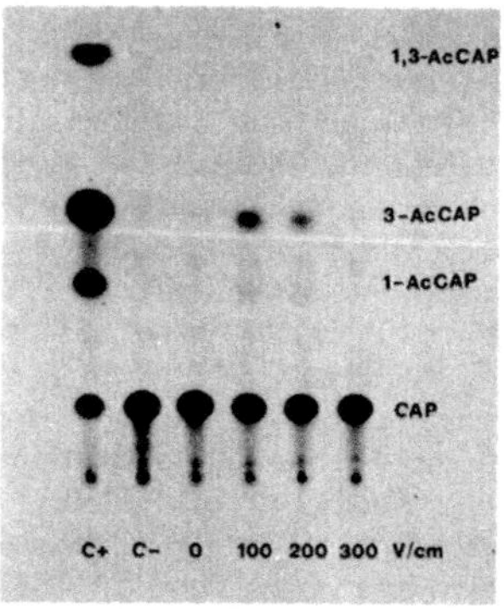

REFERENCES

1. Karp, A., Bright, S.W.J. (1985) Oxford Surveys of Plant Molecular and Cell Biology. Vol. 2, 199-325.
2. Tempelaar, M.J. and Jones, M.G.K. (1985) Planta 165, 205-216.
3. Jones, H., Tempelaar, M.J., Jones, M.G.K. (1987) Oxford Surveys of Plant Molecular and Cell Biology. Vol. 4 (in press).

INTRODUCTION OF SELECTABLE MARKERS INTO DIFFERENT DIPLOID POTATO (*SOLANUM TUBEROSUM L.*) GENOTYPES BY DIRECT GENE TRANSFER.

J. MASSON, C. BELLINI, A. CHARPENTIER*, G. PELLETIER.

Laboratoire de Biologie Cellulaire, I.N.R.A., F 78OOO Versailles.
*Laboratoire de Génétique et d'Amélioration des Plantes, I.N.R.A., F 78000 Versialles.

INTRODUCTION

The breeding strategy proposed for potato by Wenzel (1979) involves selection and crossing at the diploid level. Crosses between tetraploid potato lines and *Solanum phureja* as a pollinator gives rise to dihaploids (Hougas,1964). Desirable genes can then be introduced by crossing the dihaploids with wild species. However, tetraploid clones of potato are more vigorous and produce more tubers than diploid lines. Chromosome doubling to the tetraploid level will be required for further use of the lines. One way to accomplish this would be to fuse protoplasts from two different diploid lines and regeneration of a tetraploid hybrid combining both parental genomes.
Recovery of somatic hybrids between diploid clones could be greatly simplified by the use of markers selectable at the cellular level. Antibiotic resistance genes have been proved to be efficient for the selection of transformed cells in various species.
Diploid potato clones have been transformed using the electroporation technique described by Guerche et al. (1987). Transient expression assays allowed us to find the optimal conditions of electroporation of potato mesophyll protoplasts. Plants have been regenerated from transformed colonies selected on hygromycin or paromomycin.

PLASMIDS

-pCEL4O (C. Waldron et al., 1985). In this plasmid the bacterial APH 4 gene, conferring the resistance to hygromycin, is inserted behind the ocs promotor and followed by the nos transcription termination signal.

-pABDI (Paszkowski et al., 1984) is a 5,3 Kb plasmid containing the aminoglycoside phosphotransferase (APH 3' II) gene under the control of the cauliflower mosaic virus (CaMV) gene VI promoter and terminator. It confers kanamycin resistance.

-pCaMVCAT (Fromm et al., 1985) contains the chloramphenicol acetyl transferase (CAT) gene controlled by the CaMV 35S promoter and nos terminator.

PROTOPLAST ISOLATION

Potato mesophyll protoplasts were isolated by overnight digestion as described by Masson et al. (1987), and resuspended at final density of 2×10^6 protoplasts/ml in the medium used for electroporation (O,4 M mannitol, 2 uM MOPS, 5mM KCl, pH 7,2).

ELECTROPORATION

Protoplasts were mixed with 1O ug of circular plasmid and 5O ug of carrier DNA/ml. Samples of 2×10^6 protoplasts were pulsed in a parallelepipedic electroporation cuvette. A 16 uF capacitor charged with 25O or 3OO volts was discharged across the cuvette. Three electric pulses were delivered and monitored with an oscilloscope.

MEASURE OF THE TRANSIENT ACTIVITY AND SELECTION OF TRANSFORMANTS

CAT assays were performed as described by Fromm et al. (1985). The kanamycin resistant and hygromycin resistant transformants were selected by adding 20 ug of paromomycin (Guerche et al., 1987) or 50 ug of hygromycin B (Waldron et al., 1985) in the dilution medium C. The colonies were cultured and regenerated as discribed by Masson et al. (1987).

DETERMINATION OF OPTIMAL CONDITION OF ELECTROPORATION (Fig. 1)

Mesophyll protoplasts were electroporated at different voltages using the pCaMV CAT plasmid. Transient assays were performed on three days old cells . 250 or 300 volts gave the best results. We maintained these conditions for the transformation experiments.

SELECTION OF PROTOPLAST DERIVED COLONIES RESISTANT TO VARIOUS ANTIBIOTICS (Fig. 2)

The protoplast derived colonies obtained after twenty days of culture, following the procedure described by Masson et al. (1987), were transferred in C medium supplemented with 20 mg/l of paromomycin or 50 mg/l of hygromycin. Three weeks later transformed colonies could be transferred on regeneration medium.

REGENERATION OF TRANSFORMED PLANTS (Fig. 3 and 4)

Shoots appeared 15 days after the transfer of the selected calli on the regeneration medium supplemented with paromomycin (20 mg/l) or hygromycin (50 mg/l). Shoots were cut out and transferred on propagation medium.

PLANT CONFORMITY (Fig. 6)

Chromosome countings were done on root tips. Diploid plants were transferred in the greenhouse. Their phenotype is normal, they have normal leaves, flowers and tubers.

CONCLUSION

Potato transformation using *A. tumefaciens* (Ooms, 1987) or *A. rhizogenes* (Ooms, 1986) is a suitable method for the introduction of cloned genes in the potato genome. Nevertheless, regeneration of plants concern few potato clones, generally tetraploid clones, and most of the transformed plants exhibit somaclonal variation. As a part of somatic hybridization program we developped here an efficient method for potato transformation by direct gene transfer.
Electroporation is suitable for the transformation of the diploid clones tested. Frequency of transformation was 0,5% up to 2% (number of resistant colonies referred to the plating efficiency). Diploid normal plants can be regenerated.
In addition, we feel that direct gene transfert appears suitable for transferring monogenic characters to improve potato cultivars in preserving their heterozygoty which hold their agronomic value.

REFERENCES

Fromm M., Taylor L.P., Walbot V. (1985), Proc. Natl. Acad. Sci. USA, 82 5824-5828.
Guerche P., Bellini C., Le Moullec J.M., Caboche M. (1987), Biochimie (in press).
Hougas R.W., Peloquin S.J., Cabert A.C. (1964), Crop. Sci., 4, 593.
Masson J., Lecerf M., Rousselle P., Perennec P., Pelletier G. (1987), Plant Sci. (in press).
Ooms G., Burrell M.M., Karp A., Bevan M., Hille J. (1987), Theor. Appl. Genet, 73, 744-750.

Ooms G;, Twell D., Bossen M.E., Hoge J.H.C., Burell M.M. (1986), Plant mol. Biol., 6, 321-330
Paszkowski J., Shillito R.D., Saul M., Mandak V., Hohn B., Potrykus I. (1984), EMBO J., 3, 2717-2722.
Wadron C., Murphy E.B., Roberts J.L., Gustafon G.D., Armour S.L., Malkolm S.K. (1985), Plant Molecular Biology 5, 103-108.
Wenzel G., Schleder O., Sopory B. K., Melchers G. (1979), Theor. Appl. Genet., 55, 49-55.

ACKNOLEDGEMENTS

This relatively low cost device for electroporation of cells has been done by P. Guerche.
We thank M. Lecerf for technical assistance.
Diploid clones used in this study were created in the I.N.R.A. breeding laboratory of Landerneau.
This work was supported by a grant from A.R.P.O.T. (Association pour la Recherche sur la POmme de Terre).

DIRECT GENE TRANSFER IN PETUNIA HYBRIDA PROTOPLASTS: EVIDENCE FOR CO-TRANSFORMATION

D. TAGU, C. BERGOUNIOUX, C. PERENNES, C. CRETIN, and P. GADAL.

Physiologie Végétale Moléculaire, Université Paris-Sud, Bât 430, U.A.1128 CNRS, 91405 ORSAY Cedex France.

1. INTRODUCTION

One of the experimental opportunities of the direct gene transfer into plant protoplasts is the co-transfer of an easily selectable marker gene with non-selectable genes. Co-transformations can be obtained either by mixing the two plasmids containing the selectable marker and the non-selectable gene (1), or by constructing a vector harbouring the two genes (2). We show herein that the direct gene transfer of the neomycin phosphotransferase gene (NPT II) into petunia protoplasts via an electroporation step, is accompanied by the co-transfer of a non-selectable gene: phosphoenolpyruvate carboxylase (PEPC) cDNA from sorghum leaves.

2. RESULTS

Petunia hybrida protoplasts were transformed by direct gene transfer using electroporation as described before (3). The NPT II gene carried by the plasmid pCaMVNEO was a gift from V. Walbot (4) and the non-selectable gene was a cDNA clone of PEPC from Sorghum leaves (pCP26B: Crétin, unpublished data). As it is not a chimeric construction, its expression was not attempted in the transformed cells. Calli expressing kanamycin resistance were regenerated in plants. The average absolute transformation frequency was 7.10^{-6} (90% of treated protoplasts were viable before transformation).

DNA of calli and plants were hybridized with the NPT II coding sequence. Figure 1 shows the presence of that coding sequence among the high molecular weights un-restricted DNA of the selected calli and plants, whereas the two DNA from untransformed plants did not hybridize.

The NPT II activity was also detected on fresh leaves from these plants (5).

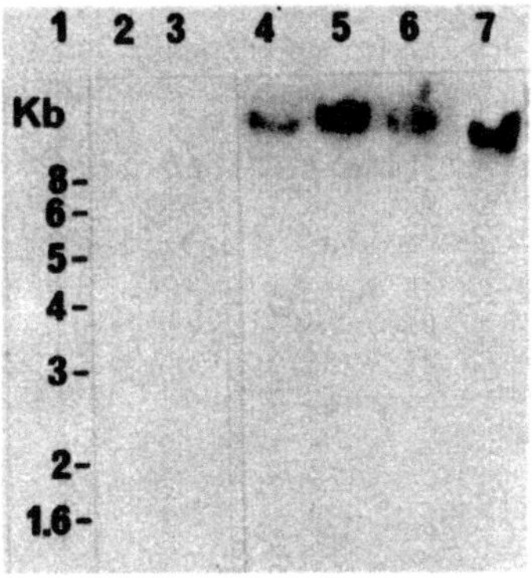

Figure 1: Southern analysis of DNA from transformants: NPT II. 10ug DNA per lane. 1% w/v agarose gel probed with the BamH1 fragment of pCaMVNEO, covering just the NPT II coding sequence. Lane 1; size markers in kilobases. Others lanes: unrestricted DNA. Lanes 2-3; control sensitive petunia and sorghum respectively. Lanes 4-5-6; randomly chosen resistant calli. Lane 7; regenerated plant.

Among the primary transformants, we tested the presence of the PEPC cDNA from sorghum leaves. The only way to spot these secondary transformants was to analyse the DNA of the primary transformants by Southern blot analysis with a probe consisting of the PEPC cDNA (fig 2).

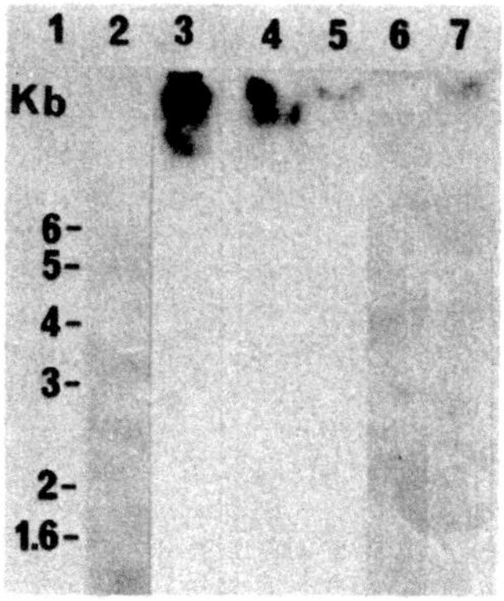

Figure 2: Southern blot analysis of DNA from transformants: PEPC cDNA. 10ug DNA per lane. 1% w/v agarose gel probed with the EcoR1 fragment of pCP26B covering just the PEPC cDNA sequence. Lane 1; size markers in kilobases. Others lanes; unrestricted DNA. Lanes 2-3; control untransformed petunia and sorghum respectively. Lane 4; resistant callus. Lanes 5-6-7; regenerated plants.

In our hybridizing and rinsing conditions, the endogenous petunia PEPC gene was not recognized by the sorghum probe contrary to the one of S. vulgare. This cDNA was incorporated among the high molecular weight DNA of the calli and the plants. Autoradiographies of restricted DNA from several lines yielded complicated patterns (6).

Out of 10 plants containing the NPT II gene, 5 plants were found to contain the PEPC cDNA.

Thus, the DNA analysis of transformed calli and plants proved the co-transfer of the non-selectable PEPC cDNA from sorghum leaves into petunia protoplasts. This preliminary work permits us to envisage further co-transformations with the entire PEPC gene from sorghum, and to study its expression in transgenic petunia.

REFERENCES

1. Schocher RJ, RD Shillito, MW Saul, J Paszkowski, I potrykus 1986. Bio/Technology 4: 1093-1096.
2. Uchimiya H, H Hirochika, H Hashimoto, A Hara, T Masuda, T Kasumimoto, H Harada, JE Ikeda, M Yoshioka 1986. Mol Gen Genet 205: 1-8.
3. Tagu D, C Bergounioux, C Perennes, S Brown, P Müller, P Gadal 1987. Plant Sci 51: 215-223.
4. Fromm ME, LP Taylor, V Walbot 1986. Nature 319: 791-793.
5. Reiss B, R Sprengel, H Will, H Schaller 1984. Gene 30: 211-218.
6. Tagu D, C Bergounioux, C Crétin, C Perennes, P Gadal. Submitted.

ELECTROPORATION OF TOBACCO PROTOPLASTS: A MODEL SYSTEM TO STUDY PLANT GENE EXPRESSION

M.T.J. DE BOTH AND M.G.K. JONES

AFRC Institute of Arable Crops Research, Rothamsted Experimental Station, Biochemistry Department, Harpenden, Herts. AL5 2JQ. U.K.

1. INTRODUCTION

The transient expression of genes introduced into plant protoplasts by direct gene transfer offers a powerful approach to the study of plant gene expression. Some advantages over study of transgenic plants are that results can be obtained in a matter of days and that the DNA is introduced in the form of free plasmids so that its expression is not influenced by position effects. It is not even necessary to use protoplasts which are capable of division, since transient expression has been obtained in non-dividing protoplast systems.

We report here on development of electroporation for tobacco protoplasts which gives reproducibly high expression of the introduced genes.

2. MATERIALS AND METHODS

Leaf mesophyll protoplasts of *Nicotiana tabacum* cv. Petit Havana SR_1 were isolated according to published procedures. They were resuspended in a 20 mM KCl solution and electroporated using a workshop-built apparatus, capable of giving electric pulses in the range of 0-400 V by the discharge of a capacitor bank of 50 - 1600 µF. The DNA used was p35SCN and pCGUS-1, carrying the genes for chloramphenical acetyl transferase (*cat*) and β-glucuronidase (*gus*) respectively under control of the CaMV 35S promoter and *nos* polyadenylation signals. In all experiments, 50 µg/ml carrier DNA was also added.

Transient expression was detected in CAT- and GUS-assays following published procedures.

3. RESULTS AND DISCUSSION

3.1. Electric pulse parameters A sharp field strength optimum was found for electroporation in the range of 175-225 V/cm (Fig. 1). This is in good agreement with theories predicting membrane breakdown in electric fields in relation to protoplast size. Optimum expression correlated with 40-50% survival of protoplasts. Pulse length, defined as the RC-value (R=Resistance, C-Capacitance) is important. Highest expression was found with C=50 µF and R=1 kΩ (RC 50 msec), with pulses of 200-225 V/cm. Longer pulses led to cell death, whereas µsec pulses gave little expression. This suggests that an electrophoretic flow of DNA into protoplasts is important (Jones *et al.* 1987). Routinely, 3 pulses were delivered at 30 sec intervals. More pulses reduced protoplast viability. Transient expression was slightly reduced using one pulse, but protoplast survival was greatly improved. For stable transformation, therefore, 1 pulse is recommended to retain good plating efficencies.

3.2. Other factors Other factors which improve levels of expression are DNA concentration and protoplast density. Optimum expression was obtained at 1 x 10^6 protoplasts per ml.

3.3. Recommended conditions These experiments led us to formulate the following conditions for optimal electroporation of tobacco protoplasts:

FIELD STRENGTH	200-225 V/cm	PROTOPLAST DENSITY	1 x 10^6/ml
RC-VALUE	50 ms	DNA CONCENTRATION	50 µg/ml
NUMBER OF PULSES	3		

Under these conditions expression of pCGUS-1 was also obtained.

3.4. Stable transformation 10^6 protoplasts were electroporated under standard conditions (1 pulse instead of 3) with pHP23, a vector containing *neo* driven by a fused CaMV 35S-19S promoter. After selection in 50 µg/ml kanamycin in 'bead' culture, 82 colonies were selected, which represents an efficiency of ~ 10^{-4}. The colonies have been regenerated and are being analysed.

3.5. Conclusion Electroporation conditions have been established for tobacco protoplasts which give high levels of transient expression. This system can be used to study plant gene expression. Stable transformants can readily be obtained.

ACKNOWLEDGEMENT The financial support of BIOSEM (Groupe Limagrain) is gratefully acknowledged. We also thank J. Paszkowski (FMI, Basel), J. Walker (CSIRO, Canberra) and J. Topping (RES) for supplying the plasmids.

REFERENCE

Jones, H., Tempelaar, M.J. and Jones, M.G.K. (1987) In Oxford Surveys of Plant Molecular Biology and Cell Biology, Vol. 4, Ed. B. J. Miflin. In press.

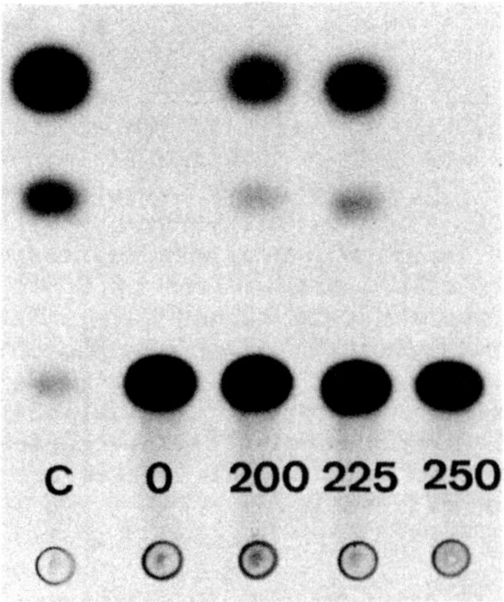

Fig. 1. Transient expression assay for CAT.
C - positive control.
Field strengths are indicated in V/cm.

INFECTION OF TOBACCO PROTOPLASTS WITH RNA OF A PEPPER STRAIN OF TOBACCO MOSAIC VIRUS BY ELECTROPORATION*

I. DIAZ, M. SERRA, B. WICKE, M.J. AVILA-RINCON, E. ALONSO, I. GARCIA-LUQUE and J.R. DIAZ-RUIZ
UEI FITOPATOLOGIA, CIB, CSIC, VELAZQUEZ, 144. 28006-MADRID. SPAIN.

1. INTRODUCTION

Tobacco mosaic virus pepper strains (TMV-ps) are becoming an important economic problem in protected pepper crops all over the world (1) since they are able to overcome all known genetic resistances to the main TMV so far found in Capsicum spp. (2). Inoculation of tobacco protoplasts with RNA of different well known TMV by electroporation has been recently accomplished (3,4,5). In order to iniciate studies on the mechanisms of pathogenesis and resistance to TMV-ps infections, RNA of a spanish TMV-ps has been used to electroporate tobacco mesophyll protoplasts with a Bio-Rad Gene Pulser apparatus. Optimal electroporation conditions for protoplasts infection are reported here.

2. PROCEDURE

Mesophyll protoplasts of Nicotiana tabacum v. Xanthi n-c were prepared from fully expanded leaves of about two month old plants grown in a chamber (25°±5°C, 12,000 lux, 16 h photoperiod). Pieces of leaves with the lower epidermis peeled were floated on an enzyme solution containing 1.5% Meicelase P, 0.05% Macerozyme R-10 in CPW 13M pH 5.8 (6), for 16 h. The isolated protoplasts were washed with CPW 13M and floated on CPW 21S (6) by centrifugation at 100g for 10 min. Floating cells were washed with CPW 13M and again with only 0.7M mannitol. They were resuspended at a concentration of $2x10^5$ cells/ml in cold 0.7M mannitol as electroporation medium containing TMV-ps RNA (10-40 µg/ml). After 5' preincubation in ince, 0.8 ml portions were transferred to Gene Pulser cuvettes (0.4 cm interelectrode distance), electroporated under different electric pulse conditions, postincubated in ice for 10 min, washed two times with 0.7M mannitol and CPW 13M respectivelly, resuspended in 1 ml of MSP1 9M culture medium (7) and cultured for 24-48 h at 25°C with artifficial light in 3 cm plastic dishes. Cultured cells were washed with CPW 13M and the percentages of survival, viability and infection were evaluated by microscopic examination and staining with Fluorescein diacetate and virus-specific fluorescent antibody respectively.Control samples with neither viral RNA nor electric pulse treatment were used. Samples of the cultured cells were pelleted in 1 ml eppendorff tubes, resuspended in 0.5 ml of 0.2M phosphate buffer pH 7.2 and used for electron microscopy examination of viral particles and local lession assay in N. tabacum v. Xanthi n-c.

Preliminary series of experiments showed that better protoplast survival were obtained with low capacitor discharges and thus low pulse times. Increasing the number of pulses in the 500-1.500 V/cm range or the voltage over 1.000 V/cm markedly decreased the percentage of survival protoplasts (no shown). Table 1 shows the effect of pulse voltage on the survival viability and infection of protoplasts after 48 h of electroporation. In the range of 250 to 1.000 V/cm (0.25 µFD, one pulse, 20 µg RNA/ml) there were not significant differences in the percentage of protoplast survival and

viability. Best percentages of protoplast infection were obtained in the range of 750-1.000 V/cm as shown by immunofluorescent assay. This date was confirmed by the electron microscopy and the local lession assay. Low RNA concentration (10 µg/ml) slightly reduced the percentage of infected protoplasts and high concentration (40 µg/ml) did not improved it.
As much as 79% of the protoplasts that survival 48 h after electroporation showed immunofluorescence in the above conditions.

TABLE 1. Effect of the pulse voltage on the survival, viability and infection of tobacco protoplasts with TMV-ps RNA (1).

	Pulse voltage (V/cm)	Pulse time (m sec.)	Protoplast survival (%)	Protoplast viability (%)	Protoplast infection (%)
Control (- RNA)	--	--	65	65	--
Control (+ RNA)	--	--	55	56	0
	250	0.6	50	60	< 1
	500	0.6	50	53	< 1
	600	0.5	55	65	5
	700	0.5	50	62	31
	750	0.5	45	55	62
	800	0.4	45	60	79
	850	0.4	45	60	63
	900	0.4	40	45	65
	1,000	0.4	25	55	72
	1,250	0.5	19	45	36
	1,500	0.6	13	22	< 1
	1,750	0.3	9	0	< 1
	2,000	0.4	7	0	< 1

(1) Electroporation conditions: 2×10^5 protoplast/ml in 0.7M mannitol containing 20 µg RNA/ml, 0.25 µ FD, one pulse. Data are average of 3 experiments after 48 h post electroporation.

3. CONCLUSIONS

TMV-ps RNA can be introduced and expresed in N. tabacum v. Xanthi n-c mesophyll protoplasts by electroporation. Optimal conditions for electroporation in 0.7M mannitol using the Bio-Rad Gene Pulser apparatus have been established.

REFERENCES

1. Wetter, C. et al. (1984) Phytopathology, 74: 405-410.
2. Rast, A.T.B. (1982) Neth. J. Pl. Path., 88: 169.
3. Hibi, T. et al. (1986) J. Gen. Virol. 67: 2037-2042.
4. Nishiguchi, M. et al. (1984) Plant Cell Reports, 5: 57-60.
5. Morikawa, H. et al. (1986) Gene 44: 121-124.
6. Frearson, E.M. et al. (1973) Dev. Biol. 33: 130-137.
7. Power, J.B. and Chapman, J.V. (1985) In: Dixon, R.A. (ed.) Plant Cell Culture. IRL Press. Oxford, pp. 37-66.

* Supported by grants from CAICYT (nº 148) and Fundación Ramón Areces, Spain.

COMPARISON OF DIRECT GENE TRANSFER INTO *NICOTIANA* PROTOPLASTS BY ELECTROPORATION AND Mg^{2+}/PEG METHODS

P.J. DIX[1], L. MARTON[2], P. MEDGYESY[2] and A. FEHER[3]
[1]Dept. of Biology, St. Patrick's College, Maynooth, Co. Kildare, Ireland.
[2]Institute of Plant Physiology and [3]Institute of Genetics, Biological Research Center, Hungarian Academy of Sciences, H-6701 Szeged, PO Box 521, Hungary.

1. INTRODUCTION

Direct transfer of DNA into plant protoplasts (1) represents a valuable addition to *Agrobacterium*-mediated transformation as a technique for genetically engineering crop plants. Simplification of the procedures required would further increase the value of the approach. The method recently developed (2) which gives efficient transformation without a requirement for electroporation, therefore merits close attention. We report here a comparative study using electroporation or the Mg^{2+}/PEG method in the transformation of diploid *Nicotiana tobacum*, and haploid and diploid *N. plumbaginifolia* protoplasts.

2. PROCEDURE

2.1 Materials and methods

2.1.1 Shoot culture. *N. tabacum*: SR1 (streptomycin resistant) and *N. plumbaginifolia*: LR400 (lincomycin resistant) and HNP28 (haploid).

2.1.2 DNA. Isolated plasmid pGA471 or pGA472 (3). Carrier DNA was calf thymus or chick erythrocyte.

2.1.3 Protoplast culture. As described in Reference 4.

2.2 Transformation

2.2.1 Electroporation and Mg^{2+}/PEG methods. These closely followed procedures described in Refs. 1 and 2 respectively, although a number of parameters were varied. Electroporation was carried out with 0.5 ml samples in plastic cuvettes. Three pulses were given, using 600 V and 56 nF (resistance 300-400 Ω).

2.2.2 Stable transformation. 50 mg l^{-1} kanamycin sulphate was added at first dilution of the protoplasts, after 8-10 days and at all subsequent steps. Resistance was checked in callus and leaf strip tests.

2.2.3 Transient expression. Measured for 3×10^6 protoplasts after 24 hours using a dot assay (5). 10 µg protein used for assay.

3. RESULTS

3.1 Comparison of transformation methods with *N. plumbaginifolia*.

Efficient transformation could be obtained using both methods. Heat shock and PEG greatly improved the electroporation procedure. More resistant colonies (about 3x) were obtained using Mg^{2+}/PEG method. Using the latter method, sucrose worked as well as mannitol as the osmoticum. For both methods higher transformation frequencies were obtained using linear plasmid DNA plus carrier DNA. Transient expression was more pronounced with circular DNA, and did not require carrier. Transient expression supported the results for stable transformation (Fig. 1).

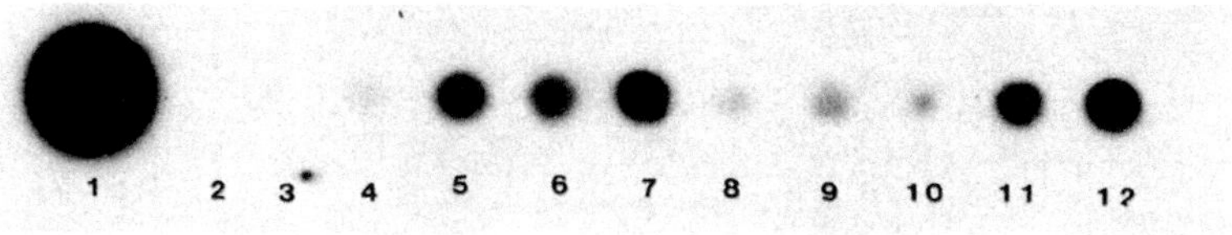

FIG. 1. Transient expression of neomycin phosphotransferase gene. 1, Leaf extract of transformed SR1 plant; 2, Empty; 3, Leaf extract of LR400 plant. Remainder, all $3x10^6$ LR400 protoplasts transformed with 60 µg circular pGA471; 4-7= Mg^{2+}/PEG method: 4, No heat shock or PEG, 5, No PEG; 6, PEG but no heat; 7, Heat + PEG; 8-12= Electroporation method: 8, No heat, PEG or electroporation; 9, Heat + PEG, no electroporation; 11, PEG + electroporation; 12, Heat + PEG + electroporation.

3.2 Comparison of haploid and diploid *N. plumbaginifolia*. Haploid protoplasts showed much poorer viability, division frequencies and colony formation after transformation, differences which were absent when DNA was left out of the treatments. Frequency of stable transformation among survivors was lower than in diploid cells.

3.3 Comparison of *N. tabacum* and *N. plumbaginifolia*. Higher frequencies of stable transformation, but not always transient expression, were obtained with *N. tabacum*. The best values for each species, using each method, were shown in Table 1.

TABLE 1. Best transformation frequencies (resistant colonies per 10^5 dividing protoplasts) obtained with *N. tabacum* (SR1) and *N. plumbaginifolia* (LR400).

Electroporation		Mg^{2+}/PEG	
SR1	LR400	SR1	LR400
84	39	394	112

4. Conclusion

Both transformation procedures are effective for these two species. *N. tabacum* responds slightly better than *N. plumbaginifolia*, and haploid protoplasts show poor recovery after transformation.

REFERENCES

1. Shillito R, Saul M, Paszkowski J, Muller M, and Potrykus I: Biotechnology 3, 1099-1103, 1985.
2. Negrutiu I, Shillito R, Potrykus I, Biasini G, and Sala F: Plant Molecular Biology 8, 363-373, 1987.
3. An G, Watson BD, Stachel S, Hordon MP, and Nester EW: EMBO J. 4,277-284, 1985.
4. Maliga P.: In: Cell Culture and Somatic Cell Genetics of Plants (Ed. IK Vasel) pp. 552-562, 1984.
5. Platt and Yang: Anal. Biochem. 162, 529-535, 1987.

QUANTIFICATION OF TRANSIENT EXPRESSION LEVELS OF GENES TRANSFERRED TO PLANT PROTOPLASTS BY ELECTROPORATION.

J. Denecke, A. Reynaerts, K. D'Halluin and J. Botterman.

PLANT GENETIC SYSTEMS N.V., J. Plateaustraat 22, B-9000 Gent Belgium.

A simple and fast procedure for the quantitative analysis of gene expression directed by chimeric constructs in plants has been developed. The problem of 'inter-experiment' variability has been reduced by the inclusion of an internal standard. Different chimeric constructs have been evaluated in <u>Nicotiana tabacum</u> protoplasts using this system and compared with expression in transformed tobacco plant tissues. The system has also been tested in <u>Brassica oleracea</u> protoplasts.
The constructions analysed contain two marker genes: 1) The cat gene encoding chloramphenicol acetyl transferase fused to the CAMV35S- or the TR2' promoters served as an internal reference. 2) The <u>bar</u> gene (1) encoding phosphinothricine acetyl transferase (PAT) was used as a reporter in THE chimeric constructs inserted in the vector containing the internal reference as shown (fig.1).

Both gene products can be extracted by the same procedure allowing simultaneous analysis in a single extract. PAT activity is calculated relative to the CAT activities. In this way direct comparision of different promoter fusions is possible. Chimeric <u>bar</u> constructs containing the TR2'-, CAMV35S-, CAMV19S-, and the <u>Arabidopsis</u> SSU promoter were inserted in the vector. The chimeric constructs can easily be transferred to appropriate vectors for plant transformation after analysis in the transient expression system.

Fig.1: Vector system

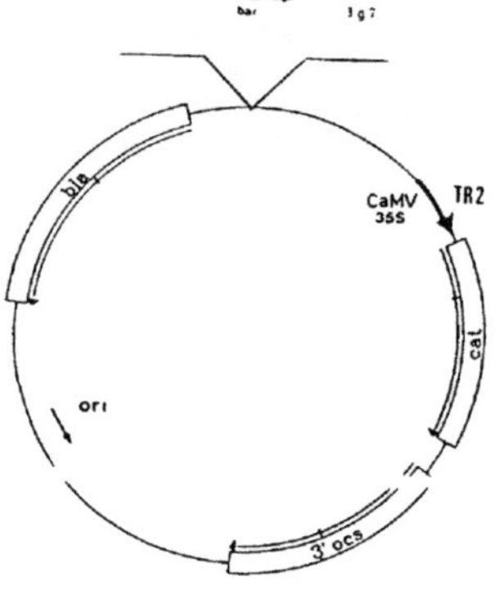

RESULTS

Expression directed by different chimeric constructs was studied in three different ways: transient expression in tobacco leaf protoplasts, in pools of transformed calli, and in individual and pooled transgenic plants.

Transient expression was measured by enzymatic and immunological methods after normalisation with respect to the internal marker. Expression of the internal marker gene showed a 4 fold variation depending on the plasmids used for transient gene expression. Low expression correlated with high molecular weights of the plasmids. The TR2'-bar construct showed the highest expression levels after normalisation with respect to the CAT activities in the same extract.

To compare these values with expression levels in transformed cells, tobacco protoplasts were cocultivated with the Agrobacterium strain C58C1 (pGV2260) carrying the appropriate plasmids. From independently selected calli transgenic tobacco plants were regenerated. In addition calli were propagated for subsequent analysis. Pools of 20 randomly chosen calli were analysed for PAT expression. Expression levels for all constructs were comparable with those obtained in transient expression experiments.
To analyse expression in regenerated plants, 15 individual plants for each construction were studied in parallel with 8 pools of 5 individuals each. The SSU promotor fusion gave the highest level of expression while the CAMV19S- and the TR2' constructs showed the lowest expression levels.

To determine whether the vector system is suitable for the analysis of gene expression in other plants, similar transient expression experiments were done with mesophyl protoplasts prepared from Brassica oleracea. Quantification of the internal CAT marker by spectrophotometric analysis and detection of the TR2'-bar chimeric construct after normalisation with respect to the internal standard was still feasible. The data show a similar variability in the internal CAT expression to that seen in tobacco.

CONCLUSIONS

1) The system presented allows the quantitative determination of expression levels of different chimeric genes. Problems of variation are overcome by the use of an internal reference. This was shown in both Nicotiana tabacum and Brassica oleracea.

2) The variation seems to be inherent to the kind of the plasmid and does not depend on the kind of the plant protoplasts used for electroporation. A correlation between increasing plasmid size and decreasing expression level was observed. We propose that high molecular weight of the plasmids hinders the uptake of the DNA into the cells.

3) The transient expression state seemed to mimic the situation in callus tissues for the analysed chimeric constructs.

REFERENCES

(1) Thompson C.J., Movva N.R., Tizard R., Crameri R., Davis J.A., Lauwereys M. and Botterman J.: Characterization of the herbicide - resistance gene bar from Streptomyces hygroscopicus. EMBO journal 6, 2519 - 2523, 1987.

GENE TRANSFER BY ELECTROPORATION IN *ALNUS INCANA* PROTOPLAST.

A. Séguin* and M. Lalonde.
Centre de Recherche en Biologie Forestière Faculté de Foresterie,
Université Laval, Ste-Foy Québec, G1K7P4, Canada.

INTRODUCTION

Actinorhizal plants are economically important in forestry programs such as land reclamation, and reforestation. They can establish symbiosis with the actinomycetes *Frankia* resulting in nitrogen-fixing root nodules (1). *Alnus* (alder) is a well-known genera of actinorhizal plants, composed of perennial dicots trees found in high latitude countries. The understanding of this association at the molecular level is vital in order to identify the specific genes involved. An efficient transformation method being a prerequisite for most molecular genetic works and for the eventual ability to transfer symbiotic genes to other woody plants.

We developed an efficient transformation method for protoplasts obtained from cell suspension of *Alnus incana* (2) using electroporation mediated gene transfer. The E. coli beta-glucuronidase (GUS) marker gene was used in this study (3). We monitored the success of gene transfer into *Alnus* protoplasts by a transient gene expression system using a fluorogenic substrate 4-methyl umbelliferyl glucuronid. Conditions for efficient transformation are presented. These results motivated us to investigate this method as a means of achieving the stable transformation of alder.

MATERIALS AND METHODS

Plasmids and Plant Materials. The plasmid pBI221 was obtained from Clontech Laboratories, Inc. (Palo Alto, CA). This is a vector of 5.7 kb which contains the 3.0 kb Hind III - Eco RI fragment of pBI121 (Jefferson et al., in press) containing the CaMV 35S promoter, the GUS gene and the NOS poly A site of the *Agrobacterium tumefaciens* Ti plasmid in pUC19. The *Alnus incana* cell suspension line was grown in the Nagata and Takebe's (NT) basal medium as described by Tremblay *et al.* (1985).

Protoplast isolation. Protoplasts were isolated as previously reported (2).

Electroporation. A 1-ml aliquot of 1.5 X 10^6 protoplast in CPW 13M was placed in an electroporation chamber (Bio-Rad, Richmond CA) consisting of a 1 cm plastic spectrophotometer cuvette (narrow portion of 0.4cm) with aluminum foil electrodes glued onto opposite faces. The supercoiled plasmid DNA was added to the protoplasts prior to a heat-shock treatment at 45°C for 5 min, and the sample was held at room temperature for 5 min. A 0.5-ml aliquot of a 28% solution of PEG 8000 (Sigma) dissolved in CPW 13 M medium containing 15 mM $MgCl_2$ was added. The sample was then subjected to electroporation with a Bio-Rad Gene Pulser™ transfection apparatus. The Gene Pulser apparatus generates an exponential decay pulse. After electroporation, the samples were placed at room temperature for 15 min. Protoplasts were diluted with CPW 13M media and collected by centrifugation for 10 min at 120x *g*, resuspended in 4.0 ml of NT 8 plating media (Tremblay *et al.* 1985) and incubated in the dark at 23°C.

β-glucuronidase assays. Protoplasts collected by centrifugation were resuspended in a lysis buffer (3) consisting of 50 mM sodium phosphate (pH 7.0), 10 mM 2-mercaptoethanol, 0.1% Triton X-100, 0.1% Sarkosyl, and 1 mM 4-methyl umbelliferyl glucuronide (MUG). Each samples were sonicated on ice for 5 sec. Glucuronidase was assayed in the lysis buffer in a 300µl volume at 37°C. The reaction was terminated by the addition of 3.0 ml of 0.2 M Na_2CO_3. Fluorescence is then measured with excitation of wavelength 365 nm, and emission at 455 nm in a Perkin Elmer fluorimeter.

RESULTS

Our initial experiments were focused on the determination of the optimal voltage for transient expression of GUS gene in protoplasts, using the expression plasmid pBI221. Fig. 1 illustrates the behavior of this expression system against the variation of the voltage. Electroporation performed at 200 V appears to achieve a reasonable compromise between maximum cell viability and transient gene expression. When protoplasts electroporated in the absence of pBI221 DNA were assayed for GUS, no activity was detected (fig. 2). The intensity of the fluorescence was positively correlated with DNA concentration when the amount of DNA present during electroporation was between 5 and 40µg (fig. 2). We did not observe a sharp optimum with respect to the use of linear DNA or supercoiled DNA and the use of carrier DNA do not resulted in increasing GUS expression (data not shown). Furthermore, we observed increase in GUS activity in heat shocked protoplasts.

Effect of increasing voltage

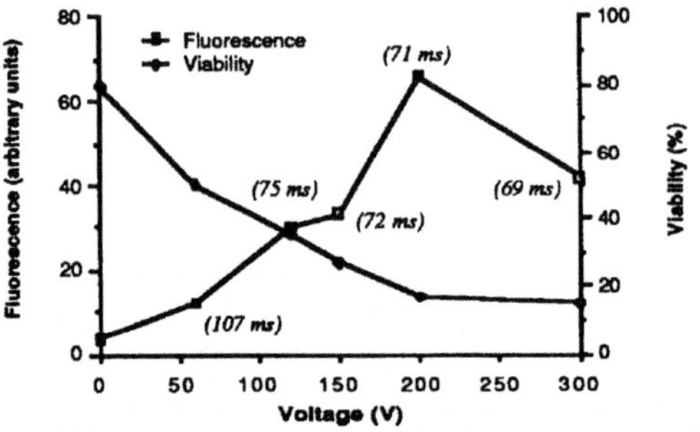

Fig. 1. Effect of increasing voltage of electroporation on protoplast viability and GUS gene expression. GUS activity is related to fluorescence in the extract. Viability of protoplasts was measured by FDA staining(4). Values are given as a percent of protoplasts remaining intact after electroporation. The time of pulse (in italic) varied between 69 and 107 msec for 37% decay (capacitance, 960 µFD). The level of GUS activity in extracts made from protoplasts 24 hrs after electroporation are shown.

Effect of the concentration of plasmid DNA

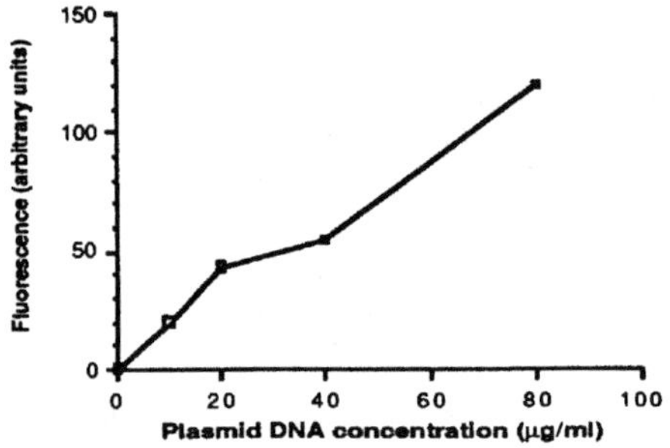

Fig. 2. Dependence of GUS gene expression on the DNA concentration used during electroporation. Protoplasts were treated with an electric pulse of 120 V (capacitance, 960 µFD).

DISCUSSION

These results show that electroporation is an effective method for the introduction of a marker gene in *Alnus* protoplasts. Different parameters of electroporation have been tested. Among these we can see that higher concentration of DNA and higher voltages up to 200 V electric pulse produce more GUS activity from fewer surviving protoplasts. The next step in this work will be to obtain stable transformants with the use of vector containing a selectable gene. Unfortunately it will not be possible to use kanamycin resistant gene because preliminary experiments have shown endogenous resistance of this antibiotic in *Alnus* species.

Finally the advances of culture technology in the area of *Alnus* protoplast isolation and callus regeneration (2) combined with an efficient transformation method highlight the need for an *Alnus* whole plant regeneration system. However the production of transformed cell lines of *Alnus incana* will be of great interest to look for induced promotors after cocultivation with *Frankia* and expression of symbiotic genes.

REFERENCES

1. Normand, P. and Lalonde, M. (1986) Plant and Soil, **90** 429-453.
2. Tremblay, F.M., Power, J.B. and Lalonde, M. (1985) Plant Sciences, **41** 211-216.
3. Jefferson, R.A., Burgess, J.M. and Hirsh, D. (1986) Proc. Natl. Acad. Sci. USA, **83** 8447-8451.
4. Widholm, J.M. (1972) Stain Tech **47**: 189-194

TRANSFORMATION OF RICE PROTOPLASTS BY DIRECT DNA UPTAKE

H. YANG, H.M. ZHANG, M.R. DAVEY, B.J. MULLIGAN and E.C. COCKING
Plant Genetic Manipulation Group, Department of Botany, University of Nottingham, Nottingham NG7 2RD, UK.

1. INTRODUCTION

Plant regeneration has been established from rice protoplasts (1), and kanamycin resistant callus obtained following PEG-induced uptake of plasmid into isolated protoplasts (2). This paper summarises studies on the transformation of rice protoplasts by direct DNA uptake using various delivery methods.

2. MATERIALS AND METHODS

2.1. Plasmid delivery to protoplasts

Protoplasts isolated from cell suspensions of *Oryza sativa* v. Taipei 309 line LB1 were treated with pCaMVNEO carrying a chimaeric gene consisting of the CaMV 35s promoter, the neomycin phosphotransferase II (NPTII) gene from Tn5 and the *nos* polyadenylation region (3). Plasmid was delivered to protoplasts by electroporation, treatment with PEG, and PEG combined with electroporation. In the case of electroporation, 1.7 x 10^6 protoplasts in 0.4 ml of modified electroporation medium (3) with a resistance of 900 Ω, were mixed with 4 µg of pCaMVNEO and 20 µg of calf thymus DNA. Protoplasts were electroporated (DIA-LOG, G.m.b.H., 4000 Düsseldorf 13, FRG) at 500 to 2,500 V with 20 to 50 nF capacitance. Samples were incubated on ice for 10 min. after electroporation; protoplasts were cultured in KPR agarose medium (1). The PEG treatment was essentially as described (4), as were the conditions for PEG combined with electroporation (5). After two weeks, the liquid medium bathing the agarose sectors was replaced with medium containing 100 µg/ml kanamycin. The liquid kanamycin-containing medium was replaced every 7 days over a period of 4 weeks. Kanamycin resistant colonies were transferred to AA2 liquid medium (1) containing 50 µg/ml kanamycin for callus proliferation.

2.2. DNA hybridisation and NPT II assay

DNA was isolated from callus using a published method (6). Ten µg of DNA digested with BamH1 were electrophoresed in a 0.8% agarose gel. The DNA was blotted to nitrocellulose (7) and hybridised with the ^{32}P-labelled 1.0Kb BamH1 fragment from pCaMVNEO containing the NPTII gene.
NPTII activity was assayed as described (8,9).

3. RESULTS AND DISCUSSION

3.1. Optimum conditions for electroporation

Kanamycin resistant colonies were selected following electroporation at 1,000 to 2,500 V with 20 to 50 nF capacitance. The number of resistant colonies from protoplasts exposed to 2,000 to 2,500 V with these capacitances was more than with other conditions, with an optimum at 2,000 V with 40 nF.

3.2. Transformation frequencies

Of 1.7×10^6 protoplasts electroporated with plasmid, 338 colonies greater than 1mm in diameter were selected on kanamycin-containing medium. These continued to proliferate in the presence of the antibiotic during subsequent transfers, giving a transformation frequency of 2.0×10^{-4} when expressed in terms of the number of protoplasts treated with plasmid. In comparison, the transformation frequency following PEG treatment was 9.1×10^{-5}, and from PEG combined with electroporation, 2.7×10^{-5}.

3.3. Confirmation of transformation

DNA from 20 kanamycin resistant clones selected from the three plasmid uptake experiments hybridised with the ^{32}P-labelled 1.0 Kb BamH1 fragment from pCaMVNEO containing the NPTII gene, indicating the presence of plasmid DNA sequences in the plant genome. A positive NPTII assay confirmed the activity of this enzyme in the kanamycin resistant tissues.

In conclusion, stable transformation of rice callus has been obtained using electroporation, PEG and PEG combined with electroporation as methods of plasmid delivery, the transformation frequency with electroporation being higher than with the other methods. Future studies will attempt to combine the plant regeneration and transformation protocols to produce transgenic rice plants.

4. ACKNOWLEDGEMENTS

This work was supported, in part, by the Rockefeller Foundation and Overseas Development Administration. HY is a Royal Fellow supported by the Royal Society of London; HMZ holds a fellowship from the Chinese Government.

5. REFERENCES

1. Abdullah R, Cocking EC, Thompson J: Bio/Technology 4: 1087-1090, 1986.
2. Uchimiya H, Fushimi T, Hashimoto H, Harada H, Syono K, Sugawara Y: Mol. Gen. Genet. 204: 204-207, 1986.
3. Fromm ME, Taylor LP, Walbot V: Nature 319: 791-793, 1986.
4. Krens FA, Molendijk L, Wullens GJ, Schilperoort RA: Nature 296: 72-74, 1982.
5. Shillito RD, Saul MW, Paszkowski J, Muller M, Potrykus I: Bio/Technology 3: 1099-1102, 1985.
6. Dellaporta SL, Wood J, Hicks JE: Plant Molec. Biol. Reporter 1: 19-21, 1983.
7. Southern E: J. Mol. Biol. 98: 503-517, 1975.
8. Reiss B, Sprengel R, Will H, Schaller H: Gene 30: 211-218, 1984.
9. Schreier PH, Seftor EA, Schell J, Bohnert HJ: EMBO J. 4: 25-32, 1985.

ISOLATION, CULTURE AND MORPHOGENESIS FROM WHEAT PROTOPLASTS, AND STUDY OF EXPRESSION OF DNA CONSTRUCTS BY DIRECT GENE TRANSFER

B. T. Lee, K. Murdoch, J. Topping, M. T. J. de Both, Q. S. Wu, A. Karp, S. Steele, C. Symonds, M. Kreis and M. G. K. Jones

AFRC Institute of Arable Crops Research, Rothamsted Experimental Station, Harpenden, Herts. AL5 2JQ, U.K.

1. INTRODUCTION

The regeneration of plants from protoplasts of some monocots has been achieved, but so far the list excludes wheat. We report here on the culture of protoplasts obtained from an inflorescence-derived suspension cell line, which when cultured gives rise to green callus showing limited morphogenesis.

The introduction of genes into protoplasts by direct gene transfer allows transient expression of introduced genes. We have examined the expression of the reporter gene β-glucuronidase (GUS) following electroporation of mesophyll and suspension protoplasts. In order to study tissue specific expression of such gene constructs, methods have been developed to isolate protoplasts from specific tissues, in particular from aleurone layers and endosperm.

2. MATERIALS AND METHODS

The green cell suspension culture was derived from immature inflorescence tissue of cultivar Maris Huntsman. Protoplasts from young wheat mesophyll cells and cell suspension lines (*T. monococcum* and C82d) were isolated following published procedures (1). Protoplasts were resuspended in KCl solutions (20 or 150 mM) and electroporation was carried out with a workshop built machine (equipped with a variable range of capacitances 50 to 2000 uF, and voltage range 0 to 400V). Plasmids used were p35SCN and pCGUS with the reporter genes chloramphenicol acetyl transferase (CAT) and β-glucuronidase (GUS) respectively. Both were under the control of the CaMV 35S promoter and NOS terminator sequences. GUS was assayed according to (2).

Developing wheat grains, 10 to 15 days post anthesis, were used to isolate endosperm protoplasts, and 19 to 25 days post anthesis grains for aleurone protoplasts. Whole endosperms were removed from grains and digested in enzyme solution (cellulase RS and pectinase). Protoplasts were freed from contaminating starch by gently spinning on a Nycodenz gradient. Aleurone protoplasts were isolated using a modified procedure (3) and freed from starch using a Nycodenz gradient.

3. RESULTS

3.1. Morphogenesis in protoplast derived callus

The cell suspension line derived from the immature inflorescence of cv. Maris Huntsman was characterised by compact green cell aggregates. Many of the isolated protoplasts contained chlorophyll and the yield from 1 ml of packed cells was 1.0×10^6/ml. Optimal plating density was at 1.5×10^5 protoplasts/ml. First divisions occurred 5 to 6 days after isolation and by 4 weeks microcolonies were visible (50% of the colonies were green at this stage). At the microcolony stage, protoplasts were plated onto solid media containing 2,4-D 1 mg l^{-1}. Addition of zeatin (1 mg l^{-1}) resulted in the production of darker green callus. Irrespective of a range of hormones tested in the media, differentiation of recognised structures (shoot meristems and leaves) occurred. Further growth beyond these stages was not obtained. The formation of such structures suggests that it should be possible to regenerate intact plants from protoplasts of wheat. Cytological examination of the inflorescence suspension culture showed that it had fewer structural chromosome changes than those of the immature embryo-derived suspension (C82d). The results suggest that control of genetic stability *in vitro* of cultures is a prerequisite for plant regeneration from protoplasts of wheat.

3.2. Electroporation and transient expression

A number of pulse types, including high voltage rectangular and capacitor discharge pulses have been used for transient expression studies. Best results were obtained with the workshop built high capacitance discharge machine which was set to give pulse lengths in the 50 to 100 ms range. Despite the use of a range of wheat protoplast types, and electroporation in a range of salts (KCl, NaCl, $MgCl_2$ and CH_3COOK), transient expression levels were extremely low. A comparison of equal numbers of mesophyll protoplasts from wheat and *Solanum brevidens* showed that levels of expression were in the order of 15 fold higher in *S. brevidens* than in wheat (Fig. 1).

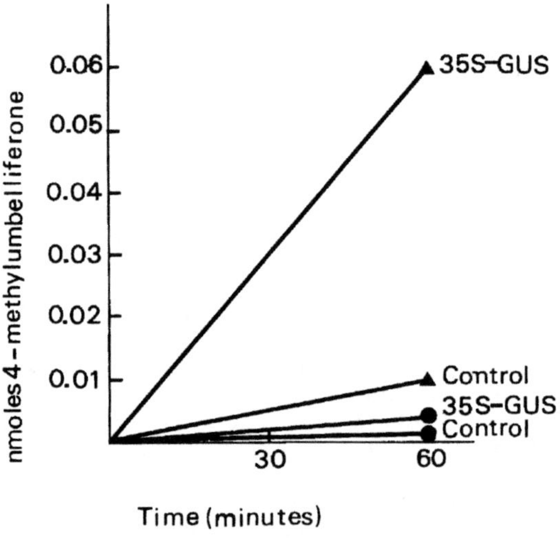

FIGURE 1. Transient expression of electroporated DNA in mesophyll protoplasts of wheat and *S. brevidens*. Mesophyll protoplasts, 1 x 10^6/ml, were electroporated in 20 mM KCl at 225V, 50 µF with 30 µg/ml pCGUS, 50 µg/ml carrier and assayed for GUS activity after 48 hours. ▲—▲ *S. brevidens*; ●—● wheat.

A similarly low level of transient expression was found in wheat mesophyll protoplasts with the p35SCN construct, whereas expression of this construct in potato and tobacco mesophyll protoplasts are consistently high (4). In principle, use of GUS as a reporter gene should be simpler than when CAT or NPTII are used as reporter genes. However, use of GUS as a reporter gene in transient expression studies with protoplasts presents some difficulties. The main problem is background activity of GUS, which comes from two sources: (1) the enzymes used in protoplast isolation and (2) endogenous tissue GUS activity. The former problem can be solved by choice of protoplast isolation enzymes and extensive washing of the protoplasts. The latter problem can be controlled by ensuring that negative controls are always pulsed, for each pulse condition. Mesophyll protoplasts of wheat are relatively free from endogenous GUS activity, whereas embryonic tissues have quite high endogenous levels of GUS. The levels of transient expression following electroporation of *T. monococcum* have also been quantified using GUS (Fig. 2).

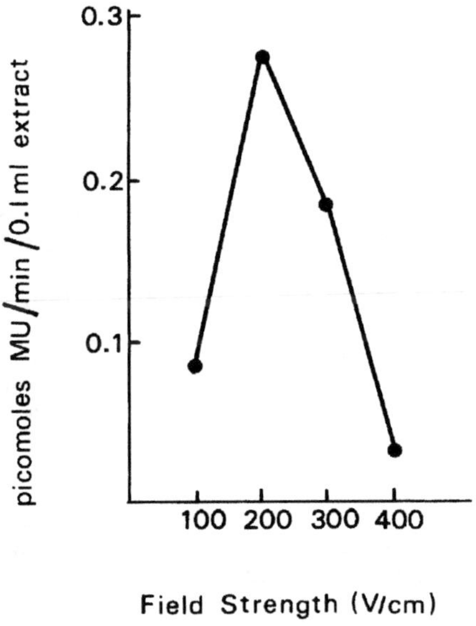

FIGURE 2. GUS activity in electroporated *T. monococcum* protoplasts. Protoplasts 2 x 10^6/ml, were electroporated in 150 mM KCl at varying voltages, 500 μF with 30 μg/ml pCGUS, 50 μg/ml carrier and assayed after 40 hours for GUS activity. (4-methylumbelliferone (MU) released from the substrate 4-methylumbelliferyl-β-D-glucuronide.)

The low levels of expression following electroporation of these reporter genes into cereal protoplasts suggest that fundamental differences exist between wheat and the dicot systems. The nature of these differences is under investigation.

3.3. Tissue specific protoplasts

Endosperm protoplasts have been isolated from developing grains of wheat (Fig. 3a). Their viability has been shown using FDA staining. Endosperm protoplasts are delicate because they contain numerous starch granules. It was only possible to obtain protoplasts during a specific developmental period (9-16 days post anthesis). At stages later than 16 days post anthesis, the presence of starch granules caused the protoplasts to rupture. Yields of endosperm protoplasts were approximately 1 x 10^5 - 1 x 10^6 protoplasts per 300 grains. Yield increases may be achieved by culturing ears before protoplast isolation.

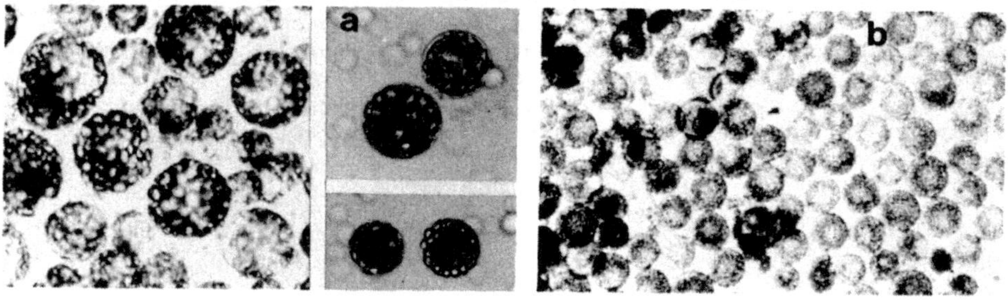

FIGURE 3. a. Light micrograph of endosperm protoplasts.
b. Light micrograph of aleurone protoplasts isolated from developing wheat grains

Aleurone protoplasts, free from contaminating endosperm, can be isolated from developing grains (Fig. 3b). Yields are between 1 to 1.5 million per 100 grains (19-25 days post anthesis). Higher yields should be possible in barley which has 3 layers of aleurone cells.

Both the endosperm and aleurone tissues are triploid and arise from the one cell. Their isolation from developing tissues in a highly purified form now allows them to be used for tissue specific studies using electroporation.

ACKNOWLEDGEMENT

The financial support of B. Lee, K. Murdoch and J. Topping by AGC/CIBA GEIGY is gratefully acknowledged. We also thank S. Maddock for the initial work on suspension cultures.

REFERENCES

1. Maddock SE: Suspension and protoplast culture of hexaploid wheat (*Triticum aestivum* L.). Plant Cell Rep, 6, 23-26 (1987).
2. Jefferson RA: GUS gene fusion system user's manual. Richard A. Jefferson, PBI Cambridge, England, March 1987.
3. Norman HA, Black M, and Chapman JM: The induction of sensitivity to gibberellin in aleurone tissue of developing wheat grains III. Sensitization of isolated protoplasts. Planta 158, 264-271 (1983).
4. Jones H, Tempelaar MJ, and Jones MGK: In Oxford Surveys of Plant Molecular and Cell Biology, Vol. 4. Ed. B. J. Miflin, 1987 (In press).

UV INDUCIBLE TRANSIENT EXPRESSION OF CHIMERIC CHS-NPTII IN PARSLEY PROTO - PLASTS

S. Lipphardt[1], F. Kreuzaler[2], J. Dangl[1] and J. Schell[1]

1. Max-Planck-Institut für Züchtungsforschung, D-5000 Köln 30

2. Institut für Biologie I, RWTH, D-5100 Aachen

1. INTRODUCTION

Light is a very important effector for differentiation in plant development. One of the most intensivly studied regulatory effects of (UV B)-light on specific metabolic pathways is the induction of flavonoid biosynthesis (Kreuzaler et al., 1983). Flavonoids are abundant in higher plants and function as flower pigments and UV protective compounds. Cultured parsley cells respond to UV irradiation by cellular accumulation of potentially UV protective flavonoids. The UV irradiation stimulates signal specific transcription and synthesis of chalcone synthase (CHS), the key enzyme of flavonoid pathway, and several other specific enzymes. We demonstrated that isolated parsley suspension protoplasts maintain fully the ability to be stimulated by UV light or other environmental factors as fungal elicitor (Dangl et al., 1987). This made the protoplasts an ideal system of reduced complexity to study inducible endogenous genes as well as exogenous transfected genes rapidly studied via transient expression.

2. PROCEDURE

Parsley suspension cultures are subcultivated once a week. A two or five days old culture is harvested, protoplasted in 0.24 M $CaCl_2$ 0.25% cellulase RS and 0.05% macerozyme for 14 hours and protoplasts are purified by pelleting, floating in B5 media (680mOs with sucrose) and a subsequent wash in $CaCl_2$ drops of 10^6 protoplasts in 200 µl are mixed with 200 µl 25% PEG in $CaNO_3$ and 20 µg DNA. After 20 minutes incubation, the mixture is washed with 5 ml $CaNO_3$, pelleted and plated in 3 ml of B5-680 mOS containing 1 mg/l 2;4-D. After a recovery period of 12 or 24 hours the transformed protoplasts are UV-irradiated for 15 hours, harvested and assayed for NPTII-activity (Schreier et al., 1985).

3. RESULTS

To study the Antirrhinum (CHS) promoter (Sommer and Saedler) a chimeric gene consisting of 1.2 kb of the Antirrhinum CHS promoter, including the TATA box and RNA-leader, were linked to the NPTII reporter gene and a genomic parsley CHS 3'end and integrated in tobacco (Kaulen et al., 1986). We assayed transient expression of previously described and newly constructed promoter deletion mutants (Fig. 1) in parsley protoplasts. The wild type promoter (714) showed background NPTII expression in the dark which could be increased 3-5 fold after UV induction with a time cours of induction similar to the endogenous parsley CHS gene (Fig. 2).
Removal of a 158 bp fragment in position -197 to -39 (1) led to loss of induced expression after UV stimulation. A larger deletion expanding from -39 to -357 (3) also did not respond to UV irradiation and the dark expression drops down as well. Trimming the 1.2 kb promoter to maintain only this 357 basepairs in their original position (2) gave rise to an inducible con-

struct. Shortening 2 to 158 basepairs in front of the TATA box (which are deleted in 1) led to lower levels of dark and UV induced expression than the wild type promoter, but, however, was sufficient to induce NPTII expression after UV irradiation. Removal of the next fragment at -197 to -357 (7) did not abolish UV induction whereas this upstream element fused to the TATA box (6) was unresponsive to UV irradiation.
These data indicate that the light responsive sequence is located within this 158 bp fragment. Interestingly internal inversion of the 158 bp fragment (714R) maintains inducibility.
Upstream sequences showed enhancing activity in background expression as well as in UV light induced expression. The region between -357 to -197 and the far upstream parts were involved in that.

4. DISCUSSION

We present a cis analysis of the UV inducible *Antirrhinum majus* CHS-NPT II gene measured via UV inducible transient expression. The data obtained in the transient assays fit to induction data observed in transformed tobacco teratoma tissue (Kaulen, 1987). The induction factor of 3-5 reached by the input gene is much lower than the increase of parsley endogenous CHS. This might be due to the transient expression system indicating that we do not mimic the *in vivo* situation but, however, may identify some *cis* acting DNA elements by transient expression.

REFERENCES

1. Dangl JL, Hauffe KD, Lipphardt S, Hahlbrock K, Scheel D: EMBO J. 6: 2551-2556 (1987).
2. Kaulen H, Schell J, Kreuzaler F: EMBO J. 5: 1-8 (1986).
3. Kaulen H: PhD-thesis (1987).
4. Sommer H, Saedler H: Mol. Gen. Genet. 202: 429-434 (1986).
5. Kreuzaler F, Ragg H, Fautz E, Kuhn DN, Hahlbrock K: Proc. Natl. Acad. Sci. USA 80: 2591-2593 (1983).

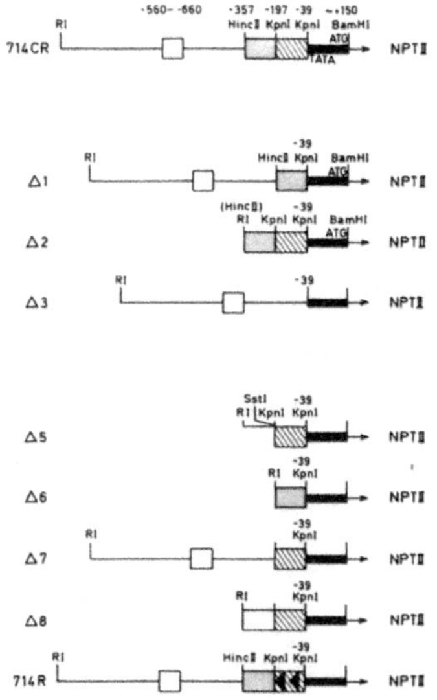

Figure 1

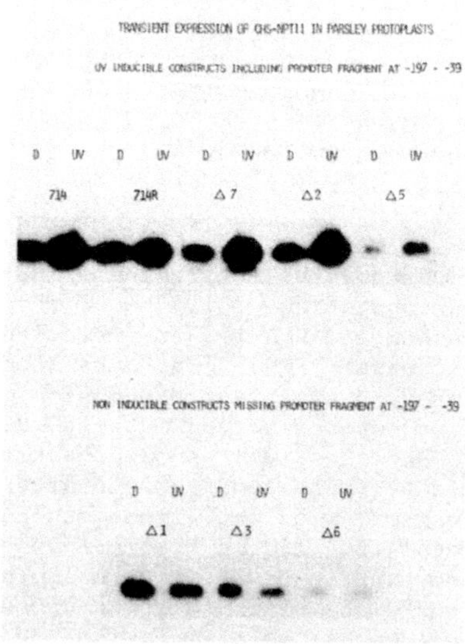

Figure 2

MODIFICATION OF PLANTS USING INTRACELLULAR ASSOCIATIONS OF L-FORM BACTERIA

E.J. ALLAN, C.M. INNES and A.M. PATON
Division of Bacteriology, Department of Agriculture,
University of Aberdeen, Scotland AB9 1UD

INTRODUCTION

Almost any bacterium can be converted into the so-called L-form by removal of the cell wall (1) and can be maintained indefinitely on nutritive media, supplemented with osmotic stabilisers. L-form bacteria differ from their parent cell walled form, in their size, vacuolation and pleomorphic nature (Fig. 1). Here we describe the longterm association of L-form bacteria with plants and show that bacterial secondary metabolites are produced by the L-form in association with the plant.

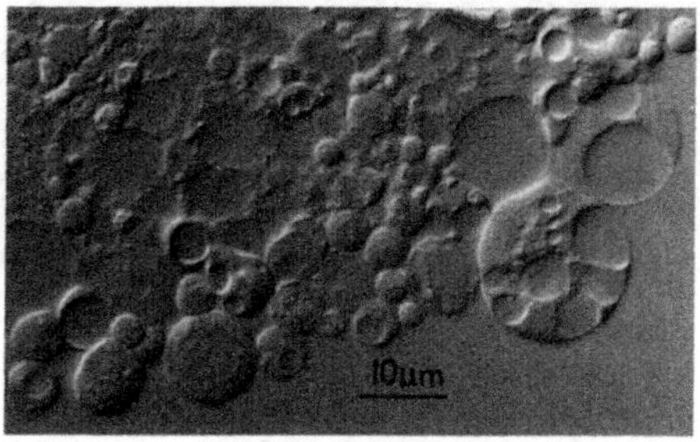

Fig. 1 L-forms of *Bacillus subtilis*

METHODS

Initiation of L-forms into cucumber

L-forms of *Bacillus subtilis* NCIMB 8054, ATCC 6634 were initiated using the *in vivo* method (2), whereby actively growing cells were incubated in a buffered saline solution containing 10% L-form Medium (sucrose 20%, glucose 0.5%, yeast extract 0.5%, peptone 0.5%, $MgSO_4 7H_2O$, 0.1%) and 10% horse serum supplemented with 1000 units benzyl penicillin (sodium) BP (Glaxo Ltd.). After a 24 h incubation at 25°C, the suspension was introduced into fresh cucumber fruits (*Cucumis sativus*). Control samples were prepared using the same bacterial suspension which had been boiled for 15 mins. The presence of L-form bacteria was confirmed by pectolysing the tissue in 0.5% (w/v) Pectinase Rohamet P5 (Serva Feinbiochemica) and examined using a Reichert-Jung Polyvar microscope. Interference, phase contrast and epifluorescence was used to identify the L-form bacteria within the cells. Cucumber samples were homogenised in a Waring Blender and passed through a 0.45μm filter. The filtrates were stored at -20°C.

Initiation of L-form Associations in Chinese Cabbage

Seeds of chinese cabbage (*Brassica campestris spp. pekinensis* cv. *Granaat*) were surface sterilised and germinated on water agar. The seedling roots were placed in a well containing filtrates of *B. subtilis*/cucumber associations. After an overnight incubation the roots were washed in sterile water and examined for the presence of L-forms prior to growing in greenhouse conditions.

Antibiotic Production in Plants

Seedlings of chinese cabbage were macerated and the sap passed through a 0.45µm filter. Bioassay plates of glucose agar seeded with the sensitive organism, Pseudomonas syringae var. phaseolicola (a pathogen of proved virulence) were used with sap being added to central wells. Antibiotic produced by the L-form bacteria and contained in the plant cell sap was thus detected by the formation of inhibition zones.

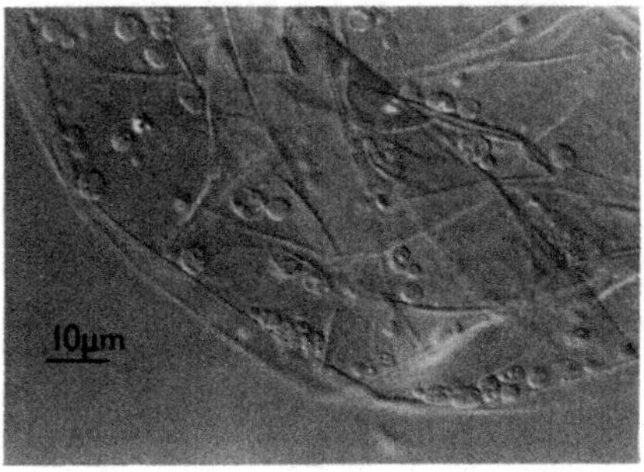

Fig. 2 L-form bacteria in cucumber cells

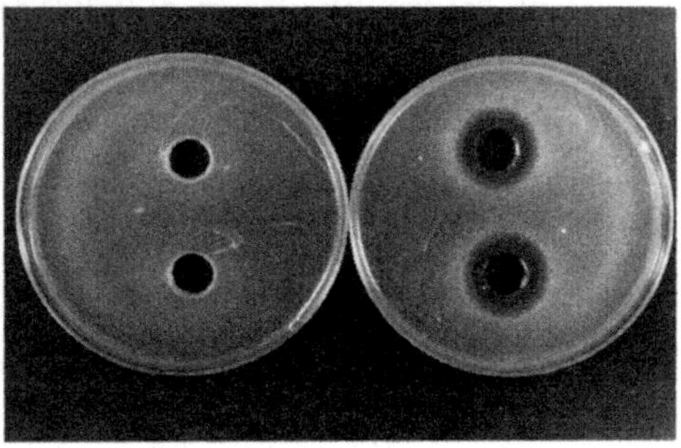

Fig. 3 Bioassay plates showing production of antibiotic in control samples compared to test samples shown on the left hand side.

RESULTS AND DISCUSSION

The association of L-forms in cucumber cells (Fig. 2) and chinese cabbage root hairs was evident compared to control samples. The subsequent association of L-form bacteria in chinese cabbage that had been inoculated with a cucumber filtrate illustrates that these unusual bacteria can pass through microbial filters. Moreover the bioassay (Fig. 3) illustrates the active longterm association of B. subtilis L-forms in chinese cabbage and demonstrates the transfer of bacterial expression into plants without the manipulation of plant DNA.

RELEVANT REFERENCES

1. Madoff, S M (1986). The Bacterial L-forms. Marcel Dekker Inc., New York and Basel.
2. Paton A M (1984). Associations of prokaryotic organisms and eukaryotic plant cells. International Patent Classification C12 15/00. Patent Number WO 85/00828.
3. Aloysius, S K D and Paton, A M (1984). Artificially induced symbiotic associations of L-form bacteria and plants. Journal of Applied Bacteriology, 56, 465-477.
4. Jones, S M and Paton, A M (1973). The L-phase of Erwinia carotovora var. atroseptica and its possible association with plant tissue. Journal of Applied Bacteriology 36, 729-737.

Acknowledgement

We would like to thank the Venture Research Unit of British Petroleum International Ltd for supporting this research.

TRANSFORMATION OF BRASSICA NIGRA THROUGH PROTOPLAST COCULTIVATION WITH AGROBACTERIUM TUMEFACIENS

M.D. SACRISTAN, M. GERDEMANN-KNÖRCK, and O. SCHIEDER

Institut für Angewandte Genetik, Freie Universität Berlin, Albrecht-Thaer-Weg 6, 1000 Berlin 33, FRG.

In this paper we report on experiments leading to the introduction of a dominant selectable marker in B. nigra lines possessing a double disease resistance (against Phoma lingam and Plasmodiophora brassicae). Three disarmed Agrobacterium tumefaciens strains, two of them carrying a gene for kanamycin resistance and the third one a gene for hygromycin resistance, were assayed for transforming B. nigra. Previous work has demonstrated that it is possible to transform all basic Brassica species by protoplast cocultivation with different wild-type and T-DNA deletion ("shooter") mutant strains of A. tumefaciens (1).

Protoplast isolation and cocultivation procedure

Mesophyll protoplasts of B. nigra were isolated from leaves of sterile shoot cultures and plated on MI medium (2) at a concentration of 2-5 x 10^5 protoplasts/ml. Four to 6 days after isolation the protoplast cultures were inoculated with agrobacteria at an initial concentration of 100-300 bacteria per protoplast. After 48 h cocultivation, bacterial growth was inhibited by adding Claforan to a final concentration of 500 ug/ml. Four days later, protoplast cultures having growing colonies were embeded in MI-seaplaque-agar and the agar blocks were transferred to plates containing liquid K3 medium (3) with the respective toxic compound for selection.

RESULTS

In a series of 22 experiments, two disarmed A. tumefaciens strains carrying a gene for kanamycin resistance (NPT II) under the control of the NOS promoter were used for protoplast cocultivation: C58CI pGV3850:1103 neo and pTiA6:pMON 501. In no case kanamycin resistant colonies could be recovered.

Alternative systems for transformation using differentiated plant tissue (hypocotyls, epicotyls, leaf discs, stems) gave also negative results with respect to transformation with the kanamycin strains, although transformants could be obtained both after inoculation with wild-type (C58) and with a shooter mutant (3851). Also in co-transformation experiments using wild-type as "helper", the obtained transformants showed to have been transformed by only the wild-type strain (Table 1).

The stable introduction of a dominant selectable marker in the B. nigra genome was possible by using the Agrobacterium strain C58 pGV3850:Hygro^R, which carries a gene for hygromycin resistance. Hygromycin showed to be lethal at a concentration of 25 ug/ml, both for B. nigra protoplast colonies and seedlings. Protoplast-derived cells were cocultivated as described, and growing colonies on selective medium were already visible 10-20 days after the cocultivation.

The results of six transformation experiments are presented in Table 2. Only part of the colonies growing during the first selection phase (in agar blocks on liquid medium with hygromycin) survived after they were individually removed and transferred to agar-hygromycin medium.

Transformed calli grow on hygromycin as vigorously as control calli on non-selective medium; however, not all of them express the nopaline gene (Table 2).

TABLE 1. Transformation experiments on differentiated tissue of sterile B. nigra plants with wild-type, "shooter", and disarmed kanamycin^R strains of Agrobacterium tumefaciens

A.t. strain	No. inoc.	Tumors develop.	%	Calli grown on selective medium* km 100mg/l	hormone-free	%	Nop$^+$ calli/ total tested
C58	149	89	59.7		84	94.4	27/32
pGV3851	152	0	0		11+(10)	7.2	6/12
pGV3850:km^R	248	0	0	(24)		0	0/15
pMON501:km^R	247	0	0	(26)		0	0/13
3850 + C58	75	31	41.3	(7)	29	0/93.5	31/36
MON501 + C58	68	29	42.6	(4)	22	0/75.8	15/19

*In brackets: calli initially growing on selective medium; growth ceases in further transfers.

TABLE 2. Transformation experiments by coculture of protoplasts from B.nigra with a disarmed A. tumefaciens strain carrying hygromycin resistance (pGV3850:Hygro^R)

Exp. no.	B.nigra line	Plated ppl.	Division rate at day 4	Survival after cocult.*	Growing colonies on hygr.	Transform. rate	Nop$^+$calli/ tot. test.
30/4	461	$7.8x10^6$	30%	10%	3	$1.3x10^{-5}$	0/3
7/5	460	$7.2x10^6$	22%	30%	27	$6x10^{-5}$	19/27
4/6	460	$3.6x10^6$	17%	15%	5	$5x10^{-5}$	1/1
19/6	460	10^7	20%	67%	12	$9x10^{-6}$	5/20
6/8	16220	10^7	20%	10%	0	0	
3/9	2051	$4.8x10^6$	30%	23%	7	$2x10^{-5}$	**

*Percentage of growing colonies from cocultured cells on non-selective medium with respect to controls.
**Not yet tested.

The reason why transformation was possible only with the hygromycin strain but not with the kanamycin strains is unclear. Similar results using the same Brassica material have been obtained in transformation experiments with plasmid DNA (Golz et al., this volume).

REFERENCES

1. Gerdemann M, Sacristán MD, Schieder O (1987). XIV Internat. Botanical Congress, Abstr.
2. Kohlenbach HW (1983). Vortr. Pflanzenzüchtg. 2, 35-50.
3. Glimelius K (1984). Physiol. Plant. 65, 38-44.

Acknowledgment
This work was supported by the Deutsche Forschungsgemeinschaft.

TRANSFORMATION OF BRASSICA SPECIES VIA DIRECT GENE TRANSFER

C.Golz, F.Köhler, M.D.Sacristan and O.Schieder
Institut für Angewandte Genetik; Freie Universität Berlin;
Albrecht-Thaer-Weg 6; 1000 Berlin 33; F.R.G.

INTRODUCTION

Two approaches may be relevant for the transfer of disease resistances using two plant species, which can not be crossed sexually: asymmetric protoplast fusion and direct gene transfer. Important prerequisites in both techniques are that the plant species used possess a good regeneration capacity and that high transformation rates can be obtained using an antibiotic marker gene. Such marker genes can also be used to select hybrids in fusion experiments. Direct gene transfer and co-cultivation experiments (Sacristan et al.,this volume) were started with

a) Brassica nigra lines possessing resistances against Phoma lingam and Plasmodiophora brassicae;

b) Brassica napus varieties, which are sensitive against both pathogens.

MATERIAL AND METHODS

Leaf material of four-week-old sterile shoot cultures of four B.nigra (2051, 2058, 16220, 461) and three B.napus lines (H1, Liropa, Garant) was incubated overnight in a solution of 0.25 % cellulase, 0.05 % macerozyme and 0.5 M mannitol. Protoplasts were purified using 0.6 M sucrose and seawater as washing solution. Per 10^6 protoplasts/ml, 20 - 50 ug plasmid DNA (pABD1 carrying the NPT II gene or pGL2 containing a Hyg-R gene), 50 - 100 ug calf thymus DNA and PEG were added resulting final PEG concentrations of 8 - 20 %. After 7 days culture a selection pressure was established by adding 75 - 100 mg/l kanamycin or 25 - 60 mg/l hygromycin to the culture media. Protoplasts were cultured in M1 medium (Li and Kohlenbach, 1982) and in K3 media (Glimelius, 1984).

RESULTS AND DISCUSSION

B.nigra: Protoplast cultures and transformation experiments with four lines showed that only in the B.nigra lines "2051" and "16220" the survival rates of protoplasts and the production of colonies after transformation with plasmid DNA/PEG were high enough to select transformants. The plating efficiencies and callus production rates of protoplasts of lines "2058" and "461" were only about 10^{-4}. In the lines "2051" and "16220" colony formation per protoplast treated was in the range of 10^{-2} - 10^{-3}. The disadvantage of line "16220" based on the very low frequency of green colonies, which is likely important for differentiation into plants.

More than 90 % of the colonies from "16220" remained "white". Using the transformation method of Negrutiu et al. (1987) transformants could be observed by treating protoplasts with pGL2 plasmid DNA conferring hygromycin-resistance (Tab.1). A high $MgCl_2$ concentration (95 mM) seems to result in higher transformation rates in comparison to the use of original $MgCl_2$ concentration (15 mM) in the incubation medium. Per 10^6 protoplasts of line "2051" about 8-12 transformed colonies could be regenerated.

Table 1: Direct gene transfer in B.nigra "2051". Protoplasts were treated with pGL2 plasmid DNA, carrier DNA and PEG (final conc.: 13,3 %). Transformation rates in dependence of the $MgCl_2$ concentration used are shown from three independent experiments.

	15 mM $MgCl_2$:	95 mM $MgCl_2$:
% Hyg^R - colonies:	0.0/0.02/0.15	0.04/0.11/0.32
mean value:	0.06	0.16

The treatment of B.nigra protoplasts with plasmid pABD1 containing the NPT II gene did not yield any kanamycin-resistant colony although identical conditions were applied like in the successful transformation experiments using pGL2 plasmid. Now, we are testing another kanamycin-confering plasmid (pHP23), which has a higher expression rate than pABD1 (Paszkowski, pers. communication).

B.napus: Like in B.nigra the treatment of protoplasts with pABD1 plasmid DNA did not result in transformed colonies. But according to the first results with the B.napus lines H1 and Liropa, hygromycin-resistant lines could be selected using plasmid pGL2, 20 % final PEG concentration and the transformation method described above for B.nigra.

In contrast to the successful experiments using pABD1 plasmid for the direct gene transfer in N.tabacum and Vignaaconitifolia (Köhler et al.,1987), the application of this kanamycin-resistance confering plasmid was unsuccessful in B.nigra and B.napus. The experiments with these two plant species clearly showed that appropriate plant genotypes and gene vectors (here: pGL2) are decisive factors in direct gene transfer.

REFERENCES

Glimelius, K. (1984), Physiol.Plant. 61: 38-44.
Köhler et al. (1987), Plant Science (in press).
Li, L. and H.W.Kohlenbach (1982), Plant Cell Rep. 1: 209-211.
Negrutiu et al. (1987), Plant Mol.Biol. 8: 363-373.

GENETIC TRANSFORMATION OF DIPLOID SOLANUM TUBEROSUM BY DIRECT DNA TRANSFER AND MEDIATED BY AGROBACTERIUM TUMEFACIENS

S.KNAPP, H.UHRIG AND F.SALAMINI

Max-Planck-Institut für Züchtungsforschung, Egelspfad, 5000 Köln 30, FRG

1. INTRODUCTION

Using two different plant transformation techniques -direct DNA transfer into protoplasts and cocultivation of leaf discs with A. tumefaciens-the bacterial gene coding for neomycin-phosphotransferase II (NPT II) was introduced into a dihaploid (2n=2x=24 chromosomes) potato clone. The final goal of the project is the establishment of a set of diploid potato clones with individual chromosomes marked by the selectable NPT II gene, facilitating genetic analysis of the potato genome.

2. MATERIAL AND METHODS

2.1. Direct DNA transfer into protoplasts

Mediated by PEG as fusing reagent, protoplasts isolated from axenic shoot cultures of the dihaploid potato clone H 81.1506/60 were incubated with plasmid DNA pLGV11neo, pLGV1103neo (L. Herrera-Estrella, Gent) and pKU2 (B. Baker, USDA, Albany), carrying the NPT II gene as selectable marker. Ten to twelve days after treatment with plasmid DNA and culture in liquid medium (1) the protoplasts were plated on solid medium allowing callus induction. Kanamycin resistant colonies were regenerated in the presence of 100 ug/ml Km.

2.2. Leaf disc transformation via A. tumefaciens

Leaf explants of 3-4 week old shoot cultures of the dihaploid potato clone H 81.1506/60 were incubated with an overnight culture of A.tumefaciens harbouring the plasmids pGV2260::I170 (A.Depicker, Gent), pVU1011 (P.v.d.Elzen,Leiden) or pGV3850::pKU2 (B.Baker, USDA, Albany),all carrying the NPT II gene as selectable marker and cocultured during 2 days. Immediately after cocultivation the leaf pieces were placed on MS supplemented with 2 mg/l zeatin-riboside, 0.02 mg/l NAA, 0.02 mg/l GA_3, 500 mg/l Claforan and 100 mg/l Km.

3. RESULTS

3.1. Direct DNA transfer into protoplasts

Kanamycin resistant colonies appeared after 8 weeks with a frequency fo 0.1 % with respect to surviving unselected calli. First shoots were obtained 6 months after protoplast isolation. Shoots were rooted in the presence of 100 mg/l Km and expressed the enzyme NPT II (Fig.1a). Southern analysis revealed the presence of the NPT II ocding sequence in Km-resistant plants, but also modifications of the input DNA (Fig.2bi). The clonal origin of shoots arising from the same callus is observed(Fig.

2b, lane 7-9). All transformants tested so far (8) were tetraploid.

3.2. Leaf disc transformation with *A. tumefaciens*

After treatment of potato leaf pieces with *A.tumefaciens* and selection on Km first shoots appeared after 6 weeks. 10-50% of the incubated leaf pieces gave rise to Km-resistant shoots which expressed the NPT II gene (Fig.1b). Southern analysis of transformed plants revealed a band of 2.9 kb using EcoRI-Hind III digests and as probe a 1.6 kb PstI fragment of plasmid pKU2, covering the coding region of the NPT II gene (Fig.2a). The 2.9 kb band represents the uninterrupted EcoRI-HindIII fragment of the input DNA (2). Four of five regenerants tested were still diploid.

Fig.1. NPT II activity in Km^R plants a) after direct DNA-transfer into protoplasts (1-4) and b) after leaf disc transformation (5-10). C^- untransformaed plant C^+ transformed tobacco

Fig.2. Southern analysis of Km^R plants a) after leaf disc transformarion (3); b) after direct DNA transfer into protoplasts (4-9). 7-9: shoots from the same callus. 1: plasmid pKU2;2:untransformed plant

4. CONCLUSION

The comparison between the two methods of transformation adapted (2,3) is summarized in the following table:

	Direct DNA-transfer into protoplasts	Leaf disc transformation via A.t.
transformation frequency	0.1 % of surviving unselected calli	10-50 % of incubated leaf discs
regeneration time	6-10 months	6-8 weeks
ploidy level after regeneration	2n=4x=48	2n=2x=24 (4/5)
fate of integrated marker DNA	modifications and rearrangements	no modifications observed

5. REFERENCES

1. Shahin EA (1984) In: Cell culture and somatic cell genetics of plants. Vasil IK (ed) Vol 1. Academic press: 381-390
2. Czernilofsky AP et al (1986) DNA 5: 101-113.
3. Horsch RB et al (1986) PNAS 83: 2571-1575.

OPTIMIZATION OF DIRECT GENE TRANSFER IN TOMATO

CORNÉ PIETERSE and MAARTEN KOORNNEEF
Department of Genetics, Agricultural University,
53 Generaal Foulkesweg, 6703 BM Wageningen, The Netherlands

1. INTRODUCTION

The successful transformation of callus derived protoplasts of the tomato genotype MSK93 was recently described (2, 4). Transformants were obtained with plasmid DNA containing the kanamycin resistance gene from Tn5 with expression signals from the nos gene as a selectable marker (plasmid pneo△18). In addition a (0.8 kb) tomato sequence which promoted autonomous replication in yeast was cloned into pneo△18 resulting in pCTW22 (2). Because thus far only a limited number of transformants were obtained (exclusively with pCTW22) and no diploid plants could be regenerated from these callus derived protoplasts, the transformation procedures were further optimized with leaf derived protoplasts of MSK93.

2. MATERIAL AND METHODS

The origin and properties of the easily regenerable tomato genotype MSK93 has been previously described (3, 4). The plasmids used and the DNA isolations are described by Jongsma et al. (2).

DNA transformation of protoplasts was performed according to the modified Krens procedure (2) and the $MgCl_2$/PEG procedure of Negrutiu et al. (5). Alliquots of 5.10^5 protoplasts (1 ml) were used with 30 µg plasmid DNA in all experiments. The plasmids were linearized by cutting the plasmids at their unique HindIII site. In some experiments a heat-shock treatment (5) was applied. In addition the transformation procedure of Hain et al (1) with the PEG treatment and dilution scheme of the modified Krens procedure and electroporation with the media of respectively Riggs and Bates (7) and Shillito et al. (8) were tested.

Protoplasts were isolated and cultured in 2 ml liquid TMP (5 cm dishes) as described previously (4). On the 3th day after transformation the protoplasts were embedded in 0.6% Sea Plaque agarose (4 ml). Agarose parts were placed into liquid medium 3 days later. Selection started on day 10 when the liquid medium was replaced with 6 ml dilution medium (4) containing 100 mgl^{-1} kanamycin.

Growing microcalli were subsequently transferred to callus growth, shoot induction and rooting media all containing 100 mgl^{-1} kanamycin.

3. RESULTS AND DISCUSSION

The modified Krens procedure and the method of Negrutiu et al. (1987) gave fairly reproducible frequencies of 0.1% (range 0-0.6%) kanamycin resistant calli among the colony forming units (RTF) (see Table 2).

The modified Hain procedure resulted in comparable RTF values, although survival was very much reduced. A heat-shock treatment and the use of linearized plasmid DNA appeared to enhance the transformation frequency. The two electroporation procedures (7, 8) yielded transformants, although at a lower frequency than the chemical methods.

TABLE 2. Plating efficiencies and transformation frequencies obtained with MSK93 leaf protoplasts.

Plasmid	Linea-rized	Transform. procedure[1]	Number of exp.	Average PE[2]	RTF($x10^{-4}$)[3]	ATF($x10^{-5}$)[4]
pneo △18	–	K	5	1.9	5.2	1.0
	+	K	4	2.7	7.4	2.0
	–	N	7	2.2	10.3	2.3
pCTW 22	–	K	5	2.7	8.1	2.2
	+	K	4	2.8	13.1	3.7
	–	N	7	2.8	10.8	3.0
control		untreated	8	7.9	0	0

1) K: Modified Krens procedure N: Negrutiu procedure
2) PE: Plating efficiency
3) RTF: Relative transformation frequency
4) ATF: Absolute transformation frequency

Plasmids pCTW22 and pneo△18 were used in all experiments and comparison of transformation frequencies showed that pCTW22 gives a 1.5 higher transformation frequency than pneo 18. Tested with a sign test this was significant at P = 0.05.

In 9 transformants that were tested, the kanamycin gene was integrated into the nuclear genome as shown by Southern blotting.

The slightly higher transformation frequency of pCTW22 might be explained by a (temporary) more stable or more abundant presence within the cell before integration takes place and not by an improved chance for integration since Jongsma et al (2) found no indication for homologous recombination in 7 pCTW22 transformants.

For 20 independent transformants (10 of each plasmid were tested) resistance to kanamycin at the callus level was still present after 4 weeks on non-selective medium.

Regeneration was tested for 67 resistant calli. After 3 months on shoot induction medium 66 transformants had formed shoots. Out of 51 plants regenerated from 24 transformants 8 plants were diploid, 42 tetraploid and 1 probably octaploid as determined by counting the number of guard cell chloroplasts. Five of the 24 transformants regenerated diploid plants.

REFERENCES

1. Hain R, Stabel P, Czernilofsky AP, Steinbiss HH, Herrera-Estrella L, Schell J: Mol. Gen. Genet. 99: 161–168 (1985).
2. Jongsma M, Koornneef M, Zabel P, Hille J: Plant Mol. Biol. 8: 383–394 (1987).
3. Koornneef M, Hanhart C J, Martinelli L: Theor. Appl. Genet. 74: 633–641 (1987).
4. Koornneef M, Hanhart C, Jongsma M, Toma I, Weide R, Zabel P, Hille J: Plant Science 45: 201–208 (1986).
5. Negrutiu I, Shillito R, Potrykus I, Biasini G, Sala F: Plant Mol. Biol. 8: 363–373 (1987).
6. Potrykus I, Saul M, Petruska J, Paszkowski J, Shillito RD: Mol. Gen. Genet. 199: 183–188 (1985).
7. Riggs CD, Bates GW: Proc. Natl. Acad. Sci. USA 83: 5602–5606 (1986).
8. Shillito RD, Saul MW, Paszkowski J, Müller M, Potrykus I: Biotechnology 3: 1099–1104 (1985).

TRANSFORMATION STUDIES IN TOBACCO, MOTH BEAN AND BRASSICA SPECIES

F.Köhler, K.Afsar, C.Golz, S.Eapen and O.Schieder
Institut für Angewandte Genetik; Freie Universität Berlin
Albrecht-Thaer-Weg 6; 1000 Berlin 33; F.R.G.

INTRODUCTION

Besides the *Agrobacterium tumefaciens* mediated gene transfer, the direct gene transfer is an established technique for transforming single cells of higher plants. Transformation rates are influenced by factors like heat shock, PEG-concentration, carrier-DNA and incubation time (Shillito et al. 1985). Here, the influence of plant genotype on the transformation rates was investigated using several varieties of four plant species. The importance of plant genotype on the regeneration potential in cell and tissue culture is recognized since many years. Other important factors, incubation media and vectors carrying different antibiotic marker genes and/or promotors, also influenced the transformation rates obtained.

MATERIAL AND METHODS

Plant material and protoplast isolation. The isolation of mesophyll protoplasts in tobacco and *Vigna aconitifolia* is described by Köhler et al. (1987a). Brassica mesophyll protoplasts (*B.nigra*; *B.napus*) were isolated after an overnight incubation in 0.25 % cellulase, 0.05 % macerozyme and 0.5 M mannitol using a standard method.

Plasmid DNA. Two kanamycin-resistance confering plasmids, pLGV neo2103 (Hain et al.,1985) and pABD1 (Paszkowski et al., 1984) and one hygromycin-resistance confering plasmid, pGL2, were used in the direct gene transfer experiments. Plasmid pGL2 was obtained by cloning the coding sequence of hyg.-gene as a BamHl fragment into plasmid pDH51 (Paszkowski, pers.communication; Pietrzak et al.1986).

Protoplast transformation. Tobacco and moth bean protoplasts were treated with plasmid DNA, carrier DNA and PEG according to the method of Shillito et al.(1985) modified as described by Köhler et al. (1987 a,b). For the transformation of Brassica protoplasts the incubation media according to Negrutiu et al. (1987) were used.

Protoplast culture and selection of transformed lines. Tobacco protoplasts were cultured in K3 medium (Nagy and Maliga, 1976) and moth bean protoplasts were suspended in V47 medium (Binding, 1974). Both culture media contained 1.5 mg/l NAA and 0.4 mg/l BAP. Brassica protoplasts were cultured in media according to Li and Kohlenbach (1982) and Glimelius (1984). After one week a selection pressure was established by adding media supplemented with 75 - 100 mg/l kanamycin or 25 - 40 mg/l hygromycin to the bead type culture system.

RESULTS

Nicotiana tabacum. Protoplasts of N.tabcum cvs. "SR1" and "W38" were treated with carrier DNA, PEG (8 %) and pLGVneo2103 DNA or pABD1 DNA. Significant differences in transformation rates were obtained (Tab. 1) in dependence of the plant genotype used. 1.54 % (Standard Deviation: 0.69) of the colonies deriving from "SR1"-protoplasts expressed the kanamycin resistance, whereas only 0.23 % (S.D.: 0.16) of the "W38"- calli were transformed. The application of pLGVneo2103 plasmid DNA for transformation instead of pABD1 DNA yielded distinctive lower levels of transformants in "SR1" (0.03 % kan.-R colonies). Using "W38"- protoplasts transformed lines were not detected. Resistant colonies could be regenerated into plants and transformation was confirmed by Southern blot analysis (Golz et al., 1987).

Table 1: Influence of plant cultivar and plasmid DNA upon transformation rates in N.tabacum. Selection pressure was 75 mg/l kanamycin. In control cultures, treated in the same manner but without adding plasmid DNA, resistant colonies were not obtained. Results from 5×10^6 protoplasts treated.

Cultivar:	"SR1"		"W38"	
Plasmid:	pABD1	pLGV	pABD1	pLGV
No. kan.-R colonies:	54	1	6	0
% transf. colonies :	1.54	0.03	0.23	0.0

Vigna aconitifolia. Direct gene transfer proved to be an efficient transformation method for legumes. Using three varieties, "560", "909" (Köhler et al., 1987b) and "GP-4" (Golz, unpublished) of the grain legume Vigna aconitifolia, a well-established commercial crop of the Indian subcontinent, transformation rates comparable with the frequencies in tobacco were obtained (Tab. 2). The two varieties "560" and "909" showed similar callus production rates (1.1×10^{-3}; 1.3×10^{-3}) but the percentage of transformed colonies in line "560" was 8-fold higher than in lines "909" and "GP-4".
Like in tobacco, the use of pLGVneo2103 containing the nopaline synthase promotor yielded transformation rates which were lower in comparison to the rates obtained after transformation with plasmid pABD1 carrying a cauliflower mosaic virus promotor. Southern blot analysis showed the integration of functional copies of the marker gene into the plant genome.
The cultivar dependence of transformation rates was also significant in transformation experiments using the co-cultivation method with Agrobacterium tumefaciens (Tab. 3). The rate of kanamycin-resistant colonies was 3.3 % in line "560" and only 0.04 % in line "909" on average of three experiments (Eapen et al.,1987). In each experiment the influence of plant cultivar on transformation rate was evident.

Table 2: Influence of plant cultivar and plasmid DNA upon transformation rates in V.aconitifolia. Selection pressure was 75 mg/l kanamycin. In control cultures (same treatment without plasmid DNA) resistant calli did not grow.

Variety	Plasmid	Protoplasts ($\times 10^6$)	Total no. of colon.	Kan.-R. colon.	% transf. colon.
560	pABD1	4.50	5500	54	0.98
909	pABD1	3.75	4850	6	0.12
GP-4	pABD1	2.00	10000	10	0.10
560	pLGVneo	4.50	4650	27	0.58
909	pLGVneo	4.00	5000	4	0.08

Table 3: Transformation rates in V.aconitifolia obtained after co-cultivation with A.tumefaciens strain C58C1 containing pGV 3850:1103neo. Results from 10×10^6 protoplasts treated per variety (Eapen et al.,1987).

Cultivar	Kan.-R. colonies	% resist. colonies
560	414	3.3
909	5	0.04

Brassica species. Transformation experiments with several B.napus and B.nigra varieties were started in order to test the "transformation capacity" of these plant species. From a large collection of B.nigra, double disease resistant lines against Phoma lingam and Plasmodiophora brassicae were selected by Sacristan (unpubl.data). We want to use these varieties for the transfer of disease resistances and other traits into B.napus protoplasts. For that, an optimal transformation system has to be established. In a first step transformation rates were estimated using antibiotic marker genes.

Table 4: Protoplast culture and direct gene transfer in four B.nigra lines. Protoplasts were treated with pABD1 DNA or with pGL2 DNA. Mean values of three experiments.

Lines :	2051	16220	461	2058
No. protoplasts :	1×10^6	1×10^6	1×10^6	1×10^6
Total No. colonies after transform. :	8000	6000	120	140
Frequency :	8×10^{-3}	6×10^{-3}	1.2×10^{-4}	1.4×10^{-4}
% green colonies :	90 %	10 %	n.d.	n.d.
No. kan.-resist. colonies :	0.0	0.0	n.d.	0.0
No. hyg.-resist. colonies :	12	2	n.d.	n.d.
% resist. colonies:	0.15	0.03	---	---

n.d. - not determined

Brassica nigra: Using the W5 incubation medium (Negrutiu et al.,1987) and plasmid pGL2, hygromycin-resistant colonies (selection pressure: 25 mg/l hyg.) were observed deriving from protoplasts of line "2051" (Tab. 4). The increase of the $MgCl_2$ concentration from 15 mM to 95 mM and a final PEG-concentration of 13 % instead of 20 % led to a transformation rate of 0.15 % (Golz et al.,this volume). In regeneration experiments using untransformed protoplasts, line 2051 also seems to possess a good "regeneration capacity" in comparison to other varieties tested (Gerdemann, pers. communication). Resistant colonies are now on regeneration media with or without hygromycin supplement.

Brassica napus: According to the first results using the variety "Liropa", hygromycin-resistant colonies (selection pressure: 30-40 mg/l hyg.) were obtained after treatment of protoplasts with pGL2 DNA and PEG (final conc.: 20 %). However, the application of plasmid pABD1 containing the NPT II gene instead of plasmid pGL2 did not result in transformed colonies. The same observation was made in B.nigra where a kanamycin-selection was also unsuccessful using plasmid pABD1 in direct gene transfer.

DISCUSSION

The results of direct gene transfer experiments in both tobacco and moth bean show that the plant cultivar possesses an significant influence on the number of stably transformed colonies. Using three varieties of the grain legume Vigna aconitifolia, transformation frequencies of 0.98 % ("560"), 0.12 % ("909") and 0.10 % ("GP-4") were observed. After co-cultivation of protoplasts with A.tumefaciens, the percentages of kanamycin-resistant colonies of lines "560" and "909" were clearly different. The application of the kanamycin confering plasmid pABD1 in B.nigra and B.napus was unsuccessful, while a hygromycin selection led to transformed colonies after treatment of protoplasts with plasmid pGL2. On the other hand, Paszkowski et al. (1986) and Guerche et al. (1986) demonstrated that in Brassica species kanamycin-resistant colonies can be observed after direct gene transfer. These opposing results may be a consequence of using different plant genotypes and/or culture conditions.

REFERENCES

Binding H. (1974), Z.Pflanzenpysiol. 74: 327-356.
Eapen S. et al. (1987), Theor.Appl.Genet. (in press).
Glimelius K. (1984), Physiol.Plant. 61: 38-44.
Golz C. et al. (1987), IAPTC, Berlin, Abstract p. 89.
Guerche P. et al. (1986), Cruciferae Newsl. 11: 98-99.
Hain R. et al. (1985), Mol.Gen.Genet. 199: 161-168.
Köhler F. et al. (1987a), Plant Science (in press).
Köhler F. et al. (1987b), Plant Cell Rep. 6: 313-317.
Li L. and H.W. Kohlenbach (1982), Plant Cell Rep. 1: 209-211.
Nagy J. and P. Maliga (1976), Z.Pflanzenpysiol. 78: 453-455.
Negrutiu I. et al. (1987), Plant Mol.Biol. 8: 363-373.
Paszkowski J. et al. (1984), EMBO J. 3: 2717-2722.
Paszkowski J. et al. (1986), Plant Mol.Biol. 6: 303-312.
Pietrzak M. et al. (1986), Nucl.Acid Res. 14: 5857-5868.
Shillito R.D. et al. (1985), Bio/Technol. 3: 1099-1103.

TRANSPOSITION AND REGULATION OF THE EN ELEMENTS OF MAIZE ASSAYED IN TRANSFORMED TOBACCO

S.R. Grant*, A. Pereira, H. Cuypers, W.H. Martin, D.J. Lydiate, and H. Saedler

Max-Planck-Institut für Züchtungsforschung, D 5000 Köln 30
Federal Republic of Germany

The En transposable element of maize has great potential as a versatile gene delivery and tagging system. En-I is a two component system composed of the autonomous En element which encodes transposase, and receptor elements, called I. "I" represents any deletion derivative of En that looses transposase function but can transpose when an autonomous element supplies transposase in trans.

I elements remain stably integrated once the autonomous element is lost by segregation. En is unique among maize transposons in that expression of loci interrupted by En or I insertions can be suppressed or activated by functions encoded by En transposon (Nevers et al. 1986).

Our group has compared the transcripts and DNA structures of naturally occuring mutant En-I elements as a first step in the investigation of mechanisms of En-I action. (Gierl et al. 1985, Pereira et al. 1986, Tacke et al. 1986, Schwarz-Sommer et al. 1987). We are now using recombinant DNA techniques to engineer new mutants of En in order to define regions encoding transposase and control functions as well as the cis-acting signals they recognise.

We chose to assay the activity of our mutants in tobacco since tobacco protoplasts are easily transformed and regenerated and because, unlike maize, tobacco has no endogenous En-I elements that could affect the phenotypes of our introduced mutant elements. In addition, Baker et al. (1987) have used a phonotypic assay to show that another maize transposon, Ac, can transpose when introduced into tobacco.

Our first task was to determine if En could also excise and transpose in tobacco. Tobacco protoplasts synchronized in S-phase by procedure of Meyer et al. (1986) were simultaneously transformed with 2 plasmids by PEG-calcium nitrate precipitation of uncut, naked DNA. The first plasmid contained an I element separating the neomycin phosphotransferase gene of Tn5 (NPT) from the TR1' promoter of Ti plasmids. The NPT gene can only be expressed, conferring kanamycin resistance, if the I element is removed. The second plasmid carries an autonomous En element and a constitutively expressed hygromycin resistance gene. Regenerated protoplasts were grown in medium with either kanamycin or hygromycin. In this way the relative efficiency of each transformation could be estimated by the number of hygromycin resistant calli.

In a typical experiment, 3 calli resistant to 200 mg/l of kanamycin were recovered from an estimated 99 calli. A control transformation in which the autonomous En was replaced by an I element gave no kanamycin resistant calli although the estimated number of hygromycin resistant transformants was 66. Repeated trials gave similar results.

DNA from three regenerated kanamycin resistant plants was subjected to genomic southern analysis. The foreign DNA was present in multiple and rearranged copies as is characteristic of DNA introduced into S-phase synchronized protoplasts (S. Kartzke and P. Meyer, personal communication). One plant had the restriction patterns expected if the I element had excised. After cloning the appropriate DNA fragment, we found a deletion that removed the entire I element along with 9 base pairs of flanking DNA. Such a deletion is characteristic of En mediated excision events in maize (Schwarz-Sommer et al, 1985). En homologous polyadenylated transcripts from this plant have been compared to those of maize. The major 2.4 kb En transcript of maize is not found in this plant. However, two transcripts are found in tobacco that are similar in size to minor En transcripts of maize, one 6.5 kb in size and the other 3.4 kb. In maize and tobacco, both hybridize to a probe from the 3' end of the major maize transcript and from the long open reading frame of En. The abundance of the various RNA's suggests that En is transcribed at a similar frequency in both maize and tobacco but the relative amount of the processed products differ considerably.

In conclusion, we have an example of a transformed tobacco plant in which an En mediated excision seems to have occured and En is transcribed in that plant. This suggests that En can function in tobacco, however the assay system we have used to detect excision is not satisfactory. Kanamycin resistance is caused more frequently by an aberrant NPT integration than by excision of an I element.

We are presently attempting to use Ti plasmids to deliver our constructs into tobacco as we expect that this will reduce the frequency of rearrangement of the transforming DNA.

Once a good assay system is available we will compare the effects of in vitro mutations on En activity in order to obtain a clear understanding of En-I action.

REFERENCES

1. B. Baker, G. Coupland, N. Fedoroff, P. Starlinger and J. Schell, EMBO J. 6 (1987) 1547-1554.
2. A. Gierl, Z. Schwarz-Sommer and H. Saedler EMBO J. 4 (1985) 579-583
3. P. Meyer, E. Walgenbach, K. Bussmann, G. Hombrecher and H. Saedler Mol Gen Genet 201 (1985) 513-518.
4. P. Nevers, N. Shepherd and H. Saedler in Advances in Botanical Research (1986) 10 459-472
5. A. Pereira, Z. Schwarz-Sommer, A. Gierl, I. Bertram, P.A. Peterson and H. Saedler EMBO J. 4 (1985) 17-23.
6. E. Tacke, Z. Schwarz-Sommer, P.A. Peterson and H. Saedler Maydica. 23 (1986) 83-91
7. Z. Schwarz-Sommer, A. Gierl, H. Cuypers, P.A. Peterson and H. Saedler EMBO J. 4 (1985) 591-597.
8. Z. Schwarz-Sommer, N. Shepherd, E. Tacke, A. Gierl, W. Rohde, L. Leclercq M. Mattes, R. Berndtgen, P.A. Peterson and H. Saedler EMBO J. 6 (1987) 287-294

Initiation of Cell Suspensions, Protoplast Culture, and Direct Gene Transfer to Protoplasts of Secale cereale L. (rye)

J. Zimny[2], B. Junker[1], H. Lörz[1], [1]Max-Planck-Institut für Züchtungsforschung, D-5000 Köln 30, Federal Republic of Germany, [2]Plant Breeding and Acclimatization Institute 05-870 Blonie Radzikow, Poland.

1. INTRODUCTION

Genetic manipulation such as somatic hybridization or introducing foreign genes into plant cells are limited due to several reasons. Firstly, it is difficult to achieve sustained divisions of protoplasts and in most cases it is not possible to regenerate fertile plants from cereal protoplasts. Regeneration of rice plants from protoplasts, confirmed by several laboratories, makes us confident that it will also become possible with other cereal species. Secondly, obtaining cell suspensions which contain small clusters of rapidly dividing cells still remain a problem, and it is a necessary stage towards the isolation and culture of totipotent cereal protoplasts (Ozias-Akins and Lörz 1984). Embryogenic cell suspensions are most suitable. And yet, this characteristic often disappears with time. Dividing protoplasts enable to carry out research on transformation and to obtain stably transformed cell lines (Lörz et al. 1985, Fromm et al. 1986). Non-dividing protoplasts (e.g. cereal mesophyll protoplasts) are suitable for studying different plasmid constructions and gene promotors in transient gene expression studies (Werr and Lörz 1986, Junker et al. 1987). In this study we have elaborated a method for the initiation of cell suspensions, a method of isolating and culturing protoplasts of rye (Secale cereale L.), and introduction of genes into rye protoplasts.

2. PROCEDURE

2.1 Materials and Methods: Callus cultures were initiated from immature embryos of winter rye line, S1D4 83, following the method described previously (Zimny and Lörz 1986). The embryo-derived callus was passaged several times on CC-10-2 medium (Potrykus et al. 1979) in order to select embryogenic tissue. This callus was moved into 20 ml liquid medium CC-2 (without coconut water) in 125 Erlenmeyer flasks onto a rotary shaker at 140 rpm at 26^{0}C and 16h light 2500 lux. Suspensions were subcultured every 5 days by removing 15 ml of suspension and adding the same quantity of fresh medium. Only a fraction of clusters and cells passing through 250 um was propagated and used for protoplast isolation. Protoplasts were isolated 2-5 days after subculture. In order to optimize conditions of isolation we used various combinations of enzymes. Protoplasts were cultured in liquid CC-10-2 as well as in agarose solidified medium (Lörz et al. 1983) at densities of $4x10^{4}$ $-6x10^{4}$ protoplasts/ml. Protoplasts isolated from mesophyll of rye leaves were used in the experiments for transient gene expression. The plasmid DNA used carryed a chimeric gene construction consisting of the promotor of the nopaline synthase gene (NOS') fused to the coding region of the neomycin phosphotransferase (NPT II) gene of Tn 5. This DNA was added to the freshly isolated protoplasts in the presence of Ca^{2+} and PEG (Krens et al. 1982). After 2 days gene activity of NPT II was analysed in the extract of the treated protoplasts.

3. RESULTS

3.1 Cell Suspensions

Suspensions of embryogenic callus subcultured 5 days contained after 6 months a mixture of meristematic cell clusters (Fig.1a), fragments of callus and non-dividing cells. Fraction 50-250 um was isolated mechanically. After a few subcultures it contained single, dividing cells (spherical or elongated) as well as small cell clusters (Fig.1b). As the clusters were getting bigger they had to be seperated on sieves.

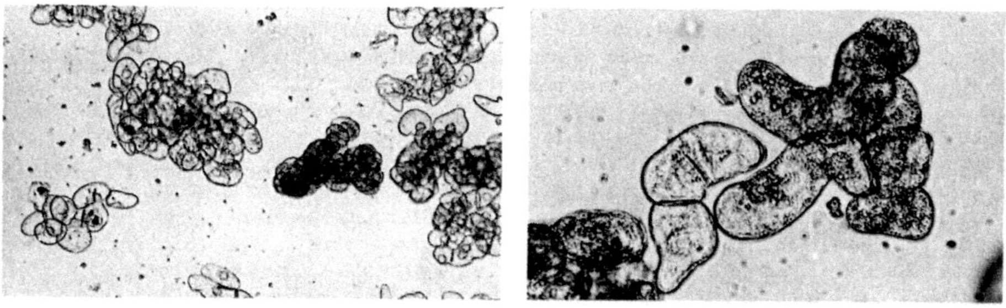

Fig.1: a)Meristematic cell aggregates separating from callus and non-dividing cells. b)Highly cytoplasmic dividing cells in an established suspension.

3.2 Protoplast Culture

From among studied mixtures of enzymes only the combination of Cellulase Onuzuka RS (2%), Macerozyme R10 (0.5%) and Pectolyase Y-23 (0.1%) gave positive results in protoplast isolation. Divisions were observed in CC-10-2 agarose solidified medium after 10 days of culture with an efficiency of ca. 0.02%.

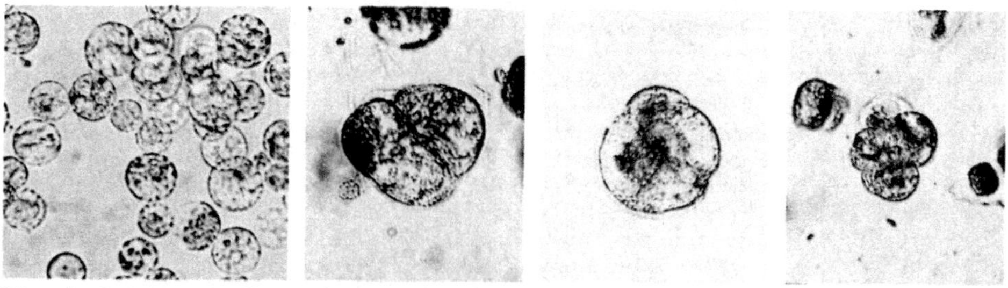

Fig.2 Culture of protoplasts isolated from the suspension of rye

3.3 Transient Gene Expression

In our experiments only non-dividing rye protoplasts were used. We have chosen a medium ensuring long survival of protoplasts, i.e. C-8 medium (see Lörz et al.1985). On this particular medium survival reached 20 days or more. Adding glucose instead of mannitol in order to increase osmotic pressure up to 720 mOs/kg H_2O increased the survival of mesophyll protoplasts significantly. Extracts of 2 days old protoplasts showed NPT II activity which means that after gene transfer, transcription and translation processes in protoplasts occured, and as a result we obtained the product of a given gene.

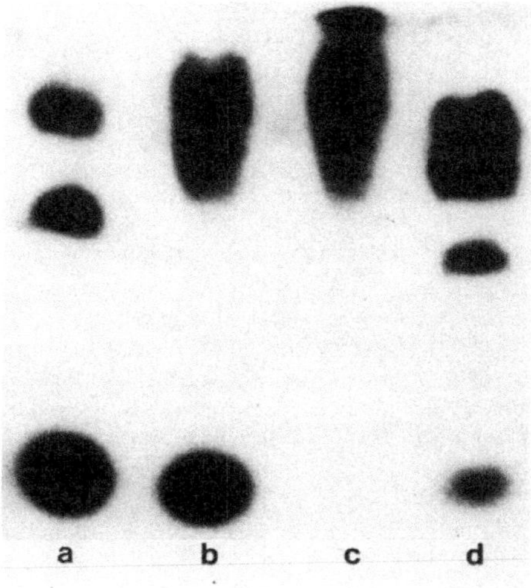

Transient gene expresion of the gene pNOS - NPT in protoplasts after PEG induced DNA uptake..br

a)Transformed Nicotiana tabacum plant (positive control)

b)Suspension derived protoplasts of Hordeum vulgare L.

c)Suspension derived protoplasts of H.vulgare L. (no DNA treatment)

d)Mesophyll protoplasts of Secale cereale L.

References

Fromm M.E., L.P. Taylor and V. Walbot (1986) Nature 319: 791-793.
Junker B., J. Zimny, R. Lührs and H.Lörz (1987) Plant Cell Rep 3: 329-332.
Krens F.A., L. Molendijk, G.J. Wullems and R.A. Schilperpoort (1982) Nature 296: 72-74.
Lörz H., P.J. Larkin, J. Thompson and W.R. Scowcroft (1983) Plant Cell Tiss Org Cult 2: 217-226
Lörz H, Baker B, Schell J (1985) Mol. Gen. Genet.199:178-182.
Ozias-Akins P.M. and H. Lörz (1984) Trend Biotechnol 2:119-123
Potrykus I., C.T. Harms and H. Lörz (1979) Theor Appl Genet 54: 209-214
Werr W. and H. Lörz (1986) Mol Gen Genet 202: 471-475
Zimny J. and H. Lörz (1986) Plant Cell Rep 5: 89-92

PROTOPLAST CULTURE AND DNA UPTAKE IN ECHINOCLOA

D.F. CHEN, A. TASSIE, P.J. DALE, A. GOLDSBROUGH, W. LIANG, M.W. BEVAN, R.B. FLAVELL and Z.A. XIA*
IPSR, Cambridge Laboratory, Trumpington, Cambridge CB2 2LQ, England
*Shanghai Institute of Plant Physiology, 300 Feng Lin Road, Shanghai, China

1. INTRODUCTION

Although protoplasts of cereals and grasses have proved difficult to manipulate in culture, mature plants have recently been regenerated from protoplasts of rice (Abdullah et al., 1986), maize (Chai et al., pers. comm.) and Polypogon (Chan and Xia, 1987). These results are encouraging, not only for protoplast morphogenesis but also for the transformation of cereals and grasses. We report protoplast culture and transient expression of foreign genes in the grass Echinocloa colonum.

2. MATERIALS AND METHODS

2.1. Protoplast isolation and culture

A finely dispersed and rapidly growing suspension culture initiated from young inflorescences was used for protoplast isolation. Protoplasts were obtained from an exponentially growing suspension culture (3-4 days after subculturing) in MS medium plus 0.5 g l^{-1} casein hydrolysate, 0.2 g l^{-1} myo-inositol and 1 mg l^{-1} 2,4-D, by using the enzyme mix: Onozuka cellulase Rs 1%, Onozuka cellulase R10 0.5%, Macerozyme R10 0.2% and Pectolyase Y-23 0.1% in cpw salts (Frearson et al., 1973) with 0.5 M mannitol. Protoplasts were purified by resuspending in 0.5 M sucrose solution and cultured in MS medium plus 0.5 g l^{-1} casein hydrolysate, 0.2 g l^{-1} myo-inositol, 0.5 mg l^{-1} 2,4-D, 0.5 mg l^{-1} NAA, 0.2 mg l^{-1} BAP, 0.1 mg l^{-1} Zeatin and 5% coconut milk, at density of 1.2×10^5 ml, 25°C in dark. Agarose, liquid layer and hanging drop methods in petri dishes were used for culture.

2.2. DNA uptake

The DNA constructs used were the β glucuronidase (GUS) gene with the NOS terminator and the following 35S cauliflower mosaic virus regulatory sequences: DNA 1, full length 35S (c850 bp); DNA 2, promoter region without putative enhancer; DNA 3, promoter with duplicated enhancer; DNA 4, promoter with triplicated enhancer. All constructs were in the plasmid pUC19. Both Ca^{++}-PEG (Kren et al., 1982) and Mg^{++}-PEG (based on Negrutiu et al., 1987) methods were used for DNA uptake treatment. The DNA samples (30-50 μg ml^{-1} plasmid DNA or 50 μg ml^{-1} herring DNA) were added to aliquots of 0.37 ml protoplast suspension ($\approx 6 \times 10^6$ cells), followed by addition of 0.63 ml PEG solution. After 30-35 min incubation at room temperature, 10 ml F medium (Ca^{++}-PEG method) or 0.2 M $CaCl_2$ (Mg^{++}-PEG method) was added stepwise in 2 ml amounts at 5 min intervals. Washing with MS medium twice by centrifugation at 90 g for 5 min the protoplasts were cultured in the same medium. Protoplasts were harvested after 1, 2, 3, 5, 10 day culture for GUS assay.

3. RESULTS

3.1. Protoplast isolation and culture

Ten to 20x10^6 protoplasts were released from 30 ml of exponentially growing cells. About 30% of the protoplasts were lost during the washing and purification procedure. First cell divisions were observed after 2 days in culture and reached a maximum at 7 days. The frequency of first cell divisions (7 days) and callus formation (4 weeks) was, respectively, 34% and 6% for agarose, and 12% and 1% for liquid layers. The hanging drop method showed a large variation in response with 0-29% of protoplasts with first divisions and some individual drops with 23% callus formation.

3.2. Transient expression of GUS gene

Although PEG-DNA treatment damages about 40% protoplasts, the remaining protoplasts are healthy enough to show expression of the introduced DNA. Ca^{++}-PEG and Mg^{++}-PEG methods were compared by detection of the accumulation of 4-methylumbelliferone (MU), which is produced from 4-methylumbelliferyl-β-D-glucuronide (MUG) when the GUS enzyme is present in cells. The fact that Ca^{++}-PEG gave 22 nM MU/μg protein/min whereas Mg^{++}-PEG gave 120 nM MU/μg protein/min suggests that Mg^{++} ions favours DNA uptake.

With the Mg^{++}-PEG method, and the full length CaMV promoter (DNA 1) MU level was calibrated to 100 (120 nM MU/μg protein/min). DNA 2 gave 32%, DNA 3 56% and DNA 4 gave 209% expression. This suggests that an enhancer does exist in the full length CaMV 35S promoter.

With both Ca^{++}-PEG and Mg^{++}-PEG treatment, GUS activity can be detected after 24 hours of culture (38 nM MU/μg protein/min) and reaches a maximum at 48 hours (120). Although the enzyme activity declined after 3 days (to 88) it was still detectable at 5 and 10 days (65 and 62 respectively) during which there was active cell division.

Similar transient expression results have been achieved in protoplasts of *Triticum monococcum* and *Polypogon fugax*.

4. CONCLUSIONS

1. Callus was formed from protoplasts of *Echinochloa colonum* on MS medium. The highest first division frequency was from the agarose method and the highest frequency of callus formation was from some individual hanging drops.
2. Both Ca^{++}-PEG and Mg^{++}-PEG methods can deliver foreign DNA into protoplasts. The latter method gave better transient expression.
3. CaMV 35S promoter can be recognised by monocot (at least *Echinochloa*, *Triticum*, *Polypogon*) plant cells and there is evidence for the existence of an enhancer in the natural 35S promoter which operates in monocot cells.
4. GUS activity is detectable in PEG-DNA treated protoplasts at 24 hours and reaches a peak at 48 hours. GUS activity was detectable after 10 days.

5. REFERENCES

1. Abdullah R, Cocking EC and Thompson JA: Biotech. 4, 1080-1090, 1986.
2. Chen DF and Xia ZA: Scientia Sinica (Series B) 30, 698-705, 1987.
3. Krens FA *et al.*: Nature 296, 72-74, 1982.
4. Negrutiu I *et al.*: Plant Mol. Biol. 8, 363-373, 1987.
5. Frearson EM, Power JB and Cocking GC: Develop. Biol. 33, 130, 1973.

TRANSIENT EXPRESSION OF FOREIGN GENES IN WHEAT PROTOPLASTS

C. BROWN, D.M. GILMOUR, R.J. MATHIAS, R.B. FLAVELL.

Cytogenetics and Molecular Biology* Departments, IPSR (Cambridge Laboratories), Maris Lane, Trumpington, Cambridge, CB2 2LQ, UK.

1. INTRODUCTION

Plant transformation depends on the introduction, integration and expression of foreign genes. Where the recovery of stable transformants is not yet possible 'transient' assays may be useful in assessing the effect of promoter, enhancer and terminator sequences on gene expression.

2. PROCEDURE

2.1. Materials and methods

2.1.1. Protoplast isolation. Leaves of 7-d old, aseptically grown, etiolated wheat (*Triticum aestivum*) seedlings were sliced longitudinally and incubated in 2.5% w/v cellulase R10, 1.0% w/v macerozyme R10 in wash medium (0.7 M mannitol, 0.01 M $CaCl_2$, pH 5.8) for 4-5 h. Protoplasts were released by squeezing the leaf pieces, filtered through a 45 µm sieve and washed twice.

Cells from 4-d old suspensions of *T. monococcum* were incubated in 1.0% w/v cellulase RS, 0.5% macerozyme R10, 0.05% pectolyase Y23, 0.6 M mannitol, 5 mM $CaCl_2$. After 3-4 h protoplasts were sieved (45, 28 µm sieves) and washed twice.

2.1.2. PEG mediated DNA uptake into protoplasts. Method 1, after Meyer *et al* (1). Method 2, after Negrutiu *et al* (2), used a $MgCl_2$, MES, 0.4 M sucrose buffer and 3350 MW PEG in $Ca(NO_3)_2$. DNA (20 µg) was added to freshly isolated mesophyll protoplasts (2.5 x 10^5) or 5.0 x 10^5 cell suspension protoplasts. Following PEG-DNA uptake, protoplasts were plated in liquid Kao medium (3) and cultured in the dark at 25°C for 2-d.

2.1.3. Plasmids. Plasmid vectors were constructed containing the *E. coli* β-glucuronidase (GUS) gene (4) with the nopaline synthase gene terminator sequence and three versions of the CaMV 35S promoter: 1) a full length promoter, 2) a 'deleted' version with only 100 bp of the promoters 3' end, (cut at the *Eco*RI site), 3) an 'enhanced' version with the putative CaMV 35S promoter enhancer (a 325 bp *Eco*RV-*Hinc*II fragment) duplicated at the *Hinc*II site (6, 7).

2.1.4. Detection of GUS activity. The GUS assay was performed as described in the GUS gene fusion system users manual (5).

3. RESULTS

3.1.1. Comparison of PEG methods. Transient GUS expression was obtained in dividing (*T. monococcum*) and non-dividing (*T. aestivum*) protoplasts. The level of GUS expression was dependent on the condition of the protoplasts and the method of PEG-DNA uptake (no GUS expression in *T. monococcum* using method 1).

3.1.2. Time course of GUS activity in wheat seedling leaf protoplasts. Protoplasts were assayed for GUS activity 15, 39, 63, 87 and 111 h after PEG transformation. The specific activity of GUS increased over 111 h (figure 1).

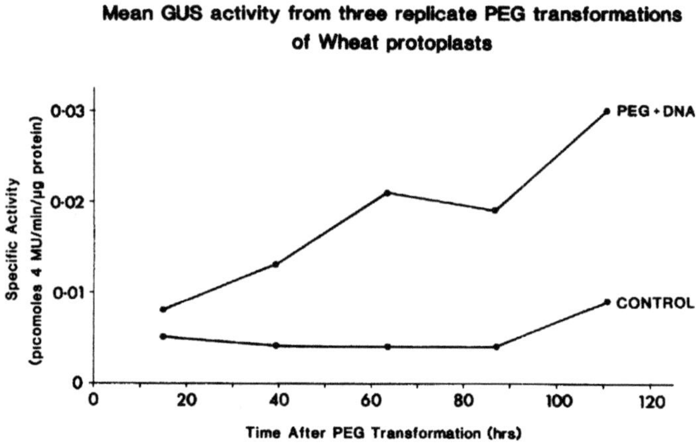

FIGURE 1. GUS activity in PEG treated wheat seedling leaf protoplasts and in PEG+DNA treated protoplasts.

3.1.3. The effect of modifying the 35S promoter on GUS specific activity in protoplasts of T. monococcum and T. aestivum (picomoles 4MU/min/µg protein).

	T. monococcum		T. aestivum
	Expt. 1	Expt. 2	
Promoter			
control	0.001	0.0004	0.004
a	0.047 +/- 0.01	0.208 +/- 0.02	0.011 +/- 0.001
b	0.051 +/- 0.01	0.140 +/- 0.03	0.007 +/- 0.001
c	0.059 +/- 0.01	0.203 +/- 0.02	N/A

In comparison to the intact 35S promoter, enhanced versions increase levels of gene expression (6, 7). In tobacco the truncated promoter reduced transient GUS expression by 30-fold (A. Goldsbrough, Pers. comm.). However in these experiments there were no differences in GUS expression from the different promoters. In interpreting these results, the following points should be considered:- a) introduction of large amounts of DNA, in transient assays, may saturate the regulatory system, b) although the 35S ("dicot") promoter is functional in monocots, the "enhancer" regions may not be recognised as in dicots.

REFERENCES

1. Meyer P, Walgenbach K, Bussman G, Hombrecher G, Saedler H: 1985, Mol. Gen. Genet., 201:513-518.
2. Negrutiu I, Shillito R, Potrykus I, Biasini G, Sala F: 1987, Plant Mol. Biol., 8:363-373.
3. Kao KN: 1977, Mol. Gen. Genet., 150:225-230.
4. Jefferson R, Burgess S, Hirsh D: 1986, Proc. Natl. Acad. Sci., 83:8447-8451.
5. Jefferson R: 1987, GUS gene fusion system users manual.
6. Kay R, Chan A, Daly M, McPherson J: 1987, Science, 236:1299-1302.
7. Ow DW, Jacobs JD, Howell SH: 1987, Proc. Natl. Acad. Sci. USA, 84:4870-4874.

Protoplast culture and transformation studies of Triticale (x Triticosecale Wittmack).

A. Stolarz , H. Lörz , Max-Planck-Institut für Züchtungsforschung , D-5000 Köln 30 , Federal Republic of Germany.

1. INTRODUCTION

Plant regeneration from protoplasts offers a unique and essential technology for genetic engineering studies of higher plants. The major cereal crops however are still considered to be difficult for protoplast culture. There are very few reports on plant regeneration from protoplasts of the Gramineae. Fertile plants have been regenerated from protoplasts of *Oryza sativa* L. (Fujimura. et al.,1985, Yamada et al., 1986, Abdullah et al.,1986) sterile plants of *Zea mays* L. (Rhodes et al., 1987) and albino plantlets of *Hordeum vulgare* L.(Lührs and Lörz 1988) for more details see reviews (Vasil, 1987; Lörz et al., 1988).

Stable integration of defined genes in cereal species has been achieved by transformation of protoplasts with naked DNA in the presence of polyethylene glycol (Potrykus et al., 1985, Lörz et al., 1985, Uchimija et al., 1986) by electroporation (Fromm et al., 1986) or by in situ injection of plasmid DNA into floral tillers of *Secale cereale* L. (De la Pena et al., 1987).

Non-stable transformation so-called transient gene expression, in the Gramineae family has been reported in tissue culture derived protoplasts of *Zea mays* L. (Fromm et al.,1985),*Triticum monococcum* L., *Sorghum bicolor* L. and leaf derived protoplasts of *Oryza sativa* L.(Ou-Lee et al. 1986). More recently PEG-facilitated plasmid DNA uptake has been shown in dividing and non-dividing protoplast culture systems of *Zea mays* L., *Hordeum vulgare* L., *Oryza sativa* L. and *Secale cereale* L. (Junker et al.,1987).

In this report we present a protoplasts culture system of *Triticosecale* used for transformation studies with different plasmid constructions containing a selectable chimeric gene coding for kanamycin resistance.

2. PROCEDURE

2.1 Materials and Methods : Establishing cell suspension cultures from embryogenic callus of *Triticosecale* has been described previously (Stolarz and Lörz, 1986). In order to select fast growing cell suspension sublines macroplates (Greiner No 657102) with 6 units were used as culture vessels. The cell suspensions were maintained by subculturing 0.2 ml aliquots to 2 ml of fresh medium (Kao 1977) into each unit at 4 days intervals and cultured at 26 ^{0}C in light on a gyratory shaker at 120 rpm. After 10-15 subcultures each subline used for further experiments was characterized by growth curves (fresh and dry weight). Samples for measurements have been collected at daily intervals in 6 replication. Protoplasts were isolated from cell suspension cultures 2-10 days after subculture. The cell aggregates from a complete macroplate were collected and fresh weight was determined. 0.5 g. of the cells was mixed with 10 ml of an enzyme mixture consisting of 2% Cellulase RS Onozuka, 1% Macerozyme R 10 and 0.05% Pectolyase Y23 with osmolarity adjusted to 780 mOsm/kg H_2O with glucose. The cells were incubated on a gyratory shaker at 50 rpm for 2-2.5 h at 27^{0}C in dark. Protoplasts were then passed through a series of sieves of 250,100,50 and 25 µm diameter mesh size to remove undigested cell clusters and cell debris. The protopsts were washed twice by centrifugation in seawater. Isolated protoplasts we resuspended in culture medium and plated at densities of 1-5 x 10^5 protoplasts/ml in 2 ml or 6 ml of liquid medium in Petri dishes (5 or 10 cm diameter) or directly embedded in agarose. The medium used for protoplast culture was identical to the suspension medium but with different concentrations of 2,4-D (0.5-8 mg/l). Plating efficiency was evaluted after 2 weeks. For transformation experiments 4 different plasmid constructions containing the NPT II gene have been used namely pLGVneo1103 and pLGVneo2103 (Hain et al., 1985), pRT99 and pRT100neo (Töpfer, 1987). Protoplast transformation was performed according to the method described by Krens et al. (1982). After treatment with DNA the protoplasts were washed and resuspended in culture medium at density 2.5 x 10^5 protoplasts/ml. The NPT II assay followed the procedure described by Reiss et al. (1984).

3. RESULTS

3.1 PROTOPLASTS ISOLATION AND CULTURE :Protoplasts of the triticale cell suspension (Fig.1) can be isolated with different efficiency depending from the time after subculture and osmolarity of the enzyme solution used for protoplast isolation.

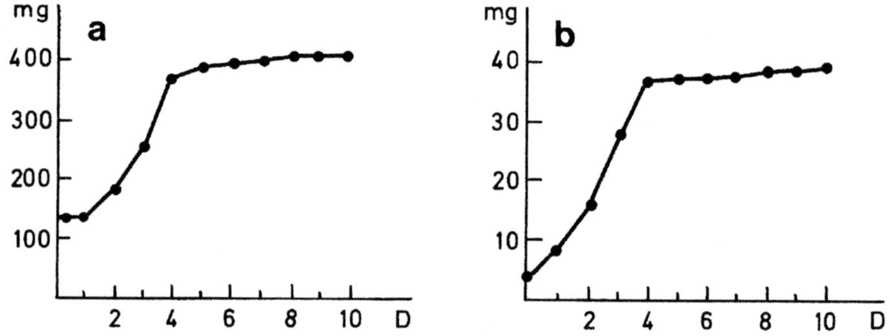

Fig.1 Growth curves of <u>Triticosecale</u> cell suspension
a - fresh weight, b - dry weight

Protoplast isolation at day 2 or 3 after subculture yields a low number of protoplasts (1-2x10^6) and these are capable to form colonies with efficiencis of 3-6%. The highest yield of 7x10^6-3x10^7 protoplasts per gram of fresh weight could be obtained from cell suspensions at day 7 or 8 after subculture.

In that case, plating efficiency evaluated after 2 weeks in liquid medium was high and ranged from 36% to 54%. The highest plating efficiency (60%) has been achieved from protoplasts isolated at day 10 after subculture. First divisions occurred after 3-5 days in protoplast cultures isolated 7 or 8 days after subculture. In protoplast cultures isolated after 2-3 and 10 days first divisions were observed only after 8 days or later. During the course of the experiments we found that embedding protoplasts in agarose redused plating efficiency significantly. Variation of the concentration of 2,4-D in the culture medium between 2-6mg/l had no effect on plating efficiency and time of the first divisions. Further experiments are aiming towords plant regeneration from protoplasts.

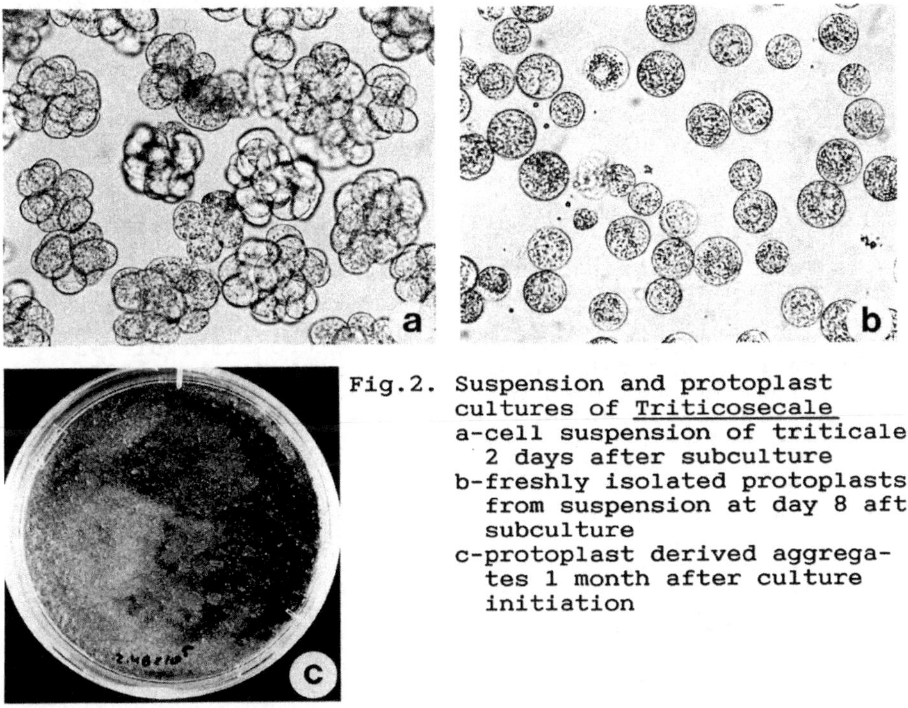

Fig.2. Suspension and protoplast cultures of Triticosecale
a-cell suspension of triticale 2 days after subculture
b-freshly isolated protoplasts from suspension at day 8 after subculture
c-protoplast derived aggregates 1 month after culture initiation

3.2 Transformation studies: Different DNA constructions carrying the NPT II gene as a marker gene have been used to transform suspension protoplasts of triticale. Positive results in transient gene expresion have been obtained by using the plasmid pRT99 and pLGV1103. The activity of the gene product neomycin phosphotransferase has been measured in the extract of treated protoplasts from day 2 to day 10 after transformation. The suspension protoplasts of triticale were treated with identical amounts of plasmid DNA (40µg). Intensity of expresion of the NPT II gene seems to be influenced by the promoter (p35S or pNOS) fused to the marker gene NPT II (Fig.3). Meanwhile protoplasts treated with plasmid DNA have been transferred to selection medium containing 100mg/l geneticin. Further experiment are needed to optimize DNA uptake and the selection scheme.

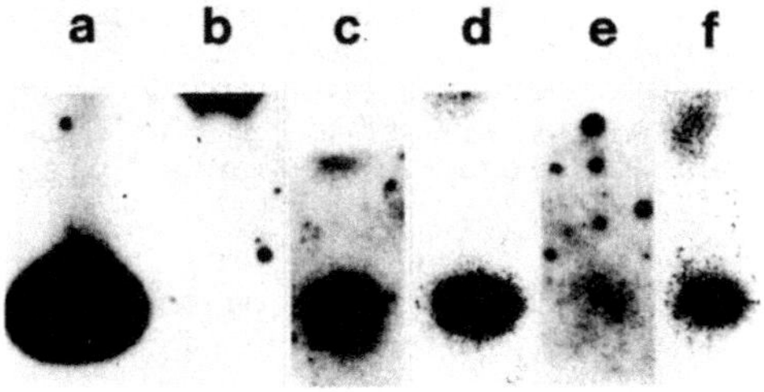

a-transformed tobacco (positive control)
b-triticale protoplasts without DNA
c-10^6 protoplasts treated with pLGVneo1103 assayed after 4 days
d-10^6 protoplasts treated with pLGVneo1103 assayed after 10 days
e-10^6 protoplasts treated with pLGVneo2103 assayed after 4 days
f-10^6 protoplasts treated with pRT99assayed after 2 days

Fig.3.Transient expresion of the NPT II gene constructions in triticale suspension protoplasts.

REFERENCES

1.Abdullah R.,E.C.Cocking,J.A.Thompson (1986) Bio-Technol.4: 1087-1090
2.De la Pena A.,H.Lörz,J.Schell (1987) Nature 325 :274-276
3.Fromm M.,L.P.Taylor,V.Walbot (1985) Proc.Natl.Acad.Sci.USA 82 : 5824-5828
4.Fromm M.,L.P.Taylor,V.Walbot (1986) Nature 319 :791-793
5.Fujimura T.,M.Sakurai,H.Akagi,T.Negishi,A.Hirose (1985) Plant Tissue Culture Lett. 2 :74
6.Hain R.,P.Stabel,A.P.Czernilofski,H.H.Steinbiss,L.Herrera -Esterella,J.Schell (1985) Mol.Gen.Genet. 199
7.Junker B.,J.Zimny,R.Lührs,H.Lörz (1987) Plant Cell Rep. 6 : 329-332
8.Krens F.A.,L.Molendijk,G.J.Wullems,R.A.Schilperoort(1982) Nature 296 : 72-74
9.Lörz H.,B.Baker,J.Schell (1985) Mol.Gen.Genet. 199 : 178-182
10.Lörz H.,E.Göbel,P.Brown(1988) Plant Breeding 100 (in press)
11.Lührs R.,H.Lörz (1988) submitted
12.Ou-Lee T.M.,R.Turgeon,R.Wu (1986) Proc.Natl.Acad.Sci. USA 83 : 6815-6819
13.Potrykus I.,M.W.Saul,J.Petruska,J.Paszkowski,R.Shillito (1985) Mol.Gen.Genet. 199 : 183-188
14.Reiss B.,P.Sprengel,H.Will,H.Schaller (1984) Gene 30 :217 -223
15.Rhodes C.A.,D.A.Pierce,I.J.Mettler,D.Mascarenhas,J.J.Detmer (1987) submitted
16.Stolarz A.,H.Lörz (1986) in : Genetic Manipulation in Plant Breeding, W.Horn (ed) 499-501
17.Töpfer R. (1987) Ph.D.thesis
18.Uchimija H.,T.Fushimi,H.Hashimoto,H.Harada,K.Soyono, Y.Sugawara (1986) Mol.Gen.Genet. 205 :461-468
19.Vasil I.K. (1987) J.Plant.Physiol. 128 :193-218
20.Yamada Y.,Y.Q.Yang,D.T.Tang (1986) Plant Cell Rep.5 : 85-88

ANALYSIS OF THE ZEA MAYS SUCROSE SYNTHASE PROMOTOR BY TRANSIENT GENE EXPRESSION IN MAIZE PROTOPLASTS

C.Maas*, W.Werr and P.Starlinger

Institut für Genetik, Universität zu Köln, Weyertal 121, D-5000 Köln 41, F.R.G.

INTRODUCTION:

The Shrunken gene of Zea mays, encoding the enzyme sucrose synthase, has been cloned and characterized in detail (Werr et al., 1985). The expression of this gene is regulated on the transcriptional level. High transcript levels accumulate in the endosperm of developing maize kernels. Reduced levels of transcript are detected in roots and shoots of etiolated seedlings. The transcription rate of this gene can be increased 10-20 fold upon anaerobic stress (Springer et al., 1986).
To analyse the Shrunken gene promotor for regulatory elements responsible for tissue specific and anaerobic induced expression we established a transient gene expression system in protoplasts of a maize suspension cell line (BMS, kindly provided by Horst Lörz). By northern experiments we verified that the internal Shrunken gene is expressed in this cell line. The level of transcript can be increased by anaerobic treatment of the cell line. Therefore it is suitable to test for the influence of promotor deletions on the transcription rate of the plasmid pSKAN 1 (Werr and Lörz, 1986, Figure 1). This plasmid contains a 2 kb promotor fragment of the sucrose synthase gene, extending upstream to position -2000 and including the transcriptionstart, fused to the NPT II-coding region from Tn 5 and a polyadenylation signal from the octopine synthase gene.

RESULTS:

We constructed a set of deletion mutants shortening the 2 kb promotor fragment from its 5´-end towards the transcription start (Figure 2). These plasmids were cotransfected in maize protoplasts with CAT-encoding chimaeric gene driven by the CaMV-35 s promotor (kindly provided by Reinhard Töpfer). This cotransfection enabeled us to minimize differences in the rate of individual transfections by arithmetical adjustment to the same CAT activity in each individual transfection . After this equalization procedure for the transfection rate, differences in transient expression of the chimaeric gene between the NPT II encoding deletion constructs can be observed (Figure 2).

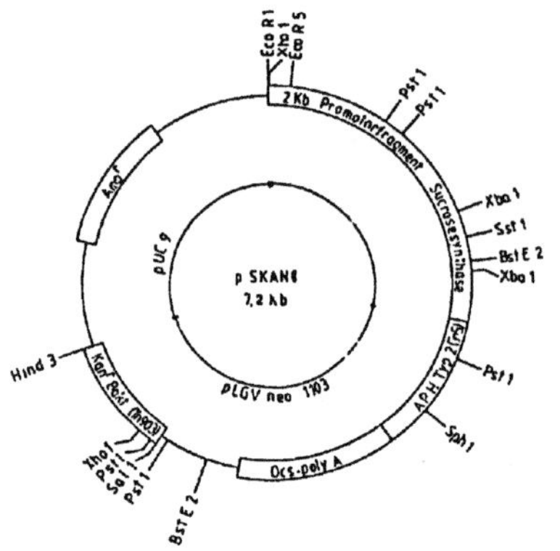

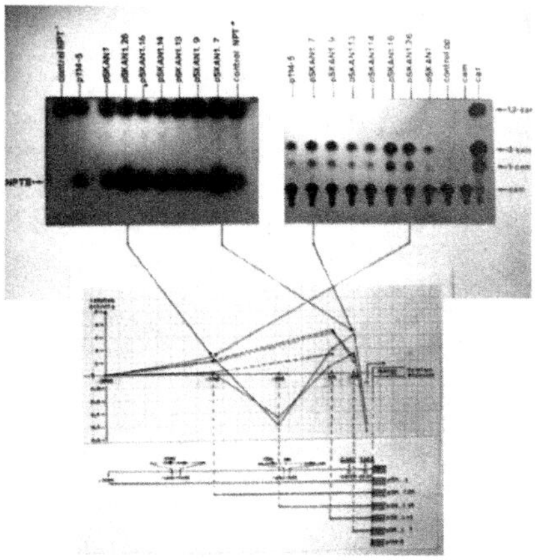

Figure 1: Figure 2:

Figure 2: The NPT II and CAT autoradiogram of one transfection series is shown. Resulting data from this series are indicated by arrows in the diagram, which shows the results of five transfection series. The relative activity of the deletion plasmids, leveled to the activity of pSKAN 1, is shown. Deletion endpoints of the plasmids are indicated below the diagram, where the 2 kb promotor fragment of the sucrose synthase gene used in pSKAN 1 is presented.

CONCLUSIONS:

High levels of transient gene expression are obtained in maize protoplasts after $Ca(NO_3)_2$-PEG mediated transfection of pSKAN 1. Differences in individual transfections were minimized by cotransfection with a second marker gene (CAT). After this equalization procedure differences in transient expression between the promotor deletion constructs were obtained.
This differences indicate, that regulatory sequences, spread over a long distance in the promotor of the sucrose synthase gene, can influence transcription in a positive and negative manner.

REFERENCES:

1. B.Springer, W.Werr, P.Starlinger, D.C.Bennett, M.Zokolica and M.Freeling Mol.Gen.Genet. 205 (1986) 461-468
2. W.Werr, W.-B.Frommer, C.Maas and P.Starlinger Embo Jour. 4 (1985) 1373-1380
3. W.Werr and H.Lörz Mol.Gen.Genet. 202 (1986) 471-475

PROTOPLAST CULTURE OF 2n S. TUBEROSUM AS A SOURCE OF MUTANT PHENOTYPES

A. SADANANDAM, H. UHRIG, and F. SALAMINI
Max-Planck-Institut für Züchtungsforschung, D-5000 Köln 30, FRG

Protoplasts of 2n S. tuberosum were used as a basis for selecting mutant phenotypes after a treatment with mutagenic agents EMS, HA and U.V. irradiation. The search of mutants was specifically related to

1. tolerance to herbicide glufosinate
2. mitotic and meiotic behaviour.

Mesophyll protoplasts were isolated and cultured essentially following Shahin's method (1984). Tolerance to the herbicide was induced by selection in the presence of glufosinate. Six lines were selected by increasing the concentration of the chemical compound from 2mg/l to 10mg/l, all were regenerated in absence of the selection pressure and data are now available for UV R line tested in vitro and R line in vivo. The selected lines show resistance to concentration of the compound of 1mg/l, a value to which the wild type is not able to resist (Fig.1).

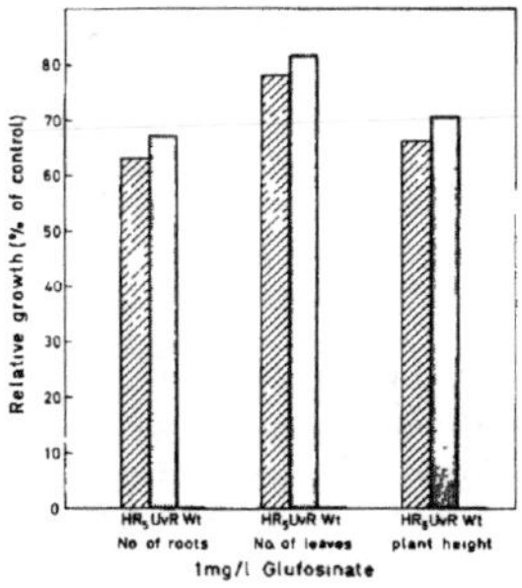

Tolerance to glufosinate was also tested in random plants regenerated after the EMS (0.025%) treatment of protoplasts. Out of 171 considered, one line (HR_5) did show a tolerance to 1.5 mg/l. for in vitro tolerance shoot segments from wildtype and tolerant lines were cultivated using selective medium. Data were collected on different parameters and analysed statistically using 't' test (Wersuhn et al., 1987). The selected lines retained their tolerance after repeated vegetative propagation. Attempts are being made to reselect the trait from protoplast cultures of the selected lines with a view to make use of tolerance as a selectable marker. Two mutant phenotypes with chromosomal anomalies were individuated among 200 plants regenerated from protoplasts treated with 0.025% EMS. One mutant shows a very abnormal meiotic pairing. This results in an abnormally high frequency of univalents at diakinesis and metaphase I. (Fig. 2a and b).

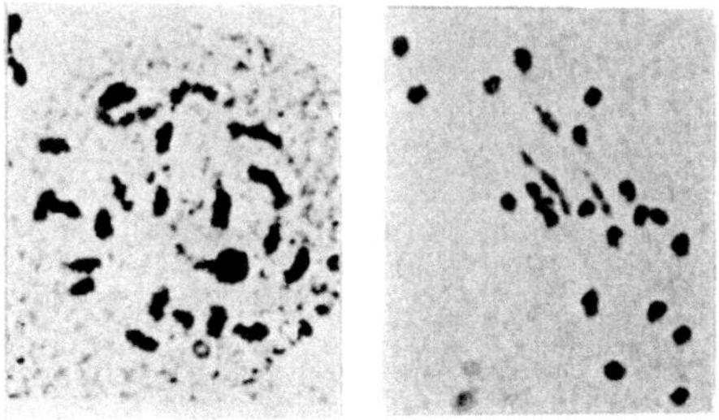

Fig. 2a and b) Univalents at diakinesis and metaphase I in the mutant.

As a consequence of univalent formation, the chiasma frequency was greatly reduced and the later stages of meiosis were highly irregular, resulted in high pollen sterility. Mutants affecting meiotic pairing are rather frequent in potato (Iwanaga, 1984). An apparent meiotic mutant from potato protoplast cultures (Gill et al., 1987) and our observation clearly documents the possibility of isolating specific mutants and emphasizes on further search for such mutant phenotypes which are useful in potato breeding.
The second show a very intimate mixoploidy at somatic level, results in finding of different level of ploidy in very restricted leaf regions. Pollen mother cells of this phenotype are triploid, tetraploid and aneuploid is also very common. Mixoploids were frequently described in potato from tissue culture regenerated plants (Sree Ramulu, 1986). However, the present phenotype shows intimate mixoploidy in all three histogenic layers.

ACKNOWLEDGEMENTS
The first author is grateful to DAAD for the award of fellowship and to Mr. W. Lehmann for his skillful assistance. Thanks are also due to Dr. G. Donn (Hoechst) for providing the chemical compound and to M. Pasemann for typing the manuscript.

REFERENCES

Gill BS, Kam-Morgan NW, Shepard JF: An apparent meiotic mutation in a mesophyll cell protoclone of the Russet Burbank potato. J. of Heredity 1985, 76:17-20.
Iwanaga M: Discovery of a synaptic mutant in potato haploids and its usefulness for potato breeding. Theor. Appl. Genet. 1984, 68:87-93.
Shahin EA: Isolation, culture and regeneration of potato leaf protoplasts from plants preconditioned in vitro. Cell culture and somatic cell genetics of plants. Vol.1:381-390.
Sree Ramulu K: Case histories of genetic variability in vitro Potato; Cell culture and somatic cell genetics of plants. Vol. 3:449-473.
Wersuhn G, Kirsch K, Gienapp R: Herbicide tolerant regenerates of potato. Theor. Appl. Genet. 1987, 74:480-482.

CYTOPLASMIC OR NUCLEAR CONTROL OF MALE STERILITY IN PROTOPLAST-DERIVED PLANTS OF NICOTIANA SYLVESTRIS

R. DEPAEPE[1], D. PRAT[2], X.Q.LI[3], P. CHETRIT[3], E. COUSIN[3] AND F. VEDEL[3]
1 Génétique et Physiologie du Développement des Plantes, CNRS, 91198 Gif sur Yvette, FRANCE.
2 Génétique des Populations d'Arbres Forestiers, ENGREF, 54042 Nancy, FRANCE.
3 Génétique Moléculaire des Plantes, UA 115, Bât. 360, Université Paris-Sud, 91405 Orsay, FRANCE.

1. PROTOPLAST CULTURE IN NICOTIANA SYLVESTRIS

Different lines of Nicotiana sylvestris, a diploid autogamous species (2n = 24), were used to initiate protoplast culture from leaf mesophyll cells:
- The original line (T line) is a pure line from the Institut des Tabacs, Bergerac, France, morphologically very homogeneous ("flat" phenotype).
- Diploid androgenetic lines were obtained by consecutive pollen cultures from the T line. These lines, theoretically completely homozygous (Doubled-Haploid or DH) show typical alterations of leaf morphology that we previously called "crumpling" (FR: fripé) (De Paepe et al, 1977; De Paepe et Pernès, 1978).

Protoplast cultures were carried out from T and DH second generation progenies, and the diploid regenerated plants, called PR1, were selfed and crossed for genetic analyses. A second cycle of protoplast culture was performed from the second generation of selfing of a non mutated PR1 plant, issued from T. Diploid regenerated plants and their progenies were called PR2.

```
T------>DH1------>DH2------>DH3------>DH4------>DH5
|        |         |         |         |         |
↓        ↓         ↓         ↓         ↓         ↓
PR1      PR1       PR1       PR1       PR1       PR1
|
↓
PR2
```

2. GENETIC ANALYSIS OF THE PR PLANTS.

- Protoplast culture induced a high mutation rate, 40 to 60% after the first cycle of culture, more than 90% after the second cycle (Prat, 1983; Li et al, 1985).
- Abnormalities are observed from germination up to adult stage, and frequently concern leaf shape or colour. They are generally under recessive nuclear control, however some of them are maternally inherited.
- Male sterile plants (ms) were obtained independently after one cycle of culture from T and from two DH lines. After the second cycle of culture, two ms plants were regenerated from the same callus, together with a fertile plant.

3. ANALYSIS OF THE MALE STERILE PLANTS.

a. Two modes of transmission of male sterility were observed. After one cycle of protoplast culture, ms plants appeared in the offspring of some fertile PR1 plants regenerated from T or DH lines. These PR1 ms plants did not transmit maternally the sterility to their progenies; ms plants reappeared in the following selfed generation, suggesting that sterility is nuclearly controlled (NS plants).
After two cycles of culture, two ms plants were regenerated from the same callus; these PR2 ms plants transmitted the sterility maternally during at least two generations: these plants are the first cytoplasmic male sterile (CMS) obtained by in vitro culture without somatic hybridization (Li et al, 1987).
b. However, outside the apparent discrepancy of their genetic control, NS and CMS plants present several morphological and molecular similarities:
- typical curvature of leaves, that is different from the "crumpling" of the androgenetic plants: the PR1 NS issued from a DH3 line is curvated and crumpled.
- absence of 40 kD mitochondrially encoded protein, that is present in all fertile plants, either parent or fertile PR plants.
c. Sac I and Bgl I mitochondrial DNA restriction profiles are changed in four PR families:
- both PR2 CMS and the fertile plant PR2F regenerated from the same callus have new patterns, different from one another. These changes are stably inherited.
- fertile and sterile plants of one PR1 family show an identical new pattern. In this case, sterility is not directly related to apparent mt DNA variation.
Interestingly, some changes are shared by different PR lines:
- loss of a 12.6 kb Sac I fragment in NS and CMS plants.
- occurence of a new 16.4 kb Bgl I fragment in NS and PR2 F plants.

4. DISCUSSION

The results reported here strongly suggest the existence of some common mechanism governing sterility in both NS and CMS plants. One possible hypothesis would be the presence in some PR plants of a nuclear gene acting on mitochondrial DNA organisation: during the first cycle of culture, partial substitution of normal mt molecules by abnormal ones would produce heteroplasmic regenerated plants, and complete substitution during the second cycle would produce homoplasmic plants (CMS).
Experiments are in progress to check the validity of this model.

REFERENCES

DE PAEPE R., NITSCH C., GODARD M. and PERNES J. 1977.In: Plant tissue culture and its biotechnological application,pp.341-352. Berlin, Heidelberg, New York: Springer.
DE PAEPE R. et PERNES J. 1978. Physiol. Vég. 16,195.
PRAT D. 1983. Theor.Appl.Genet.64:223,230.
LI.X.Q., PRAT D., DE PAEPE R. and PERNES J. 1984. GMICN.1(1):81,84.
LI.X.Q., CHETRIT P., MATHIEU C., VEDEL F., DE PAEPE R., REMY R., AMBLARD-BRETTEVILLE F. Curr.Genet.(sous presse).

SELECTION AND CHARACTERIZATION OF NITRATE REDUCTASE-DEFICIENT MUTANTS OF *NICOTIANA PLUMBAGINIFOLIA*

F. Pelsy, J. Gabard, I. Cherel, C. Meyer, A. Marion-Poll, A. Müller °and M. Caboche.
Laboratoire de Biologie Cellulaire. INRA. F 78000 Versailles. FRANCE
° Zentralinstitut für Genetik und Kulturpflanzenforschung der Akademie der Wissenschaften der DDR. DDR-4325 Gatersleben. DDR

Nitrate reductase mutants are a tool to study nitrate reductase (NR), a key enzyme of nitrogen assimilation in higher plants (Review by Kleinhofs, 1985).

Mutants deficient in nitrate reductase (NR) activity were obtained either spontanously or after mutagenesis with Gamma rays of haploid mesophyll protoplasts derived from cell culture of *Nicotiana plumbaginifolia*. Selection for chlorate resistance was carried out by growing p-cells in diammonium succinate for 10 days. Under these conditions, the depletion of ammonium succinate from the medium by cell utilisation induces nitrate reductase without nitrate in the culture medium. Then, the addition of chlorate as a selective agent for a 2 day period kills wild type cells very efficiently and allows recovery of most of the chlorate resistant colonies.

Opimal regeneration conditions were found to be a hight concentration of the reduced nitrogen source and a low concentration of sucrose in the regeneration medium. During regeneration, diploidisation occured spontanously in most cases and resulted in homozygote diploids.

The regenerated plants were multiplied by shoot cultures and then tested for their ability to grow on nitrate as sole nitrogen source.

It was found very difficult to grow the mutants on their roots in the greenhouse, therefore they were routinely grafted on a wild type *Nicotiana tabacum* . Typically, the NR mutants were chlorotic and showed crinked leaves. Their fertility is variable but we were able to characterize most of them by sexual crosses.

These mutants have been analysed by genetic crosses, biochemical and immunological methods.

Initial classification of *nia* and *cnx* types was based on xanthine dehydrogenase measurements and was subsequently confirmed by genetic analysis. Among 211 mutants isolated and tested, about 34% have been classified as *cnx*, impaired in the synthesis of the molybdenum cofactor or in its insertion in the enzyme (Gabard et al., 1985, 1987).76% have been characterized as *nia* , i.e.unable to synthetize a functional apoenzyme but able to synthetize an active molybdenum cofactor. These mutants have been shown to be affected in the structural gene of apoenzyme.

Among *nia* mutants, 70% of them were obtained spontanously and 30% after mutagenesis. Half of the *cnx* mutants were obtained spontanously. It seems that the nia locus mutates more easily.

All the mutants we obtained were affected at a single locus for the nitrate reductase-deficiency and segregated the mutation as a recessive Mendelian character.

All the *nia* mutants belonged to a single complementation group. However, at this locus, intragenic complementation was observed, essentially when some of the *nia* mutants were crossed with the *nia* mutants E56 and F19 (Table 1). Whether the complementation was total (+) or partial (§) could be determined by observing the subsequent development of the hybrid plants. When total complementation had occured nitrate reductase activities were fully restored and plants grew normally in the greenhouse under standard conditions. Partial complementation resulted in poor growth.

Seven complementation groups have been identified by genetic analysis for the *cnx* mutants: *cnx* A, B, C, D, E, F and G. The last three complementation groups have never been described previously for higher plants. (Table 2)

	D61	D80	E23	E56	E83	E113	F1	F19	F68	G152	I2	I19	K25	K27	K32
D61	–														
D80	–	–													
E23	–	–	–												
E56	+	§	–	–											
E83			–	+	–										
E113		–	–	–		–									
F1	–			+			–								
F19	+	§	–	+		–	+	–							
F68	–	–	–	–		–	–	–	–						
G152	–	–		–	+			–		–					
I2	–	–	–	+	–	–		+		–	–				
I19			–	§		–	–	–	–		–	–			
K25	–	–		+	–		–	–	–	–	–	–	–		
K27	–		–	–				–		–	–		–	–	
K32	–	–		–				–	–				–		–

Table 1

	NIA		A		B		C		D		E		F		G
	F68	K25	B25	D72	D134	D128	E137	F8	E110	K4	D16	F104	D86	I22	D93
F68	-	-													
K25	-	-													
B25		+	-	-											
D72	+		-	-											
D134	+	+	+	+	-	-									
D128			+	+	-	-									
E137	+	+	+	+		+	-	-							
F8	+		+		+		-	-							
E110	+	+	+		+	+	+	+	-	-					
K4	+	+	+	+	+	+		+	-	-					
D16		+	+	+	+		+				-	-			
F104	+			+	+	+		+	+	+	-	-			
D86			+			+		+		+	+	+	-	-	
I22	+	+	+	+	+		+	+	+		+		-	-	
D93	+	+	+	+	+	+	+	+		+	+	+	+	+	-

Table 2

This classification has been confirmed by somatic hybridisation with already characterized mutants described in the literature (Dirk *et al.*,1985) This work was done by Andreas Müller and his collegues at Gatersleben.

Nia mutants were characterized by ELISA measurements using a monoclonal antibody 96 (9) 25 specific for NR (Cherel *et al* ,1986). A broad range of immunoreactive responses were detected in the mutants: from at least one mutant which overproduces an inactive enzyme (D51) to others in which the NR protein was not detected (E23). About half of the mutants in our collection do not show any immunoreactive material.

Some of the *nia* mutants, although they are nitrate reductase-deficient, still display nitrate reductase terminal activities using reduced methyl viologen, $FMNH_2$ and bromphenol blue as the electron donors (Gabard *et al*, 1987; Meyer *et al*, 1987). This data shows that the nia complementation group corresponds to the structural gene of nitrate reductase

The *cnx* A class mutants were previously shown to synthesize an inactive cofactor repairable *in vivo* and *in vitro* by high concentrations of molybdate, and still able to effect the dimerisation of the two subunits of the enzyme; on the contrary the other *cnx* mutants can not be repaired *in vivo* by molybdate in the same conditions (Mendel *et al*, 1986). ELISA tests show that *cnx* A mutants overproduce an inactive apoenzyme, while the *cnx* mutants belonging to classes B, C, D, E and F also produced an inactive protein, but in amount lower than the wild type.

As we have recently cloned a c-DNA coding for nitrate reductase, we are now beginning to study the expression of the apoenzyme mRNA in these different classes of mutants.

REFERENCES

Kleinhofs A., Warner R.L. and Narayanan K.R.,1985. Current progress towards an understanding of the genetics and molecular biology of nitrate reductase in higher plants. In Oxford Surveys of Plant Molecular and Cell Biology, 2, 91-121.

Gabard J., Goujaud J., Marion-Poll A. and Caboche M., 1985. Nitrate reductase-deficient mutants of *Nicotiana plumbaginifolia*. In:"Somaclonal variations and crop improvement". J. Semal ed. Martinus Nijhoff publishers. Dordrecht pp. 264-270

Dirk R., Negrutiu I., Sidorov V. and Jacobs M., 1985. Complementation analysis by somatic hybridisation and genetic crosses of nitrate reductase-deficient mutants of *Nicotiana plumbaginifolia*. Evidence of a new category of *cnx* mutants. Mol. Gen. Genet., 201, 339-343.

Grafe R., Marion-Poll A.and Caboche M. , 1986 . Improved in vitro selection of nitrate reductase-deficient mutants of *Nicotiana plumbaginifolia*. Theor. Appl. Genet., 73, 299-304.

Cherel I.,Marion-Poll A., Meyer C. and Rouzé P., 1986. Immunological comparisons of nitrate reductases of different plant species using monoclonal antibodies.Plant Physiol. 81, 376-378.

Gabard J., Marion-Poll A., Cherel I., Meyer C., Muller A. and Caboche M., 1987. Isolation and characterization of *Nicotiana plumbaginifolia* nitrate reductase-deficient mutants: genetic and biochemical analys of the *nia* complementation group. Mol. Gen. Genet. 209, 596-606.

Meyer C., Cherel I., Moureaux T., Hoarau J. , Gabard J. and Rouzé P., 1987. Bromphenol blue: nitrate reductase activity in *Nicotiana plumbaginifolia*: an immunochemical and genetic approach. Biochimie, 69, 735-742.

Mendel R., Marton L. and Muller A., 1986. Comparative biochemical characterization of mutants at the nitrate reductase/molybdenum cofactor loci cnx A, cnx B and cnx C of *Nicotiana plumbaginifolia*. Plant science, 43, 125-129.

ISOLATION OF TEMPERATURE-SENSITIVE AUXIN-REQUIRING VARIANTS OF NICOTIANA PLUMBAGINIFOLIA

Yvan Fracheboud, Patrick J. King
Friedrich-Miescher-Institut, P.O.Box 2543, 4002 Basel, Switzerland

Introduction

The metabolism and mode of action of auxin is still very poorly understood. We think that mutants would be very useful for research in this field.

Temperature-sensitive auxotrophic variants will only show a requirement for exogenous auxin at the higher temperature (33 °C), and thus plant regeneration from these clones should occur in the normal way at low temperature (26 °C).

Isolation

Auxin-auxotrophic clones were isolated from populations of haploid protoplasts after mutagenesis with UV or MNNG using 5-bromo-2'-deoxyuridine (BUdR) and 5-fluoro-2'-deoxyuridine (FUdR) to counter select wild-type cells. Under restrictive conditions (at 33 °C in the absence of auxin) the selective agents killed about 99% of the wild type-colonies (table1).

Yield of variants

Previous total isolation experiments by Anne D. Blonstein in this lab led to the isolation of auxin-requiring clones with a frequency of 10^{-4}, but no temperature-sensitive auxin auxotrophs were found amongst 60000 colonies tested. In contrast, in the present experiments using BUdR enrichment, both auxin auxotrophs and temperature-sensitive auxin auxotrophs were found, the temperature-sensitive variants being present at a frequency of ca. 10^{-3} amongst the surviving colonies (table 2).

Characterization

We are presently studying these clones more closely by defining dose-response relationships, characterizing the phenotype at the whole plant level and analysing free and total indole-3-acetic acid content by radio-immuno-assay.

Table 1: Results of 5 different selection experiments. Colony forming efficiency represents the % of protoplasts forming colonies in cultures not mutagenised and not treated with the selective agents. Differences in colony forming efficiencies and number of survivals are due to different techniques of culture and concentrations of the selective agents.

starting protoplasts	% colony forming efficiency	% colonies killed by mutagenesis	surviving colonies	interesting clones
16 10^6	0.3	80	69	1
3.5 10^6	1.44	72	2746	3
9.4 10^6	2.33	47	3985	1
11 10^6	1.12	50	337	0
20 10^6	2.68	43	3301	2

Table 2: The phenotype of the most interesting clones. For most of these lines, temperature-sensitivity and auxin auxotrophy and resistance were only studied at the tissue culture level.

name	temperature sensitivity	auxin auxotrophy	auxin resistance	regenerated plants	seeds
IVA3	+	+	+	yes	no
XIIB21	+	+	+	no	-
VIIB2	+	+	-	yes	yes
XIXB5	+/-	+	-	no	-
10C6	+/-	+	-	yes	?
B8	-	+	?	?	?
Y11	+/-	+	?	?	?

The use of phosphinothricin resistance as a selectable marker in tobacco protoplast transformation.

M. De Block*, K. D'Halluin, J. Botterman and J. Leemans

* Plant Genetic Systems, J. Plateaustraat 22, 9000 Gent

1. The herbicides

Phosphinotricin (PPT) is a new non-selective contact herbicide. It is an analogous of glutamate and acts as an irreversible inhibitor of the plant glutamine synthetase. The compound is obtained by chemical synthesis (Basta^R, Hoechst AG) or naturally produced by some microorganisms as part of a tripeptide antibiotic - e.g. the tripeptide bialaphos from Streptomyces hygroscopicus (Herbiace^R, Meiji Seika Ltd).

2. Cloning of the herbicide resistance gene

A genomic bank of the bialaphos producing strain Streptomyces hygroscopicus ATCC 21705 was made and transformed in Streptomyces lividans 66 (Murakami et. al., Mol. Gen. Genet. (1986) 205, 42-50). Transformants resistant to bialaphos were selected. By an in vitro coupled transcription-translation system using S. lividans extracts one could identify a 22 kDal polypeptide encoded by the resistance gene. The total protein extract of S. lividans expressing the gene product was blotted on polybrene coated glass fibre (PCGM-1 membranes) and the aminoterminal aminoacids were sequenced. Together with the information of the DNA sequence we could subclone the gene precisely. The resistance gene is called bar.

The bar gene codes for an enzyme which transfers an acetyl group from acetyl-CoA to the amino group from PPT yielding acetyl-PPT. The enzyme is called phosphinothricin acetyltransferase (PAT).

3. Transfer to plants

The bar gene was placed under control of the strong CaMV 35S promoter and is followed by a 3' end poly adenylation signal sequence. A chimeric kanamycin resistance gene is included as an additional selection marker for plant transformation. The expression units are inserted between terminal T-DNA border repeats in an intermediated vector and cointegrated in the Agrobacterium strain C58C1 (pGV2260) (Deblaere, et al., Nucl. Acids Res. 1985, 13, 4777-4788), yielding C58C1 (pGSFR1280).

Transformants were selected in medium containing 50 mg/l kanamycin or various concentrations of PPT (0.5 to 50 mg/l). After one month, no growth was observed on the selective plates of the C58C1 (pGV2260) co-cultivation. Thus, the low dose of 0.5 mg/l PPT is sufficient to inhibit growth of sensitive cells. Calli transformed with C58C1 (pGSFR1280) grew undistinguishable from their growth in non-selective medium under all selective conditions. Plants were regenerated from the calli selected on 25 mg/l PPT. The regenerated plants contained high PAT-activity and showed to be completely resistant towards high doses of the commercial formulations of phosphinothricin and bialaphos. Protoplasts made from these transgenic plants showed to be resistant to concentrations up to 1250 mg/l PPT.

Studies on the S1 and the S2 progenies obtained by selving the transformed tobacco and tomato plants proved that the herbicide resistance is inherited as a stable dominant mendelian trait.

4. Conclusion

This work illustrates a successful approach to obtain phosphinotricin resistant transgenic plants by detoxification of the herbicide. The gene conferring herbicide resistance provides a selectable marker for transformation of tobacco in leaf disc and in protoplasts cocultivation experiments. Probably the PPT-marker can also be used in other protoplast transformation systems where markers like kanamycin resistance are difficult to use. Experiments are in progress to test this possibility.

SELECTION FOR SALT/DROUGHT TOLERANCE USING ISOLATED PROTOPLASTS AND PROTOPLAST-DERIVED CALLI OF COLT CHERRY (PRUNUS AVIUM X PSEUDOCERASUS)

S.J. OCHATT AND J.B. POWER
Plant Genetic Manipulation Group, University of Nottingham, Nottingham NG7 2RD, UK

1. INTRODUCTION

The successful cultivation of rosaceous fruit trees relies heavily on the quality and performance of the rootstocks employed. It would, therefore, hold special interest for the fruit grower to have general purpose rootstocks, adapted for a wide range of environmental stresses. This may be achieved by exploiting somaclonal variation and, in this respect, isolated single cells or protoplasts should be considered as a suitable starting point provided plants can be regenerated following selection for a specific phenotype/genotype. Plant regeneration from Colt cherry protoplasts has been described (1), and *in vitro* selection experiments, were undertaken using this genotype.

2. MATERIALS AND METHODS

Isolated Colt cherry (*Prunus avium* x *pseudocerasus*) mesophyll and root cell suspension protoplasts were cultured under conditions so as to give callus (1). These calli were assessed, alongside direct explant-derived calli of comparable origin (leaf and root), for their responses to salt and water stress. Callus was transferred to a basal proliferation medium, which consisted of MS medium (2) with 2.0 mg/l NAA and 0.5 mg/l BAP, supplemented with 0, 25, 50, 100 or 200 mN of NaCl, KCl or Na_2SO_4 (low to high salinity) or iso-osmotic concentrations of mannitol (mild to severe water-stress).

Cell lines were subjected to a direct recurrent selection strategy whereby they were progressively enriched in putatively tolerant cells whilst those that were merely physiologically adapted were progressively discarded. This method consisted of a sequence of alternate two-subculture passages (3 weeks each) on media with and without the stress-inducing agents, and was initiated once a callus line itself was established under the various stress conditions. After applying at least three cycles of recurrent selection, plants were regenerated using the media as reported earlier (1), but supplemented with the same concentration of the stress-inducing agent as tolerated at the callus level. The stability of the induced stress-tolerance was assessed by reinitiating callus lines from leaves and roots of the first generation of regenerants and under the corresponding stress conditions (medium plus the salts or mannitol as shown above).

All cultures were maintained at 25°C (continuous illumination of 1000 lux) and subcultured every 3 weeks. Each treatment included at least 25 replicates with an initial inoculum of ca. 50 mg fresh weight of callus. Subsequent responses were measured in terms of the extent of callus growth at the end of each subculture (fresh and dry weight), % cell viability (using fluorescein diacetate, FDA), and callus survival (the percentage of surviving calli after each selection cycle).

A separate series of experiments was initiated so as to investigate the cellular basis of salt-tolerance. Protoplasts, of both sources, were cultured in the respective media containing the three salts, with or without the addition of 2,6-dichlorobenzonitrile (DBN); an inhibitor of cell wall synthesis (3). Protoplast responses were evaluated in terms of viability (assessed twice weekly during the first two weeks of culture) and plating efficiency, measured after 14 days and expressed as the percentage of the initially plated protoplasts that had regenerated a cell wall and divided at least once.

3. RESULTS AND DISCUSSION.

Salt-tolerant cell lines were obtained at all concentrations above 25 mN and for all three salts. With increasing concentrations of the salts callus growth, cell viability and survival diminished. The nature of the tissue source affected the responses of cells in relation to salinity, with root-derived calli appearing to be more tolerant to salts than those from leaves. In turn, calli from cell suspension protoplasts were more tolerant than the root-explant calli but in contrast calli of leaf protoplast origin were more sensitive to the salts than those of direct explant origin. Sodium ions were primarily responsible for cell damage as compared to potassium ions. No differences were detected between Na^+ supplied either as chloride or sulphate. The anionic effect on cells was therefore negligible.

In relation to water-stress of tissues growth was progressively inhibited with increased mannitol concentration with protoplast-derived calli exhibiting a greater inhibition than explant calli. Growth inhibition with mannitol was always more pronounced than with ion toxicity when iso-osmotic saline media were used for the culture of tissues. Cell lines tolerating up to 66.6 g/l mannitol (iso-osmotically equivalent to 200 mN of salts) added to the culture medium were isolated. Plants in turn were regenerated from all the salt and water-stress tolerant cell lines.

The culture, of leaf discs and root segments taken from the regenerated plants of such tolerant cell lines, on the corresponding stress-induced media, showed that they too were able to produce friable, fast-growing calli. This could be regarded as an indirect assessment of the stability of the acquired tolerance since tolerance was retained from cell to plant and vice versa.

The results using isolated protoplasts confirmed that sodium ions were the most detrimental to culture. A dramatic drop in cell viability was also observed concomitant with cell wall regeneration followed by a progressive decline thereafter as assessed by plating efficiency measurements. When cell wall regeneration was inhibited, in the presence of DBN, protoplast viability showed no significant difference to that of the controls. Thus, the precise mechanism conferring increased salt-tolerance seems to be clearly associated with cell wall resynthesis.

REFERENCES

1. Ochatt SJ, Cocking EC, Power JB: Plant Sci. 50: 139-143, 1987.
2. Murashige T, Skoog F: Physiol. Plant. 15: 473-497, 1962.
3. Galbraith DW, Shields BA: Physiol. Plant. 55: 25-30, 1982.

FIELD TESTS OF PROGENY FROM PROTOPLAST REGENERATED OILSEED RAPE

Kristofer Vamling[1] and Waheeb Heneen

Department of crop genetics and breeding, Swedish University of Agricultural Science, S-268 00 Svalöv, Sweden.

[1]Present adress:"MARIBO" SEED, P.O. Box 29, DK-4960 Holeby, Denmark.

1. INTRODUKTION

To investigate the "agronomic" consequences of a simple protoplast - plant regeneration cycle, we chose two commercial oilseed rape varieties for evaluation of their products after such a cycle. The regenerated plants were selfed and the resulting seeds were used for field trials.

2. MATERIALS AND METHODS

Hypocotyl protoplasts were isolated from two commercial varieties of double low summer type oilseed rape (Brassica napus, cv. Topas, Svalöf AB and cv. Hanna, Weibulls AB). Protoplast isolation, culture and plant regeneration techniques were with slight modifications as described by Glimelius (1984).

The chromosome numbers of the regenerated plants were determined (meiosis, anaphase I analysis) in 69% of the material presented here (148 out of 214).

Based on the seed yield from the self pollinated regenerated plants, the material was devided into four groups for the field trials. Progenies from plants with high seed yields were tested in large and small plots (10.5 m^2 and 1.2 m^2) while progenies from plants with low seed yields were tested in small plots only. Plots with the two cultivars Topas and Hanna acted as controls.

The material was characterized for several agronomic traits: flowering date, duration of flowering, plant morphology, and maturity date. The results of seed yield and chemical analysis of the seeds were not ready at the time this poster was prepared. The material grown in the largest plots was also rated in terms of "breeders impression of potential".

3. RESULTS

The regenerated plants, grown under greenhouse conditions, showed a large variation in morphology e.g. abnormal leaves and flowers. Most plants with abnormal morphology were aneuploid or tetraploid and had a low seed yield. The variation documented in the greenhouse was less pronounced in the second generation. Mainly the progeny from tetraploid and aneuploid plants continued to show an altered morphology compared to diploid plants. The majority of the tetraploids also showed a later date of flowering and a later maturity date.

Progeny from high yielding and diploid regenerants had

largely a normal development and appearance. But there was also a clear variation in the various characters studied, especially when comparing offsprings of different plants. This variation was more or less gathered around the average value for respective cultivar showing that the cultivar "characters" were maintained in the offspring. Only as regards the character "duration of flowering", there was a tendency towards a longer duration than in the controls (Fig. 1). The variation within each plot was lower than in the control in most of the material originating from high yielding and diploid plants. However, in certain lines, a clear segregation was observed, e.g. in plant height, chlorophyll content and flower morphology and color. The variation in "breeders impression of potential" is shown in fig. 2.

FIGURE 1. <u>Duration of flowering</u>

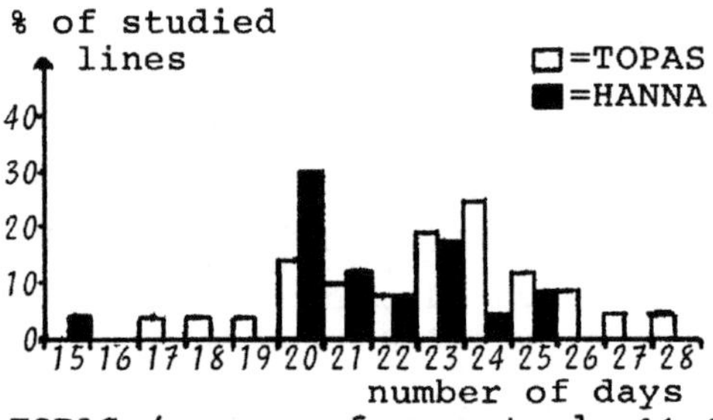

TOPAS (average for controls 21.3, average for studied lines 23.3)
HANNA (average for controls 20.0, average for studied lines 21.0)

FIGURE 2. <u>Breeders impression of potential</u>

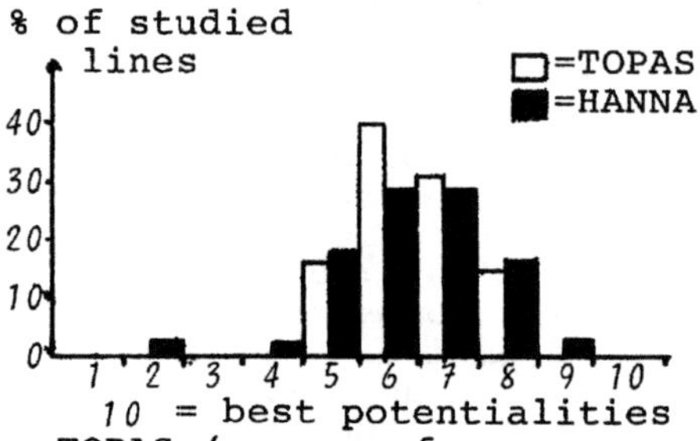

TOPAS (average for controls 7.6)
HANNA (average for controls 6.7)

4. DISCUSSION

The main cause of the variation between progenies from diploid and high yielding plants is probably the heterogeneity already existing in the cultivars tested. The majority of the observed abnormalities, e.g. bright-yellow flower color, narrow petals and haploid or male sterile flower morphology, have also been observed in the control material. This does not exclude that there are progenies expressing somaclonal variation.

The very low variation in field characters within progenies from diploid high yielding plants/shoots indicates that these have a homozygotic origin.

5. CONCLUSION

(1) The progeny of diploid regenerated plants maintain the high breeding value characteristic of the original material.
(2) The "character" of a cultivar is maintained relatively well through such a cycle.

REFERENCE

Glimelius K: High growth rate and regeneration capacity of hypocotyl protoplasts in some <u>Brassicaceae</u>. Physiol. Plant. 61:38-44, 1984.

Studies of lettuce plants regenerated from protoplasts.

ROGUET Y.

Laboratoire d'amélioration des plantes- Bt 360- Faculté d'Orsay,
Paris XI- 91405 Orsay cedex- FRANCE.

Protoplasts is a method wich induced much variability. Plants regenerated from tissue culture or protoplast cultures are often quite different from mother plants. Such variation may be beneficial by providing new disease resistances or improvements in horticultural characteristics.

In lettuce,calli and protoplasts regeneration display somaclonal variations (1-2).

Few morphological observations were made on the regenerated plants Fo, because their development has been asynchronous in a green house over a period of eight months so it was difficult to separate the genotypic effect from the environemental one on the phenotype. Observations on the selfed-progenies of the Fo plants (F1) allowed simultaneous replicated comparisons of the protoplasts regenerated clones in similar environements and also provided the opportunity to observe recessive mutations masked in the Fo.

For phenotype comparisons, only the Fo plants wich had produced more than 200 F1 seeds are used. The control was made by seeds of selfing parental plants in the same conditions of culture than the Fo plants. All the Fo plants included in the trail has no changes in chromosome number. The pollen fertility of the Fo plants was not significantly different from the control. F1 seeds were very variable and different from the control in size, color and form.

The germination curves of the F1 seeds are regroup in 3 types, all the curves are quite similar in form but their values :

1- are the same ;
2- are lower ;
3- are higher than the control values.

Except for two families whose curves are different in form and values. The slope of the curves was higher than the control one. The curves are similar for all plants arriving from the same callus.

Seedlings from 4 clones derived from protoplasts segregated for mutant phenotypes in approximatively 3:1 ratios suggesting single-gene recessive mutations.

The F1 plants were measured in 4 field trials on morphological characters such as leaf number and heart diameter at different stages, and, at maturity, axe diameter, shoot number, heart weight and also percentages of seed germination, pollen fertility and seed production.

All the F1 plants have normal chromosome number (2n=18), normal pollen fertility and normal seed production. The percentage of seed germination was different : the F1 pe̊rcentage of germination was superior to the control one.

The heart diameter of the F1 plants was lower than the control one, the heart height of F1 plant was higher than the control one but the weight was no significantly different. The F1 plant was more egg-shaped and the control plants are round.

The leaf number was not significantly different at early stage but became highly different after.

The F1 plants have also more lateral shoots than thecontrol. This may be due to a decrease of apical dominance.

The F1 plants can be regroup in 4 types :
1- Albinos, died at cotyledonary stage ;
2- Frilly and non-heading plants with lateral shoots ;
3- Darker and frilly heading plants with lateral shoots ;
4- Darker and moderatly frilly plants with lateral shoots (between the the control and type 3).

The statistical tests confirm the fields observations that describe the F1 plants as with more lateral shoots, a higher and more confined heart, darker and more frilly leaves than the control.

F1 progenies are quite different from the control but the variation is of the same kind comparatively to the control (despite of some intensity fluctuations). This may be due to the fact that all plants Fo have the same history, the correlation break-up are the same and the disruption will be identical for all plants.

The further studies on F2 progenies and cross-breeding F1 progenies will be select between two hypothesis :
1- Somaclonal variation ;
2- Physiological variation.

REFERENCES :

1-SIBI M Ann. Am. Pl. 26 (1976) 523;

2- ENGLER D.E., GROGAN R.G. J. Heredity 75 (1984) 426.

SOMACLONAL VARIATIONS FOR INDOLE ALKALOID ACCUMULATION IN PROTOCLONES DERIVED FROM AN HABITUATED STRAIN OF CATHARANTHUS ROSEUS

C. Perronnet*, C. Denécheau, M. Rideau, J.C. Chénieux

Plant Biotechnology Laboratory, Faculty of Pharmaceutical Science of Tours
2bis, Bd Tonnellé - 37042 Tours - France.

1 - INTRODUCTION

It is well established that in vitro tissue culture can accumulate secondary metabolites with pharmaceutical and/or industrial potential interest. However, in subcultured strains, it is not unusual to see cells loosing their growth substances requirements. It is admitted that habituation is an epigenetic phenomenon [Meins, 1982], but its influence on the secondary metabolite production capacity has been seldom studied, and this, only with uncloned habituated strains which could be mixture of autotrophic and non autotrophic cells [Furuya et al., 1983].

Our aim was to study the accumulation capacity of indole alkaloids in protoplast-derived clones ("true clones") of Catharanthus roseus (Apocynaceae).

2 - MATERIALS AND METHODS

Protoplasts from Catharanthus habituated cell line C20A were isolated from 3 day-old cell suspension culture by incubation in 2 % cellulase Onozuka (R10) + 1 % macerozyme Onozuka (R10) + 0.05 % pectolyase (Sigma) dissolved in osmoticum : 0.55 M Sorbitol + 1mM $CaCl_2$ (pH 5.5). After incubation for 16 to 18 h 27°C in the dark with occasionnal shaking, protoplasts were separated from undigested tissue by filtration through a 100 µm nylon cloth. Sedimentation was twice for 3 min. at 100xg. Protoplasts were purified twice by overlayering over sucrose solution 21 %. Protoplasts were finally washed twice by sedimentation in osmoticum (3 min - 100xg) and resuspended in culture medium, adjusting the population density to 5 x 10^5/ml.

Fresh culture medium was added every 5 days. After 15 days, protoplast suspension was gently mixed with semi-solid agar C20A cell line medium at room temperature. Two weeks later, proliferating clones were transferred on hormone-free solid medium, and subcultured every week. Alkaloid contents were estimated according to Merillon (1986)

3 - RESULTS AND DISCUSSION

Protoplasts were isolated from an habituated cell line (C20A), selected three years earlier from an auxin heterotrophic strain (C20) [Mérillon and al., 1986]. Protoplasts were cultured in three different culture media. Best results were obtained with : Gamborg's medium supplemented with 0.35 M glucose, 0.35 M sorbitol, 1 mg/l 2,4-D, 0.1 mg/l NAA, 0.1 mg/l Kin., 0.1 mg/l B.A. Growth of C20A cells is independent of phytohormones, whereas growth of protoplasts derived from the strain need a supply of auxin and cytokinin to initiate cell division. Then, the protoclones could be transferred on C20A plant growth substance-free culture medium. Protoclones exhibited differences in relative growth rate, pigmentation (anthocyanins and flavonoids) and friability. The indole alkaloid accumulation was studied in 200 cultivated clones. The amounts of ajmalicine and serpentine showed a large fluctuation among the clones (Fig. 1). Whereas tabersonine and vindolinine were not detectable in C20A cell line suspension, these alkaloids were shown to be accumulated in about respectively 40 % and 15 % protoclones. Frequency distributions in alkaloid contents fluctuate according to the composition of the initiation culture medium.

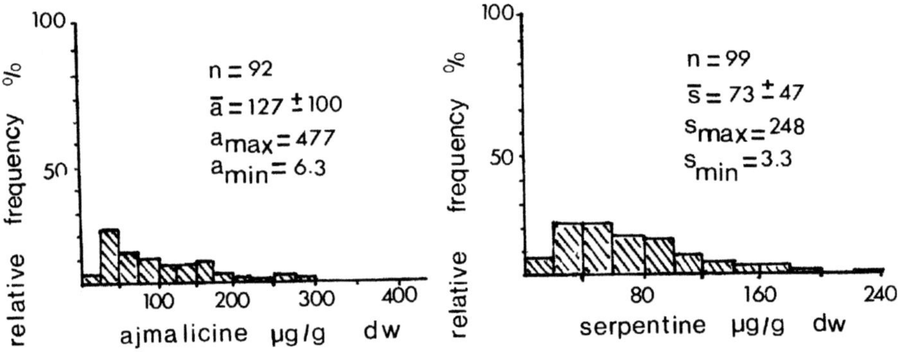

Fig. 1 : Frequency distribution of ajmalicine and serpentine contents in protoclones obtained after plating of Catharanthus roseus habituated cell-line.

REFERENCES

1. T. Furuya, T. Yohikawa, T. Ishii and K. Kajii, Planta med. 47 (1983) 183-187.
2. F. (Jr) Meins in G. Kahl and J.S. Schell, Molecular biology of plant tumors (1982) Acad. Press, New York, 3-31.
3. J.M. Mérillon, K. Ramawat, F. Andreu, J.C. Chénieux and M. Rideau, CR Acad. Sci. Fr. III, 303 (1986) 689-692.

SOMACLONAL VARIATIONS OF DIHYDROFUROQUINOLINE ALKALOIDS IN CELL LINES ISSUED FROM STRAINS AND LEAF MESOPHYLL OF CHOISYA TERNATA BY PROTOPLAST CLONING.

J. Trémouillaux-Guiller, F. Andreu, J. Crèche, B. Arbeille, J.C. Chénieux, M. Rideau

Plant Biotechnology Laboratory, Faculty of Pharmaceutical Sciences
2bis, Bd Tonnellé - 37042 Tours, France.

I - INTRODUCTION

Plant tissue cultures can accumulate secondary metabolites but unfortunately often in low amounts. To improve the productivity, cloning strategies have been devised allowing the selection of cells with high-accumulating capacities. So far, cloning has been carried-out mostly with "aggregate-clones" and only a few workers established true clones by single cell or protoplast-cloning (Constabel et al., 1981 ; Fujita et al., 1985 ; Oksman - Caldentey and Strauss 1986 ; Trémouillaux-Guiller et al., 1987).

Our aim was to compare with regard to their dihydrofuroquinoline-alkaloid capacities, Choisya ternata lines obtained by protoplast-cloning to those previously obtained by classical way (i.e. selection upon biochemical and morphological criteria during regular subcultures over a long period : Crèche et al., 1987). For that purpose protoclones were prepared, by protoplast-cloning from 4 of these well-established strains and from mesophyll cells.

2 - MATERIALS AND METHODS

Plant material : The initiation and culture conditions of the cell lines obtained from a single tree have been previously described (Crèche et al., 1987). Plant shoot cultures were originated from a fragment of stem bearing two axillary buds ; this explant, excised from the same tree as for cell line initiation, was cultivated on Murashige and Skoog medium, 0,75 % glucose and 0,8 % agar. Every 6 weeks, shoots grown from the buds were cut into several foliated stems and subcultured in the same way.

Protoclones from cell lines : They have been obtained by a procedure previously described (Trémouillaux-Guiller et al., 1987).

Protoclones from mesophyll cells : Leaves from shoot cultures were peeled off and the tissues were digested in filter sterilized solution containing 0.5 % $CaCl_2$, 2 H_2O, 0.25 % cellulase Onozuka R10 and 0.125 % macerozyme R10 in mannitol 0.6 M. Incubation was carried out in the dark for 16-18 hrs. After centrifugation (50xg, 3 min), protoplasts were washed and cultured in Murashige and Skoog medium supplemented with mannitol 0.36 M, glucose 0.25 M, 2,4-D $1 mgl^{-1}$ NAA $1 mgl^{-1}$ KIN $0.1 mgl^{-1}$ and BA $1 mgl^{-1}$.

Alkaloid analysis : Were done by a spectrofluorometric method previously described (Crèche et al., 1987).

3 - RESULTS AND DISCUSSION

1. The variability among the clones was very large, whatever the plant material used (mesophyll or cell lines). The best clones could accumulate only one of the alkaloids (platydesminium) in higher amounts than the mother plant. The other alkaloid (balfourodinium) was always accumulated in very low amounts.

2. Protoclones prepared from well-established strains (all originating from a single individual) gave very similar frequency distribution of alkaloids whatever was the alkaloid content in the parental strain used for the isolation of protoplasts (except for the lowest strain $S_1G/4$ which is otherwise very particular). As for other systems (Dougall et al., 1980) it seems that the capacities of alkaloid accumulation in these clones have an upper limit.

3. Protoclones prepared from mesophyll cells highly ressembled to the primary callus usually arising when a stem explant is cultivated *in vitro*. During subsequent passages their phenotypes were modified and alkaloid accumulation shifts were often observed. In some clones both platydesminium and balfourodinium contents increased and for the later it was possible to obtain amounts much higher than in the clones derived from well-established strains.

		Platydesminium $\mu g.g^{-1}$ dr.wt		Balfourodinium $\mu g.g^{-1}$ dr. wt	
Leaves in vivo		300 to 500		4 000 to 8 000	
Leaves (shoot cultures)		360		16 000	
Established strains					
S1G/4		37		tr.	
S1H-MS-JAO		200		29	
S1H-MS-JAL		440		35	
S1H-MS-VAL		600		155	
Cell line protoclones					
S1G/4	n = 35	tr-345	92±75	tr- 37	1,5
S1H-MS-JAO	n = 56	79-2836	984±570	tr-144	48±45
S1H-MS-JAL	n = 35	139-2087	663±451	tr-199	35±49
S1H-MS-VAL	n = 30	97-2499	857±572	tr-169	48±43
Leaf protoclones passages :					
1st	n = 57	15-873	355±189	12-192	68±48
2nd	n = 46	15-1180	468±308	13-297	89±66
3th	n = 31	138-2015	591±425	9-400	105±86
4th	n = 29	154-2201	748±551	9-690	165±156

Alkaloid levels in protoclones, leaves, cell lines and protoclones (range ; mean values ± sd)

REFERENCES

1. J. Crèche, J. Guiller, F. Andreu, M. Gras, J.C. Chénieux, M. Rideau, Phytochem, 26 (1987) 1947-1953.
2. F. Constabel, S. Rambold, K.B. Chatson, W.G.W. Kurz, J.P. Kutney, Plant cell rep. 1 (1981) 3-5.
3. Y. Fujita, S. Takahashi, Y. Yamada, Agr. Biol. Chem., 49 (1985) 1755-1759.
4 K.M. Oksman-Caldentey, A. Strauss, Planta med. (1986) 6-12.
5. J. Trémouillaux-Guiller, F. Andreu, J. Crèche, J.C. Chénieux, M. Rideau Plant cell rep. 6 (1987) 375-378.
6. D.K. Dougall, J.M. Johnson, G.H. Whitten, Planta, 149 (1980 292-297.

SOME ASPECTS OF SALINITY TOLERANCE OF CALLI DERIVED FROM LEAF TOMATO PROTOPLASTS

J. ROSSI, N. DORION and C. BIGOT*

Department of plant physiology, National Institute for Horticultural Science, 4 rue Hardy, 78009 Versailles Cedex, France

1. INTRODUCTION

Salinity represents one of the major environmental stresses influencing the productivity of agricultural systems around the world (McWilliams 1986). Beyond the agronomic improvement of soils, the selection of plants with a better tolerance, in particular by cell culture, is also a way to solve this problem (Nabors 1983). Therefore selection against salts was attempted from microcalli derived from leaf protoplasts of *Lycopersicon esculentum* var. Edkawy, a variety naturally more tolerant to salinity (Hassan and Desouki 1982).

2. PROTOPLASTS ISOLATION, MICROCALLUS PRODUCTION AND PLANT REGENERATION

Mother plants were sowed *in vitro* and maintained during 9 weeks by apex subcultures (each 3 weeks). They were kept in the dark for 48 h. to improve subsequent protoplast development. Leaves were cut into small strips and incubated first during 6 h. at 10 °C in a preplasmolysis medium, then during 16 h. at 25 °C (8 h. under 21 Wm^{-2}) in enzymatic solution. The more effective enzyme mixture contained cellulase Onozuka RS 0.2 %, Macerozyme R 10 0.02 %, Driselase 0.05 % ; the cellulase concentration might be doubled for plants more than 7 weeks old. In these conditions yield of protoplasts averaged 10^7/g f.w. and mean survival was 87 %. Protoplasts were washed and cultured in 5 ml liquid medium at a final density of 3 x 10^5/ml. The suspension was diluted (x 1/2) after 4 days, then 2.5 ml of the medium were renewed each week during 3-4 weeks. Formation of cell clusters was improved by dilution (x 1/2) of mineral salts (MS without NH_4NO_3), use of NAA 0.3 mg/l, 2.4 D 0.6 mg/l, BAP 0.75 mg/l. Moreover the addition of polyvinylpyrrolidone (PVP 10, 20 g/l) prevented agglutination and browning of protoplasts and clusters (Figure 1) so that plating efficiency reached commonly 10 % sometimes 18 %. For plant regeneration the microcalli must be cultured individually in test tubes ; if subcultured in Petri dishes they became brownish and friable without organogenesis. Shoots appeared after 2 subcultures in tubes with IAA 0.2 mg/l, Z. 1 mg/l and GA_3 0.2 mg/l or IAA 0.2 mg/l and 2 IP 4 mg/l. After 7 subcultures, by mean of the selection of organogenetic parts, calli were always producing shoots, some of them being rooted and acclimatized.

3. DETERMINATION OF SALT TOXIC CONCENTRATIONS AND SELECTION ATTEMPTS

In a first experimentation, the concentrations leading to 50 % of growth inhibition and to death were determined for 3 salts. 12 concentrations were tested 0-207 mM for NaCl and KCl, 0-103 mM for Na_2SO_4 10 H_2O. Fresh and dry weights of 60 microcalli were measured after 6 weeks (2 subcultures). Inhibition increased with salt concentration except for KCl which showed an optimum for 34 mM (2.5 g/l). Letal concentration was the same for all salts, equivalent to 207 mM of NaCl (12 g/l), on the other hand the 50 % of growth inhibition varied with salts, it was about 121 mM for KCl, 103 mM for NaCl (7 g/l) and 43 mM for Na_2SO_4 (Figure 2). Nevertheless these results depended on experimental conditions as a calcium protection was observed.

Selection attempts were made by 2 ways (Rossi 1987) :

1 - by hardening with a reduced number of calli

2 - by confrontation of numerous calli (3.6×10^6) with various concentrations of NaCl 7, 9, 12 g/l.

1 - 120 calli subcultured on agar medium containing 9 and 11 g/l of NaCl necrosed after 6 weeks. On the other hand calli (600) on NaCl 7 g/l could be subcultured after 9 weeks (3 subcult.) on a medium containing 9 g/l of NaCl. During the first 2 weeks growth reduction was observed, then the calli developed too fastly as previously. After 8 subcultures, rescued calli were transfered on a medium with 11 g/l of NaCl (results are in progress).

2 - In this system, after 3 weeks calli developed well on control medium and more slowly with 7 g/l. No growth was observed with 12 g/l whereas with 9 g/l 13 calli could be subcultured. 5 protoclones seemed to retain there growth ability on the same saline medium for several subcultures (> 6).

4. CONCLUSION

Studies are in progress to separate habituation from resistance in the last 2 methodologies, to start organogenesis process on these protoclones, and to test regenerated plant against salinity. Experiments are also developed to connect saline resistance ability with natural enhancement of proline biosynthesis.

FIGURE 1. Microcalli derived from leaf protoplasts after 7 days.
a) on medium without PVP 10, b) on medium containing PVP 10 20 g/l.

FIGURE 2. Determination of salt toxicity.

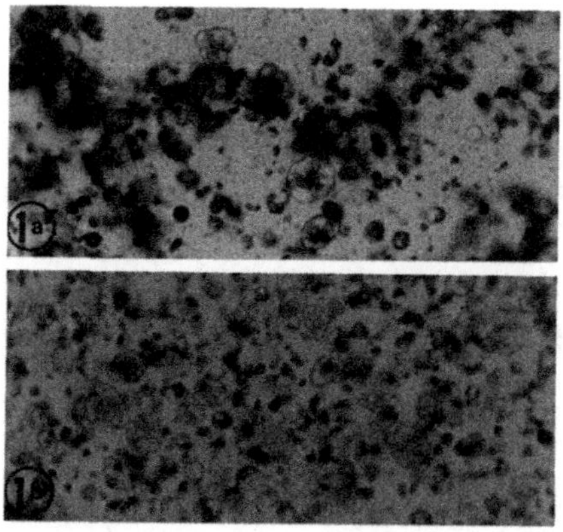

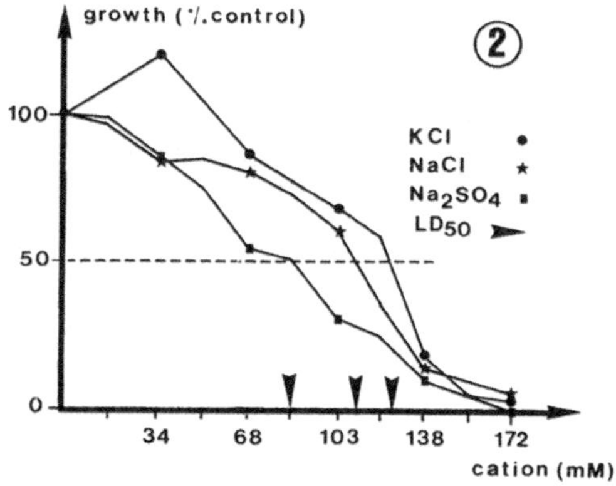

REFERENCES

1. Hassan AA and IAM Desouki : Egypt. J. Hort., 9, 153-162, 1982.
2. McWilliams JR : Aust. J. Plant Physiol., 13, 1-13, 1986.
3. Nabors WM : Current topics in Plant Biochemistry and Physiology, Vol. 2, Ed. DD Randall, 165-184, 1983.
4. Rossi J : Mémoire de fin d'études, ENSH, Versailles, 69 p., 1987.

PROTO-VARIATION IN PROTOPLAST DERIVED <u>BRASSICA</u> <u>NAPUS</u> PLANTS

S.M. JAIN and R.J. NEWTON
Department of Forest Science, Texas Agricultural Experiment Station, Texas A&M University System, College Station, TX 77843, USA.

TAES Article #23183

INTRODUCTION

Plant regeneration from protoplasts facilitates gene transfer, somatic cell fusion, and mutation. Genetic variability in plants regenerated from tissue culture is termed "somaclonal variation" (Larkin and Scowcroft, 1981), and we propose that variation observed in protoplast derived plants be referred to as "proto-variation". Our aims were 1) to regenerate plants from *Brassica napus* L. protoplasts; and 2) to select desirable proto-variants (p-variants) with herbicide resistance and high seed quantity.

PROCEDURE

Materials and Methods

Growth chamber-grown *B. napus* L., amphidiploid, 3-week-old seedlings were kept in dark for 48 h before isolation of protoplasts. Leaves were sterilized in 10% chlorox for 15 min, washed in sterile water 4 times, and incubated in an enzyme mixture (pectolyase Y23 0.025%, cellulase RS 0.4%, $CaCl_2$ 100mM and sorbitol 10%, pH 5.8) for 4-5 h in dark, at 25°C, and agitated at 50 rpm. Released protoplasts were washed in B5 medium (Gamborg et al. 1968) containing 12% sucrose and cultured (50,000 protoplasts/ml) on 1.5 B5 liquid medium having 1mg/l of each kinetin and 2,4-D, 4.5% sucrose and 4% sorbitol. Ethyl methane sulfonate (EMS 0.0005%) was added in protoplast cultures for 48 h. Protoplast calli (p-calli, 2-3mm) were cultured on shoot regeneration medium (B5+ 3mg/l BAP, 1mg/l zeatin and 1% sucrose). Shoots were rooted on either B5 medium containing 0.2mg/l IBA or in the soil under 100% humidity. All p-variants were self-pollinated and seeds were collected from each plant in the R1 generation, and were evaluated for total seed weight and herbicide resistance. Seedlings (10-12 day old) in R_2 generation were sprayed twice, after every 15 days, with 10^{-5}M each phenmedipham (Phen) and atrazine (Atr) herbicides and their sensitivity was indicated by bleaching of green leaves after 10-15 days. Seeds were soaked on a wet filter paper for 16 h and were treated with 1% EMS for 6 h. All seeds were collected from each mutagenized self-pollinated plant and screened for variation to compare with p-variants.

RESULTS AND DISCUSSION

For improving protoplast division and growth of p-calli, it was necessary to a) precondition protoplasts in the culture medium before culture; b) give dark treatment, initially, for 7-10 days; c) add fresh medium in protoplast cultures by removing old medium; and d) reduce p-calli density by 1/3 in each petri dish. After 4-5 days, caulogenesis occurred first in p-calli before development of shoot buds, which elongated into shoots, and eventually were rooted. Total seed weight per plant was 0-5gm in 98% p-variants(EMS), in 67% of the p-variants, and in

98% of the seedlings (EMS), as compared to 11-20 gm with 83% in the control seedlings (Table 1). However, 0.44% p-variants yielded highest total seed weight/plant (31-35 gm) whereas 1% p-variants(EMS) produced 21-25 gm of total seed weight/plant. EMS was deleterious to plant cells and p-variants yielded a wider total seed weight range as compared to p-variants(EMS) and seedlings(EMS). Three percent of the p-variant seedlings survived in the ratio of 3:1 (Resistant:Sensitive) after spraying with Phen and 1% of the p-variant seedlings survived after spraying with Atr (Table 2) indicating that Phen and Atr resistance are due to a single gene mutation. Evans and Sharp (1983) reported single gene mutations in tomato somaclonal variants. In conclusion, desireable traits can be selected in p-variants and use of mutagens should be avoided.

REFERENCES

1. Evans, D.A. and W.R. Sharp: Science 221 (1983) 949-951.
2. Gamborg, O.L., R.A. Miller, and K. Ojima: Exp. Cell Res. 50 (1968) 151-158.
3. Larkin, P.J. and W.R. Scowcroft: Theor. Appl. Genet. 60 (1981) 197-214.

TABLE 1. Variation in total seed weight of _B_. _napus_ p-variants and seedlings in R1 generation. Numbers in brackets indicate percentage.

Total Seed wt. per plant(gm)	No. of p-variants (EMS)	No. of p-variants	No. of seedlings (EMS)	No. of seedlings
0 less than 1	40(56)	88(39)	50(91)	0
1-5	30(42)	62(28)	4(7)	0
6-10	0	40(18)	0	0
11-15	1(1)	22(10)	0	20(33)
16-20	0	7(3)	0	30(50)
21-25	1(1)	3(1)	0	10(17)
26-30	0	0	1(2)	0
31-35	0	1(0.44)	0	0

TABLE 2. Variation in phenmedipham and atrazine resistance among _B_. _napus_ p-variants in R2 generation. Numbers in the brackets indicate percentage.

Phenmedipham resistance			Atrazine resistance		
No. of p-variants	No. of p-variants (EMS)	Percent ratio R:S	No. of p-variants	No. of p-variants (EMS)	Percent ratio R:S
22(30)	15(45)	1:1	20(24)	15(44)	1:1
15(21)	8(24)	1:2	15(18)	10(29)	1:2
8(11)	4(12)	1:3	10(12)	5(15)	1:3
5(7)	3(9)	1:4	10(12)	3(9)	1:4
2(3)	1(3)	1:5	5(6)	1(3)	1:6
4(5)	1(3)	1:6	5(6)	0	1:8
5(7)	0	1:12	5(6)	0	1:16
2(3)	0	3:1	1(1)	0	3:1
8(11)	1(3)	2:1	8(10)	0	2:1
2(3)	1(3)	4:1	4(5)	0	4:1
30(100)*		0	30(100)*		0

*Control seedlings, R-resistant, S-sensitive

INDEX OF ORGANISM

INDEX OF AUTHORS

GPSR Compliance

The European Union's (EU) General Product Safety Regulation (GPSR) is a set of rules that requires consumer products to be safe and our obligations to ensure this.

If you have any concerns about our products, you can contact us on ProductSafety@springernature.com

In case Publisher is established outside the EU, the EU authorized representative is:

Springer Nature Customer Service Center GmbH
Europaplatz 3
69115 Heidelberg, Germany

Batch number: 10815974

Printed by Printforce, the Netherlands